CHEMISTRY RESEARCH AND APPLICATIONS

TRENDS IN POLYANILINE RESEARCH

CHEMISTRY RESEARCH AND APPLICATIONS

Additional books in this series can be found on Nova's website
under the Series tab.

Additional e-books in this series can be found on Nova's website
under the e-book tab.

POLYMER SCIENCE AND TECHNOLOGY

Additional books in this series can be found on Nova's website
under the Series tab.

Additional e-books in this series can be found on Nova's website
under the e-book tab.

CHEMISTRY RESEARCH AND APPLICATIONS

TRENDS IN POLYANILINE RESEARCH

TAKEO OHSAKA
AL-NAKIB CHOWDHURY
MD. AMINUR RAHMAN
AND
MD. MOMINUL ISLAM
EDITORS

New York

For permission to use material from this book please contact us:
Telephone 631-231-7269; Fax 631-231-8175
Web Site: http://www.novapublishers.com

Additional color graphics may be available in the e-book version of this book.

Library of Congress Cataloging-in-Publication Data

Trends in polyaniline research / [edited by] Al-Nakib Chowdhury, Takeo Ohsaka, Aminur Rahman and Mominul Islam (Department of Chemistry, Bangladesh University of Engineering and Technology, Dhaka, Bangladesh, and others).
pages cm
Includes bibliographical references and index.
ISBN: 978-1-62808-424-5 (hardcover)
1. Aniline--Synthesis. 2. Conducting polymers. I. Chowdhury, Al-Nakib, editor of compilation.
QD341.A8T86 2013
668.9'2--dc23

2013024360

Published by Nova Science Publishers, Inc. † New York

CONTENTS

PREFACE

This book presents an overview on recent "Trends in Polyaniline (PAni) Research" covering synthesis, properties and applications such as capacitors, electronics, sensors, composites, adsorption, biomedical and membrane technology. Scientists and researchers from various disciplines including Physics, Chemistry, Materials, Nanoscience and Engineering contributed chapters for the book based on their expertise in these fields. PAni, an interesting conducting polymer (CP), has attracted great attention over the last decades owing to its tunable properties and potential applications in multidisciplinary areas. PAni is found to be the most promising because of its ease of synthesis, low cost monomer, tunable properties, wide range of application possibilities and higher thermal stability compared to other CPs. Therefore, it is the appropriate time to comply a book with a comprehensive review of the recent trends on PAni synthesis, properties and application. This book presents the latest research on PAni from around the world.

Chapter 1 describes the synthesis, characterization and gas sensing applications of PAni. The sensor films prepared by spin coating technique are characterized for structural, morphological, optical and electrical properties. The room temperature gas sensing properties of PAni sensor films are tested for volatile and non-volatile gases. Chapter 2 discusses the synthesis of PAni using electrochemical methods. Electrochemical syntheses by galvanostatic, potentiostatic or potentiodynamic methods are adopted, most often on inert metal and carbon-based electrodes, to produce films directly adherent on the electrode surface. PAni is electrochemically synthesized on aluminium and its alloys through the electrochemical polymerization. The antibacterial and antifouling properties and corrosion resistance behavior of the PAni films are also described. In chapter 3 sonochemical synthesis and application of functional hybrid nanomaterials containing PAni are presented. Besides, ultrasound assisted mini-emulsion synthesis of functional conducting latex (PAni latex) and various applications such as anticorrosion, coatings, photoelectrochemical cell and sensor are also reported.

The structure control synthesis of PAni is very interesting. Chapters 4 and 5 provide an introduction and examples of structure control syntheses of PAni using electrochemical and chemical polymerizations in unique reaction fields and media. Ionic liquids and supercritical fluids are now used as electrolytic media for controlling the physical structure of the materials.

Chapter 6 describes the recent advances of PAni as an electrode material for supercapacitor. Literature data on historical background of capacitor, classification of supercapacitors and status of PAni and its composites with carbon and/or metal oxide for supercapacitor work are summarized in this part.

Chapter 7 provides a comprehensive review on the PAni based composite materials and addresses the perspective of the use of PAni in composites with variety of materials, characterization techniques, and multi-facet applications in different areas. Chapter 8 describes the fabrication and electrical behaviour of electrospun PAni-carbon black composite nanofibers. The role of carbon black in PAni matrix, which can be beneficial as conductive electrode applications in electronic and photovoltaic storage devices is discussed. In chapter 9, the designing of conducting ferromagnetic PAni composites for EMI shielding is described. It also focuses on synthesis, conduction mechanism, magnetic and dielectric properties of conducting ferromagnetic PAni nanocomposites and associated phenomenon to EMI attenuation.

In chapter 10, electrochemical synthesis of PAni nanocomposite is described with a novel biphasic electro-polymerization technique. Multiwalled carbon nanotube/PAni nanocomposite formation based on biphasic electro-polymerization technique provides a general method for the formation of uniform nanocomposites for other polymers and fillers.

PAni has also played a significant role in membrane technology and applications. Chapter 11 deals with the PAni based membranes field and focuses on their gas/vapor separation applications.

Chapter 12 focuses the gas sensing ability of PAni. In this chapter, the fabrication of various morphological PAni gas sensors, PAni/metal nanocomposite gas sensors, other PAni nanocomposite gas sensors and PAni based hetero-junction type gas sensors are discussed.

Chapter 13 furnishes the anticorrosive properties of PAni and focuses the mechanism of corrosion protection of coatings based on PAni. Here, a short review of application of nanomaterials to improve the anticorrosive property of PAni coating is also reported.

Chapter 14 describes the application of PAni nanocomposites as a potential material for the removal of pollutant metal ions and dye molecules from wastewater. The mechanism of removal of pollutants by using PAni and its composites and nanocomposites are elucidated.

Chapter 15 accumulates the biomedical application of PAni in tissue engineering, biosensors and drug delivery devices and also several modification manners of PAni in order to improve its technological usage.

The book "Trends in Polyaniline Research" is a sincere attempt to make available the recent developments in PAni synthesis, properties and its application in a single issue which we believe to bring comfort to the readers worldwide. We hope, this book would be a piece of latest information on PAni researches.

Finally, we would like to thank all of our family members for their understanding, strong support and encouragement. If, any technical errors exist in this book, all editors and chapter contributors would deeply appreciate the reader's comments for further improvement.

Edited by

Dr. Takeo Ohsaka
Professor
Department of Electronic Chemistry
Interdisciplinary Graduate School of Science and Engineering
Tokyo Institute of Technology Mail box: G1-5, 4259 Nagatsuta, Midori-ku,

Yokohama 226-8502, Japan
Tel: +81-45-924-5404, Fax: +81-45-924-5489
ohsaka.t.aa@m.titech.ac.jp

Dr. Al-Nakib Chowdhury
Professor
Department of Chemistry
Bangladesh University of Engineering and Technology
Dhaka-1000, Bangladesh
Tel: +880 2 966 5614, Fax: +880 2 861 3046
nakib@chem.buet.ac.bd

Dr. Md. Aminur Rahman
Associate Professor
Graduate School of Analytical Science and Technology (GRAST)
Chungnam National University, 79 Daehangno,
Yuseong-gu, Daejeon 305-764, Korea
Tel: +82-042-821-8546, Fax: +82-042-821-8541
marahman@cnu.ac.kr

Dr. Md. Mominul Islam
Assistant Professor
Department of Chemistry, Faculty of Science
University of Dhaka
Dhaka 1000, Bangladesh
Tel: +88-01947-558235, Fax: +880 2 861 5583
mominul@du.ac.bd

In: Trends in Polyaniline Research ISBN: 978-1-62808-424-5
Editors: T. Ohsaka, Al. Chowdhury, Md. A. Rahman et al.

Chapter 1

POLYANILINE: SYNTHESIS, CHARACTERIZATION AND GAS SENSING APPLICATIONS

Vikas B. Patil
Principal Investigator, DST FTP, DAE-BRNS, Materials Research Laboratory,
School of Physical Sciences, Solapur University, Solapur,
Maharashtra, India

This chapter describes the synthesis of polyaniline (PAni) by oxidative chemical polymerization method. The characterization concerning structural, morphological, optical and electrical properties of the sensor films prepared by spin coating technique have been discussed. The room temperature gas sensing properties of PAni sensor films that have been tested for volatile and non-volatile gases are described with examples. Gas sensing mechanism for PAni films is also described. The research work describing the detailed study of selective PAni based sensors to ammonia gas is also presented.

1. INTRODUCTION

Polyaniline (PAni), polypyrrole (PPy), polythiophene (PTh) and their derivatives, conducting polymers, have been widely investigated for the development of room temperature (RT) gas sensors. The sensors made of conducting polymers shows improved characteristics such as high sensitivities and short response time at room temperature. However, a major disadvantage of conducting polymers is their lack in specificity towards the target gas molecules since they may sense all the gases present in the system. The selectivity and enhance sensitivity of a particular gas sensor obviously depends on the characteristics of sensors with conductive polymers. As a result, a lot of attention has been paid to fabricate smart gas sensors with conducting polymers [1]. This chapter represents the summary of works that have been carried out on the synthesis and characterization of PAni films used as gas sensors operating at room temperature.

1.1. Conducting Polymers

Conducting polymers are polymers with conjugated structures [1]. Chemical structures of some of the commonly known conducting polymers are shown in Figure 1.1.

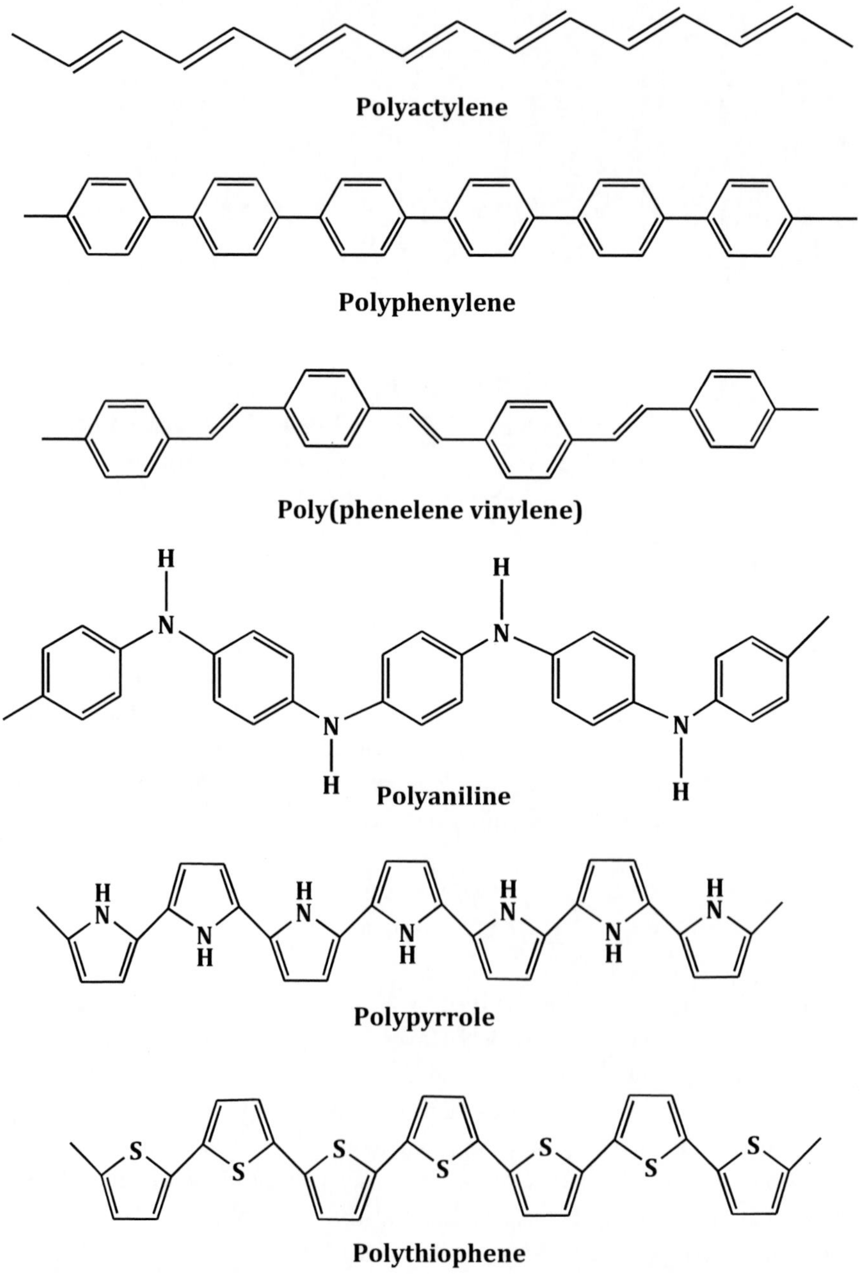

Polyactylene

Polyphenylene

Poly(phenelene vinylene)

Polyaniline

Polypyrrole

Polythiophene

Figure 1.1. Chemical structures of some of the most significant conjugated organic conducting polymers.

In the structure of the polymers, a common feature, i.e., the occurrence of double bonds alternating with single bonds along the polymer chain known as conjugated bonds can be seen. In fact, such an arrangement of conjugated bonds in the organic polymers is the origin of conduction that can be understood as follows. In each repeated unit of polymers, three of the four electrons in the outer shell of every carbon atom occupy hybridized states formed from one 's' and two 'p' states (sp^2 hybridization) [2]. These electrons form three strong 'σ bonds' that play a key role in forming the polymer structure. In polyacetylene each carbon atom forms these σ bonds with one hydrogen and two neighboring carbon that leaves one valence electron left over (the π-electron), which occupies a p orbital. The π-electron wave functions from different carbon atoms overlap to form a π -bond. π-electrons are delocalized over large segments of the polymer chain, which is responsible for the electronic properties of the conjugated polymers. The addition of heteroatoms (atoms other than carbon and hydrogen) and side chains allows fabricating a larger variety in this class of materials.

Conducting polymers in their neutral states are insulators and exhibit a strong UV-visible absorption characteristic. Neutral conjugated polymers with a small conductivity, typically in the range 10^{-10}-10^{-5} S cm^{-1} can be converted into semi-conductive or conductive states with conductivities of 1-10^4 S cm^{-1} through chemical or electrochemical redox reactions. Conjugated polymers are treated as a quasi one-dimensional (1D) system, wherein the polymer chains are assumed to behave independently. Their physical and chemical properties depend on interactions within the single chains. The π-bonding scheme of conjugated polymers decreases the gap between highest occupied molecular orbital (HOMO) and lowest unoccupied molecular orbital (LUMO) states. The band gap of these polymers lies between 1.5 and 3 eV, in the same range as of inorganic semiconductors [3].

Most conducting polymers possess conductivities comparable with those of traditional metal at room temperature. However, the temperature dependence in the conductivity of polymers is mostly non-metallic, especially at lower temperature. In order to achieve high conductivity in conducting polymers the so-called doping is necessary [4, 5]. The study of the conductivity of polyacetylene shows that the conductivity could be increased by more than seven orders of magnitude upon doping of iodine or arsenic pentafluoride (AsF_5) [6]. This draws a great interest of the researchers and similar works lead to the award of Nobel Prize for chemistry in 2000.

When an electron is added to or withdrawn from a conducting polymer, a chain deformation takes place around the charge, which costs the elastic energy and puts the charge in lower electronic energy state. The competition between elastic and electronic energies determines the size of the lattice deformation that sometimes could be the order of 20 units. As a result, the volume of polymer changes and simultaneously, the absorption bands related to the neutral conjugated polymers reduce as well as new absorption bands associated with charge carriers appear at longer wavelengths.

Based on composition-driven redox behavior and accompanying fundamental characteristics such as optical, electrical, electrochemical and mechanical properties the conducting polymers have been generally considered as potential candidates for a range of technological applications. These applications include biomimetics (biosensors, electronic noses, artificial nerves etc.), medical prosthetics (artificial muscles and limbs), battery technology, corrosion inhibition, field-effect transistors, light-emitting diodes and electrochromic display devices [7-11]. Electronic noses and biosensors specifically utilize the change in the recognition pattern of an array of different conducting polymers. Artificial muscles and other prosthetic medical devices harness the ability of conducting polymer to change shape and size during redox switching and film compositional dynamics. For redox switching devices, the conducting polymer is deposited on metal nanowires and /or can be grown as freestanding films.

The principle of electrochromic display devices is the use of change in color associated with different oxidation states exhibited by conducting polymers upon the application of voltage. Using the light emitting properties that can be tuned by changing chemical structure, the conducting polymers have been commercially used as light emitting diodes (LEDs) and displays. Moreover, the conducting polymers of which electron-hole pairs could be generated upon illumination of light have been employed as materials for photovoltaic devices. The conducting polymers are being investigated as a candidate for organic/molecular electronics due to their unique combination of properties that make them an attractive, alternative or a complement to the Si based microelectronics.

1.1.1. Electronic Conduction Mechanisms in Polymers

1.1.1.1. Highly Anisotropic ('quasi-1D') Metallic Conduction

A key feature of polymer conductivity is its high anisotropicity, being much greater along the polymer chains. This quasi-1D nature leads to a mechanism that avoids the usual limitation on conductivity due to scattering of carriers by thermally excited lattice vibrations. Basic idea can be understood with reference to the idealized situation depicted in Figure 1.2, where a highly anisotropic (quasi-1D) polymeric metal in which Fermi surface consists of sheets perpendicular to the chain direction (k_F is the electronic Fermi wavevector) is presented in which the charge carriers are taken to have wavevectors k_F or $-k_F$ parallel to the direction of polymer chains.

It may be stated that to scatter these carriers and create resistance, phonons would need to have a wave vector $2k_F$ spanning the Fermi surface [12]. The energy of these phonons is large leading to excitation of few at ordinary temperatures and the resistivity is suppressed in conventional isotropic metals.

Such quasi-1D metals could be expected to possess the conductivities much higher than those of conventional metals of which the scattering of electrons by phonons limits the conductivity to about 0.6 x 10^6 $S.cm^{-1}$ at room temperature even in the best metals like copper and silver.

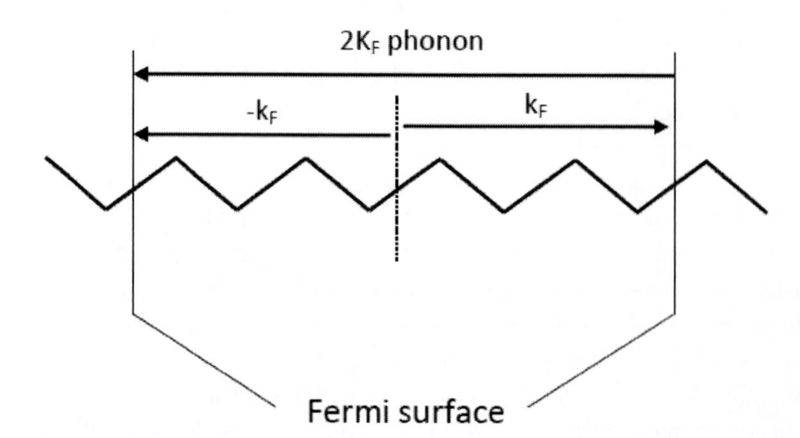

Figure 1.2. A sketch showing backscattering along the polymer chain direction by phonons of wave vector $2k_F$.

1.1.1.2. Hoping in Disordered Semiconductors

For a large variety of disordered materials on the semiconductor side of a metal-semiconductor transition, the conductivity is well-described by Mott's law for variable-range hoping [13]. In disordered semiconductors with localized states in the band gap, conduction occurs by hopping (phonon-assisted tunneling between electronic localized states centered at different positions). As the thermal energy k_BT (k_B is Boltzmann constant and T is temperature in Kelvin) decreases with temperature, there exists fewer nearby states with accessible energies, resulting in increased mean range of hopping. This can be expressed mathematically as the expression for the temperature dependent conductivity σ (T) [14] as follows:

$$\sigma(T) = \sigma_o e^{-(\frac{T_O}{T})\gamma} \tag{3.1}$$

For hopping in 3D, the exponent has the value γ = 1/4. If the electronic wave functions decay with distance 'r' as exp(-r/L_{loc}), where L_{loc} is the localization length, the constant T_0 is characteristic Mott temperature and given approximately by k_BT_0~16/N (E_F)L^3_{loc}, where N(E_F) is the density of localized states at the Fermi level. The prefactor σ_0 is also temperature dependent, with different authors deriving different power laws [14], but this temperature dependence is often neglected compared to the stronger temperature dependence of the exponential term. At sufficiently high temperatures the hopping occurs to the nearest neighbors and the conductivity would instead show a simple activated form given by eqn. (3.1) with γ = 1. This form also applies for electron transport by thermal excitation into a conduction band of extended states in crystalline semiconductor where the conductivity is approximately proportional to the number of electrons excited across the semiconductor gap. For 2D hopping, γ = 1/3, and for 1D hopping γ = 1/2, although according to Efros et al. [15], latter exponent could arise for hopping when electron-electron interactions are considered.

It has been emphasized [16] that quasi-1D hopping could play a key role in conducting polymers where polymer chains traverse disordered regions to connect 'crystalline islands'. Charge carriers would diffuse along such electrically isolated disordered chains as a part of the conduction path, but would readily localize owing to the 1D nature of the chains. In such a case, quasi-1D variable-range hopping along the disordered chains with γ = 1/2 could dominate the overall resistance of the polymer.

1.1.1.3. Tunneling Between Metallic Regions

According to calculations by Sheng and Klafter [17] and Sheng [18], if conduction is by electronic tunneling through non-conducting material separating mesoscopic metallic 'islands' rather than localized states, the expression for the tunneling conductivity approximately follows eqn. (3.1) with γ = 1/2 (same form as for 1D variable range hopping). Such a picture should be appropriate for a granular metal in which small metallic grains surrounds by non-conducting shells. These calculations apply when the conductivity is limited by the electrostatic charging energy when an electron is transferred from one island to the next. The inhibition of tunneling when the thermal energy k_BT is less than the charging energy referred to as a 'Coulomb blockade' [19]. However, the metallic regions are large enough that the electrostatic charging energy is much smaller than k_BT for accessible temperatures. Tunneling can occur between metallic states of the same energy on different sides of the barrier without thermal excitation, provided that the wave functions overlap. Fluctuations in the voltage across the tunneling junction can greatly increase the tunneling current as the temperature increases [20]. The conductivity due to this fluctuation-assisted tunneling for a simple parabolic barrier shape can be written with several assumptions in the simple form:

$$\sigma(T) = \sigma_t e^{-\left(\frac{T_t}{T+T_s}\right)} \tag{3.2}$$

where T_t represents the temperature at which the thermal voltage fluctuations becomes large enough to raise the energy of electronic states to the top of the barrier, and the ratio T_t/T_s determines the tunneling in the absence of fluctuations. The prefactor σ_t is approximated as independent of temperature.

1.1.1.4. Solitons, Polarons and Bipolarons

The elementary excitations for the single polyacetylene chain illustrated in Fig 1.1 are solitons rather than electron-hole pairs as observed in 3D metals. Solitons represent discontinuities in the pattern of alternating single and double bonds that arise when one carbon atom has single bonds to both of its neighboring carbons [21-23]. Additional electronic states are created at the centre of the semiconductor-like band gap in polyacetylene, as shown by optical absorption measurements at low doping levels [21]. Motion of a simple soliton discontinuity along a polymer chain does not lead to charge transport but an inter-chain soliton hopping mechanism could contribute to conduction with conductivity varying as power of temperature [24]. The agreement of a vast amount of conducting polymer data with the hopping laws over a wide range of temperature suggests that once localized states are formed in the gap by any mechanism, conduction takes place predominantly by the usual variable range-hopping processes for lightly or moderately doped samples [25, 26].

In heavily doped samples, the overlap of wave functions means that a metallic picture is likely to be more appropriate at least in the ordered regions. For conducting polymers other than polyacetylene, different solitonic states possess different energies and hence solitonic conduction mechanisms are not expected. On the other hand, states appear in the band gap [21, 27] owing to the formation of polarons having charge and spin and their motion can contribute to charge transport. Polarons of opposite spin often pair up to form bipolarons with zero spin, which can lead to charge transport without spin.

Inspired by the work of Sheng [18] on granular metals, Zuppiroli et al. [28] proposed that conduction in disordered conducting polymers takes place via the correlated hopping between polaronic clusters. In this case, the charging energy for charge-limited tunneling between metallic grains could be defined, and the calculated conductivity has been reported to follow the same temperature dependence as given in eqn. (3.1) with $\gamma = 1/2$.

In the recent years, conducting polymers such as PAni and PPy are studied as gas sensors essentially due to their operation at room temperature and ease of processing for sensor element [29, 30].

1.2. Literature Survey

This review is written with purpose to obtain general understanding of the preparation of conducting polymer, PAni and its use for gas sensing application. We start with an overview of recent accomplishments in the methods of PAni synthesis followed by characterization and gas sensing its application.

There are mainly two methods used in fabricating gas sensors based PAni films namely electrochemical and chemical methods. Electrochemically PAni can be synthesized either by potentiostatic or galvanostatic techniques. Films deposited potentiostatically and galvanostatically produce uniform deposit but not adherent to the substrate. On the other hand, electro-oxidation of aniline by continuous cycling using cyclic voltammetric technique between the predetermined potentials produces a polymeric film which adheres strongly to the electrode surface [31].

Zhao et al. [32] reported highly ordered PAni nanofibrils arrays that have been fabricated within the pores of porous anodic aluminum oxide (AAO) template membrane by electrochemical polymerization. The electrochemical properties of the PAni nanofibrils have been investigated using cyclic voltammetry technique. The results showed that the PAni nanofibrils are of good orientation and uniform.

Cyclic voltammetry result showed that oxidation of PAni occurs in the potential range of +1.0 to +1.2 V and reduction occurs in the range + 0.2 to + 0.8 V. Chloride doped PAni film with a better adhesion to the substrate, less porosity and uniform deposition have been prepared in HCl acid using potentiodynamic method by Hussain et al. [34]. The studies on effect of concentration and substrate resistance on redox properties of PAni prepared using HCl acid is investigated by Bedekar et al. [35]. Thin polymeric films were deposited by cyclic voltammetry, potentiostatic or galvanostatic techniques on the Fe-disc electrode from aqueous oxalic acid solutions [36]. Also the films deposited on iron substrate electrode were studied as enzyme- modified electrode [37].

The comparison between PAni coatings synthesized chemically and electrochemically using a pulse potentiostatic procedure was made by Ivanov et al. [38]. The chemical coatings, if obtained under suitable synthesis conditions, can compete with the electrochemical ones concerning their surface homogeneity and stability. This finding is important from a practical point of view such as cost effectiveness of polymerization. It may be noted that for a large scale polymerization the chemical method is cheaper and more appropriate than the electrochemical one.

Huang et al. [39] have examined the morphological evolution of PAni during its chemical polymerization. It has been revealed that polymerization process preferentially produces nanofibers in aqueous solution. It has been mentioned that pure nanofibers can be obtained by preventing secondary growth of PAni. Two facile approaches, e.g., interfacial polymerization and rapidly mixed reactions, have been developed for a readily production of high-quality nanofibers. In fact, the pure nanofibrillar morphology of PAni, in general, results in significantly improved interactions of PAni with its environment. This leads to an enrich properties of the nanofibers such as excellent water processibility and much faster and more responsive chemical sensing that ultimately assist to design new inorganic/PAni nanocomposites and ultra-fast nonvolatile memory devices. On the other hand, flash welding of the highly conjugated, polymeric nanofibers demonstrates a new nanoscale phenomenon that is not accessible with current inorganic systems. The advantages of "nanostructures + conducting polymers" have been demonstrated.

Delvaux et al. [40] prepared micro- and nanostructures composed of PANi formed by chemical and electrochemical oxidative polymerization of aniline within the pores of particles track-etched membranes (PTM). In both cases, it has been shown that the polymerization starts at the pore walls and is regulated by the diffusion of monomer into the pores. Such an polymerization ultimately leads to the formation PAni tubules. Conductivity measurements have been carried out on this PAni tubules and it has been observed that when tubule diameter decreases, the electrical conductivity increases. Such an increase in conductivity has been attributed to the presence of a larger ratio of oriented polymer chains within narrow PAni tubules.

A variety of PAni based gas sensors have been reported in literature. Nanostructural polyanilines such as nanowires, nanofibers and nanorods have been proven to show a great promise as chemical sensors. This result of PAni has been considered to be originated due to their high surface area and small diameters that facilitate the fast diffusion of gas molecules into the structures. Yan et al. [41] have demonstrated for the first time that the sensor based on PAni nanofibers prepared simply by the interfacial polymerization has the advantages of sensitivity, spatial resolution, and rapid time response for NO_2 gas at room temperature.

Liu et al. [33] prepared a PAni film with a loose 2D nanowire network structure which showed high sensitivity upon exposure to a low concentration ammonia gas. They have investigated the electrochemical formation of nano-PAni on the insulating gap area of an interdigitated electrode with large gap width. The advantage of deposition on the insulating substrate is to avoid the formation of a compact bottom layer structure. Therefore, the obtained 2D loose nano-network structure has been considered to be more suitable for the application as chemi-resistive sensors. It may be noted that the thus-developed method can be used to fabricate cost-effective nanostructured conducting polymer gas sensors with high sensitivity.

Humidity sensor based on PAni nanofibres has been fabricated by Zeng et al. [42] and its response towards humidity has also been investigated. The study showed that PAni nanofibres based humidity sensor behaves very differently to conventional conducting polymer based sensors. At low RH (relative humidity) where proton effect is believed to dominate in decreasing the electrical resistance has been observed. At higher RH, the so-called polymer swelling occurs causing a reverse response, i.e., an increase in resistance has been observed. This peculiar behavior has been believed to be due to the distortion of the nanostructure as the consequence of absorption of water molecules. It has been confirmed by IR spectroscopy that excess water absorption occurs and that a change in polymer oxidation state might have taken place.

Li et al. [43] have recently developed a technique based on UV-irradiation of an aqueous precursor solution. By this technique photo-patterning of thin films of PAni nanofibers has been successfully carried out in an one-pot, single-step synthesis. It has been shown that this technique can be applied to fabricate sensors by growing nanofibers in the active area of an interdigitated electrode array. In this method, it has been emphasized that the sensors prepared are ready for operation as soon as the polymerization is completed, and no additional processing steps are necessary. Due to their higher surface area, the response of PAni nanofibers has been found to be considerably faster and more intense than that of bulk

PAni. The results also showed that nanofiber-based devices can be produced by our bottom-up lithographic technique.

PAni-chemically coated the electrode of quartz-crystal microbalance (QCM) has been developed for the determination of phosphoric acid (H_3PO_4) in the liquid phase. The QCM sensor demonstrated a rapid response to the acid with an excellent reversibility. Although much work has been performed on the ES form of PAni exposed to ammonia, little work has been reported on the use of the EB form of PAni as a sensor for acids in aqueous media. Ayad et al. [44] reported use of PAni coated electrode of QCM as a sensor for H_3PO_4 in aqueous medium. This can be performed during the subsequent redoping–dedoping in acid and ammonia solutions, respectively, and the successive redoping in acid solutions with different concentrations. It has been shown that the pH dependence on the electronic absorption of PAni phosphate film are comparable with the corresponding dependence based on PAni-sulfate films.

The development and optimization of ammonia gas sensor composed of inkjet-printed PAni nanoparticles have been demonstrated by Crowley et al. [45]. The conducting films have been assembled on interdigitated electrode arrays and characterized with respect to their layer thickness and thermal properties. The sensor has been further combined with heater foils for operation at a range of temperatures. When operated in a conductometric mode, the sensor has been shown to exhibit temperature dependent analytical performance to ammonia detection. At room temperature, the sensor has been found to respond rapidly to ammonia ($t50$ = 15 s). Sensor recovery time, response linearity and sensitivity could all significantly be improved by operating the sensor at temperatures up to 80 °C. The sensor has been found to have a stable logarithmic response to ammonia in the range of interest (1–100 ppm). The sensor has also been observed to be insensitive to moisture in the range from 35 to 98 % RH.

During last two decades, the research in PAni thin films synthesized chemically or electrochemically showed that films could be synthesized as doped PAni (by adding protonic acids such as HCl, H_2SO_4, HNO_3 and $HClO_4$) on various substrates. It could be mentioned that the bronsted acids or inorganic dopants are very common dopant [46]. Three sulphonic acid-doped PAni's have been synthesized through chemical oxidation at low temperature (0-5 ^{o}C) and the potential of these polymers as sensing agent for O_2 gas detection in terms of fluorescence quenching has studied been by Draman et al. [47]. Sulphuric acid, dodecylbenzene sulphonic acid (DBSA) and camphor sulphonic acid (CSA) have been used as doping agents. Spectroscopic and thermal properties of the polymers have been found to be affected by the type of the dopants used. Fluorescence characteristics of polymer solution/film upon exposure to O_2 gas showed a good potential to be exploited as sensing material for O_2 gas detection. A good repeatability and reproducibility of measurement have been obtained. A complete regeneration cycle took about 13 minutes and 4 complete cycles have been observed in 60 minutes of testing.

Manigandan et al. [48] explored the Langmuir Blodgett (LB) technique to produce dopant induced nanostructures, especially nanorod and nanoparticles of conducting PAni emeraldine salt. PAni emeraldine base is doped using camphor sulfonic acid (CSA), *p*-toluene sulfonic acid (PTSA) and perfluoro-octanoic acid (PFOA). TEM analysis reveals the formation of dispersed nanorods of less than 100 nm diameter of PAni-CSA. Dispersed nanoparticles of less than 80 nm diameter forms for PAni doped with PTSA, while the formation of connected nanoparticles occurs in the case of PAni-PFOA. Among the three different systems the rod-like nanostructure of PAni-CSA exhibits the highest conductivity.

Study of dynamic response (conductivity variation) of these PAni nanostructures with ammonia vapor showed that the dispersed rod-like PAni-CSA exhibits the fastest response characteristics (shortest response time and time constant) but the connected nanoparticles of PAni-PFOA show the most sluggish response.

Jiang et al. [49] have demonstrated the facile route for the synthesis of mono-disperses Au nanoparticles in 2D ordered, liftable, and patterned PAni monolayer sheet by using tetrachloroauric acid as the oxidizing agent. The gas-sensing performance based on the obtained product for NH_3 has been investigated. The sensor exhibits an ultra-fast response (5 s) and recovery (7 s) behavior, which are of great importance in gas detection and control. Most importantly, this method can offer a powerful platform to design and incorporate functional metal nanoparticles in 2D pattern conducting polymer sheet for desirable properties.

An organic–inorganic hybrid material PAni/SnO_2 has been prepared by Geng et al. [50] through a hydrothermal method for sensing ethanol and acetone. The PAni/SnO_2 hybrid material is sensitive to ethanol and acetone when operated at 60 or 90 °C and shows a good reversibility. They have suggested that the sensing mechanism is related to the existence of p–n hetero-junctions in the PAni/SnO_2 hybrid material. When operated at 90 °C, the PAni/SnO_2 hybrid material exhibits a short response and recovery time (1 min). It has been concluded that the materials developed can overcome the shortcomings of long response time of PAni and the high operation temperature of SnO_2 that are very important features for practical uses.

The NH_3 sensing capability of PAni/TiO_2 thin film sensors has been investigated by H. Tai [51]. The polymerization temperature has been optimized. The PAni/TiO_2 thin film prepared at 10 °C exhibits a stable, reproducible and reversible resistance change in the presence of NH_3 in the range of 23–141 ppm. It has been found that when the response time is kept to be 2 s, then the recovery rate becomes very fast in the range of 20–60 s depending on the NH_3 concentration. In addition, the sensor has also been reported to possess a high selectivity and long-term stability. The difference of gas-sensing property among sensors prepared under different polymerization temperatures has been characterized with UV–vis spectra and SEM. In this study, a simple schematic energy diagram has been presented to reveal the gas sensing mechanism. The in-situ self-assembly approach has been proposed to be easy-adoptable and feasible to produce a PAni/TiO_2 thin film in large scale. It has been emphasized in this study that the PAni/TiO_2 gas sensor could be preferred in operation at room temperature. This route could be followed to develop a practical NH_3 gas sensor at low cost. In short, extensive literature is available on the PAni based gas sensors. Most of the PAni sensors are prepared by doping protonic acids or Bronsted acids or organic/ inorganic nanoparticles. In the present work, we deal with the ammonia sensor based on undoped PAni (EB) thin film sensors.

1.3. Preparation and Characterization of Polymer Films

The experimental set up and methodologies used for polymerization of aniline are discussed below. The preparative parameters have been optimized in order to get uniform and useful PAni films for gas sensing application. Furthermore, the prepared films have been characterized using various techniques. Detailed synthesis and deposition procedure of PAni thin films using spin coating technique are as follows:

1.3.1. Experimental Set up for Polymerization and Deposition of Thin Films

The schematic diagram of experimental set up employed for the polymerization of aniline by oxidative chemical reaction is as shown in Figure 1.3. In fact, this the cross-section of the experimental setup used for polymerization as shown in Figure 1.4. It consists of a freezing mixture bath kept on the magnetic stirrer. The ice-salt freezing mixture surrounding a beaker used to maintain the polymerization temperature below 0 °C. The PUF (Poly Urethane foam) insulation is used in freezing mixture bath to prevent heat exchange between system and surrounding. The digital display indicates the temperature of the solution.

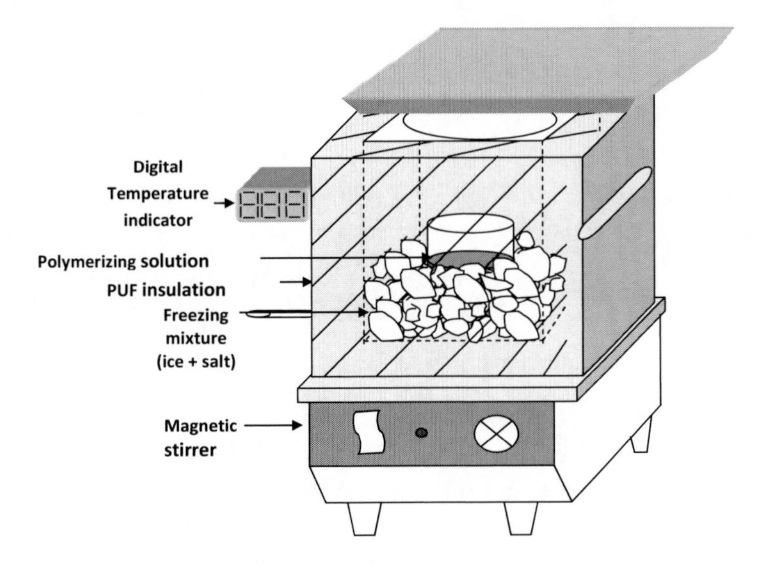

Figure 1.3. Schematic of experimental set up for polymerization of aniline.

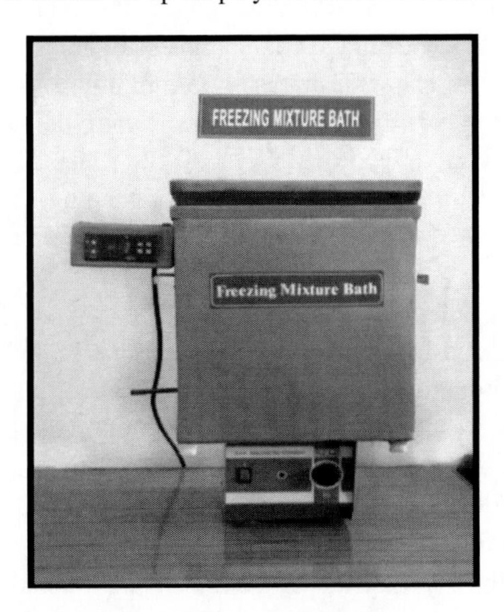

Figure 1.4. Actual experimental set up for polymerization of aniline.

The spin coating unit employed for the deposition of PAni thin film is as shown in Figure 1.5.

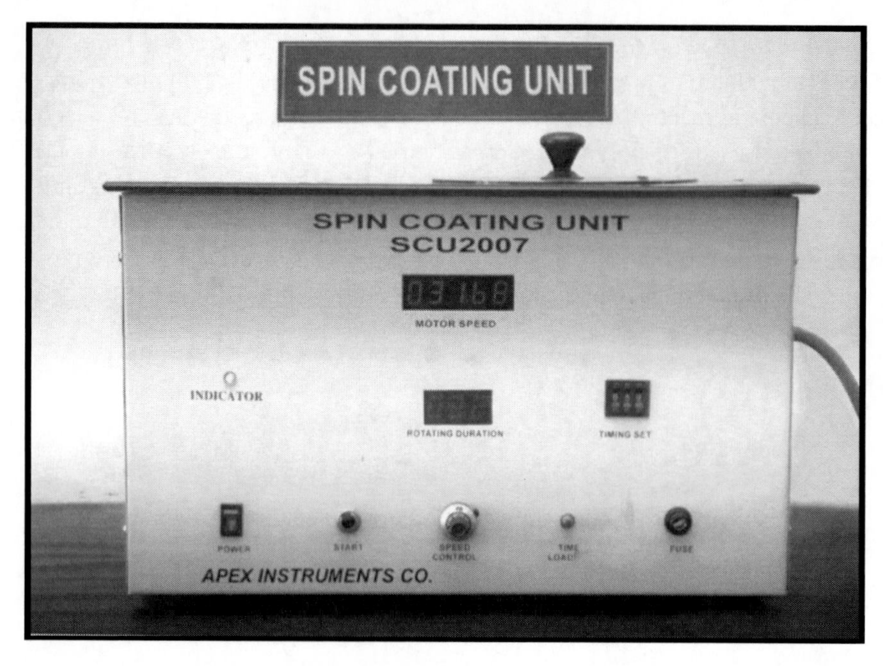

Figure 1.5. Spin coating unit for thin film deposition.

1.3.2. Substrate Cleaning

Substrate cleaning is the process of breaking the bonds between substrate and contaminants without damaging the substrate. In thin film deposition process, substrate cleaning is an important factor to get films with reproducible characteristics such as smoothness, uniformity, adherence and porosity and so on. The substrate cleaning process obviously depends on the nature of the substrate; degree of cleanliness required and nature of contaminates to be removed. The common contaminates are grease, adsorbed water, air borne dust, lint, oil particles etc. The glass substrates (1.35 mm thick) were used for the deposition of films. The following steps have been adopted for cleaning the substrate: (i) the substrates have been washed with AR grade hydrochloric acid, (ii) it has been washed with detergent and double distilled water, (iii) it has been subjected to cleaning ultrasonically for 15 min, and (iv) the substrates have dried, degreased in AR grade acetone and kept in dust free chamber.

1.3.3. Synthesis of PAni

PAni has been synthesized by polymerization of aniline in the presence of hydrochloric acid as a catalyst and ammonium peroxidisulphate as an oxidant. One method may be described as follows: 1 M 50 mL HCl and 2 mL of aniline have been added into a 250 ml beaker equipped with an magnetic stirrer. Then 4.9984 g ammonium peroxy-disulphate ($(NH_4)_2S_2O_8$) has been suddenly added into the above solution.

The polymerization temperature has been maintained to be ca 0 °C and the reaction has been allowed to continue for 7 hrs for its completion. Then the precipitate obtained has been filtered and washed successively by 1 M HCl followed by distilled water until the washed solution turned colorless.

For the preparation of emeraldine salt (ES) form of PAni, the obtained mass has been re-filtered and washed once again by distilled water and then dedoped with 0.1M NH_4OH solution by stirring for 3 hrs. Furthermore, the obtained mass has been filtered and dried at 60 °C in vacuum dryer for 24 hrs. This powdery material obtained is the insulating PAni (EB form) [52]. The pH values of both PAni (ES) and PAni (EB) have been determined and found to be 2 ± 0.1 and 11 ± 0.1, respectively [52, 53, 54].

1.3.4. Fabrication of PAni Thin Film Sensor

One of the fabrication methods of PAni thin film may be described shortly [52]. Figure 1.6 shows the flow diagram for synthesis and fabrication of PAni sensor film using the spin-coating method. The PAni (EB) powder has been put in *m*-cresol and stirred for 11 hrs to get a casting solution. By spin coating method, thin films of PAni have been coated on glass substrate by a single wafer spin processor (Apex Instrument Co. Model SCU 2007).

After setting the substrate on the disk of the spin coater, the coating solution of approximately 0.2 mL has been dropped and spin-coated by spinning the substrate at 3000 rpm for 40 s in air. The silver paste strips of 1 mm wide and 1 cm apart from each other were made on top of films for contacts. The obtained thin film has been dried on hot plate at 100 °C for 10 min. Figure 1.7 shows PAni (EB) sample and spin coated PAni thin film on glass substrate.

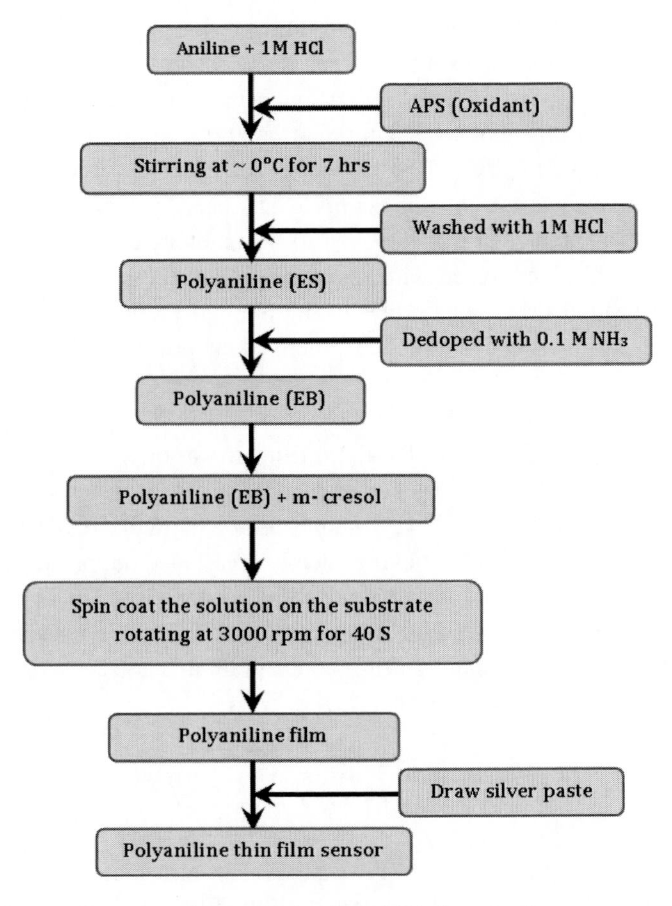

Figure 1.6. Flow diagram for synthesis and fabrication of PAni sensor film.

Figure 1.7. PAni (EB) sample and thin film.

1.3.5. Structural, Morphological, Optical and Electrical Characterization

X-ray diffraction (XRD) study has been carried out in 2θ range of 10–70° using an X-ray diffractometer (Model: Philips PW3710). High resolution transmission electron microscopy (HRTEM) and selected area electron diffraction (SAED) pattern have been measured to investigate the morphology and structure of PAni (EB) thin films. The HRTEM images have been recorded with a Hitachi Model H-800 transmission electron microscope (TEM). Morphological study the films has been carried out using scanning electron microscopy (SEM Model: JEOL JSM 6360) operating at 20 kV. Atomic force microscopy has been done using Nanoscope (Digital Instruments) IIIa. Fourier transform infrared (FTIR) spectroscopy (Model: Perkin Elmer 100) of PAni (EB) has been studied in the frequency range of 400–4000 cm^{-1}. A UV-vis spectrum of the sample has been recorded over 200-1000 nm wavelength range on a Simandzu 100 UV-vis spectrophotometer. The electrical transport properties viz, variation of resistance with temperature and thermo-emf of PAni (EB) thin films have been measured using four probe technique and thermoelectric power unit. The thickness of the film has also been determined using Dektak profilometer.

1.3.6. Gas Response Characteristics

For monitoring the gas response of the PAni films to various gases, the films have been mounted in the 250 mL air-tight container and contacts were made on silver strips. The known gas (NH_3, CH_3-OH, C_2H_5-OH, H_2S and NO_2) of particular concentration has been injected through a syringe. After achieving steady resistance in presence of a gas, the container has been exposed to atmosphere for the recovery of sensor.

The sensitivity has been estimated by measuring the difference in resistance of film in air (R_a) and in the presence of particular gas (R_g) concentration. The sensitivity of film may be calculated using the following equation:

$$S\,(\%) = \frac{|R_g - R_a|}{R_a} \times 100 \tag{3.3}$$

The response-recovery times have been calculated from a plot of resistance with time after exposure to gas as well as during the recovery of the film.

1.4. Results and Discussion

1.4.1. Structural, Morphological, Optical and Electrical Characterization of PAni (EB)

1.4.1.1. XRD Analysis

Figure 1.8 shows the XRD pattern of pure EB form of PAni in the 2θ range of 10-70°. A broad peak at 2θ = 25.30° which corresponds to (110) plane of PAni [52, 53, 54] has been observed. This broad peak indicates that the synthesized PAni is semi-crystalline as obtained by the other group for PAni synthesized by electrochemical method [55]. Similar amorphous/semi-crystalline nature of XRD analysis for conducting polymers (polypyrrole/PAni) has also been reported earlier [56].

1.4.1.2. Fourier Transform Infrared (FTIR) Spectroscopy

The chemical structure of PAni has been studied by measuring FTIR spectrum in the frequency range of 400-4000 cm^{-1} as typically shown in Figure 1.9. Tabular representation of different peaks observed in FTIR spectrum of PAni is shown in Table 1.1. The band at 3225–3451 cm^{-1} is resulted from the NH stretching of aromatic amines, whereas those observed at 2845–2914 cm^{-1} are due to the stretching of aromatic CH. The band at 504 cm^{-1} is observed for the CH out-of plane bending vibration. The CH out-of-plane bending mode has been used as a key to identifying the type of substituted benzene. The bands at 1572 and 1489 cm^{-1} are attributed to the C=N and C=C stretching mode of vibration for the quinonoid and benzenoid units of PAni, respectively. The bands at 1296 and 1239 cm^{-1} are assigned to the C–N stretching mode of benzenoid ring. The peak at 1239 cm^{-1} is the characteristic of the conducting protonated form of PAni. The bands in the region 1000–1115 cm^{-1} are due to in-plane bending vibration of C–H mode. The band at 797 cm^{-1} originates from out-of-plane C–H bending vibration [57].

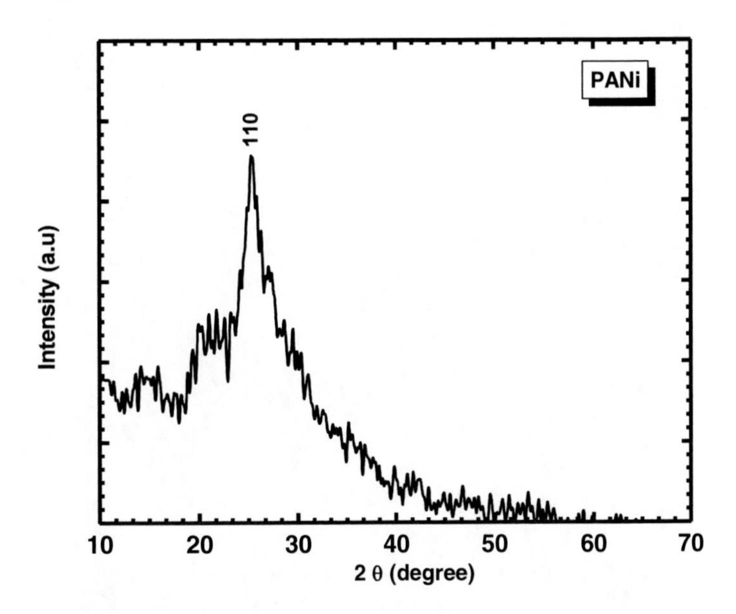

Figure 1.8. XRD pattern of PAni (EB).

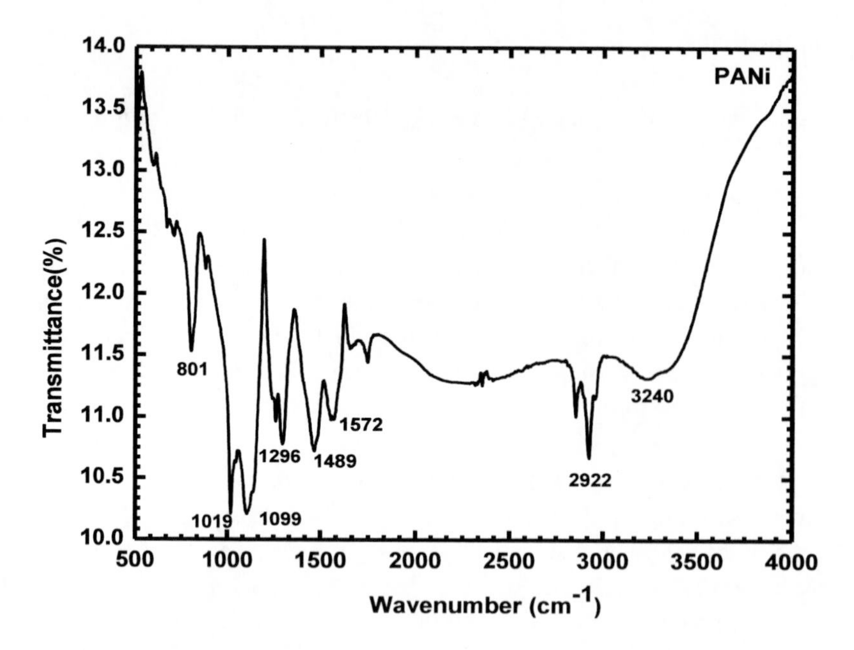

Figure 1.9. FTIR spectrum of PAni (EB).

Table 1.1. FTIR assignments of PAni (EB) [23]

Wave number (cm^{-1})	Assignments
3225–3451	NH stretching of aromatic amines
2845–2914	aromatic CH-stretching
504	CH out-of plane bending vibration
1572	C=C stretching mode of vibration
1489	C=N stretching mode of vibration
1296 and 1239	C–N stretching mode of benzenoid ring
1000–1115	in-plane bending vibration of C–H mode
801	out-of-plane C–H bending vibration

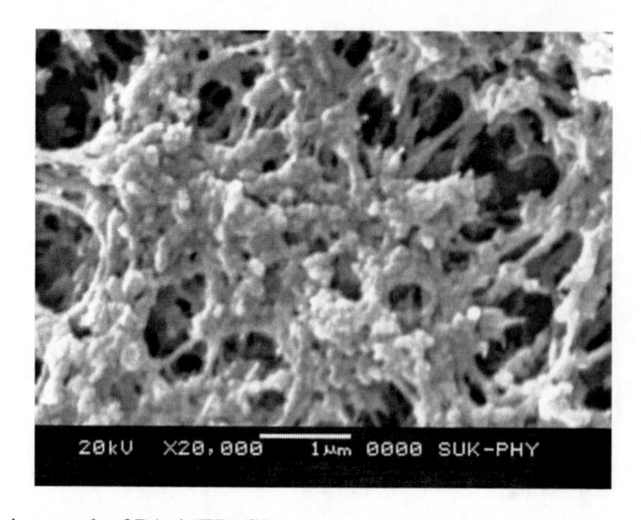

Figure 1.10. SEM micrograph of PAni (EB) film.

1.4.1.3. Scanning Electron Microscopy (SEM)

Figure 1.10 shows the SEM image of PAni (EB) thin film for magnification x 20,000. The SEM image of the PAni film exhibits a fibrous web-like structure. The contrast in the SEM image demonstrates that the film has hollow cavities which are highly porous. Practically, such a porous nature of the PAni makes it a potential candidate for various surface related applications. Such a porous surface of thin films of PAni (EB) is obviously suitable of gas sensing application, since fibrous morphology of PAni offers a high surface area, a large number of gas accommodation sites and a suitable pathway for a rapid diffusion of analytes into the film.

1.4.1.4. Atomic Force Microscopy (AFM)

The two- (2D) and three- (3D) dimensional surface morphologies of the PAni thin films have been investigated using AFM. Figure 1.11 shows the 2D and 3D AFM micrographs of PAni thin films.

From the micrograph (2D), total coverage of the substrate with fine grains of average size ~ 40-50 nm has been seen. On the other hand, it is seen that the film consists of distributed cuboidal shaped grains with some visible voids in 3D micrograph (Figure 1.11 (b)), which is consistent with the SEM image (Figure 1.10).

1.4.1.5. Transmission Electron Microscopy (TEM)

The morphology of strongly interconnected grains has been further studied with HRTEM as shown in Figure 1.12 (a). It shows that the film composed of interconnected nanograins of average diameter around 50 nm, which is consistent with grain size determined from SEM studies.

Figure 1.12 (b) shows corresponding selected area electron diffraction (SAED) pattern of PAni nanograin. The blurred bright electron diffraction rings show that the PAni film is amorphous or poorly crystalline, supporting to XRD result (Figure 1.8).

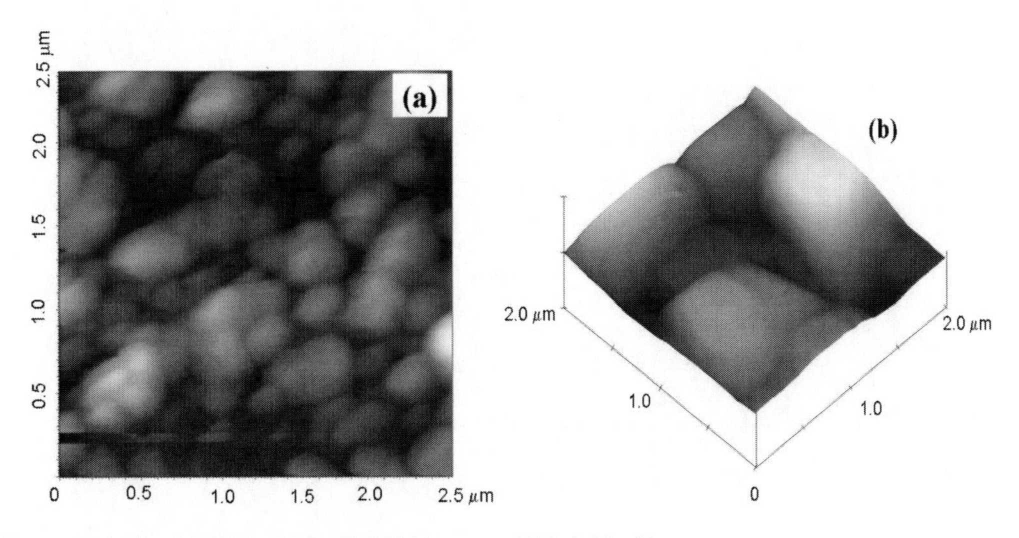

Figure 1.11. The 2 D (a) and 3 D (b) AFM images of PAni thin film.

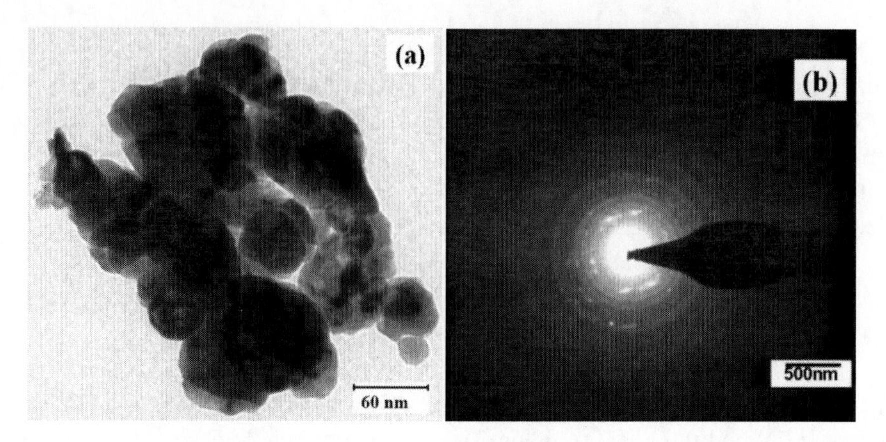

Figure 1.12. (a) TEM image of PAni nanograins and (b) SAED pattern of corresponding PAni nanograin.

1.4.1.6 UV-vis Analysis

Figure 1.13 shows the variation of optical absorbance (αt) with incident photon wavelength (λ) of the PAni (EB). It appears three distinct peaks of PAni at about 336, 451 and 924 nm, which are attributed to the π–π*, polaron-π* and π-polaron transition, respectively [58, 59]. A steadily increasing free carrier tail, starting from < 850 nm confirms the extended coil conformation of the PAni [60-62]. The band gap of PAni (EB) has been estimated from a plot of $(\alpha h\upsilon)^2$ vs. photon energy (hυ). The intercept of the tangent to the plot on x-axis gives a good approximation of the band gap energy for a direct band gap material (see inset of Fig 1.13). The optical band gap of PAni has been compared with the band gap of bulk PAni (E_g = 2.5 eV).

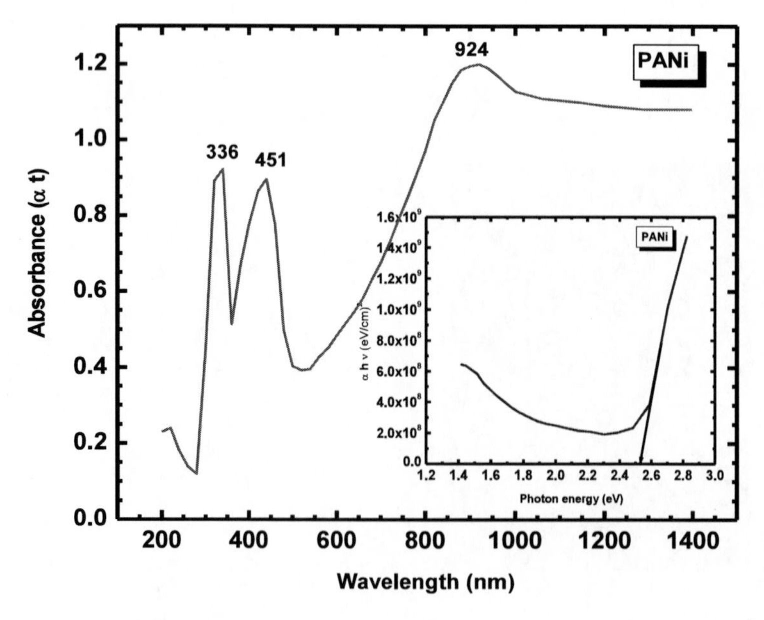

Figure 1.13. Variation of optical absorbance vs. wavelength of PAni. Inset: Plot of $(\alpha h\upsilon)^2$ vs. photon energy to determine the direct band gap energy.

The band gap has been found to be 2.54 eV that is slightly greater than the bulk PAni. It is well known that the band gap energies for the well-crystallized thin films are comparable to those of crystallized bulk materials, whereas in amorphous and/or nano-crystallized forms, the band gap energies are higher than those of the corresponding bulk materials [63]. Thus, the observed slightly larger band gap energy may be attributed to the size quantization of amorphous PAni thin films.

1.4.1.7. Electrical Resistance Measurement

Figure 1.14 shows the variation of electrical resistance as a function of temperature of pure PAni. It was observed that as temperature increases the electrical resistance decreases and hence conductivity increases. This suggests that the thermally activated behavior of conductivity has been confirmed.

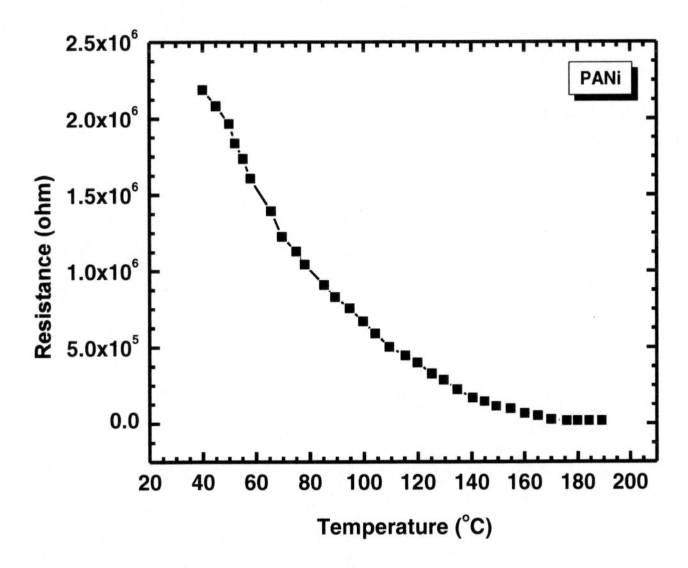

Figure 1.14. Plot of resistance versus temperature of PAni (EB).

It is suggested that the thermal curling affects the alignment of chain in the polymer, which leads to the increase of conjugation length and hence decrease the resistance. Also, heating results in molecular rearrangement that makes the molecules to be at the favorable stage for electron delocalization [64]. The temperature dependence of the conductivity for conducting polymers is expressed by a variable range hopping (VRH) model that has been proposed by Mott [65].

According to this model, the behavior of electronic conduction in disordered and non-metallic materials is controlled by the thermally assisted hopping of electrons between localized states near randomly distributed traps and the resistance. Mathematically, this can be expressed as follows:

$$R(T) \,\alpha\, \exp[-T_0 / T^{1/(n-1)}] \quad (3.4)$$

$$k\,T_o = \lambda\, d^3/\rho_0 \quad (3.5)$$

where, d is the coefficient of exponential decay of the localized states, ρ_0 is the density of states at the Fermi level and λ is a dimensional constant. However, many models have predicted that $R = T^{-1/2}$.

1.4.1.8. Thermo-emf Measurement

The temperature difference causes a transport of carriers from the hot to cold end and thus creates an electric field, which gives rise to a thermally generated voltage. The thermo-emf property of PAni has been measured as a function of temperature and is shown in Figure 3.15. The thermo-emf developed between two ends showed that the PAni shows a p-type electrical conductivity and that the holes contribute to TEP [67]. The thermo-emf increased linearly with increasing temperature. Yakuphanoglu et al. [68] have also reported the thermoelectric power measured for HCl-doped organo-soluble PAni increases with the increase in temperature. However, the TEP results indicate that the conductivity mechanism of the polymer is controlled by the large polaron hopping model.

1.4.1.9. Thermo-Gravimetric Analysis (TGA) and Differential Thermal Analysis (DTA)

TGA and DTA analyses of powder form PAni have been carried out at a heating rate of 10 K/min in air atmosphere from 273 to 1273 K as shown in Figure 1.16. The thermal evolution in air atmosphere takes place in four consecutive stages corresponding to weight losses in which the inflection point coincides with the temperature of the endotherms and exotherms in DTA trace. The weight loss of PAni begins at 323 K. The weight loss commencing at around 373 K has been assigned to the loss of initially present water molecules [69]. Rapid weight loss has been found in temperature range of 493–743 K due to the consequence of structural decomposition of the polymer and elimination of dopant molecules. After 743 K, the DTA trace has been found to be stable with no further weight loss has been observed. This indicates that the PAni is stable up to 323 K and then PAni starts to undergo degradation process slowly. The smooth thermogram shows only one exothermic peak at 668 K, where thermal decomposition of PAni takes place. Ansari and Keivani [70] have reported to observe the similar behavior of PAni that has been prepared by electropolymerization using cyclic voltametric technique. The authors have reported that the PAni prepared by electrochemical method shows a high thermal stability than that of PAni prepared by potentiosatic mode.

1.4.2. Gas Sensing Studies

Gas sensing study of PAni films fabricated by the method described in the section 1.3.4. The actual gas sensor setup used is as shown in below Figure 1.17.

1.4.2.1. Selectivity of PAni Thin Film Sensor

The ability of a sensor to respond to a certain gas in the presence of other gases is known as selectivity. The selectivity of a sensor in relation to a definite gas is closely associated with its operating temperature. Here, the selectivity measured in terms of selectivity coefficient/factor of a target gas to another gas is defined as $K = S_A/S_B$, where S_A and S_B are the responses of a sensor to a target gas A and an interference gas B, respectively. Figure 1.18 shows the histogram of maximum gas response to different gases for PAni (EB) film.

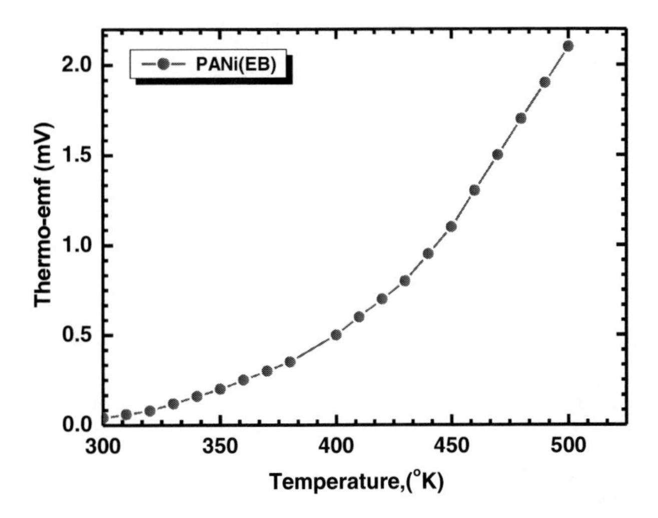

Figure 1.15. Variation of thermo-emf with temperature of PAni thin film.

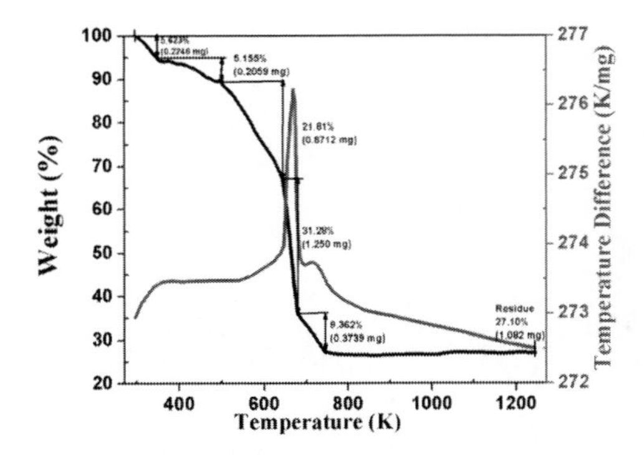

Figure 1.16. TGA - DTA spectra of PAni thin film in air atmosphere.

Figure 1.17. Actual gas sensing measurement set up.

The histogram revealed that the PAni (EB) sensor offered maximum response to NH_3 (23.66 %), NO_2 (0.8 %), CH_3OH (0.1 %), C_2H_5OH (0.15%) and H_2S (0.307 %) at room temperature for 100 ppm. This may be due to the different gases have different energies for reaction to occur on the surface of PAni (EB) sensor. Thus PAni (EB) sensor is highly selective to the NH_3 over CH_3OH compared to H_2S, C_2H_5OH, NO_2 (S_{NH3}/S_{CH3OH} = 236.66,

S_{NH3}/S_{C2H5OH} = 157.73, S_{NH3}/S_{H2S} = 77.06 and S_{NH3}/S_{NO2} = 29.58) at room temperature. Therefore, the further dependence of NH_3 response has also been studied for various concentration of NH_3 at room temperature.

1.4.2.2. Ammonia Sensing Properties of PAni Thin Film Sensor

As ammonia is more selective to PAni; the gas sensing studies have been carried out for ammonia (NH_3) at room temperature. Therefore the sensitivity of PAni towards different concentrations of NH_3 has been tested.

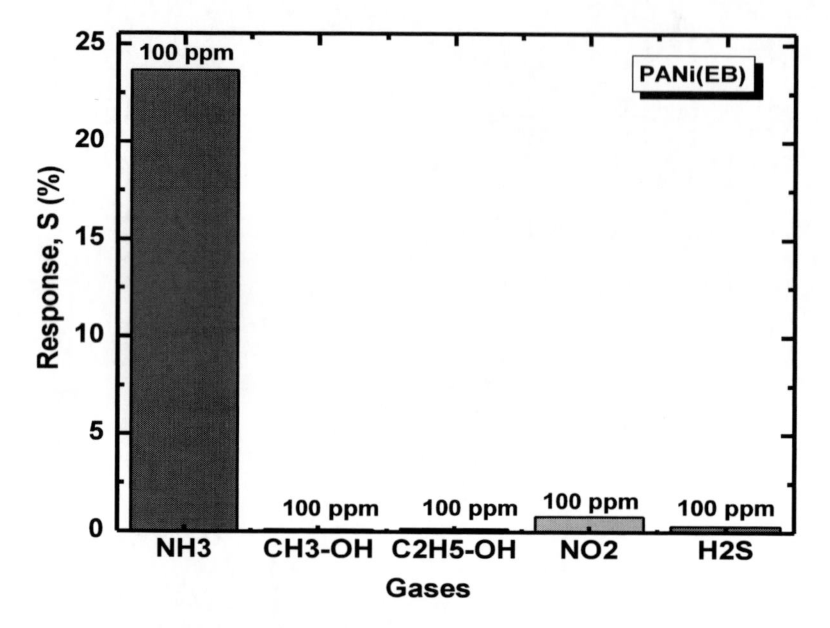

Figure 1.18. Selectivity bar chart of PANi (EB) thin film sensor.

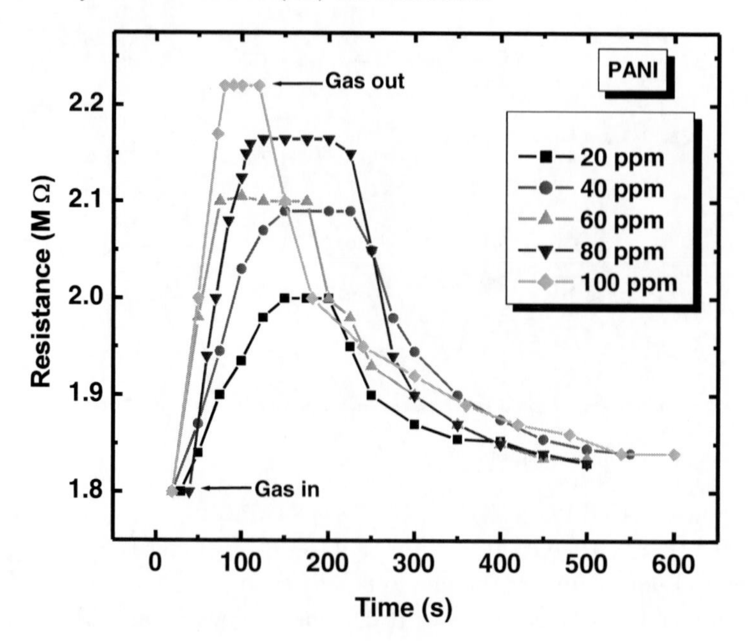

Figure 1.19. Gas responses of PANi film to NH_3 (20-100 ppm).

Figure 1.19 shows electrical response of PAni (EB) to 20, 40, 60, 80 and 100 ppm of NH_3. It has been observed that when ammonia vapor has been injected into the gas testing chamber, the resistance of PAni film has been found to increase initially from its base resistance value (R_a) and then to attain a stable value (R_g). When the chamber has been flashed with a clean air, the response has been found to decay exponentially as shown in Figure 1.19. The transient response has been attributed to be resulted for the high surface area offered by nanofibrous morphology and the ease of diffusion of NH_3 in the nanofibrous film.

1.4.2.3. Gas response Transient of PAni Thin Film

The responses of PAni thin film sensor to different concentrations of ammonia (20-100 ppm) is as shown in Figure 3.20. The interaction process between the thin film and the adsorbed gas is a dynamical process. When the thin film is exposed to NH_3 gas, the adsorption and desorption processes simultaneously takes place and thus, the thinner the films, the quicker the gas desorption is. Consequently, the resistance initially increases and attains a stable value when dynamic equilibrium is attained. However, further interaction mechanism study needs to be carried out to verify this speculation [71]. The response values of PAni (EB) thin film sensor is plotted as a function of NH_3 concentration as shown in Figure 1.21. It has been generally observed that the value of response increases as the concentration of NH_3 increases. The rate of increment of response with concentration has been found to be large for a low concentration range of 20-40 ppm, whereas it has been to be relatively small for a high concentration range of80-100 ppm. The highest response value (23.66 %) has, however, been obtained for 100 ppm of NH_3. The rapid response may be attributed to higher surface area offered by nanofibrous morphology and the ease of diffusion of NH_3 in the nanofibrous film. Moreover, the relative response tends to increase with increasing concentrations and this phenomenon may possibly be due to the availability of a large number of reactive species in the sensing layer [72].

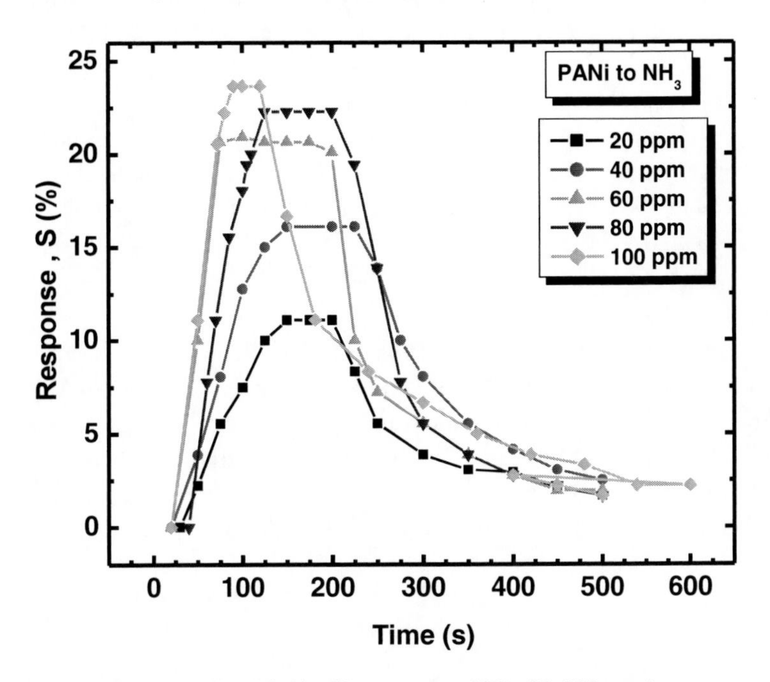

Figure 1.20. Dynamic responses of PANi thin film sensor to NH_3 (20-100 ppm).

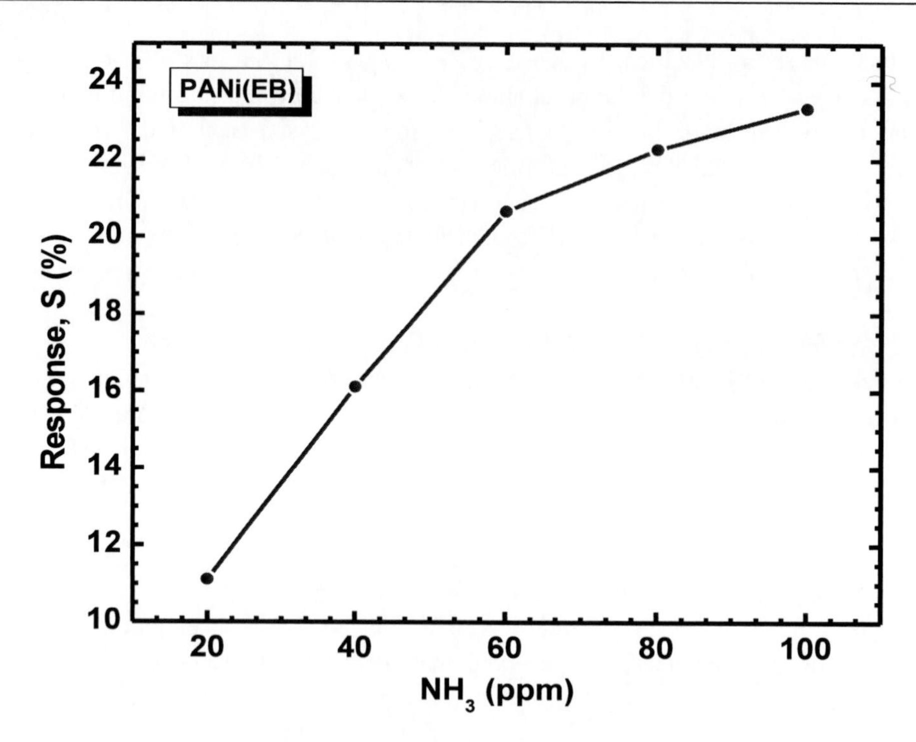

Figure 1.21. Response of PANi thin film sensor to NH_3 (20-100) ppm.

1.4.2.4. Response and Recovery Characteristics of PAni Films

The response and recovery times are important parameters in gas sensing. The time taken for the sensor to attain 90 % of the maximum increase in resistance on exposure of the target gas is known as response time. Similarly, the time taken by the sensor to get back 90 % of the maximum resistance when the film is exposed to clean air is known as recovery time. After exposure to clean air, the recovery time has been measured till the resistance nearly dropped to the original value.

Figure 1.22 represents variation of the response and recovery times with different concentration of NH_3 at room temperature. It has been revealed that as the NH_3 concentration increases, response time decreases from 110 to 72 s and corresponding recovery time increases from 270 to 480 s. The response time is the order of 72 s and recovery time is 8 min for 100 ppm of NH_3.

1.4.2.5. Stability of PAni Sensor

In order to check the stability of PAni sensor element, the electrical resistance measurements have been performed at room temperature upon exposure of 100 ppm NH_3 for 45 days, where the interval between two measurements has been 5 days. The typical results of gas response with time are illustrated in Figure 1.23. Initially, PAni gas sensor shows relatively high response and drops from 23 to 14 %. The stable sensitivity has been found to attain after 15 days. This has been considered to the interface modification of PAni at the initial stage.

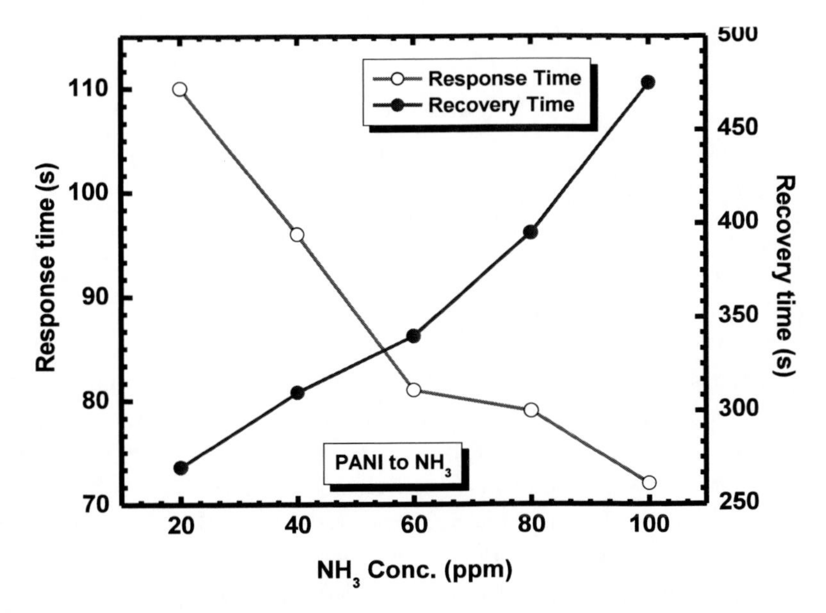

Figure 1.22. Variation of response and recovery time of the PANi thin film sensor with NH3 concentration.

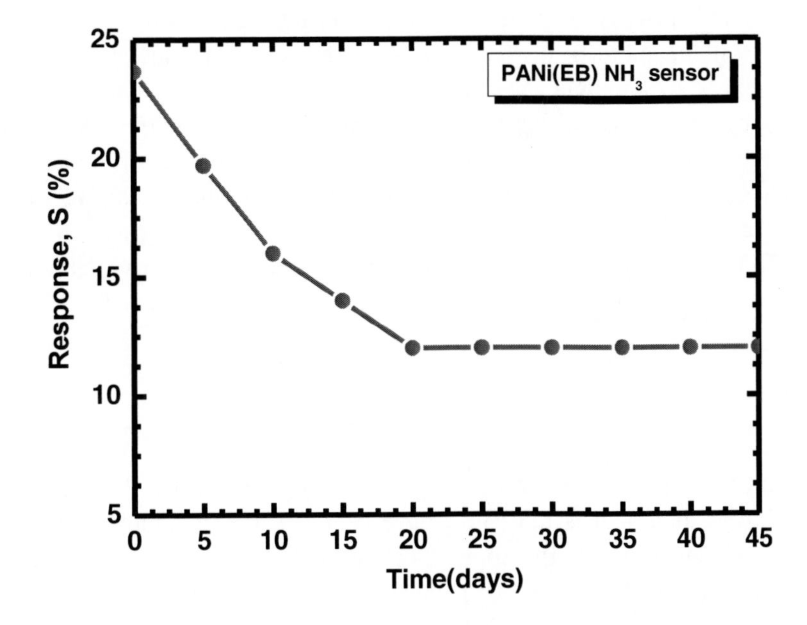

Figure 1.23. Stability studies of PAni sensor upon exposure of 100 ppm NH_3.

1.4.2.6. Discussion of Sensing Mechanism and Response Characteristics

The higher sensitivity towards NH_3 can be explained on the basis of different interactions between sensing film and adsorbed gas. The physical properties of PAni strongly depend on their doping levels.

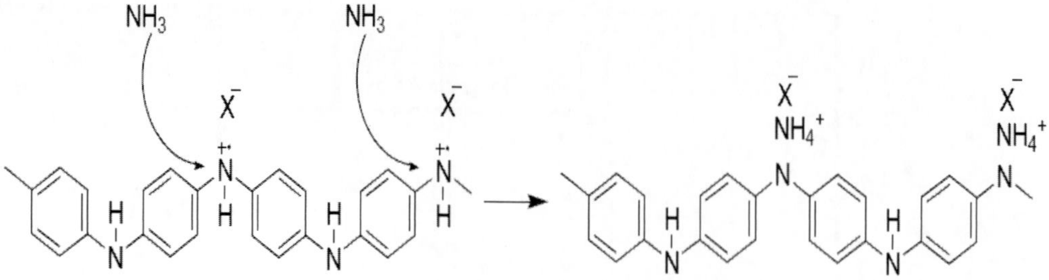

Figure 1.24. The deprotonation of PAni by ammonia.

Fortunately, the doping levels of PAni can be easily changed by chemical reactions with many analytes at room temperature, and this provides a simple technique to detect the analytes. The doping level can be altered by transferring electrons from or to the analytes. Electron transferring can cause the changes in resistance and work function of the sensing material. The work function of a conducting polymer is defined as the minimal energy needed to remove an electron from bulk to vacuum energy level.

When exposed to ammonia gas, PAni undergo dedoping by deprotonation [73-78]: The protons on –NH– groups has been propose to transfer to NH_3 molecules to form ammonium ions while PAni itself turned into its base form as shown in Figure 1.24. This results in decreasing the conductivity of material and as a result, the film resistance increases. This process is reversible, and in fact, when ammonia atmosphere is removed, the ammonium ion can be decomposed to ammonia gas and proton.

In case of oxidizing gas like NO_2, there is an increase in charge carrier concentration, since NO_2 is an electron accepting in nature. Hence, it increases the conductivity of the material. Thus the sensing mechanism of polymers essentially deals with modulating the redox levels of polymers. Other than redox reactions, partial charge transferring also leads to alteration of conductivity. The direction of transfer is determined by electronegativity and work function of vapor [79].

CONCLUSION

PAni has been synthesized by polymerization of aniline by the chemical oxidative polymerization method and its thin films have been prepared by spin coating technique. A detailed study of structural, morphological, optical and electrical properties has been carried out. The XRD pattern of PAni shows amorphous structure and broad peak at $2\theta = 25.30°$, which corresponds to (110) plane. The typical morphology observed for the films has been found to be porous and interconnected network of fibers. Based on TEM analyses, the formation of the PAni nanograins with average diameter of about 50 nm has been inferred. The AFM images show the less aggregation of the polymer with a surface roughness of ~ 7 nm. UV-vis spectra of PAni shows three distinctive peaks at about 341, 441 and 924 nm, which is attributed to the π-π*, polaron-π* and π-polaron transition respectively. The band gap has been found to be 2.54 eV, which is slightly greater than the bulk PAni. The high band gap energy has been attributed to the size quantization of semi-crystalline PAni thin films. It has also been observed that as temperature increases the electrical resistance decreases and hence conductivity increases. These observations suggest the thermally activated behavior of conductivity of PAni film. The p-type electrical conductivity of PAni has been confirmed from thermo-emf studies. The thermal stability study has revealed that the PAni is stable only up to 323 K.

The gas sensing studies for different reducing and oxidizing gases like ammonia, ethanol, methanol, hydrogen sulfide and nitrogen dioxide at room temperature have been carried out. The gas sensing studies shows that PAni films are selective to ammonia gas. The response curves obtained indicate higher sensitivity as well as fast response and recovery for ammonia gas. The response time is the order of 72 sec and recovery time is of 8 min for 100 ppm of NH_3. The sensitivity of sensor has been found to be linear between 20-100 ppm of ammonia. It has been proposed that PAni film undergos dedoping by deprotonation when it exposes to ammonia gas. The protons on –NH– groups are proposed to transfer to NH_3 molecules to form ammonium ions, while PAni itself turns into its base form. This results in the reduction of majority charge carrier density.

REFERENCES

[1] Kaiser A.B. *Rep. Prog. Phys.* 64 (2001) 1.
[2] Salaneck W.R *Prog. Phys.* 54 (1991) 1215.
[3] Kamloth K.P. *Chem Rev.* 108 (2008) 367.
[4] Hatchett D.W, Josowicz M. *Chem. Rev.* 108 (2008)746.
[5] Mac Diarmid A.G, Angew. *Chem. Int.* Edit. 40 (2001)2581.
[6] Chiang C.K, Fincher C.R, Park Y.W, Heeger A.J, Shirakawa H, Lious E.J, Gau S. C, MacDiarmid, A.G. *Phys. Rev. Lett.* 39 (1977) 1098.
[7] Naarmann H. *Polymers to the year 2000 and beyond*, John Wley & Sons (1993).
[8] Margolis J. *Conductive polymers and plastics*, Chapman and Hall (1989).
[9] Alcacer L. *Conducting Polymers Special Applications*, D. Reidel Company (1987).
[10] Salaneck W.R, Clark D.T, Samuelsen E. J. *Science and Application of Conducting Polymers*, IOP Publishing (1991).
[11] Heeger J.A. *Current Appl. Phys.* 1 (2000) 247.
[12] Pietronero L. *Syn. Met.* 8 (1983) 225.
[13] Hauser J.J. *Phys. Rev.* B 9 (1974) 2623.
[14] Mott N.F, Davis E.A. *Electronic processes in Non-Crystalline Materials* 2nd Edn (Oxford: Clarendon) (1979).
[15] Efros A.L, Shlovskii B. I. J. *Phys. C: Solid State Phys.* 8 (1975) L49.
[16] Epstein A.J. *Advances in Synthetic Metals P Brnier*, S Lefrant and G Bidan (1999) 349.
[17] Sheng P, Klafter *J. Phys. Rev.* B 27 (1983) 2583 .
[18] Sheng P. *Phil. Mag.* 65 (1992) 357.
[19] Ferry D.K, Goodnick S.M, *Transport in nanostructures* (Cambridge: Cambridge University press) (1997).
[20] Sheng P. *Phys. Rev. B* 21 (1980) 2180.
[21] Roth S. *One dimensional Metals* (Wienheim: VCH) (1995).
[22] Roth S, Bleier H. *Adv. Phys.* 36 (1987)385.
[23] Heeger A.J, Kivelson S, Schrieffer J.R, Su W.P. *Rev. Mod. Phys.* 60 (1988) 751.
[24] Kivelson S. *Phys. Rev. Lett.* 46 (1981) 1344.

[25] Epstein A.J, Joo J, Kolhman R.S, Du S, MacDiarmid A.G, Oh E.J, Min Y, Tsukamoto J, Kaneko H and Pouget J. P. *Syn. Met.* 65 (1994) 149.
[26] Ehinger k, Roth S. *Phil. Mag.* B 53 (1986) 301.
[27] Stafstrom S, Breadas J. L. *Phys. Rev. B* 38 (1988) 4180.
[28] Zuppiroli L, Bussac M. N, Paschen S, Chauvet O, Forro L. *Phys Rev. B* 50 (1994) 5196.
[29] Ameer Q, Adeloju S. B. *Sens. & Actu.* B 106 (2005) 541.
[30] Ramanathan K, Bangar A, Yun M, Chen W, Myung N V, Mulchandani A. *J. Am. Chem. Soc.* 127 (2005) 496.
[31] Syed A. A, Dinesan M. K. *Talanta* 38 (1991) 815.
[32] Zhao Y, Chen M, T Xu T, Liu W. *Colloids and Surfaces A* 257 (2005) 363.
[33] Liu C, Hayashi K, Toko K. *Electrochemistry Communications* 12 (2010) 36.
[34] Hussain A. M.P, Kumar A. *Bull. Mater. Sci.* 26 (2003) 329.
[35] Bedekar A.G, Patil S.F, Patil R.C, Vijaymohanan K. *Mater. Chem. Phys.* 48 (1997) 76.
[36] Sazou D. *Synth. Met.*118 (2001) 133.
[37] Eftekhari A, *Synth. Met.*145 (2004) 211.
[38] Ivanov S, Mokreva P, Tsakova V, Terlemezyan L. *Thin Solid Films* 441 (2003) 44.
[39] Huang J. *Pure Appl. Chem.*, Vol. 78, No. 1, pp. 15–27, 2006.
[40] Delvaux M, Duchet J, Stavaux P, Legras R, Champangne S. *Synthetic metals* 113 (2000) 275.
[41] Yan X.B, Han Z.J, Yang Y, Tay B.K. *Sens. and Actu. B* 123 (2007) 107.
[42] Zeng F.W, Liu X. X, Diamond D, King K. T. *Sens. and Actu. B* 143 (2010) 530.
[43] Li Z.F, Blum F.D, Bertino M.F, Kim C.S, Pillalamarri S.K. *Sens. and Actu. B* 134 (2008) 31.
[44] Ayad M.M, Salahuddin N.A, Alghaysh M.O, Issa R.M. *Current Applied Physics* 10 (2010) 235.
[45] Crowley K, Morrin A, Hernandez A, O'Malley E, Whitten P.G, Wallace G.G, Smyth M. R, Killard A.*J. Talanta* 77 (2008) 710.
[46] Trivedi D. C. *Ind. J. Chem.* A33 (1994) 552.
[47] Draman S.F.S, Daik R, Musa A. *Proceedings of World Academy of Science, Engineering And Technology*, 33 (2008) 2070.
[48] Manigandan S, Jain A, Majumder S, Ganguly S, Kargupta K. *Sens. and Actu. B* 133 (2008) 187.
[49] Jiang S, Chen J, Tang J, Jin E, Kong L, Zhang W, Wang C. *Sens. and Actu. B* 140 (2009) 520.
[50] Geng L, Zhao Y, Huang X, Wang S, Zhang S, Wu S. *Sens. and Actu. B* 120 (2007) 568.
[51] Tai H, Jiang Y, Xie G, Yu J, Chen X, Ying Z. *Sens. and Actu. B* 129 (2008) 319.
[52] Pawar S.G, Patil S.L, Chougule M. A, Mane A.T, Jundale D. M, Patil V.B. *International Journal of Polymeric Material* 59 (2010) 777.
[53] Ku B. C, Lee S.H, Liu W, Kumar J, Bruno F. F, Samuelson L.A. *Mat. Res. Soc. Symp. Proc.* 708 (2002) 12.
[54] Mu. S.L, Kong Y, Wu J. *Chinese J. of Poly. Sci.* 22 (2004) 405.
[55] Lekha P. C, Subramanian E, Pathinettam D, *Padiyan. Sens. and Actu. B* 122 (2007) 274.
[56] Skotheim T.A. "*Handbook of conducting polymers*", Marcel Dekker, New York (1986) 118.

[57] Pawar S.G, Patil S.L, Mane A.T, Raut B.T, Patil V.B. *Archives of App. Sci.* Research 1 (2) 2(009) 109.
[58] Gospodinova N, Terlemezyan L. *Prog. Polym. Sci.* 23 (1998)1443.
[59] Xia H. S, Wang Q. *Chem. Mater.* 14, (2002) 2158.
[60] Pawar S.G, Patil S.L, Patil V.B, *Oriental Journal of Chemistry* 4 (2009) 25.
[61] Pawar S.G, Patil S.L, Raut B.T, Patil V.B. International Workshop on NanomaterialsAnd Characterization, 9-11 December (2008), Shivaji University, Kolhapur, India.
[62] Pawar S.G, Patil S.L, Raut B.T, Patil V.B. Homi Bhabha Centenary BRNS-GND University Workshop on Molecular and Organic Electronic Devices MOED-(2009), 22-25.
[63] Chattopadhyay K K, Banergee A N. *Introduction to Nanoscience and Nanotechnology*, PHI Learning Pg. 142 2009.
[64] Kobayashi A, Ishikawa H, Amano K, Satoh M, Hasegawa E. *J. Appl. Phys.* 74 (1) (1993) 296.
[65] Mott N. F, Davis E. *Electronic Processes in Noncrystalline Materials*, Clarendon Press, Oxford, 1979.
[66] Joshi A.C. Ph. D. Thesis 'Fabrication and characterization of Polypyrrole, ZnO and their composite based gas sensors for oxidizing and reducing gases', (2010).
[67] Joshi S. S., Lokhande C. D., *Appl. Surf. Sci.* 252 (2006) 8539.
[68] Yakuphanoglu F, Senkal, B. F. A. Sarac¸ *J. Electron. Mater.* 37 (2008) 6.
[69] Duval C., *Inorganic Thermogravimetric Analysis*, Elsevier, Amsterdam (1963) 315.
[70] Ansari R., M. Keivani E B., *J. Chem.* 3 (2006) 202.
[71] Tai H, Jiang Y, Xie G, Yu J, Chen X, Ying Z. *Sens. and Actu. B* 129 (2008) 319.
[72] Sutar D.S, Padma N, Aswal D.K, Deshpande S.K, Gupta S.K, Yakhmi J.V. *Sens. and Actu. B* 128 (2007) 286.
[73] Jin Z, Su Y.X, Duan Y.X. *Sens. and Actu. B* 72(2001) 75.
[74] Nicho M.E, Trejo M, Garcia-Valenzuela A, Saniger J.M, Palacios J, Hu H. *Sens. Actu. B* 76 (2001) 24.
[75] Hu H, Trejo M, Nicho M.E, Saniger J.M, Garcia-Valenzuela A. *Sens. Actu. B* 82 (2002) 14.
[76] Bekyarova E, Davis M, Burch T, Itkis M.E, Zhao B, Sunshine S, Haddon R.C. *J. Phys. Chem. B* 108 (2004) 19717.
[77] Hong K.H, Oh K.W, Kang T.J. *J. Appl. Polym. Sci.* 92 (2004) 37.
[78] Liu H.Q, Kameoka J, Czaplewski D.A, Craighead H.G. *Nano Lett.* 4 (2004) 671.
[79] Blackwood D, Josowicz M. *J. Phys. Chem.* 95 (1991) 493.

In: Trends in Polyaniline Research
Editors: T. Ohsaka, Al. Chowdhury, Md. A. Rahman et al.
ISBN: 978-1-62808-424-5

Chapter 2

POLYANILINE DEPOSITION ON ALUMINIUM ALLOYS: SYNTHESIS, CHARACTERISATION AND APPLICATIONS

Teresa To and Paul A. Kilmartin
Hybrid Polymers Research Group, Polymer Electronics Research Centre, School of Chemical Sciences, University of Auckland, Auckland, New Zealand

ABSTRACT

Synthesis of Polyaniline (PAni) can be undertaken either chemically or electrochemically. Electrochemical syntheses typically employ galvanostatic, potentiostatic or potentiodynamic methods, most often on inert metal and carbon-based electrodes, to produce films directly adherent on the electrode surface. More active metals can also be employed for the deposition of conducting polymers, and the case of aluminium and its alloys are the focus of the current chapter.

Electropolymerisation of PAni offers the advantage of being able to control the chemical and physical properties of the coatings formed on the substrate by changing electrochemical parameters (e.g. voltage), from a monomer-electrolyte solution. The electrochemical deposition of PAni on an electrode surface also involves oxidation of the substrate prior to electropolymerisation, including the formation of a passive layer on the electrode surface. The protective oxide layer on aluminium alloys is well-known to exhibit excellent resistance to general corrosion. The passive film plays two important roles: protection of the electrode against dissolution, and the establishment of a suitable surface for deposition of the conducting polymer. However, aluminium often suffers from localised corrosion when exposed to an environment containing aggressive chloride ions. The deposition of PAni onto aluminium provides the opportunity for protection of aluminium alloys in more aggressive environments.

PAni coated on aluminium and its alloys has been the subject of a significant number of studies over the past 15 years. The PAni layer is typically characterised by both electrochemical techniques, such as cyclic voltammetry and AC impedance, and by spectroscopic and surface science analyses, such as FTIR, SEM and XPS. The easily synthesised and air stable PAni-coated aluminium has been considered for a range of applications, owing to the antibacterial and antifouling properties of the PAni films, in addition to their corrosion-resistance.

INTRODUCTION

Polyaniline (PAni) is one of the most promising conducting polymers, and has considerable potential for many applications due to its good processability, ease of preparation [1, 2], environmental stability and relatively low cost [3]. PAni exists in three different oxidation states: leucoemeraldine (fully reduced form), emeraldine (partially oxidized form) and pernigraniline (fully oxidized form). The emeraldine salt form of PAni is the only conducting state of the polymer, which also relies on acid-doping for high conductivity and electroactivity. The ability of the polymer to easily adjust conducting oxidation states creates great interest for industry [4]. Given its versatility and the various forms available, PAni is suitable for many applications, including the following: secondary batteries, biosensors [1], corrosion protection [3, 5], conducting coatings [1, 3, 6-15] films [16-19], composites [20] and antistatic packing materials [6]. At the same time, conducting polymers can have drawbacks due to their intractable nature in certain solvents [21].

PAni can be prepared by various methods: electrochemical polymerisation, chemical polymerization, photochemically initiated polymerization, enzyme-catalyzed polymerization and polymerization employing electron acceptors. The easily synthesised and air stable PAni-coated aluminium has been considered for a range of applications, owing to the antibacterial and antifouling properties of the PAni films, in addition to corrosion resistance. In this chapter, the focus is placed upon chemical and electrochemical oxidative polymerization of aniline monomers in acidic solutions, for their application to the coating of aluminium surfaces.

1. CHEMICAL POLYMERIZATION OF ANILINE

Chemical polymerization of aniline can be carried out under different synthesis conditions [22], with a variety of oxidizing agents, preparation media, dopant concentrations, resulting in variations in the chemical and redox state of the PAni. An example of a conventional method of preparation is the use of ammonium persulfate as an oxidant in an acidic medium such as hydrochloric acid or sulfuric acid, where a precipitate is formed that can be washed and dried [6]. The PAni powders can then be dispersed in a coating system for application on aluminium surfaces, as discussed below. An advantage of chemical polymerisation is that this is a simple process capable of producing bulk quantities of PAni, and is thus well suited for commercial purposes.

The chemically synthesized PAni in either the doped (emeraldine salt) or undoped (emeraldine base) form can be incorporated into coatings for use as a primer or top coat for the protection of active metals against corrosion and fouling. PAni can be used a primer alone, by adding a relatively small concentration (~10 wt %) of chemically synthesised PAni in a coating system [12, 13]. The PAni primer coating can also be followed by a conventional epoxy and polyurethane resin topcoat. In relation to obtaining a good epoxy formulation with the incorporated PAni, it needs to be noted that PAni prepared under normal conditions renders the epoxy powder less electrically charged, and due to the amine terminal groups in PAni, and so the curing of the epoxy becomes more rapid [23]. However, this can be overcome by the use of an appropriate PAni chemical synthesis procedure to produce a successful epoxy coating formulation [23]. PAni can also be used as an additive in paint formulations [8], and can replace commonly used inorganic anticorrosive inhibitors. Modification with a quite low concentration of conducting polymer (less than 1 wt %), usually shows better corrosion resistance than unmodified paints [14, 15]. However, for long term protection, a higher pigmented level is required to allow slow release over time of corrosion inhibitors, particularly in a marine environment [8].

There are many reports that dispersing PAni in both the doped and undoped form in a coating can display corrosion-prevention behaviour [22]. However, the selection of solvent has been challenging in this area, as PAni in the doped or undoped form is insoluble in most solvents commonly used with coatings. Methods of incorporating PAni in a coating system involve dissolving in xylene [8], *N*-methylpyrrolidinone [24-28], DMSO [29], THF [29] or using an alkyl phenol as co-solvent and blending with an epoxy resin [30]. Commercially formulated versions are available, including VERSICON [31-33], PANDA [15], CORREPAIR [34] or CORRPASSIV [35] which have reported to lower rates of corrosion on various steels [36].

Studies by Armelin et al. showed that epoxy with PAni in the base form showed better corrosion resistance than the conducting form when used as a system pigment [8]. The authors suggested that the protection was based on the fact that the base form of PAni has the ability to store charge, acting as a molecular condenser, and that the mechanism was based on the electroactivity of partially oxidized polymers more so than direct conductive properties [8]. Epstein et al. studied the corrosion protection properties of PAni in both the undoped and doped form, when cast-deposited on aluminium AA 3003 and AA 2024-T3 alloys. When exposed to chloride environments, the coatings were effective in lessening corrosion. It was suggested that the PAni coating facilitated the extraction of copper from the surface of the AA2024-T3 alloy, which lessened the galvanic coupling between aluminium and copper, and thereby lowered the corrosion rate [37]. Ceccetto et al. [38] suggested that an emeraldine base coating promotes proton activity and delays local acidification and its contribution to corrosion [39]. The conversion between emeraldine base and the PAni salt form is believed to occur on the aluminium-coating interface via an acid-base interaction with a hydrated surface oxide, leading to substrate ennoblement, which in turn is thought to improve corrosion protection [40]. The dopant anions associated with PAni can also play an important role in determining the level of corrosion protection afforded to aluminium alloys, as these can act as inhibitors in their own right. In this sense, the PAni layer can operate as a "smart-release" coating, where corrosion reactions occurring at a coating defect drive the release of inhibitor anions, which in turn can stifle cathodic oxygen reduction on copper-rich inter-metallic particles [39].

PAni has been dispersed in numerous types of binders, e.g. polyvinyl butyral-co-vinylalcohol-co-vinylacetate, prepared by dispersion with various volume fractions doped with paratoluene sulphonic cid (PAni-pTS) and applied on aerospace aluminium alloy AA2024-T3 for corrosion protection [39]. The process of initiation and propagation of chlorine-induced filiform corrosion was monitored using an in-situ scanning Kelvin probe technique as a quantitative measure of the influence of the PAni-pTS and PAni EB coatings on the AA2024-T3.

Although a topcoat is the final line of protection, the primer is also very important for antifouling properties. For example, an excellent antifouling effect was observed when the primer was zinc-rich coated [41]. Wang et al. evaluated DBSA doped polyurethane coatings as a primer for corrosion protection and found that it provided a useful antifouling effect [41]. The presence of unknown electron transfer processes between the conducting coating and the primer are expected. PAni used as a pigment in a coating system has provided a non-toxic coating that has attracted attention in the coatings field as a step towards an ideal marine antifouling and corrosion prevention system [42].

There have also been reports on using an overcoat, often an epoxy top-coat on top of a PAni-coated surface, mainly for the purpose of protecting the PAni-coating and to provide an additional barrier against corrosion. There are many examples of concentrated PAni solutions being applied onto iron or steels [36]. When designing the PAni pigment for coating systems as an antifouling and anticorrosion agent, the agglomeration of the intrinsically conducting polymers should be minimised, along with the achievement of well-dispersed nanoparticles of uniform size and with superior adhesion [21].

Wang et al. reported the use of this method with PAni as a marine antifouling and corrosion-prevention agent [41]. In their studies, the coatings consisted of different forms of conducting PAni that were cast on to mild steel and then immersed into the sea to evaluate their antifouling effect. Both the base and doped form of PAni were tested and after the first week the conductive PAni of an epoxy coating system and p-dodecylbenzensulfonic acid doped PAni of a polyurethane coating system displayed good antifouling properties, while the base form of PAni in an epoxy system displayed no antifouling behaviour. The antifouling effect of HCl-doped PAni was not as good as DBSA-doped PAni in an epoxy coating system after two weeks. This could be explained by the change of electrical conductivity of the doped PAni. Due to the alkaline nature of seawater at pH 8, an ideal pH value for organism growth, the conductivity of the PAni decreases with immersion time. The electrical conductivity of HCl-doped PAni dropped sharply from 5 to 10^{-6} S/cm after 8 weeks while DBSA-doped PAni remained stable along with its antifouling properties. It was reported that the DBSA-doped PAni polyurethane coating was effective as an antifouling coating for at least two months. Although this period is not long enough for real applications, the synergetic antifouling effect between the doped PAni and a biocide, such as cuprous oxide or dichlorodiphenyltrichloroethane (DDT), are an encouraging step forward. DDT epoxy coating was effective for only 1 month, while with the addition of 5 wt% of DBSA-doped PAni, an extra 5 month period of protection was achieved. Similar observations were made with cuprous oxide-epoxy coating systems, as the cuprous oxide alone showed only 2-3 months of antifouling protection, while with the addition of doped PAni this was extended to 9-12 months. This behaviour was explained by the special electron transfer occurring between cuprous (I)-cupric (II) ion and the conducting PAni, which has two redox processes at around 0.10 – 0.15 V (SCE) and 0.6 – 0.7 V (SCE). Another proposal was that the conducting coating surface had a pH value range from 4 to 5, which may provide a weak acidic micro-environment beneficial to enhance the redox process of the cuprous (I)-cupric (II) ion couple and to improve the antifouling behaviour.

One method of PAni preparation for coatings without altering the polymer is to simply create a blend [6]. Such a blend would combine the properties of the two components without chemical interactions, and add value to the overall coating, i.e. the electrical conductivity of PAni can be provided along with the physical and mechanical properties of the polymeric matrix [17, 20]. Blends of PAni [6] with conventional resins afford corrosion and fouling protection, and the presence of PAni in polyurethane or alkyl resins has been shown to improve the corrosion protection of iron and steel [43, 44]. Research was undertaken by Chen et al. using water-based polyurethane dispersions as the polymer matrix with DBSA-doped PAni [6]. The purpose of the study was to combine the electrical conductivity of DBSA-doped PAni with the physical and mechanical properties of water-based polyurethane and make a superior coating that could replace current coatings. For this reason polyurethane was chosen due to its versatility and a wide range of excellent properties, which include abrasion resistance, safety in handling, impact strength, low temperature flexibility and an ability to be applied to numerous substrates [6]. DBSA was selected due to the ability of the dopant to provide a proton acid and to act as a surfactant [6]. Additionally, DBSA-doped PAni has many H-donors and H-acceptors on the backbone structure, which can create various H-bonding types with polyurethane [19, 45, 46]. This coating was then cast on polyethylene terephthalate substrates to form water-based polyurethane-DBSA PAni blend films [6].

2. Electrochemical Polymerization of Aniline

Electrochemical synthesis of PAni can involve galvanostatic, potentiostatic or potentiodynamic methods, most often on inert metal and carbon-based electrodes, to produce films directly adherent on the electrode surface. More active metals can also be employed for the deposition of conducting polymers, and the case of aluminium and its alloys are the focus of the current chapter.

Electrochemical polymerization offers the advantage of simultaneous formation and deposition of polymer coatings on the substrate from a controlled environment, that guarantees reproducibility of polymer in chemical and physical properties, which can be altered simply changing the electrochemical parameters (e.g. the applied potential) [1]. The electrochemical synthesis of PAni on aluminium alloys has proven to be much more difficult than on inert electrodes such as gold and platinum. For many years it was thought that electropolymerization of PAni could only be achieved on inert metals, but over the past two decades it has been shown that more active substrates can be employed [36], including the likes of Ti, Cr, Al, Zn, brass, stainless steel [47], iron [48] and mild steels [30, 49]. It is important to note that during electropolymerization of conducting polymers on active electrodes, the dissolved ions of the substrate may be inserted into the polymer [1], as nickel ions have been reported in the film formation on a Ni substrate electrode when electropolymerizing a poly(3-methylthiophene) film [50].

In general the use of conducting polymers on active metals by electrochemical means requires high positive potentials to form the polymers. At the same time, for potentials greater than 1 V, many of these metals begin to corrode rapidly, which limits the formation of many conducting polymers on their surfaces. During a typical electrochemical polymerisation, an acidic aqueous solution of aniline is kept at a low pH to solubilize the aniline to generate a conducting emeraldine salt form of PAni in the reaction. The proposed mechanism for electrochemical polymerisation of aniline initiates with the formation of the radical cation of aniline by electrode surface oxidation, followed by the elimination of two protons via a radical coupling, after which the formed dimer or oligomer undergoes further oxidation with aniline on the electrode surface [42, 51].

A classical method of electrochemical PAni film formation on active metals is to use an oxalic acid solution to provide both the acidity and also passivation of the active metal. This allows the conducting polymer to grow on the electrode before electrode dissolution occurs. Electrochemically synthesised PAni coated on an aluminium surface has great potential for replacing the currently used toxic chromate-based treatments [2], as chromate has been listed as hazardous to the environment and to human health [2]. For antifouling applications, there is the potential for conducting polymers to replace the toxic and/or carcinogenic inhibitors currently used in several protective coating systems [39].

3. ALUMINIUM

Aluminium is an abundant metal in nature in its oxide forms. Once processed from its ores, aluminium is easy to handle, and is widely used due to its high technological value in industrial applications, ranging from household industries to aerospace [3]. It is attractive due to its low price, high electrical capacity and high energy density [3], but is prone to corrosion in seawater [2, 52]. The aluminium metal is protected from the environment by the formation, on the metal surface, of a thin, but highly stable barrier hydroxide layer that blocks electron transfer. Aluminium is available in different alloy compositions to provide different grades for various applications. However, the alloying elements are believed to play a role in the corrosion of aluminium [53]. For example, the presence of 4-5% copper in the 2024-T3 alloy is a particular problem as this introduces a galvanic couple between the aluminium and the copper, thereby accelerating the rate of corrosion [53].

A protective oxide layer on aluminium alloys forms naturally, but a thicker layer can be produced in the process of anodising [3], and is well-known to exhibit excellent resistance to general corrosion [5]. The aluminium oxide film growth at the aluminium-coating interface [39] can be followed using a combination of secondary ion mass spectrometry and atomic force microscopy [39]. The anodisation process is often followed by a sealing process, and can involve different methods such as boiling water sealing, dichromate sealing and nickel fluoride sealing. Sealing treatments enhance the protective and mechanical properties of the aluminium alloy [53-60]. Although aluminium forms a protective aluminium oxide layer on the surface, this coating could breach in aggressive environments causing corrosion [53]. In a marine environment with the presence of NaCl, the formation of aluminium chlorides can occur, which in turn reduce the effectiveness of the oxide layer in preventing corrosion [52]. Certain grades of aluminium corrode faster than others due to the content of the alloy, i.e. Aluminium 2024-T3 alloy contains 4-5% copper and copper introduces a galvanic couple between the aluminium thereby accelerating the rate of corrosion [53]. However, it is important to develop a coating for this grade of aluminium as this is an alloy used in aircraft applications where its increased strength adds value to the aluminium structures [53].

Aluminium with a protective oxide layer is quite stable in the pH range from 4-10 [1]. On the other hand, dissolution reactions can occur at higher and lower pH extremes:

$$Al \rightarrow 3e^- + Al^{3+} \qquad pH < 4$$

$$Al + 4OH^- \rightarrow 3e^- + AlO_2^- + 2H_2O \qquad pH > 10$$

These reactions are limited by the presence of the protective oxide layer. The ionic conductor properties of the aluminium oxide film then feature in the electrochemistry of the surface in solution, which involve aluminium and oxygen vacancies acting as charge carriers within the film [61, 62]. Moreover, the movement of point defects through the anodic oxide film on the aluminium electrode contributes to the film formation [51, 63-66]. As reported by Eftekhari [1], and using the point defect model, the electrochemical reactions at the aluminium/oxide film interface can be written as follows [67]:

$$Al(m) = Al_{Al}(ox) + 3/2V_o^{2+}(ox) + 3e^-$$
$$Al(m) + V_{Al}^{3-}(ox) = Al_{Al}(ox) + 3e^-$$

$Al(m)$ = normal aluminium atom in the regular site
$Al_{Al}(ox)$ = normal aluminium atom in the regular site of the oxide film
$V_o^{2+}(ox)$ = positively charged oxygen vacancy in the oxide film
$V_{Al}^{3-}(ox)$ = negatively charged aluminium vacancy in the oxide film

The reactions at the film/solution interface can also be written as follows [1]:

$$H_2O + V_o^{2+}(ox) = O_o(ox) + 2H^+(aq)$$
$$H_2O = O_o(ox) + 2/3V_{Al}^{3-}(ox) + 2H^+(aq)$$
$$Al_{Al}(ox) = Al^{3+}(aq) + V_{Al}^{3-}(ox)$$

$O_o(ox)$ = normal oxygen ion in the regular site of the oxide film

PAni coated on aluminium and its alloys has been the subject of a significant number of studies over the past 15 years. One important procedure prior to coating or electropolymerizing of the aluminium alloy is cleaning and pretreatment. The surface must be pretreated in a similar manner to that indicated below (Figure 1) to remove residues and to provide a clean surface for good adhesion of the conducting polymer:

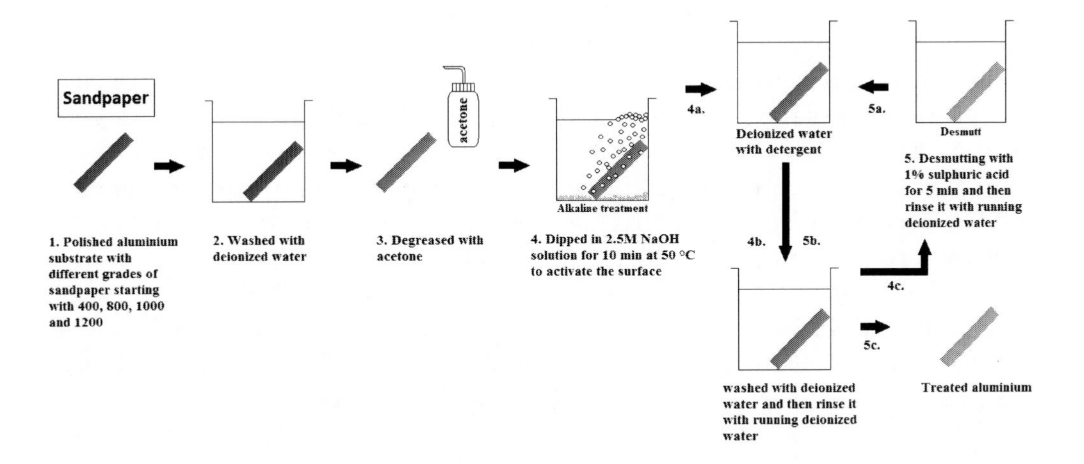

Figure 1. Cleaning and pretreatment of aluminium alloys.

The aluminium surface is typically polished with different grades of sand paper, prior to rinsing and degreasing with acetone. To activate the surface, aluminium strips can also be alkaline treated in a strong solution of sodium hydroxide for several minutes at elevated temperatures.

As a final pretreatment stage, the aluminium can be desmutt in 1% sulfuric acid, an etching process, for few minutes at room temperature before rinsing. Such a cleaning process is important to achieve good adhesion of the conducting polymer, while being aware that dissolving precipitated impurities can potentially limit the growth of a conducting polymer on an aluminium substrate [1]. With aluminium, pretreatments prior to electrochemical synthesis of a conducting polymer can also include anodic [68] or cathodic [39] pretreatment steps. Some authors have used anodic activation of aluminium in 0.1 M HNO_3 aniline-containing solution as pretreatment of the electrode prior to electrodepositing of PAni from solution of aniline and sulfuric acid [68].

4. Electrochemical Polymerization of Aniline on Aluminium Alloys

Electropolymerization of PAni offers the advantage of control over the chemical and physical properties of the coating formed on the substrate by changing the electrochemical parameters (e.g. voltage and current) [1]. The electrochemical deposition of PAni on an electrode surface also involves oxidation of the substrate electrode prior to electropolymerization, including the formation of a passive layer on the electrode surface [1]. The passive film plays two important roles: protection of the electrode against dissolution, and the establishment of a suitable surface for deposition of the conducting polymer [1]. However, aluminium often suffers from localised corrosion when exposed to an environment containing aggressive chloride ions. The deposition of PAni onto aluminium provides the opportunity for protection of aluminium alloys in more aggressive environments. Camelet et al. [69] reported that passivation of oxidizable metals prior to electropolymerization is necessary to obtain adequate adhesion of a PAni film [4].

PAni has been grown on numerous aluminium alloys. Tallman et al. studied the corrosion protection of aerospace industry AA 2024-T3, AA 6061 and AA 7075 grades of aluminium [70] with electrochemically synthesised PAni [71]. The electrochemical polymerisation of aniline on aluminium alloys can be undertaken either on top of the outer oxide layer (Al_2O_3), or immediately following an intermediate step whereby the initial oxide layer is stripped off the surface under acidic conditions. This later approach allows more ordered pore formation and greater adhesion between the coating and the substrate.

The voltammograms presented in Figure 2 for the formation of PAni on an aluminium alloy show two pairs of redox waves, I/I' and II/II'. These are commonly assigned to inter-conversion between the leucoemeraldine and emeraldine conducting form (I/I') and between the emeraldine conducting form and pernigraniline (II/II'). An increase in the current intensity of these redox waves points to a thickening of the conducting polymer [2].

During electropolymerisation of PAni, the formation of aluminium oxides will in fact continue throughout the entire potential range employed. This process allows the formation of a passive layer, which can inhibit the dissolution of the oxidizable metal without blocking access of the monomer and its further oxidation [1].

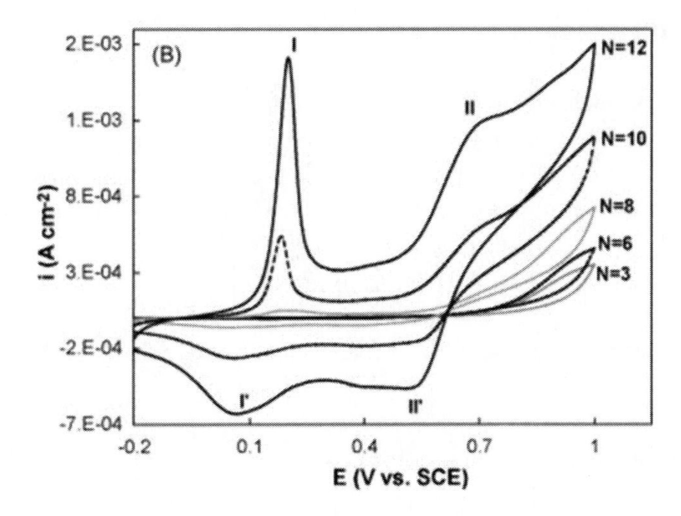

Figure 2. Cyclic voltammograms for the aluminium alloy 6061-T6 in a 0.5 M H_2SO_4 + 0.5 M aniline solution, recorded at 0.5 V s^{-1} at cycle 3, 6, 8, 10 and 12 (reproduced with permission from [2]).

This will have the effect of lowering the current density at the electrode as a passive oxide layer is produced. Studies undertaken by Eftekhari showed that a very stable PAni film was formed on an oxidized aluminium electrode, and even more so than for a comparable electrochemical synthesis of PAni on a platinum electrode [1]. This could be explained by the good connection of the PAni to the electrode surface with an oxide film present [72]. The presence of the oxide film on aluminium also provides protection against dissolution of the metallic substrate [1]. A smooth and adherent film can be formed on the electrode surface in an acidic medium. The good adhesion of the conductive coating and improvement of corrosion resistance to the metal surface has attracted much attention from industry. A galvanic interaction can occur between aluminium and coatings containing the oxidised and conducting forms of PAni. Cathodic reactions such as oxygen reduction can also be suppressed by the presence of PAni [39].

The type of substrate electrode and the electrolyte solution selected are two critical parameters for the successful growth of conducting polymers on active metals. Electropolymerization of aniline on aluminium using oxalic acid [73] and tosylic acid [74] solutions have been carried out and the galvanostatic interaction between the polymer and aluminium was observed to give rise to oxidation of the aluminium substrate and reduction of the polymer [74]. The PAni deposit has then been observed to provide increased corrosion resistance for the aluminium substrate [74].

There are other approaches to electrochemically depositing PAni on aluminium. Huerta-Vilca et al. used alizarin as a chelating agent in the electrosynthesis of PAni films on aluminium, and found that the film formation induction time was lower and growth rates were increased compared to the absence of a chelating agent, although the stability of the films was weak [75, 76]. Pournaghi-Azar et al. pretreated their aluminium in an H_2PtCl_6 solution prior to electrodeposition of PAni [77]. The pretreatment provided the opportunity to lower the monomer concentration and induction time needed to initiate the polymerization process, and at the same time increased the polymer growth rate and the stability of the PAni film on the electrode [77].

4.1. Galvanostatic Polymerization Method

Galvanostatic polymerization methods for the deposition of PAni have been reported. This approach provided more uniform conducting polymer film deposition under specific conditions compared to other methods of electrochemical deposition. PAni prepared galvanostatically at current densities of 0.5, 1.5, 5 and 15 mA cm^{-2} on AA 3004 aluminium alloy was studied by Shabani-Nooshabadi et al. [78]. During the polymerisation, the potential values increased over a period of 30 to 900 s to reach a plateau value of around 15 V on the AA 3004 electrode. The combined metal oxide and conducting polymer layer eventually restricted electron transfer and resulted in a decrease in the electrical conductivity [78].

4.2. Potentiostatic Polymerization Method

Electropolymerization of 0.5 M aniline on a 6061-T6 aluminium alloy was carried out at different constant potentials, in an attempt to optimise the conditions of growth at this particular substrate [2]. The anodic current was found to generally increase due to the nucleation and growth of the polymer, and the induction time involved for this increase depended upon the applied potential. At the same time, for an applied potential from 0.7 to 0.8 V, an increase in the maximum current was seen, indicative of a higher polymerisation rate (Figure 3) [2]. However, for potentials above 0.8 V, a dramatic decrease was seen in the maximum current values produced, which points to dominance of degradation over polymerisation for the more positive potentials [79]. The presence of Si and Mg precipitates were shown to play an important role in the aniline electropolymerisation process [2]. During etching these species dissolve away and this affects the polymer growth rate, indicating that they are involved in conduction pathways.

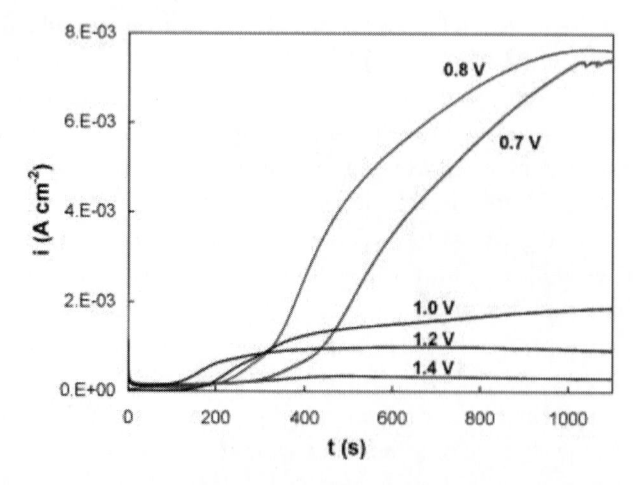

Figure 3. Potentiostatic current-time plots recorded for aluminium alloy 6061-T6 in a 0.5 M H_2SO_4 + 0.1 M aniline at different applied potentials (reproduced with permission from [2]).

In this case, as the applied current density increased, the induction time decreased. This can be explained by more Al^{3+} being produced per unit time at a higher applied current density with concomitant precipitation of more Al oxalate crystals on the substrate. In other words, the substrate could be covered by a passive Al oxalate layer in a shorter period of time, and this passive layer inhibits further dissolution of Al^{3+} without affecting other electrochemical processes, and so behaves like an inert metal [78]. It is commonly agreed that there is a galvanic interaction between aluminium and a coating containing the oxidised and conducting polymeric forms [39]. As a result, the emeraldine salt form of PAni can polarise the underlying aluminium to potentials where cathodic reactions such as oxygen reduction are greatly suppressed [39]. Without a topcoat the protection of PAni is temporary [80], and the coating failure is commonly related to the polymer changing from the doped state to the undoped state [39].

The height of the first anodic peak can also be used to provide an estimate of the thickness of the polymer film, and this has been reported by many authors as a way of calculating the PAni thickness on a metal surfaces. For example, a film formed by 55 scans to 1.1 V, which showed an anodic peak current, starting from the fully reduced film, of 4.7 mA/cm^2, would have a film thickness of ~160 nm [36, 81]. However, this method does not accommodate for polymer porosity and the counter-ion volume, therefore the values calculated only give an estimated thickness for the sake of comparison [82]. Conducting polymers electrochemically deposited on various metal substrates, such as iron and steel, have shown some effectiveness against corrosion with a doped PAni layer [25, 47, 81, 83-86]. In 1985, DeBerry [47] showed that PAni could be grown on 410 and 430 stainless steels. Ahmad, MacDiarmid, Wessling [83] reported that PAni has the oxidation capability to passivate mild steel [83], copper [83] and stainless steel [29].

Likewise, several groups have reported the use and incorporation of PAni on aluminium alloys. Racicot et al. [87] studied a double strand form of PAni, and the polyanion composition provided corrosion protection for the aluminium AA 7075 alloy. Apart from anilines there have also been several studies on ring substituted anilines such as *o*-toluidine, *m*-toluidine, *o*-anisidine and *o*-chloroaniline electrodeposited on passive surfaces [88].

5. Characterisation

Conducting polymer layers are commonly characterized by both electrochemical techniques, such as cyclic voltammetry and AC impedance, and by spectroscopic and surface science analyses, such as FTIR, SEM and XPS. These methods provide structural information about the PAni that has been deposited, including oxidation state, the presence of dopants, and degradation characteristics.

5.1. Fourier Transform Infra Red (FTIR) Spectroscopy

For PAni characterisation by FTIR, the following features are typically seen. A doublet at 3200 cm^{-1} and 3260 cm^{-1} represents the secondary amine, while 1512 cm^{-1} and 1592 cm^{-1} bands represent the benzenoid and quinoid rings of PAni [3].

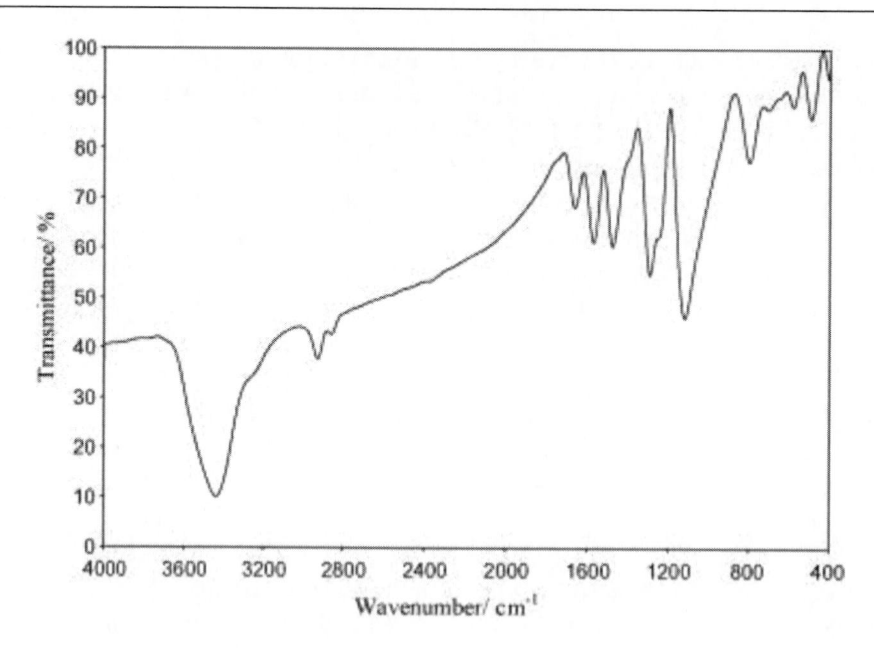

Figure 4. FTIR spectra of electrodeposited PAni on AA3004 alloy (reproduced with permission from [78]).

Peaks appearing at 2900 to 2950 cm^{-1} and 3267 cm^{-1} are due to the stretching vibration modes of the –C-H (-CH_3 or –CH_2-) and –N-H bonds [6]. The bands at 1300 and 1200 cm^{-1} are due to the C-N stretching mode, for the proton conducting state of the benzenoid ring, while a peak at 1113 cm^{-1} is due to the plane bending vibration of C-H which is formed during protonation [4]. A characteristic feature of the B-NH-Q bond or the B-NH-B bond, are out-of –plane bending vibrations of the benzenoid and quinonoid –C-H and –N-H bonds [6], which give rise to peaks in the 800 to 700 cm^{-1} range. An example of using FTIR to monitor PAni dopants, is a peak at 1100 – 1200 cm^{-1} due to the S=O stretching mode of camphorsulfonic acid when used as a dopant in PAni synthesis [3]. As a further example, DBSA doped PAni has asymmetric and symmetric O=S=O stretching peaks at 998 to 1040 cm^{-1} [1], and a band at 1669 cm^{-1} was assigned to the carbonyl group of a benzoate [89].

At the same time, bands due to the aluminium substrate and its oxide layer can also be seen with FTIR. Al-O bond bending is observed at 918 cm^{-1} and 795 cm^{-1}, and this also overlaps with a doublet at 450 to 600 cm^{-1} due to Si-O bond bending, Al-O bond stretching and a Mg-O band (468 cm^{-1}) [3]. A typical FTIR spectra of PAni prepared on an aluminium alloy is shown in Figure 4 [78].

Blending DBSA doped PAni in water-based polyurethane (WPU) and casting it in a film, leads to a peak appearing at 1730 cm^{-1} attributed to the –C=O stretching vibration mode [6]. The position of this band is affected by hydrogen bond formation between the –N-H of PANDB and the –C=O of the water-based polyurethane [6]. There are more significant shifts in the stretching vibration of –N-H bond in the urethane segments, which exhibit a more significant shift than that of a pure WPU film [6]. The peak of –N-H peak at 3348 cm^{1} for the pure WPU film shifts with different PAni concentrations from 9 wt % at 3296 cm^{-1}, 3278 cm^{-1} for 17 wt% and 3240 cm^{-1} for 33 wt% films [6].

5.2. X-ray Photoelectron Spectroscopy

X-ray photoelectron spectroscopy (XPS) has been used very effectively to identify the iron oxides Fe_2O_3 and Fe_3O_4 that lie beneath a PAni-containing phase [25].

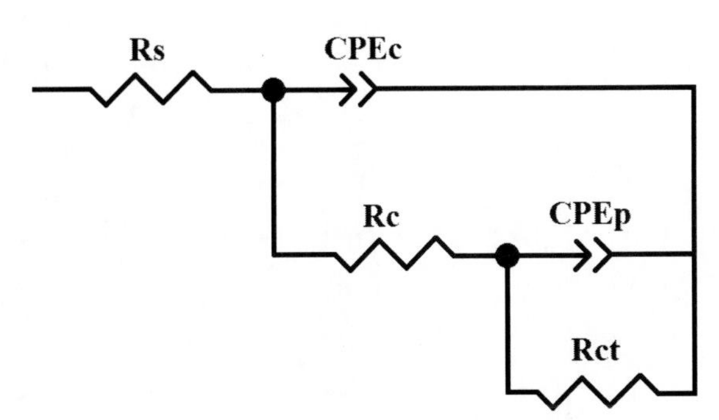

R_s = electrolyte resistance.
R_{ct} = charge transfer resistance.
R_c = coating resistance.
CPE_c = constant phase elements connected with coating capacitance.
CPE_p = electrochemical phenomena (double layer capacitance and possible diffusion processes) in the coating pores.

Figure 5. Equivalent electrical circuit for impedance measurements (adapted from [3]).

Likewise XPS can be used to examine the process of electrodeposition of PAni films on aluminium alloys, such as AA60601-T6 [2]. In this case, several samples were prepared by different numbers of cyclic voltammetry scans between -200 mV and 1250 mV and analysed by XPS [2]. The XPS peaks corresponding to the ionization of aluminium (Al 2p), carbon (C 1s), nitrogen (N 1s) and sulfur (S 2p) were analysed [2]. These results helped establish the need for several cycles before PAni electrodeposition became fully established [2].

5.3. AC Impedance Spectroscopy

Electrochemical impedance spectroscopy (EIS) is commonly used in studies involving conducting polymers and corrosion projects, as a way to evaluate the corrosion properties of a surface. The EIS output enables the coating capacitances and films resistances to be determined. This can be monitored as a function of their exposure time to various aqueous environments, including seawater for marine applications [3]. In the case of aluminium, the oxidation of aluminium is expected to take place with a PAni-coated aluminium surface, where the PAni redox reactions also take place [74].

A typical electrical equivalent circuit used for fitting EIS data is given in Figure 5 [3], and contains a number of circuit elements that can be related to physical properties of the electrochemical cell involving a PAni-coated aluminium surface. EIS studies are commonly carried out at open circuit potential with the frequencies ranging from 10^4 to 10^{-2} Hz, with the amplitude of the superimposed AC signal being around 10 mV [3].

5.4. Scanning Electron Microscopy (SEM)

Scanning Electron Microscopy (SEM) is a very effective technique to examine changes in the aluminium surface as a result of conducting polymer deposition.

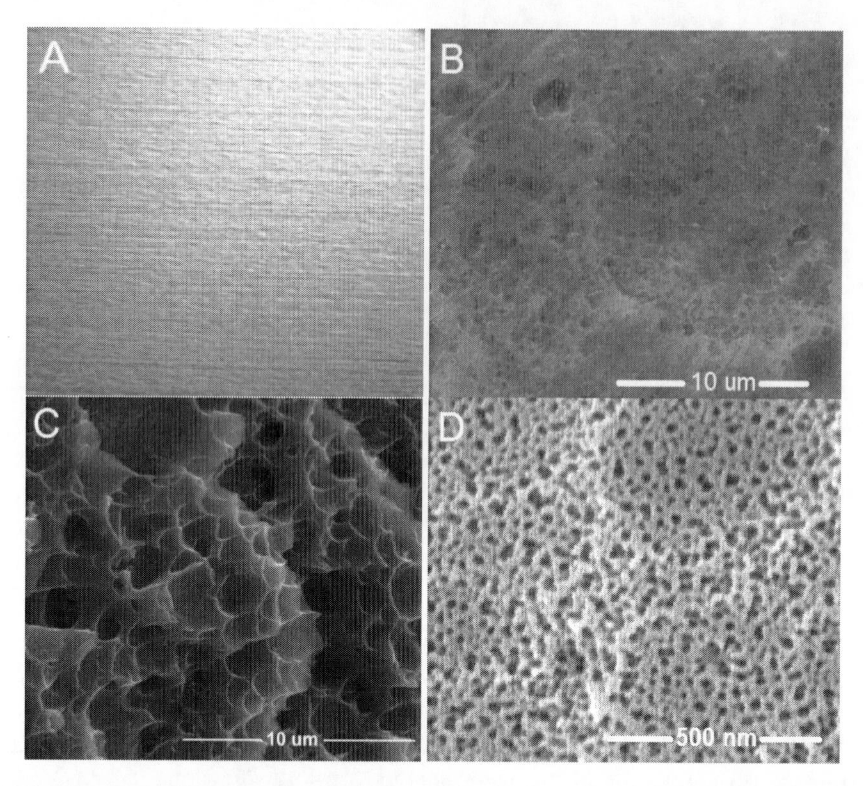

Figure 6. A, aluminium as supplied; B, chemically cleaned and pretreated aluminium; C, anodized aluminium (Al_2O_3) after being potentiostatically treated at 8 V for 30 mins in 0.5 M oxalic acid; D, potentiostatically grown PAni on top of the formed Al_2O_3 in 0.5 M oxalic acid + 0.1 M aniline at 8 V for 3 hr (results from the University of Auckland).

As illustrated in the SEM images shown in Figure 6, the initial aluminium surface is often quite smooth in appearance. As the sample is chemically cleaned, and the surface pretreated, the surface becomes more porous and rough (Figure 6B). The surface can be modified further to a honeycomb-like structure through anodisation processes (Figure 6C), while a more fluffy porous surface is obtained following the electrochemical deposition of PAni, also dependent upon the electrolyte solution employed, in this case an oxalic acid solution (Figure 6D).

6. Applications and Performance of PAni Coated Aluminium and Alloys

6.1. Anticorrosion Activity of PAni

Corrosion of metals is a serious problem throughout the world [3], with implications for many chemical and manufacturing industries [4]. Pigments based on lead oxides and chromates are most commonly used for protection of steels against corrosion. Many studies have been undertaken in this area using conducting polymers, which are promising coatings for the corrosion protection of iron, steel, zinc, aluminium and other oxidizable metallic materials [29, 83]. DeBerry [47] and Wessling [83] emphasised the important role that the passivation of the metal surface plays in the protective activity of PAni when used as an anticorrosive coating.

In general, we can consider three possible approaches to protecting metals from corrosion:

1. Modification of the environment to which the material is exposed
2. Electrical methods of control
3. Use of protective coatings

One of the most commonly used methods of preventing corrosion is to protect the surface of the metal with a coating. However, many of these layers contain toxic and environmentally hazardous materials, especially chromium compounds [2]. These considerations have led to the development of alternative protective coatings with the incorporation of conducting polymers [2]. An ideal anticorrosion coating must be pinhole-free and provide excellent corrosion resistance in aggressive environments. An example of such a testing situation would be to monitor cyclic voltammograms and Tafel plots for exposure of samples to hot saline (65°C or 75 °C) for 45 days [23]. A further requirement is good adhesion of the conducting polymer to the substrate of interest [3]. Like other polymeric coatings, conducting polymers often exhibit a self-healing ability [36], in the reconstruction of their physical barrier properties and their electroactivity, which can heal artificial defects from scratches and pinholes in the metal substrate against a corrosive reagent. In addition, the redox properties of the conducting polymers allow them to shift the potential of the original substrate to a value where the rate of corrosion is lowered, and in this sense they act like a macromolecular inhibitor.

Organic inhibitors, often rich in π bonds, have established an important role in corrosion prevention. These inhibitors adsorb onto a metal surface and lower the number of active sites, creating a barrier layer which limits the transport of corrosive agents into the metal and subsequently lowers the rate of surface corrosion [36]. Over the years, this area has been intensely investigated due to interest in finding alternatives to toxic chromium and phosphate treatments. The use of organic polymers also has the potential of being a cheaper option for industry.

Mechanisms of corrosion protection of undoped PAni on iron and cold rolled steel have been described [74]. The protection has been related to charge transfer occurring from the metal to the polymer [53], leading to the formation of a passive layer on the steel substrate consisting of Fe_3O_4 and γ-Fe_2O_3 [90]. A barrier mechanism based on a high diffusion resistance against corrosive ions has been proposed to explain the protection afforded by undoped PAni [24]. On the other hand, the protection from corrosion with the use of doped PAni has been associated with the conductivity of the organic polymers, and changes to the flow of electrons from the metal to the outer oxidizing species [13].

The unique ability of conducting polymers to store and transport charge, to anodically protect metals against rapid rates of corrosion, is one of their most attractive features [36]. The mechanisms by which PAni exerts corrosion control over aluminium and its alloys have been explained by the galvanic interaction that can occur between aluminium and coatings containing the oxidised and conducting forms of PAni, namely the emeraldine salt form, often doped with acids such as HCl, H_2SO_4 or oxalic acid [39]. As a result, the doped form of PAni can polarise the underlying aluminium to potentials where cathodic reactions such as oxygen reduction are greatly suppressed [39]. When the conducting polymer coating is used as a primer alone under immersion conditions, the protective nature of the PAni coating can be temporary [80]. Coating failure is attributed to the reduction of conducting PAni salt form to non-conducting (non-oxidising) leuco-emeraldine form [39]. However, it has been claimed by others [38, 53, 91] that the non-conductive form of PAni is also an effective corrosion inhibitor. The ability to moderate proton activity and changes in local acidification can contribute to the efficiency of undoped PAni coatings [38]. Cecchetto et al. [40] proposed that it is at the aluminium-coating interface, via an acid-base interaction with a hydrated surface oxide, that the conversion of undoped PAni to doped PAni takes place, leading to ennoblement of the substrate, which in turn is thought to improve the prevention of corrosion.

The nature of the dopant anion and its interaction with the conducting coating is also important in determining the level of corrosion protection afforded to aluminium alloys, such as AA 2024 [92]. The dopant anions arising from the doped PAni can act as inhibitors in their own right. Corrosion reactions that occur at a coating defect drive the release of inhibitor anions, which in turn stifle cathodic oxygen reduction on the likes of copper-rich intermetallic particles [39].

6.2. Performance of Chemically Prepared PAni on Aluminium and Its Alloys

Chemically synthesising PAni in camphorsulfonic acid with ammonium peroxydisulfate as surfactant and initiator with montmorillonite nanocomposite materials has been successfully prepared by in-situ emulsion polymerisation in the presence of inorganic nanolayers of clay on aluminium Al 5000 [3]. Improvements in the anticorrosion properties of epoxy coatings on treated aluminium during anodizing were investigated [3], including the PAni-polymer matrix with clay, which was shown to exhibit better coating resistance than an epoxy incorporated PAni and an epoxy alone without any clay [3]. The initial coating resistance value for the PAni-polymer matrix with clay coating on anodized aluminium was calculated to be 1.32 x 10^5 $\Omega\cdot cm^2$, while PAni in an epoxy coating without clay under the same conditions led to a lower resistance value of 15,520 $\Omega\cdot cm^2$, whereas the epoxy coating alone had a resistance of only 1400 $\Omega\cdot cm^2$. The electrolyte diffusion process was influenced by increasing the clay loading and hence an increase in the coating resistance [3]. The first sign of electrolyte diffusion took place after 8 to 23 hours of immersion with clay loadings of 0.5 to 1%. However, the first step of diffusion for 5% of clay loading was not observed, and this could be explained by greater agglomeration of the clay in the polymer matrix with an increase in clay content [3].

The addition of chemically synthesized PAni to epoxy coatings should increase the degree of crosslinking over and above the usual epoxy + hardener system, and provide better barrier properties and corrosion resistance for the epoxy-PAni system, particularly compared to an epoxy alone [23]. In addition, PAni has also shown a self-healing effect with intentionally damaged samples, and this is due to redox characteristics of PAni [23]. In many cases, the conducting polymer liberates dopant ions near the metal or alloy surface, which creates a passivating layer and limits further corrosion [93]. It must be emphasised that the PAni used in a number of these coatings was not fully doped, so could also capture ions, including the corrosion promoter Cl^- [94].

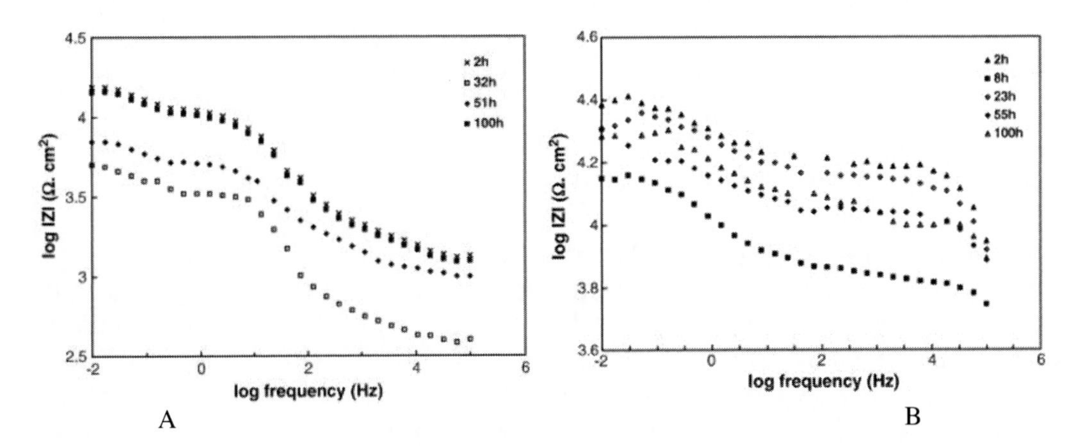

Figure 7. Bode plots (log f vs. log | Z |) for impedance spectra at various immersion times for epoxy coated, A, and PAni-epoxy coatings, B, on anodized aluminium in 3.5 wt% saline solution at 65 °C (reproduced with permission from [3]).

Emeraldine base PAni coatings have been shown to facilitate the extraction of copper from the surface of AA 2024-T3, leading to a lowering of the galvanic couple between aluminium and copper which accelerates the rate of corrosion on aluminium alloys [53]. The coating was spin coated, drop-cast and immersed in 0.1 M NaCl solution, kept at room temperature for 10 to 66 hours [53]. The corrosion current density for uncoated metal to coated metal was tenfold more, indicating the importance of the coated surface [53].

The results from Epstein et al. [53] suggest that emeraldine base and sulfonated PAni can play an important role in dissolving away the copper containing corrosion products [53]. This explains how the corrosion rate could be decreased as it eliminates the presence of galvanic coupling between the aluminium and copper [53].

Bode plots have been used in studying the corrosion protection with the epoxy coatings on aluminium after 32 hr of immersion. In Figure 7A it can be seen that the coating resistance decreased to a value of around 446 $\Omega{\cdot}cm^2$ at a 100 kHz frequency, for epoxy on its own. This decline was attributed to the influence of Cl^- ions incorporated at the surface of the aluminium oxide [3]. A more rapid decrease in coating resistance was observed with a higher water permeability. After this initial period, the coating resistance increased (after 51 hours of immersion) to a value of 1000 $\Omega{\cdot}cm^2$. This period of increase can be explained by plugging of the pores with corrosion products. After a longer immersion time the coating resistance increased further to 1050 $\Omega{\cdot}cm^2$, but the epoxy coated anodized aluminium remained unchanged and constant [3].

When PAni was included in with the epoxy coating, higher resistance values were obtained (Figure 7B), of the order of some thousands of $\Omega{\cdot}cm^2$. The penetration of the electrolyte into the coatings was observed, seen by a resistance decrease of the coating after 8 hours of immersion [3]. However, when the coating was immersed for a longer time (23 hours), the resistance increased gradually, caused by the contact between the electrolyte and the metal surface and the beginning of the electrochemical processes at the metallic surface

[3]. This is a good indication of an increased protective property of PAni incorporated epoxy coating compared to epoxy coatings alone [3]. The related Nyquist plots at various immersion times for PAni-epoxy coatings on anodized aluminium in 3.5 wt% saline solution at 65 °C, showed two capacitive loops, one at a higher and the other at a lower frequency [3]. The first capacitive loop was attributed to the coating characteristics while the second one was related to processes occurring beneath the film. Although after a long immersion time, the coating resistance decreased, and the effectiveness of the film would decrease due to the ingress of chloride ions and water [3]. Some authors have suggested that the ability of PAIN to protect against corrosion is due to the barrier properties of the reduced polymer [74].

Performance of Electrochemically Prepared PAni on Aluminium and Its Alloys

The effectiveness of a PAni coating formed electrochemically has been shown to be sensitive to the preparation conditions used. This starts with the pretreatment step prior to electrochemical synthesis, which affects the state of the surface and the extent and nature of the oxide layer present on the aluminium surface [1, 68]. It is uncertain whether a nitric acid treatment serves to activate the metal surface in some way, or whether it simply adds to the thickening of the aluminium oxide layer due to the passivating character of nitric acid [1, 68]. By contrast, in marine environments the aluminium can form aluminium chlorides which in turn lessen the effectiveness of the oxide layer in preventing corrosion [53].

A non-uniform PAni coating can also accelerate the rate of corrosion on aluminium substrates and the galvanic action between the substrate and PAni decreases with time upon exposure to an electrolyte. Upon aging in air the PAni coated aluminium would thicken due to the formation of an oxide with atmospheric oxygen [4, 39]. Moreover, PAni in its doped form would not change due to exposure to atmospheric oxygen, because the electropolymerised PAni is already in an oxidised state [4, 39].

In a further study, aluminium was subjected to anodic galvanostatic pretreatment in a 0.1 M HNO_3 aniline-containing solution, prior to electrodeposition of PAni from a solution of aniline and sulfuric acid [68]. Corrosion tests in 0.1 M NaCl showed the importance of this pretreatment for resistance to corrosion [68]. However, the reason for the improved protection was not clear, as it could be due to surface activation or from the formation of a thicker aluminium oxide film due to the passivating character of nitric acid [68].

Wang et al. [5] electropolymerized PAni on 1100 wire beam aluminium with 1 M tosylic acid solution using a potentiostatic method [5]. This was the first published article with PAni electropolymerised on wire beam electrode to study the electrodeposition process [5]. The electrode was cathodically pretreated at a potential of -900 mV prior to electrodeposition [42]. This pretreatment prevented localized corrosion of AA 1100 and ennobled the pitting potential by about 130 mV [42].

CONCLUSION

Incorporating PAni chemically or electrochemically on aluminium alloys have shown promising performance in a wide range of applications. Chemically synthesised PAni gives the flexibility of being able to be used as a pigment in a coating system. By contrast, electrochemically synthesised PAni offers the advantage of controlling the chemical and physical properties of the coatings, providing the opportunity for developing the ideal coating for a particular aluminium alloy and its end application. However, there are also limitations as different types of aluminium, of varying composition, differ in their performance, and so it is very challenging to achieve the ideal formulation that could apply to all grades of aluminium.

For more than two decades PAni coated on aluminium and its alloys have been the subject of many studies mostly focusing on anticorrosive properties. Moreover, these coatings are now revealing further properties that researchers are exploring in more detail, such as antistatic, antibacterial and antifouling properties. The success of PAni coatings on aluminium alloys in these areas can lead to the development of novel coating approaches, and less reliance on toxic coatings including chromates.

REFERENCES

[1] Eftekhari, A. *Synth. Met.* 2001, *125*, 295-300.

[2] Martins, N. C. T.; Moura e Silva, T.; Montemor, M. F.; Gernandes, J. C. S.; Ferreira, M. G. S. *Electrochim. Acta.* 2010, *55*, 3580-3588.

[3] Hosseini, M. G.; Jafari, M.; Najjar, R. *Surf. Coat. Technol.* 2011, *206*, 280-286.

[4] Kang, E. T.; Neoh, K. G.; Tan, K. L. *Prog. Polym. Sci.* 1998, *23*, 277-324.

[5] Wang, T.; Tan, Y. J. *Mater. Sci. Eng., B.* 2006, *132*, 48-53.

[6] Chen, C.H.; Kan, Y.T.; Mao, C. F.; Liao, W. T.; Hsieh, C. D. *Surf. Coat. Technol.* 2012, In press.

[7] Sathiyanarayanan, S.; Karpakam, V.; Kamaraj, K.; Muthukrishnan, S.; Venkatachari, G. *Surf. Coat. Technol.* 2010, *204*, 1426-1431.

[8] Armelin, E.; Meneguzzi, Á.; Ferreira, C. A.; Alemán, C. *Surf. Coat. Technol.* 2009, *203*, 3763-3769.

[9] Gomes, E. C.; Oliveira, M. A. S. *Surf. Coat. Technol.* 2011, *205*, 2857-2864.

[10] Rout, T. K.; Jha, G.; Singh, A. K.; Bandyopadhyay, N.; Mohanty, O. N. *Surf. Coat. Technol.* 2003, *167*, 16-24.

[11] Pereira da Silva, J. E.; Córdoba de Torresi, S. I.; Torresi, R. M. *Corros. Sci.* 2005, *3*, 811-822.

[12] Samui, A. B.; Patankar, A. S.; Rangarajan, J.; Deb, P. C. *Prog. Org. Coat.* 2003, *47*, 1-7.

[13] Sathiyanarayanan, S.; Muthukrishnan, S.; Venkatachari, G.; Trivedi, D. C. *Prog. Org. Coat.* 2005, 53, 297-301.

[14] Huh, J. H.; Oh, E. J.; Cho, J. H. *Synth. Met.* 2003, *137,* 965-966.

[15] Dominis, A. J.; Spinks, G. M.; Wallace, G. G. *Prog. Org. Coat.* 2003, *48*, 43-49.

[16] Chen, C. H.; Mao, C. F.; Su, S. F.; Fahn, Y. Y. *J. Appl. Polym. Sci.* 2007, *103*, 3415-3422.

[17] Zilberman, M.; Titelman, G. I.; Siegmann, A.; Haba, Y.; Narkis, M.; Alperstein. D. *J. Appl. Polym. Sci.* 1997, *66*, 243-253.

[18] Plesu, N.; Ilia, G.; Pascariu, A.; Vlase G. *Synth. Met.* 2006, *156*, 230-238.

[19] Ho, K. S.; Hsieh, K. H.; Huang, S. K.; Hsieh, T. H. *Synth. Met.* 1999, *107*, 65-73.

[20] Lakshmi, K.; John, H.; Mathew, K. T.; Joseph, R.; George, K. E. *Acta Mater.* 2009, *57*, 371-375.

[21] Riaz, U.; Ahmad, S. A.; Ashraf, S. M.; Ahmad, S. *Prog. Org. Coat.* 2009, *65*, 405-409.
[22] Tan, K. L.; Tan, B. T. G.; Khor, S. H.; Neoh, K. G.; Kang, E. T. *J. Phys. Chem. Solids.* 1991, *52*, 673-680.
[23] Radhakrishnan, S.; Sonawane, N.; Siju, C. R. *Prog. Org. Coat.* 2009, *64*, 383-386.
[24] Wei, Y.; Wang, J.; Jia, X.; Yeh, J.-M.; Spellane, P. *Polymer* 1995, *36*, 4535-4537.
[25] Fahlman, M.; Jasty, S.; Epstein, A. J. *Synth. Met.* 1997, *85*, 1323-1326.
[26] Santos Jr, J. R.; Mattoso, L. H. C.; Motheo, A. J. *Electrochim. Acta.* 1998, *43*, 309-313.
[27] Kinlen, P. J.; Menon, V.; Ding, Y. *J. Electrochem. Soc.* 1999, *146*, 3690-3695.
[28] McAndrew, T. P.; Miller, S. A.; Gilicinski, A. G.; Robeson, L. M. *ACS Symp. Ser* 1998, *689*, 396-408.
[29] Ahmad, N.; MacDiarmid, A. G. *Synth. Met.* 1996, *78*, 103-110.
[30] Talo, A.; Passiniemi, P.; Forsén, O.; Yläsaari, S. *Synth. Met.* 1997, *85*, 1333-1334.
[31] Lu, W.-K.; Elsenbaumer, R. L.; Wessling, B. *Synth. Met.* 1995, *71*, 2163-2166.
[32] Kinlen, P. J.; Silverman, D. C.; Jeffreys, C. R. *Synth. Met.* 1997, *85*, 1327-1332.
[33] Sitaram, S. P.; Yu, P.; O'Keefe, T.; Stoffer, J. O. *Polym. Mater. Sci. Eng.* 1996, *75*, 354-355.
[34] Li, P.; Tan, T. C.; Lee, J. Y. *Synth. Met.* 1997, *88*, 237-242.
[35] Wessling, B.; Posdorfer, J. *Electrochim. Acta.* 1999, *44*, 2139-2147.
[36] Kilmartin, P. A.; Trier, L.; Wright, G. A. *Synth. Met.* 2002, *131*, 99-109.
[37] Wang, T.; Tan, Y.-J. *Corros. Sci.* 2006, *48*, 2274-2290.
[38] Cecchetto, L.; Ambat, R.; Davenport, A. J.; Delabouglise, D.; Petit, J. P.; Neel, O. *Corros. Sci.* 2007, *49*, 818-829.
[39] Williams, G.; McMurray, H. N. *Electrochim. Acta.* 2009, *54*, 4245-4252.
[40] Cecchetto, L.; Delabouglise, D.; Petit, J.-P. *Electrochim. Acta.* 2007, *52*, 3485-3492.
[41] Wang, X. H.; Li, J.; Zhang, J. Y.; Sun, Z. C.; Yu, L.; Jing, X. B.; Wang, F. S.; Sun, Z. X.; Ye, Z. J. *Synth. Met.* 1999, *102*, 1377-1380.
[42] Laco, J. I. I.; Mestres, F. L.; Villota, F. C.; Alter, L. B. *Mater. Corros.* 2004, *55*, 689-694.
[43] Ahmad, S.; Ashraf, S. M.; Riaz, U. *Polym. Adv. Technol.* 2005, *16*, 541-548.
[44] Laco, J. I. I.; Villota, F. C.; Mestres, F. L. *Prog. Org. Coat.* 2005, *52*, 151-160.
[45] Vikki, T.; Pietilä, L.-O.; Österholm, H.; Ahjopalo, L.; Takala, A.; Toivo, A.; Levon, K.; Passiniemi, P.; Ikkala, O. *Macromolecules* 1996, *29* (8), 2945-2953.
[46] Ikkala, O. T.; Pietilä, L. O.; Passiniemi, P.; Vikki, T.; Österholm, H.; Ahjopalo, L.; Österholm, J. E. *Synth. Met.* 1997, *84*, 55-58.
[47] DeBerry, D. W. *J. Electrochem. Soc.* 1985, *132*, 1022-1026.
[48] Sazou, D.; Georgolios, C. *J. Electroanal. Chem.* 1997, *429*, 81-93.
[49] Talo, A.; Forsén, O.; Yläsaari, S. *Synth. Met.* 1999, *102*, 1394-1395.
[50] Marawi, I.; Khaskelis, A.; Galal, A.; Rubinson, J. F.; Popat, R. P.; Boerio, F. J.; Mark Jr, H. B. *J. Electroanal. Chem.* 1997, *434*, 61-68.
[51] Moon, S.-M.; Pyun Su-Ii, P. S. *J. Solid State Electrochem.* 1998, *2*, 156-161.
[52] Yin, Y.; Liu, T.; Chen, S.; Liu, T.; Cheng, S. *Appl. Surf. Sci.* 2008, *255*, 2978-2984.
[53] Epstein, A. J.; Smallfield, J. A. O.; Guan, H.; Fahlman, M. *Synth. Met.* 1999, *102*, 1374-1376.
[54] Bautista, A.; González, J. A.; López, V. *Surf. Coat. Technol.* 2002, *154*, 49-54.
[55] Snogan, F.; Blanc, C.; Mankowski, G.; Pébère, N. *Surf. Coat. Technol.* 2002, *154*, 94-103.

[56] Bartolomé, M. J.; López, V.; Escudero, E.; Caruana, G.; González, J. A. *Surf. Coat. Technol.* 2006, *200*, 4530-4537.
[57] Zuo, Y.; Zhao, P.-H.; Zhao, J.-M. *Surf. Coat. Technol.* 2003, *166*, 237-242.
[58] Moutarlier, V.; Gigandet, M. P.; Ricq, L.; Pagetti, J. *Appl. Surf. Sci.* 2001, *183*, 1-9.
[59] Moutarlier, V.; Gigandet, M. P.; Pagetti, J. *Appl. Surf. Sci.* 2003, *206*, 237-249.
[60] Liu, W.; Zuo, Y.; Chen, S.; Zhao, X.; Zhao, J. *Surf. Coat. Technol.* 2009, *203*, 1244-1251.
[61] Randall Jr, J. J.; Bernard, W. J. *Electrochim. Acta.* 1975, *20*, 653-661.
[62] Khalil, N.; Leach, J. S. L. *Electrochim. Acta.* 1986, *31*, 1279-1285.
[63] Pyun, S.-I.; Hong, M.-H. *Electrochim. Acta.* 1992, *37*, 327-332.
[64] Kim, J.-D.; Pyun, S.-I.; Oriani, R. A. *Electrochim. Acta.* 1995, *40*, 1171-1176.
[65] Kim, J.-D.; Pyun, S.-i.; Oriani, R. A. *Electrochim. Acta.* 1996, *41*, 57-62.
[66] Melendres, C. A.; Van Gils, S.; Terryn, H. *Electrochem. Commun.* 2001, *3*, 737-741.
[67] Moon, S.-M.; Pyun, S.-I. *Electrochim. Acta.* 1998, *43*, 3117-3126.
[68] Huerta-Vilca, D.; de Moraes, S. R.; de Jesus Motheo, A. *Synth. Met.* 2004, *140*, 23-27.
[69] Camalet, J. L.; Lacroix, J. C.; Aeiyach, S.; Chane-Ching, K.; Lacaze, P. C. *Synth. Met.* 1998, *93*, 133-142.
[70] Kamaraj, K.; Karpakam, V.; Sathiyanarayanan, S.; Venkatachari, G. *J. Electrochem. Soc.* 2010, *157*, C102-C109.
[71] Tallman, D.; Spinks, G.; Dominis, A.; Wallace, G. *J. Solid State Electrochem.* 2002, *6*, 73-84.
[72] Eftekhari, A. *Synth. Met.* 2004, *145*, 211-216.
[73] Akundy, G. S.; Rajagopalan, R.; Iroh, J. O. *J. Appl. Polym. Sci.* 2002, *83*, 1970-1977.
[74] Conroy, K. G.; Breslin, C. B. *Electrochim. Acta.* 2003, *48*, 721-732.
[75] Huerta-Vilca, D.; Moraes, S. R.; Motheo, A. J. *J. Appl. Polym. Sci.* 2003, *90*, 819-823.
[76] Huerta-Vilca, D.; de Moraes, S. R.; de Jesus Motheo, A. *J. Solid State Electrochem.* 2005, *9*, 416-420.
[77] Pournaghi-Azar, M. H.; Habibi, B. *Electrochim. Acta.* 2007, *52*, 4222-4230.
[78] Shabani-Nooshabadi, M.; Ghoreishi, S. M.; Behpour, M. *Electrochim. Acta.* 2009, *54*, 6989-6995.
[79] Aoki, K.; Tano, S. *Electrochim. Acta.* 2005, *50*, 1491-1496.
[80] Cogan, S. F.; MGilbert, M. D.; Holleck, G. L.; Ehrlich, J.; Jillson, M. H. *J. Electrochem. Soc.* 2000, *147*, 2143-2147.
[81] Kilmartin, P. A.; Wright, G. A. *Electrochim. Acta.* 1996, *41*, 1677-1687.
[82] Kraljić, M.; Mandić, Z.; Duić, L. *Corros. Sci.* 2003, *45*, 181-198.
[83] Wessling, B. *Adv. Mater.* 1994, *6*, 226-228.
[84] Jasty, S.; Epstein, A. J. *Polym. Mater.: Sci. Eng.* 1995, *72*, 565-566.
[85] Wei, Y.; Wang, J.; Jia, X.; Yeh, J. M.; Spelling, P. *Polym. Mater.: Sci. Eng.* 1995, *72*, 563-564.
[86] Wessling, B. *Mater. Corros.* 1996, *47*, 439-445.
[87] Racicot, R.; Brown, R.; Yang, S. C. *Synth. Met.* 1997, *85*, 1263-1264.
[88] Sazou, D. *Synth. Met.* 2001, *118*, 133-147.
[89] Shah, K.; Iroh, J. *Synth. Met.* 2002, *132*, 35-41.
[90] Schauer, T.; Joos, A.; Dulog, L.; Eisenbach, C. D. *Prog. Org. Coat.* 1998, *33*, 20-27.
[91] Ogurtsov, N. A.; Pud, A. A.; Kamarchik, P.; Shapoval, G. S. *Synth. Met.* 2004, *143*, 43-47.

[92] Kendig, M.; Hon, M.; Warren, L. *Prog. Org. Coat.* 2003, *47*, 183-189.
[93] Mišković-Stanković, V. B.; Stanić, M. R.; Dražić, D. M. *Prog. Org. Coat.* 1999, *36*, 53-63.
[94] Lux, F. *Polymer.* 1994, *35*, 2915-2936.

In: Trends in Polyaniline Research
Editors: T. Ohsaka, Al. Chowdhury, Md. A. Rahman et al.
ISBN: 978-1-62808-424-5

Chapter 3

SONOCHEMICAL SYNTHESIS OF FUNCTIONAL HYBRID NANOMATERIALS CONTAINING POLYANILINE AND THEIR APPLICATIONS

S. S. Barkade[1], B. A. Bhanvase[2] and S. H. Sonawane[3,*]

[1]Department of Chemical Engineering, University Institute of Chemical Technology, North Maharashtra University Jalgaon India
[2]Vishwakarma Institute of Technology, Upper Indira Nagar, Pune, India
[3]Department of Chemical Engineering, National Institute of Technology, Warangal AP India

ABSTRACT

Polyaniline (PANI) is a conducting polymer which has been studied extensively due to its intriguing electronic and redox properties and numerous potential applications in many fields. Due to the combination of intrinsic properties and synergistic effect of each component, PANI based nanocomposite are playing an important role in various applications. The synthesis of multi-functional PANI nanocomposite has attracted a great attention because of a number of applications in polymer based solar cell, flexible electronics devices, anticorrosion coatings, conducting latex etc due to it's superior electron transport. Synthesis of functional hybrid materials containing non-conducting/conducting polymer, inorganic-conducting polymer composite is one of the important technological challenges. Synthesis of smaller and narrow particle size distribution of PANI conducting particles is a major technological challenge in order to exhibit the desired properties. In this book chapter, ultrasound assisted miniemulsion synthesis of functional conducting latex (PANI latex) and its hybrid composites are reported. Comparative performance of conventional synthesis and ultrasound assisted miniemulsion based polymer nanocomposite is reported. Various applications of conducting functional latex such as anticorrosion coatings, latex for photoeletrochemical cell, sensor applications and role of PANI coatings in nanocontainers is also reported.

* Corresponding author: S.H. Sonawane. Tel: +91-870-2462426; E-mail address: shirishsonawane@rediffmail.com, shirishsonawane09@gmail.com.

1. INTRODUCTION

From the time when conducting polyacetylene was prepared by Shirakawa al. [1] in 1977, there has been a potential growth of research in the field of organic conducting polymers due to their excellent electrical properties and their applications in chemical/biosensor, electrochemical catalysis, battery, electrochemical capacitor, etc. [2]. Among all the conducting polymers, polyaniline (PANI) has achieved widespread attention due to its simple synthesis, low cost, environmental stability and superior electrical properties. However, poor processing ability of PANI make them insoluble in common solvents and its infusibility are the key problems related to the potential application of PANI. The most promising and viable approach to solve the problem is to prepare the PANI nanocomposite [3-5]. Also conducting polymer/inorganic hybrid materials have attracted considerable attention because they can provide new synergistic properties that cannot be attained from the individual materials alone. Hybrid materials based on conducting organic polymers shows a unique combination of useful properties which, includes electronic conductivity (e^- or h^+), ionic transport, reversible electro activity, electro optical properties typically similar to semiconductor, pH and composition-dependent properties [6]. Further PANI based hybrid materials are an important class of hybrid materials with potential applications in a variety of domains, ranging from the encapsulation and controlled release of active substances to their utilization as fillers for the paint and coating industries [7]. Different strategies and concepts are employed for synthesizing polyaniline based hybrid particles with defined shapes (core-shell, fibers, multinuclear and raspberry particles) and nanoscale dimensions [8].

In recent years, with the rapid development in the area of nanotechnology and composite material synthesis, PANI nanocomposites have attracted considerable interests. So attempts have been made to synthesize PANI nanocomposite films with better functional properties. Some examples of PANI composites are PANI/TiO_2, PANI/$CaCO_3$, PANI/Fe_3O_4, PANI/Ag, PANI/CdS, etc. [4,5,9-11]. The preparation of PANI nanocomposites is an effective route to improve the performances of PANI, aiming to obtain materials with synergic effect between PANI and fillers.

Among the different experimental strategies developed to prepare PANI based hybrids, the use of ultrasound radiation turned out to be a very attractive and alternative tool for many researchers. The main advantages to use the ultrasound waves during the polymerization process is that reactions can be initiated in the absence of external chemical initiators. Propagation of ultrasound waves through a fluid causes the formation of cavitation bubbles. Collapse of these bubbles, described as an adiabatic implosion in the hot-spot theory, is the origin of extreme local conditions: high temperature (5000K) and high pressure (1000 atm). Cooling rates obtained after collapse are greater than 1010 Ks^{-1} which leads these experimental conditions to be classified as non-conventional conditions [12]. More recently this technique has been used in the preparation of polyaniline based hybrid nanomaterials. Ultrasonic assisted polymerization allows to use a low surfactant concentrations and obviates chemical initiators. Therefore the properties of these polyaniline based hybrids synthesized via sonochemical route are so outstanding hence it is one of the attractive routes of synthesis for both industry and academia. Among conducting polymers, polyaniline based hybrids are more studied because of electrical conductivity, low cost, facile synthesis and environmental stability [13].

In the present chapter comparative study of conventional and ultrasound assisted miniemulsion polymerization for preparation of PANI, mechanism of formation process of PANI nanocomposites, role of PANI in sensors, anticorrosion coatings and role in photo anode in electrochemical cell has been discussed. Further the applications of PANI based nanocomposites are also reported.

2. Details Literature Review on Functional Hybrid Materials Containing PANI and Applications

Conducting polymer/inorganic nanoparticle composites with different combinations of components have attracted more attention, since they have interesting physical properties and many potential applications [14]. However, conducting polymers are not usually soluble in common solvents, therefore it is difficult to synthesize conducting polymer/inorganic nanocomposites by conventional blending or mixing in solution or melt form. Further doped PANIs are conducting materials of interest for use in electrochemical application that have been also satisfactorily combined with inorganic oxides [15]. Xia and Wang [9] have used a new approach, i.e. ultrasound assisted *in-situ* emulsion polymerization of aniline in the presence of nanocrystalline TiO_2 to synthesize PANI/nanocrystalline TiO_2 composite particles. During ultrasound assisted *in-situ* emulsion polymerization of aniline, they have been used nanocrystalline TiO_2 as nanofillers along with an initiator and surfactant in aqueous solution. Further, ultrasonic irradiation has been carried out with ultrasonic horn immersed into the mixture emulsion system. It has been concluded that the aggregation of nano TiO_2 in the aqueous solution can be removed under ultrasonic irradiation, which leads to improvement in the conductivity of the composites compared with the conventional stirring process. Wang et al. [16] have also used ultrasonic irradiation for the preparation of PANI and gamma-zirconium phosphate (γ-ZrP) nanocomposite by intercalation of aniline into γ-ZrP. It has been reported that the intercalation rate of aniline into γ-ZrP was improved significantly by power ultrasound irradiation, particularly in the acid environment. The polymerization and cross-linking of aniline occurred at low pH, initiated by ammonium persulfate and γ-ZrP exfoliated in PANI bulk was obtained by the aid of ultrasound irradiation. High intensity ultrasonic irradiation has also been used by researchers for preparation of polypyrrole/gold or platinum nanocomposite [17], PANI nanotubes containing Fe_3O_4 nanoparticles [10], and core-shell SiO_2 nanoparticles/poly(3-aminophenylboronic acid) composites [18].

Further, Sonawane et al. [3] have prepared metal oxide-encapsulated poly butyl methacrylate (PBMA) latex by ultrasound assisted *in-situ* miniemulsion polymerization to make stable photoelectrodes. It has also been reported that the generation of photocurrent increases with an increase in the semiconductor oxide loading, whereas a reverse effect was observed with an increase in the thickness of the latex. Haldorai et al. [19] and Barkade et al. [4] have also used ultrasound assisted *in-situ* emulsion polymerization for the preparation of electrically conducting copolymer poly(aniline-co-p-phenylenediamine)/SiO_2 nancomposite and PANI/silver nanocomposite respectively. It has been reported that this approach can resolve the issue of the dispersion and stabilization of nanoparticles in the polymer latex, which will lead to the improvement in the properties of nanocomposite.

Bhanvase and Sonawane [5] have carried out synthesis of PANI/$CaCO_3$ nanocomposite by semi-batch in-situ emulsion polymerization using indirect ultrasound (in ultrasound bath) irradiation. It has been reported that most of the myristic acid coated $CaCO_3$ nanoparticles are finely dispersed in the PANI matrix. It is attributed to the hydrophobic nature of $CaCO_3$ and micro-mixing caused by ultrasonic irradiations. Electrochemical corrosion tests at room temperature (25^0C) have been reported. The reported electrochemical current ($I_{corr.}$) decreased from 0.89 to 0.03 $\mu A/cm^2$, when neat alkyd resin and PANI/$CaCO_3$ nanocomposite was tested in NaCl electrolyte solution. Additionally, E_{corr} value shows shifts in positive side from -1.74 to -1.47 V by addition of PANI-$CaCO_3$ nanocomposite into alkyd coating. This is an indication of improvement in the anticorrosion properties of coatings with the dispersion of PANI and PANI - $CaCO_3$ nanocomposite in alkyd resin.

Also Jia et al. [20] have prepared novel functionalized hybrid PANI nanofibers with integrated Pt nanoflowers. They have employed this hybrid as the sensing platform for urea detection in a flow-injection-analysis system. Langa et al. [21] have reported three-dimensional bicontinuous nanoporous Au/polyaniline (PANI) hybrid composite films made by one-step electrochemical polymerization of PANI shell onto de alloyed nanoporous gold skeletons for the applications in electrochemical supercapacitors. Further Raut et al. [11] have also prepared camphor sulfonic acid doped PANI/CdS nanohybrid materials by chemical oxidative polymerization method for detection of hydrogen sulfide gas at room temperature (300 K). Preparation of PANI -Fe_3O_4 composite material have been carried out by Belaabeda et al. [22], the composite was dispersed in epoxy resin for modulating the electromagnetic properties to fabricate microwave absorbing and electromagnetic shielding materials with high performances. Deng et al. [23] have prepared a hybrid material of carbon nanotubes (CNTs)– PANI by in situ emulsion polymerization. The prepared hybrid composite showed high conductivity and high thermal stability due to fine dispersion of CNTs in PANI matrix. The hybrid network formed new conductive passageway, which account for high conductivity. Divya and Sangaranarayanan et al. [24] have used a very facile, simple, one step, synthesis of novel mesoporous crystalline copper–polyaniline composite with a high content of copper (46.63 wt.%) as well as high specific surface area, which has excellent environmental stability. Wang et al. [25] have synthesized PANI/graphene hybrids by in situ polymerization of aniline monomer in graphene dispersion. The reported results suggests that the PANI/graphene hybrid material is a potential alternative to Pt as the counter electrode materials for dye-sensitized solar cells. Palladium-PANI hybrid nanocomposite material as a catalyst for the coupling of phenylboronic acid with aryl halides have been prepared by Islama et al. [26] using an in-situ technique in the presence of an inorganic base which showed excellent yield. Electroactive hybrid films with cubic nickel hexacyanoferrate/PANI were synthesized on carbon nanotubes modified platinum electrodes by a facile one-step electrosynthesis method. The hybrid film prepared had good stability and reproducibility in the detection of H_2O_2, and should be useful in practical H_2O_2 sensors [27]. Further, Paulraj et al. [28] have synthesized PANI–Ag hybrid nanocomposite through interfacial polymerization method which was used for the electrocatalytic determination of hydrazine in environmental samples.

PANI/TiO_2 composite nanotubes have been prepared by Zhang et al. [29] by in situ polymerization in the presence of TiO_2 nanoparticles and b-naphthalenesulfonic acid as both the dopant and the template. Fabrication of PANI/WO_3 hybrid nanocomposite based sensor has been carried out by Parvatikar et al. [30] and it has been reported that the film conductivity increased with increasing humidity. Tai et al. [31] have also fabricated a PANI-TiO_2 nanocomposite for NH_3 and CO and reported superior response for NH_3 as compared to CO. In their study, in situ polymerization of aniline in presence of TiO_2 nanoparticles on the sensor substrate at different temperatures has been reported. Tudorache et al. [32] have reported the preparation of PANI-Fe_2O_3 and PANI-$MgFe_2O_4$ hybrid via in situ polymerization of aniline in the presence of these oxides particles. The result shows that the introduction of conducting PANI not only improves the conductivity of iron oxides particles, but also the dispersibility of ferrites. These hybrids were most sensitive to acetone vapors and can be used as gas sensor. Deshpande et al. [33] have synthesized the SnO_2-intercalated PANI hybrid through solution route technique, which showed better sensitivity to ammonia gas at room temperature than SnO_2. Jaroslav et al. [34] synthesized particles of zinc ferrite, $ZnO{\cdot}Fe_2O_3$ coated with polyaniline phosphate during the in situ polymerization of aniline in an aqueous solution of phosphoric acid. The improvement in mechanical and anticorrosion properties has been reported. Synthesis of PANI/TiO_2 hybrid nanoplates have been carried out by Katoch et al. [35] via a sol–gel chemical method. Plate like structure of the hybrid was more advantageous for the electrochemical stability. Further synthesis of PANI/mesoporous TiO_2 composite was carried out by Qiao et al. [36] and it was used as an anode for microbial fuel cells. Kowsari et al. [37] have prepared polyaniline-Y_2O_3 hybrid nanocomposite with controlled conductivity with the assistance of ultrasound and an ionic liquid. Ultrasound energy and the ionic liquid replace the conventional oxidants and metal complexes in promoting the polymerization of aniline monomer for the first time. Due to sonication a strong interaction between PANI and nanocrystalline Y_2O_3 occurred, which gives rise to changes in the surface properties and electrical conductivity of the hybrid, and also improved its thermal stability.

3. Polymerization of Aniline: Special Emphasis on Ultrasound Assisted Emulsion Polymerization

3.1. Comparative Study of Conventional and Ultrasound Assisted Miniemulsion Polymerization for Preparation of PANI

PANI and their derivatives have attracted significant attention for their wide variety of potential technological applications such as anticorrosion coatings [38], electromagnetic shielding, rechargeable polymeric batteries [39], polymer photovoltaics, and polymer actuators [40]. The intense collapse effects arising from cavitation boost the polymerization process and due to these effects polymer sonochemistry has become an active area of research. The electrical and mechanical properties of polythiophene film were improved considerably by the ultrasonic irradiation of the electrolytic solution during polymerization [41]. Xia and Wang [42] have prepared PANI nanoparticles through inverse microemulsion polymerization using ultrasonic irradiations. In this process, use of a surfactant (CTAB) and an ammonium peroxydisulfate as an oxidant in a sonication process lasting 1 h has been reported. The size of the particles obtained in case of ultrasonic irradiation has been significantly reduced and it was between 10 and 60 nm. They have also used ultrasonic irradiations for the preparation of TiO_2/PANI nanocomposite [9]. Further Atobe et al. [43] have reported use of ultrasonic irradiation for the preparation of PANI colloids using polyethylene oxide as a stabilizer and potassium iodide as an oxidant. It is interesting to know that ultrasound can be used as an oxidizing agent and a stabilizer, whereas in case conventional methods chemical oxidising agent and stabiliser has been generally used. Therefore ultrasound method for the preparation of PANI nanocomposite can be an environmental friendly alternative.

Further conventional method consists of surfactant, initiator and emulsified monomer droplets in a continuous aqueous phase (usually water), which are in the range of 1 to 10 μm [44]. The conventional polymerization can be initiated by water soluble or insoluble initiator such as azo, peroxy and persulphate compounds. Initiator is generally dissociated by thermal, photochemical method, or via a redox reaction, which is known to generate active free radicals. Conventional initiation processes have disadvantages, such as elevated temperature is required for thermal breakdown of the chemical initiator. Photochemical breakdown generates only a surface initiating radical. Radiolysis employs a radiation source, which requires specific safety precautions. Ultrasonic irradiation could be used as a source of free radicals generation for polymerization. It could be an effective method for the breakdown of water soluble initiator [45-46].

Sivakumar and Gedanken [12] have also carried out an investigation to use ultrasound for dispersion of aniline in water and its polymerization to give PANI, without any surfactant and/or stabilizer. It has been reported that the use of H_2O_2 during ultrasound assisted polymerization of aniline leads to formation of PANI within 20 min. However in case of conventional mechanical rapid stirring a small amount of PANI formation has been reported after 18 h. These results clearly show the significant effect of ultrasonic irradiation on increasing the polymerization rate of aniline to provide PANI.

3.2. Mechanism of the Formation Process of PANI Nanocomposites

3.2.1. Mechanism of Polyaniline/Ag Hybrid Nanocomposite Formation

Miniemulsion polymerization provides control over reaction rates and heat dissipation compared to conventional techniques [47]. Appropriate combination of surfactant (sodium dodecyl sulfate), initiator and monomer system leads to synthesis of nanolatex particles of small size less than 100 nm. In the first stage of aniline polymerization, radical generation occurs due to dissociation of ammonium persulphate which acts as an initiator. In case of ultrasound assisted polymerization process, cavitation generates few more radicals due to dissociation of water molecules during the collision of high energy cavities [48].

$$H_2O \xrightarrow{\text{))) Ultrasound}} \bullet H + \bullet OH$$

These radicals enter into micelles within short period of time because of microjets generated by the ultrasonic irradiations. These radicals are responsible to accelerate the rate of aniline polymerization and conversion. Solution in the ultrasonic reactor turns from black to green in color. Secondary growth of polymer occurs due to extra addition of aniline monomer. Finally it turns into irregularly shaped agglomerates containing nanosize polyaniline polymer particles. Formation of nanosize polyaniline polymer is possible because of chemical and physical effects produced by cavitation [49].

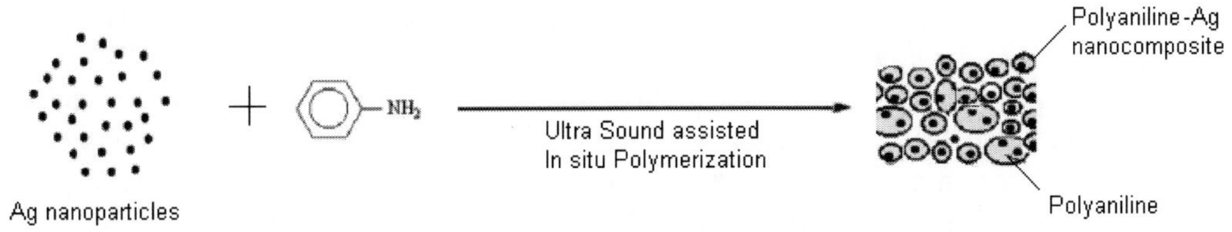

Figure 1. Ultrasound assisted synthesis of Polyaniline/silver hybrid.

The prepared colloids of silver nanoparticles as reported by Barkade et al. [4] were added into the reaction mixture. Reactor temperature was maintained at 4°C throughout the run which leads to the formation of colloidal nanocomposite. The ultrasound assisted insitu emulsion polymer nanocomposite formation process is shown in figure 1.

3.3. Mechanism and Role of PANI in Sensor, Anticorrosion, Electrochemical Cell, Etc.

Doped/undoped PANI can be synthesized based on exposure to an acidic or alkaline environment. This type of doping mechanism is due to the presence of -NH group on the polymer backbone, the protonation/deprotonation of which brings about a change in the electronic conductivity as well as change in color. Aniline basically undergoes oxidative polymerization in the presence of a protonic acid. The product formed is a simple 1,4-coupling of the monomer. Protonation induces an insulator-to-conductor transition, while the number of π electrons in the chain remains constant (Scheme 1).

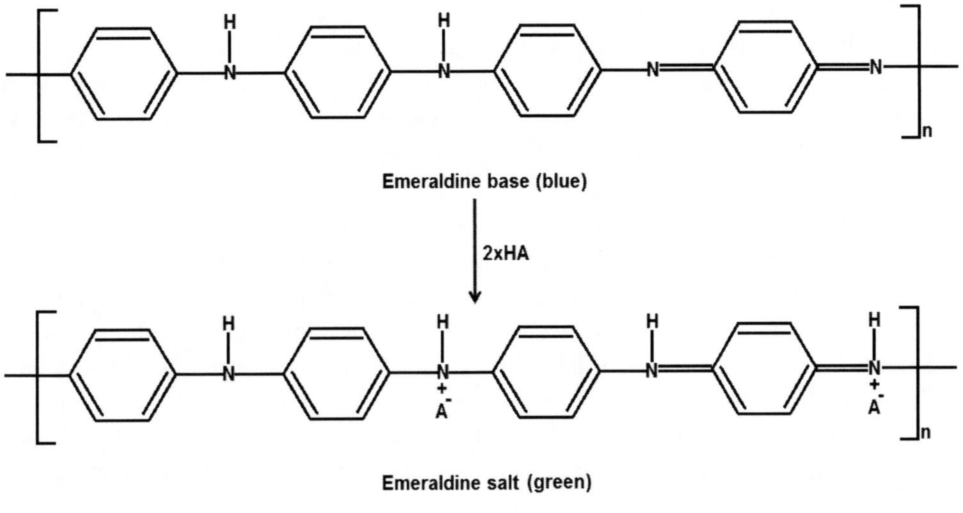

Scheme 1. Protonation of PANI emeraldine base to PANI emeraldine salt.

Polyaniline exists in different oxidation states. Leucoemeraldine base (x =1, fully reduced) is very reactive. It reacts even with minute amount of oxygen and they are environmentally unstable. Emeraldine base (x = 0.5, partially oxidized) is environmentally stable and does not undergo any change in chemical structure on prolonged storage. Pernigraniline base (x = 0, fully oxidized) is also environmentally stable and further oxidation is not possible with fully imine groups. The fully oxidized and fully reduced state of polyaniline is not in a conducting state. This simple doping/undoping phenomenon makes polyaniline and its hybrids highly suitable for sensing applications. But low process ability, low insolubility in common solvent and poor mechanical properties of polyaniline has offered obstructions in the preparation of sensing element. In order to overcome these problems incorporation/encapsulation of polyaniline with inorganic or other entities is always preferred. The incorporation of metal, metal oxide, carbon nanotubes and graphene nanomaterials with different shapes could effectively improve the electrical, optical and dielectric properties of the polyaniline hybrid composites [28]. These properties are very much sensitive to small changes in the metal content and in the size and shape of the nanoparticles. In case of sensing these incorporated nanomaterials via different innovative strategies act as conductive junctions between polyaniline chains which resulted in the change of the electrical conductivity of the hybrid composite with lower response time and better reproducibility.

Polyaniline is an excellent anticorrosive material used in coating industry. Bhanvase and Sonawane [5] have studied the preparation of PANI and PANI/$CaCO_3$ nanocomposite by ultrasound assisted insitu emulsion of polymerization method. It has been reported that the incorporation of PANI and PANI/$CaCO_3$ nanocomposite in alkyd resin coating shows significant improvement in anticorrosion properties. It is also found that with an increase in the percent loading of PANI and PANI/$CaCO_3$ nanocomposites in alkyd resin, the corrosion rate was reduced significantly in 5% HCl, NaCl and NaOH solution. PANI plays an important role in improvement in the anticorrosion properties. In case of PANI/alkyd coatings, a compact iron/dopant complex layer formation takes place at the metal-coating interface which acts as a passive protective layer. The protective behavior depends on the size and charge of the dopant i.e. as the size of the dopant increases, the strength of the iron/dopant complex film increases, which improves the anticorrosion efficiency [50-51].

Semiconductor materials have been used as photoanodes for the translation of solar energy into electrical energy in photoeletrochemical cells. Two approaches has been widely used for the preparation of thin film based photoanodes: (i) metal or semiconductor metal oxides are deposited onto a conductive glass at higher temperature using chemical vapor deposition or electrochemical deposition and (ii) semiconductors incorporating conducting polymers, such as PANI/polypyrrole/porphyrin, are coated onto the nonconducting transparent matrix such as PMMA and polyethylene terphthalate.

Conductive polymers possess poor mechanical properties; however, casting with nonconductive polymers gives better mechanical properties and adhesion along with high photocurrent. It has been also found that with an addition of small quantity of PANI during the latex synthesis to facilitate the transport of charge carriers leads to increase in the conductivity. Photocurrents of 4×10^{-5} and 1×10^{-5} mA/cm^2 have been reported by Sonawane et al. [3] in the presence and absence of polyaniline, respectively, at 0.78 V applied potential.

4. SYNTHESIS OF POLYANILINE CONTAINING HYBRID FUNCTIONAL NANO PARTICLES

4.1. PANI-Ag Nanocomposite

Hybrid nanocomposites formed by immobilization of Ag nanoparticles into the PANI matrices have received much attention in recent days due to its applications in catalysis, conducting ink, anticancer activity, sensor, fuel cell and as a potential antibacterial agent. Although many groups have involved in the synthesis of PANI incorporated silver nanoparticles by utilizing various methods, still it is a challenge to explore the various interesting properties of hybrid nanocomposites. Bedre et al. [52] have reported the synthesis of PANI and PANI - Ag nanoparticles nanocomposite via interfacial polymerization methods and the formation of PANI-Ag was accomplished using $(NH_4)_2S_2O_8$ as an oxidizing agent prior to the addition of silver nitrate solution in the aqueous phase. Chowdhari [53] have prepared PANI/Ag nanocomposites by in situ oxidative polymerization of aniline monomer in the presence of different concentrations of Ag nanoparticles. Khanna et al. [54] have synthesized PANI/Ag nanocomposite via in situ photo-redox mechanism. Deposition of a thin film of metal nanoparticles on conducting polyaniline film in both chemical reduction and electrochemical approaches has been reported by Wang et al. [55]. Barros and Azevedo et al. [56] have prepared Polyaniline (PANI) and polyaniline/silver nanocomposites by sonochemical and ionizing radiation technique. The formation of hybrid are based on the fact that both methods produce hydroxyl radical $^{\bullet}OH$ and hydrogen radical $^{\bullet}H$, which acts as an oxidizing agent for the polymerization process of aniline monomer and as a reducing agent for silver ions. Paulraj et al. [28] have used interfacial polymerization processes using aniline dimer (4-aminophenyldiphenylamine, APDA) as starting material in chloroform in contact with an aqueous solution of silver nitrate and studied its application on electrochemical oxidation of hydrazine.

4.2. PANI- Metal Oxide Nanocomposites

Organic-inorganic-metal oxide/polyaniline hybrid materials are currently of great interest for exploring enhanced characteristics, due to their synergetic or complementary behaviors that is not observed in their single counterparts. Different metal oxides like WO_3, ZnO, SnO_2, Fe_2O_3, TiO_2, Mn_2O_3, Co_3O_4, NiO, CuO and SrO are incorporated into PANI to design/generate desired property hybrid for various applications. Somani and coworkers [57] prepared highly piezo resistive PANI/TiO_2 composites by in situ polymerization of aniline in the presence of fine grade powders of anatase TiO_2 (~100 nm). Hu et al. [58] have reported PANI/SnO_2 composite material in which the nanostructured SnO_2 particles were embedded within the netlike PANI. The net like PANI provides high active surface area for the electrochemical reaction, and on the other hand, SnO_2 nanoparticles nucleated over polymer chains contributing to enhanced conductivity and stability of the nanocomposite material by interlinking the PANI polymer chains. Parvitkar et al. [59] have synthesized PANI/WO_3 composites via 'in situ' deposition technique by placing fine graded WO_3 in polymerization mixture of aniline. The hybrid composites in the pellet form exhibit almost linear behavior within a chosen range of humidity (ranging between 10 and 95% RH). Eskizeybek et al. [60] have synthesized PANI/ZnO nanocomposite in aqueous diethylene glycol solution medium via the chemical oxidative polymerization of aniline. The photocatalytic activities of PANI/ZnO nanocomposites have been investigated by the degradation of methylene blue and malachite green dyes in aqueous medium under natural sunlight and UV light irradiation. Wang et al. [61] have synthesized magnetic oleic acid modified Fe_3O_4 polyaniline nanoparticles with multicore/single-shell morphology via a novel and facile in situ miniemulsion polymerization approach. This method can be extended for fabricating magnetic/conductive core/shell composites which could find applications in catalyst supports or biomedical areas. PANI/NiO nanocomposite was prepared in the aqueous medium using polyvinyl alcohol and hydroxypropylcellulose as a surfactant by Aleahmad et al. [62]. The thermal stability of nanocomposite decreased with increasing NiO content.

5. Properties of PANI based Nanocomposites

Among conducting polymers, PANI is probably the most extensively studied due to its several unique properties [63-64]. Ease of preparation, light weight, low cost, better electronic, optical properties, high stability in air, solubility in various solvents, and good processibility [65-66] makes PANI widely usable in several applications. PANI exhibits remarkable changes in its electronic structure and physical properties at protonated state. Based on the oxidation level, PANI can be prepared in diverse insulating forms such as the fully reduced leucoemeraldine base, half oxidized emeraldine base and fully oxidized pernigraniline base. The conducting emeraldine salt form is achieved by doping with aqueous protonic or functionalized acids where protons are added to the –N= sites. This leads to an increase in the conductivity by more than ten orders of magnitude depending on the strength of the acid and the method of processing [67]. PANI composites with magnetic and conducting properties have been studied by Wan et al. [68-69]. They prepared materials with very low cohesive force and relatively high saturation magnetization. The soft magnetic spinel ferrites of the form $A^{2+}B_2^{3+}O_4$ such as Fe_3O_4, $CoFe_2O_4$, $NiFe_2O_4$, $MnFe_2O_4$, and $ZnFe_2O_4$ have been extensively used in microwave devices because of their high saturation magnetization, high permeability, high electrical resistivity and low eddy current losses [70-71]. Zhu et al. [72] have studied the electrical conductivity and dielectric permittivity of polyaniline–Al_2O_3 nanocomposites which are strongly related to both the morphology of the filler and the dispersion quality. Dey et al. [73] have also studied the dielectric properties of PANI–TiO_2 nanocomposite at room temperature. Araujo et al. [74] have studied the magnetic properties of PANI-Fe_3O_4 nanocomposite. The AC conductivity measurements reveal that the nanocomposites show conductivity of the order of 10^{-5} Scm^{-1}, two orders of magnitude higher than pure Fe_3O_4, confirming that there is increase in conductivity with the increasing amount of PANI. This increase in the conductivity is due to the long range pathways of conductivity created by polyaniline. The magnetic measurements showed ferromagnetic behavior for the nanocomposite, with high-saturated magnetization (M_S =74.30 $emug^{-1}$) and a coercive force of 93.40 Oe. Further Olad and Naseri [75] have studied the anticorrosive performance of polyaniline/clinoptilolite nanocomposite.

6. APPLICATIONS OF PANI BASED NANOCOMPOSITES

6.1. Gas Sensor Application of PANI-Ag Nanocomposite

The synthesis of hybrid nano materials with functional properties is an important area of research because of their versatile applications in gas sensors [76]. Polyaniline has been most extensively studied due to its excellent electrical properties which can be modified by oxidation state of the main chain and by protonation mechanism. Also electrical properties of polyanilne can be controlled by charge transfer doping using acids and protonation. Therefore, it acts as a conductor of electrons and shows sensitivity to the number of chemicals which helps in device fabrications of chemical and biological sensors [77-78]. It has been reported that PANI can be used in layered thin film form in sensors providing the best replacement option for metal and metal oxide used in preparation of gas sensing device [79]. Further an improvement of the sensing performance of PANI can be achieved by the incorporation of compatibilized and functional metal and metal oxide nanoparticles into polyaniline by using in situ emulsion polymerization process [33, 80]. Enhanced performance of gas sensor containing polyaniline and Ag nanoparticles at various atmospheric conditions has also been also reported by many researchers [53].

Barkade et al. [4] have attempted ultrasound assisted in-situ miniemulsion polymerization of aniline along with different loading of silver nanoparticles and have studied its application for ethanol vapor sensing. Ultrasound assisted in-situ miniemulsion polymerization of aniline in presence of different loading of silver nanoparticles for the synthesis of PANI/Ag nanocomposite have been carried out and its formation mechanism has been reported in figure 1. Surfactant solution prepared independently with an addition of sodium dodecyl sulphate in water and initiator solution by adding ammonium persulfate in water was transferred to the semi-batch sonochemical reactor (sonochemical reactor in which continuous addition of monomer is made). 1 M aqueous solution of hydrochloric acid (dopant) has been added into reactor for protonation. Aniline was added into the reactor in semibatch manner within 60 min to control the reaction rate at 4°C temperature.

Formation of cavities occurs, which contains vapor of the reaction liquid medium, when ultrasound waves passes through an emulsion polymerization reaction medium. During the collapse of cavities, these vapors have been subjected to an extreme condition of high temperature and pressure, leads to generation of highly reactive radical species due to composition of water, surfactant, monomers and oligomers. The created radical are responsible for the initiation of polymerization and it also helps in propagating polymerization. The ultrasonic irradiation also leads to decrease in the size of synthesized PANI/Ag nanocomposite particles which is reported by Barkade et al. [4] which are near to be 50 nm. It has been further reported that Ag nanoparticles (5–10 nm) are finely dispersed into polyaniline matrix without any aggregation. This uniform dispersion is attributed to strong affinity of silver for nitrogen and intense environment generated by ultrasonic irradiations. The electrical conductivity of nanocomposite is affected by the adsorption of vapor molecules on its surface. Change in resistance of polyaniline/Ag and polyaniline film sensors were characterized at room temperature in air. The sensor response (S) is defined as,

$$S = \frac{(R_g - R_a)}{R_a} \tag{1}$$

where, R_g and R_a are the resistances of the sensitive film in measuring ethanol vapor and in clean air, respectively. The initial resistance of pure polyaniline and that of nanocomposite before exposure to ethanol vapors were 240 and 150 Ω, respectively [4]. A linear response up to 100 ppm exhibited by the PANI/Ag nanocomposite has been depicted in Figure 2a. Further, the change in resistance is found independent of concentration [4]. The increase in resistance of sensor on exposure to ethanol may originate due to the interaction of –OH groups of ethanol molecules and nitrogen of polyaniline, leading to electron delocalization and charge transport through the polymer chain. In comparison to pure polyaniline, sensor response of polyaniline/Ag nanocomposite shows more stability as well as good reproducibility to ethanol vapors under the same condition. It has been also reported that ethanol vapor sensing ability of the nanocomposite sensor increases on addition of silver nanoparticles in the polymer matrix. These results show that polyaniline/Ag nanocomposite serves as an efficient sensor compared to pure polyaniline. The improvement in sensitivity of polyaniline/Ag is due to enhanced degree of interactions between the nanocomposite and the ethanol vapors.

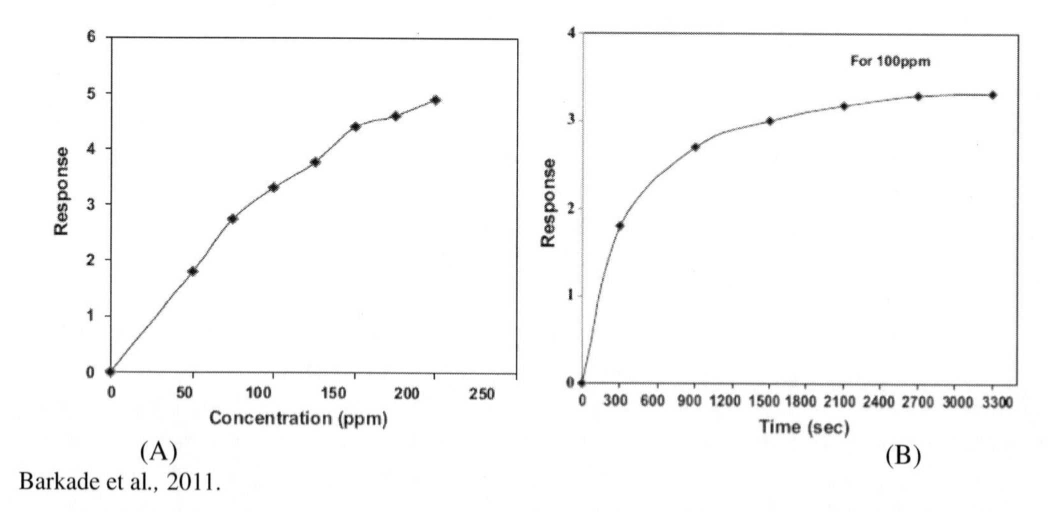

Barkade et al., 2011.

Figure 2. Response of Polyaniline/Ag film sensor (a) at different ethanol concentrations (50–200 ppm), (b) to ethanol (100 ppm).

A steady linear response up to 2100 s at 100 ppm as depicted in Figure 2b has been reported, which on further increase in time leads to saturation of the nanocomposite film. The possible reason reported is decrease in available free volume for vapor permeability into the nanocomposite [4].

6.2. Photoanode in Photoelectrochemical Cell by PBMA-PANI-TiO_2/ZrBiTiO$_4$/ZnO Nanocomposites

Semiconductor materials have been used as photoanodes for the translation of solar energy into electrical energy in photoelectrochemical cells. For effective energy conversion the devices based on polymer-semiconductor metal oxide materials offer cost effectiveness. Two approaches has been widely used for the preparation of thin film based photoanodes: (i) metal or semiconductor metal oxides are deposited onto a conductive glass at higher temperature using chemical vapor deposition or electrochemical deposition and (ii) semiconductors incorporating conducting polymers, such as polyaniline (PANI)/polypyrrole/ porphyrin, are coated onto the non conducting transparent matrix such as poly(methylmethacryalate) and polyethylene terephthalate. Wang et al. [81] have fabricated a highly transparent conductive thin film using a polypyrrole/PMMA core-shell matrix. Also Jang et al. [82] have used conductive nanospheres in the core matrix, while the shell was made up of a nonconductive polymer composite. Conductive polymers possess poor mechanical properties; however, casting with nonconductive polymers gives better mechanical properties and adhesion along with high photocurrent.

Further Sonawane et al. [3] have used ultrasound assisted insitu emulsion polymerization for the preparation of metal oxide-encapsulated poly butyl methacrylate (PBMA) latex to make stable photoelectrodes, which might find applications in solar cells. In order to have compatibility of Bi_2O_3/$ZrTiO_4$ oxide particles with polymer, they have been treated initially with myristic acid in presence of ethanol at 60 ^{o}C in ultrasound bath [3]. It has been further reported that the particle sizes, determined by TEM analysis as depicted in Figure 3, of TiO_2, ZnO, and Bi_2O_3/$ZrTiO_4$ embedded into PBMA latex were found to be 21, 15, and 7 nm, respectively [3]. It has been also reported that the particles in the case of Bi_2O_3/$ZrTiO_4$ as depicted in Figure 3d, are crystalline in nature. It is attributed to the use of ultrasound assisted methods for the synthesis of semiconductor nanocomposite oxides, which leads to the formation of fringes structured nanoparticles indicating crystalline nature.

A standard three electrode assembly consisting of a standard calomel electrode, platinum counter electrode under UV irradiation and a working electrode (semiconductor oxide-PBMA latex photoanode) has been used for electrochemical measurements. For the photo-anode preparation, the oxide-loaded latex was cast as a film on ITO (film thickness 50 μm, active area 1.0 cm^2). The use of an aqueous solutions containing 0.01 M NaOH, KI, and formic acid or oxalic acid as electrolytes has been reported. In the photo-anodes containing the semiconductor-loaded latex particles, the generation of photo-currents can be expected to occur via a similar mechanism as that in a conventional photoelectochemical cell. The advantage of using the polymer latex is that the semiconductor nanoparticles are protected from photo-dissolution, which increases the stability of the photo-anodes against photo-corrosion [83].

Photocurrent as a function of voltage curves for the ZnO/PBMA system has been depicted in Figure 4 [3]. It can be evidently seen from the photogenerated currents that the electrode is extremely photoresponsive. It has also been reported that the device is exhibited with an open-circuit voltage (V_{oc}) of -0.03 V, a short-circuit current (J_{sc}) of 0.8 × 10^{-4} mA/cm^2, and a fill factor (FF) of 0.20 (FF is defined as the ratio of the maximum power to the product of open-circuit voltage and short-circuit current). It has been found that an addition of small quantity of polyaniline during the latex synthesis facilitates the transport of charge carriers and leads to increase in the conductivity. Photocurrents of 4×10^{-5} and 1×10^{-5} mA/cm^2 have been reported in the presence and absence of polyaniline, respectively, at 0.78 V applied potential [3].

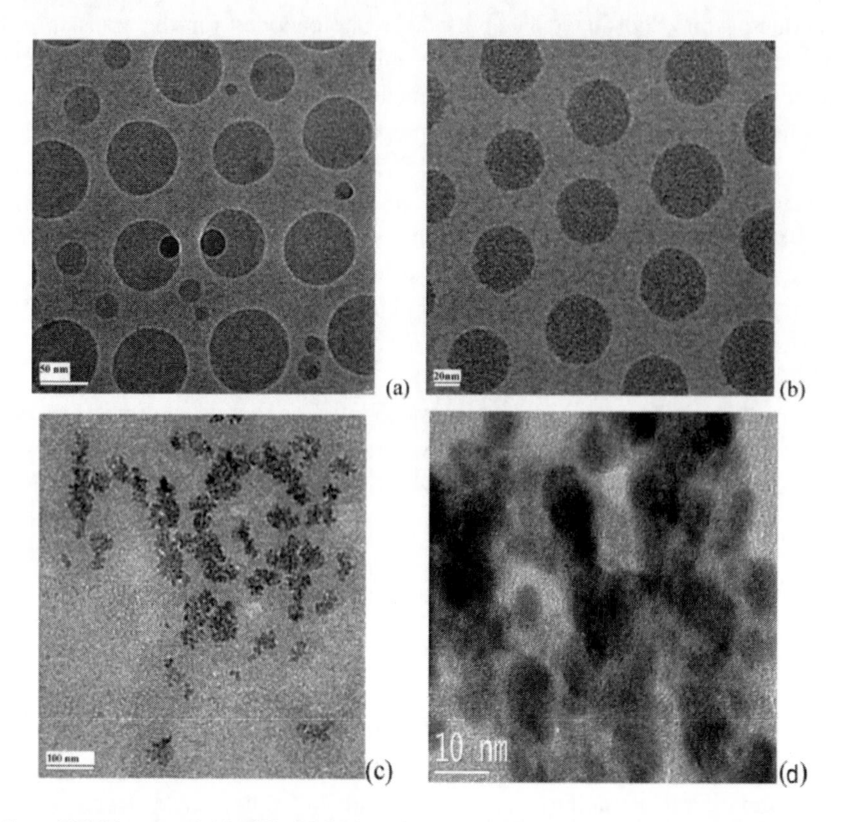

Figure 3. Cryo-TEM image of (a) TiO_2-PBMA and (b) ZnO-PBMA latex particles, TEM image of (c) $Bi_2O_3/ZrTiO_4$ nanoparticles in the PBMA matrix, and (d) $Bi_2O_3/ZrTiO_4$ nanoparticles. [3].

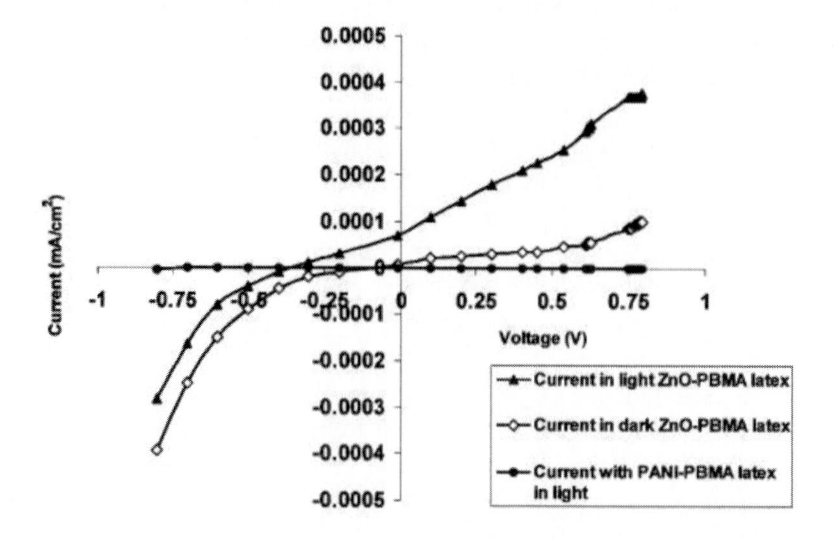

Figure 4. Representative example of the photocurrent of ZnO-PBMA latex nanocomposites (0.1 g loading) [3].

The photoelectrochemical results of the polymer latex for three different systems (ZnO, TiO_2, and $Bi_2O_3/ZrTiO_4$) in various electrolyte solutions have been reported in Table 1.

Table 1. Summary of the photoelectrochemical properties of the photoelectrodes in different electrolyte solutions

Composition	Electrolyte 0.01 Mol/dm^3	I_{Sc} Short circuit photocurrent	I_{max} Photocurrent @ 0.798 V (μA)	Open Circuit Potential (V_{oc}) (Volts)
TiO_2-PBMA 0.1g	Oxalic acid	0.05	0.06	0.9
	Formic acid	0.06	0.08	0.1
	NaOH	0.09	0.10	-0.05
	KI	0.04	0.05	-0.03
ZnO-PBMA 0.1 g	Oxalic acid	0.005	0.10	-0.26
	Formic acid	0.02	0.3	-0.18
	NaOH	0.1	0.4	-0.31
	KI	0.1	0.4	-0.33
$Bi_2O_3/ZrTiO_4$ PBMA 0.1 g	Oxalic acid	0.1	0.3	-0.18
	Formic acid	0.03	0.6	0.018
	NaOH	0.1	0.4	-0.020
	KI	0.1	0.3	-0.008
TiO_2-PBMA 0.4g	Formic acid	0.04	0.50	-0.34
	Oxalic acid	0.70	0.30	-0.68
	NaOH	0.11	0.11	-0.78
	KI	0.04	0.04	-0.62
ZnO-PBMA 0.4 g	Oxalic acid	0.07	0.14	-0.34
	Formic acid	0.07	0.70	-0.23
	NaOH	0.17	0.27	-0.61
	KI	0.10	0.28	-0.41
$Bi_2O_3/ZrTiO_4$ PBMA 0.4 g	Oxalic acid	0.10	0.26	-0.14
	Formic acid	0.10	0.90	0.02
	NaOH	0.22	0.42	-0.50
	KI	0.11	0.46	-0.20

Sonawane *et al.,* 2010.

Sonawane et al. [3] have tested three latex materials with two different loadings (0.1 and 0.4 g) and four different electrolyte solutions (NaOH, KI, formic acid, and oxalic acid). It has been reported that, at 0.1 g loading, TiO_2 showed a 0.1 $\mu A/cm^2$ maximum photocurrent at 0.798 V in NaOH electrolyte and that ZnO latex showed a photocurrent of 0.4 $\mu A/cm^2$ at 0.798 V in both KI and NaOH solutions. Bi_2O_3/ $ZrTiO_4$ have attributed to a maximum photocurrent of 0.6 $\mu A/cm^2$ at 0.798 V in the presence of formic acid under UV light. Sonawane et al. [3] have also reported the study of the effect of catalyst loading on the photocurrent. It was found that, in the case of TiO_2-PBMA latex, the short-circuit photocurrent was 0.7 $\mu A/cm^2$ in the presence of oxalic acid at a 0.4 g loading. At the same level of loading (0.4 g), 0.5 and 0.9$\mu A/cm^2$ photocurrents were observed with ZnO-PBMA and $Bi_2O_3/ZrTiO_4$ in the presence of formic acid. In overall, it has been reported that the $Bi_2O_3/ZrTiO_4$ generates the maximum current. It has also been reported that the generation of photocurrent is highly dependent on the encapsulation, particle size of the semiconductor, and their morphology. The reported size of $Bi_2O_3/ZrTiO_4$ was 7 nm, which was smaller than the size of all other photocatalysts used, and hence, the conversion was found to be higher in the case of $Bi_2O_3/ZrTiO_4$. The effect of the coating thickness without increasing the catalyst loading was also reported by Sonawane et al. [3]. The results revealed that all the electrodes were found to be inefficient with the increasing coating thickness, which might be due to the poor charge transport through thick polymer Films. Further, there has been an effect of electrolyte solution on the performance of the semiconductor oxide and $Bi_2O_3/ZrTiO_4$ shows a better performance in the presence of acid electrolytes. To conclude, it has been accomplished that the generation of photocurrent increases with an increase in the semiconductor oxide loading, whereas a reverse effect was observed with an increase in the thickness of the latex. $Bi_2O_3/ZrTiO_4$ showed photocurrents of 0.6 and 0.9 $\mu A/cm^2$ at 0.1 and 0.4 g loadings, respectively, in formic acid electrolyte. ZnO showed photocurrents of 0.4 and 0.6 $\mu A/cm^2$ in NaOH and KI, respectively, at a 0.1 g loading. The order of photocurrent generation efficiency with the different catalysts was $Bi_2O_3/ZrTiO_4$>ZnO> TiO_2 [3].

6.3. PANI–$CaCO_3$ in Alkyd Coating for Mechanical and Anti-corrosion Properties

Bhanvase and Sonawane [5] have reported the preparation of PANI/$CaCO_3$ nanocomposites by semi-batch in-situ emulsion polymerization by indirect ultrasound technique (ultrasonic bath, Sonics and Materials, 20 kHz, 600 W). For this purpose distilled aniline as a monomer, ammonium persulphate as an initiator and sodium lauryl sulphate as a surfactant has been used.

The reaction was performed in semi-batch in 90 minutes at 4 °C. In order to enhance the reaction rate and micromixing of the reaction mixture the experiments were performed in ultrasonic bath. In order to achieve compatibility of $CaCO_3$ nanoparticles with PANI, functionalization of $CaCO_3$ nanoparticles was carried out using myristic acid. During preparation of PANI/$CaCO_3$ nanocomposite myristic acid coated $CaCO_3$ percentage was varied from 2 to 8 % (Based on monomer quantity).

It has been reported that most of the $CaCO_3$ particles are finely dispersed in the PANI matrix and that exhibit the semi-crystalline nature. It is attributed to the hydrophobic nature of myristic acid coated $CaCO_3$ and micro-mixing caused by ultrasonic irradiation leads to formation of fine myristic acid coated $CaCO_3$ embedded emulsion droplets, which results into the formation of finely dispersed PANI/$CaCO_3$ nanocomposites. Fine dispersion of myristic acid treated $CaCO_3$ in PANI/$CaCO_3$ nanocomposite has been confirmed by TEM analysis [5] (Figure 5). The simultaneous enhancement in the mechanical and anticorrosion properties of PANI/$CaCO_3$/Alkyd coating has been also reported, which is attributed to the fine dispersion of myristic acid coated $CaCO_3$ nanoparticles in PANI matrix.

The reported anticorrosion properties of PANI/$CaCO_3$/alkyd coatings have been tested by using electrochemical corrosion and dip (weight loss) tests at room temperature (25 °C). The electrochemical corrosion (Tafel plot (log $|I|$ vs E)) test was carried out in 5 % NaCl solution as an electrolyte using coated panel as a working electrode.

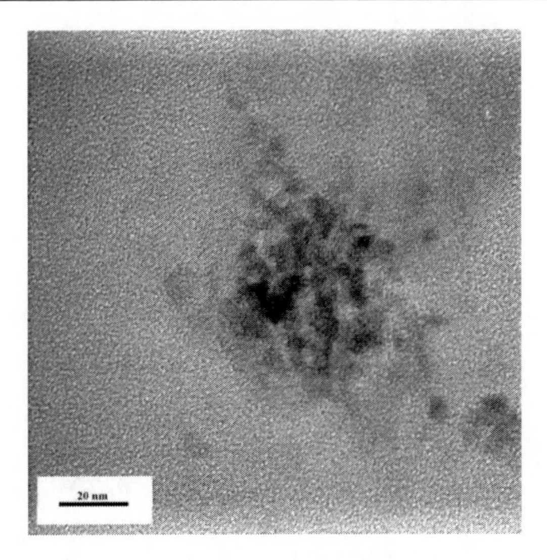

Figure 5. Transmission electron microscopic images of PANI/$CaCO_3$ composite [5].

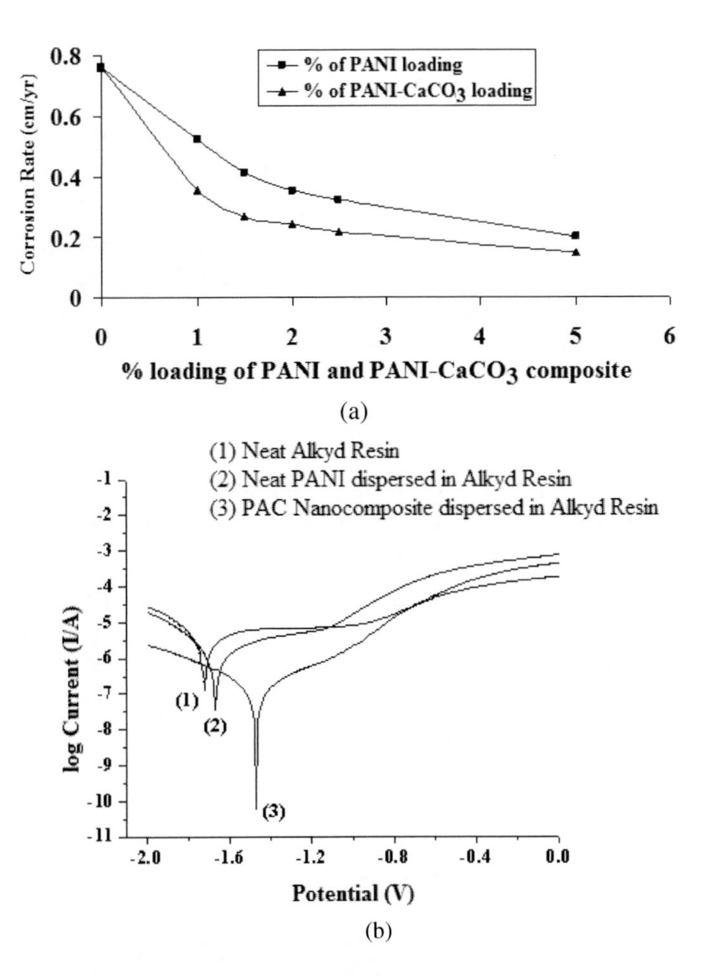

Figure 6. (a) Corrosion-protective efficiency of PANI/alkyd and PANI/$CaCO_3$/alkyd coatings in 5% NaCl, (b) Tafel plots of (1) alkyd resin, (2) PANI dispersed in alkyd resin and (3) PAC nanocomposite dispersed in alkyd resin coatings recorded in 5 % NaCl solution [5].

Further the corrosion rates of pure alkyd, PANI/alkyd and PANI/$CaCO_3$/alkyd coatings were monitored for a period of 200 h by using equation 2.

$$R_c = \frac{\Delta g}{Atd} \tag{2}$$

where Δg is the weight loss in grams, A is the exposed area of the sample in cm^2, t is the time of exposure in years, and d is the density of the metallic species in g/cm^3.

The thickness of the coating film on mild steel plate was kept close to 50 μm. The corrosion rate is reported in Figure 6a. The corrosion rate of pure alkyd coatings in different corrosive media i.e. 5% NaOH, HCl and NaCl is significant which is indicated by the disappearance of pure alkyd coatings in corrosive media. Corrosion rate of alkyd coating was found to decrease with an increase in the addition of PANI and PANI/ $CaCO_3$ nanocomposites in alkyd resin. It has been reported that for 1% loading of PANI and PANI/$CaCO_3$ nanocomposites in alkyd resin, the corrosion rate was 0.7 and 0.42 cm/yr respectively and it gets reduced to 0.35 and 0.1 cm/yr, when 5 % PANI and PANI/$CaCO_3$ nanocomposites were loaded in alkyd resin respectively in 5% HCl solution [5]. Further corrosion rate value is especially similar for 5 % PANI/$CaCO_3$ nanocomposite and PANI loading, when coated panels were dipped into NaOH and NaCl solutions (0.25 and 0.21 cm/yr for PANI and PANI/$CaCO_3$ nanocomposite in NaOH and close to 0.1 for NaCl solution). PANI and PANI/$CaCO_3$ nanocomposites for 5% loading in alkyd resin showed the best performance in the case of NaCl solution. It has been reported that 5 % loading of PANI/$CaCO_3$ nanocomposite in alkyd resin is more efficient for decreasing corrosion rate in acid, alkali and salt solutions. Further the loading of PANI/$CaCO_3$ nanocomposite or neat PANI can improve the adhesion onto the metal substrate. In case of PANI/alkyd, a compact iron/dopant complex layer formation takes place at the metal-coating interface which acts as a passive protective layer. The protective behavior depends on the size and charge of the dopant i.e. as the size of the dopant increases, the strength of the iron/dopant complex film increases, which improves the protective efficiency [50-51].

The corrosion rate of PANI/$CaCO_3$ nanocomposite/alkyd was found to decrease appreciably with an increase in the loading of PANI/$CaCO_3$ nanocomposites in alkyd. The corrosion rate of PANI/$CaCO_3$/alkyd coatings in case of 5% NaCl solution is depicted in Figure 6a. In the case of PANI/$CaCO_3$/alkyd coatings, the corrosion inhibition effect of the nanocomposite coatings is attributed to the presence of $CaCO_3$ particles. It has been also reported that the small pore size and uniform dispersion of the PANI/$CaCO_3$ nanocomposite in alkyd helps in the formation of a well-adhered, dense, and continuous network-like structure, that slows down the penetration of the corrosive ions through the metal substrate, and inhibits the MS from the attack of the corrosive species [84].

Figure 6b shows the Tafel plot for neat alkyd resin, PANI/alkyd and PANI/$CaCO_3$/alkyd coating. It has been reported that corrosion current ($I_{corr.}$) gets reduced from 0.89 to 0.03 $\mu A/cm^2$, when neat alkyd resin and PANI/$CaCO_3$ nanocomposites coatings were tested in NaCl electrolyte solution. The reported corrosion current for neat alkyd coating was 0.89 $\mu A/cm^2$, while for PANI/alkyd and PANI/$CaCO_3$/alkyd coating it was 0.50 and 0.03 $\mu A/cm^2$ respectively, which is an indication of improvement in the inhibition activity. Furthermore E_{corr} values shows shift in positive direction from -1.74 to -1.47 V by addition of PANI-

$CaCO_3$ nanocomposites into the alkyd coating. In overall, results indicate that PANI/$CaCO_3$ nanocomposites show improvement in anticorrosive properties which supports the corrosion rate data obtained by the dip test method. It has also been concluded that the adhesion of coating could be an important factor which control the corrosion rate. PANI/$CaCO_3$/Alkyd coating undergo surface crazing phenomenon due to the diffusion of oxygen, acidic, and alkaline ions. Metal iron complexes (Fe^{2+}, etc.) are formed due to the attack of acid, salt, and alkali. Also because of alkyd cross-linking and hydrogen bonding of alkyd with PANI/$CaCO_3$ nanocomposites, coating becomes more compact [85-86]. Compact cross-linking of alkyd resin and hydrophobic nature of $CaCO_3$ inhibit the moisture infiltration and moisture contact with metal through barrier mechanism leads to improvement in the corrosion efficiency, which is poorer in case of neat alkyd coatings. Also PANI has the ability to store some charges (Fe^{2+} etc.) generated by corrosion mechanism on mild steel panel (passivation of metal). Further it has been reported [5] that small quantity of PANI acts as electrochemical inhibitor and electrochemical properties of PANI nanocomposites film provides additional backup for protection. Both electrochemical and barrier mechanism communally gives the corrosion protection to mild steel [87-88].

Mechanical properties such as cross-cut adhesion (ASTM D3359-87), impact resistance test (ASTM D2794) and gloss at 45^0 evaluated as per American Society for Testing and Materials (ASTM) standards are depicted in Figure 7. The adhesion strength of alkyd coating was found to be a vital factor in the improvement of the coating performance to control the corrosion. The reported cross-cut adhesion for pure alkyd resin is poorer and it is found increased as the loading of PANI increased from 1 to 5 %. The possible reason reported is the enhanced adhesion between the PANI/alkyd coatings with mild steel plate. Also for aniline, the amino group and the aromatic ring are in the same plane. This coplanar orientation with respect to the metallic surface renders PANI a greater capacity to form more homogeneous and compact films, which leads to a better adhesion to the metal substrate. Furthermore, the presence of a lone pair of electrons in polyaniline structure enhances the electrostatic interaction between the coatings and the metal substrate, resulting in superior cross-cut adhesion. The cross-cut adhesions of PANI/$CaCO_3$/alkyd coatings were found to be higher than the PANI/alkyd coatings. The cross-cut adhesion at identical loading was found to be different because of the variation in the morphology of the PANI/alkyd and PANI/$CaCO_3$/alkyd coatings [5]. The presence of myristic acid functionalized $CaCO_3$ in PANI/$CaCO_3$ nanocomposites promotes the adhesive property as well as toughness to coatings.

The impact strength of pure alkyd is found to be 60 kg/cm^2 and it is reported to increase to 70 kg/cm^2 with an addition of 1 % PANI in alkyd resin. At 5 % PANI loading in alkyd resin the impact strength is found to be highest i.e. 78 kg/cm^2. With an addition of 4 % of myristic acid functionalized $CaCO_3$ along with varying percentage of PANI, a considerable improvement in impact strength has been reported (from 92 to 102 kg/cm^2). An enhancement in impact strength is attributed to the nano-size of both PANI and $CaCO_3$ particles, which impart positive effect on the alkyd coating. Further, the gloss values of PANI/$CaCO_3$/alkyd coatings has been reported to be increased with % loading of PANI/$CaCO_3$ nanocomposites, which attains optimum at 1.5 % loading of PANI/$CaCO_3$ nanocomposites and then decreases with an increase in % loading of PANI/$CaCO_3$ nanocomposite in alkyd resin. The decrease in the gloss value is attributed to muddiness due to higher loading of PANI/$CaCO_3$ nanocomposite. It has been concluded that the physico-mechanical properties of the

PANI/alkyd coatings are found to be significantly enhanced with the loading of the PANI/$CaCO_3$ nanocomposite in alkyd [5].

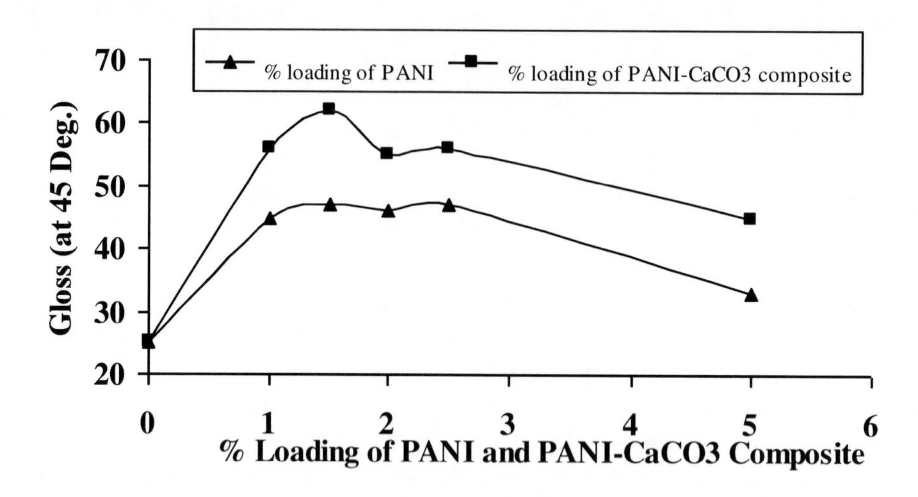

Figure 7. Variation of gloss with loading of PANI and PANI/$CaCO_3$ in alkyd [5].

6.4. Nanocontainer Containing Layer by Layer Assembly of PANI-PAA for Responsive Release

Corrosion is one of the key technological difficulties faced by all the industries such as manufacturing, chemical and petrochemical industries. Three probable approaches are being used to protect the metal surfaces from corrosion i.e. by cathodic protection, anodic protection and barrier mechanism [89]. Barrier coatings are generally used for corrosion protection because of its low permeability to the corrosive chemicals. Corrosion inhibitor additives can be simply incorporated into paint formulation, which can generate additional protective layer on the surface. Encapsulation of active materials loaded on core or in the hollow lumen of microcapsule/nanocontainer is an important concept, which offers additional advantage of selective release of the active ingredients on demand [90-92]. These polyelectrolyte containers have been used in number of applications such as biomedical, drug delivery, catalyst, textile, etc. to deliver ingredients in a sustained manner. The use of the conductive polymers as building blocks for micro and nanocontainers with the shell sensitive to the electrochemical potential has also been reported [93].

Use of polyaniline (PANI) as a polyelectrolyte layer for encapsulation of benzotriazole corrosion inhibitor on ZnO core has been reported [94]. Further it has been well known that PANI has inhibiting effect due to the formation of a compact iron/dopant complex layer at the metal-coating interface, which acts as a passive protective layer having a redox capability to undergo a continuous charge transfer reaction at the metal-coating interface, in which PANI is reduced from emeraldine salt form (ES) to an emeraldine base (EB) [95]. The resulting hybrid film has a pronounced protective efficiency for the corrosion inhibition.

Sonawane et al. [94] have used a novel approach for the synthesis of nanocontainers using layer-by-layer (LbL) assembly of oppositely charged species of polyelectrolytes and inhibitor on the surface of ZnO, which are capable of responsive release of corrosion inhibitors (benzotriazole). Sonochemical synthesis of zinc oxide nanocontainer has been using a multi-step approach has been reported which are (1) encapsulation of ZnO nanoparticles in PANI using ultrasound assisted insitu emulsion polymerization, (2) loading of benzotriazole (corrosion inhibitor) on PANI encapsulated ZnO particles in 0.1 N NaCl solution using 10 mg mL^{-1} of benzotriazole in the acidic media (pH = 3) for 20 min, and (3) deposition of polyelectrolyte layer (Polyacrylic acid) on benzotriazole loaded PANI-ZnO nanoparticles and using 2 mg mL^{-1} polyacrylic acid solution in 0.5 M NaCl for a period of 20 min. Nanocontainer/alkyd resin coatings were prepared by dispersing ZnO nanocontainers (2 to 5 wt %) in alkyd resin with the aid of pigment muller and were applied on mild steel panel in order to test the anticorrosion properties of the coatings.

The reported zeta potential and average particle size analysis confirms the formation of the layer-by-layer assembly of ZnO nanocontainer. Initially ZnO nanoparticles were coated with the sodium dedocyl sulphate surfactant. The reported surface charge on ZnO nanoparticles was near to the value of ≈−3.37 mV and with an deposition of PANI layer shows the slight increase in the negative value of Zeta potential (≈−3.73 mV). Further, addition of benzotriazole shows a drastic increase in zeta potential value which was near to ≈−22 mV and finally addition of polyacrylic acid decreases the zeta potential value near to ≈−12.8 mV. The reported particles size distribution analysis shows average particles size of ZnO nanocontainer near to 950 nm. Also transmission electron microscopic (TEM) analysis show average size of about 900 nm with agglomerated morphology of ZnO nanocontainer.

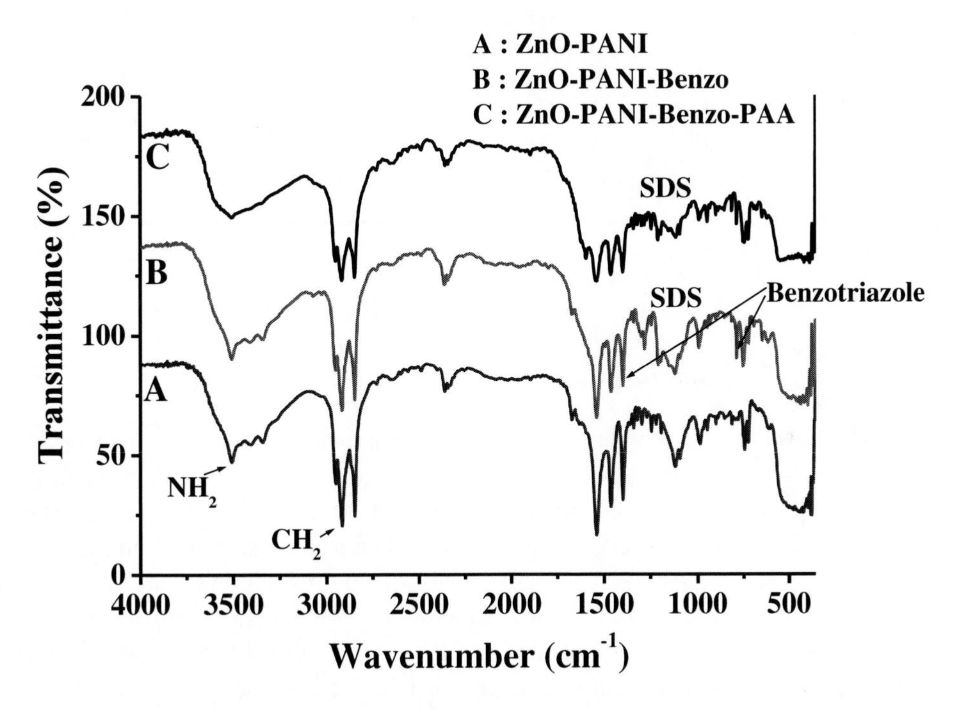

Figure 8. FTIR spectra of: (a) ZnO loaded with PANI, (b) ZnO loaded with PANI and benzotriazole, (c) ZnO loaded with PANI, benzotriazole and PAA (Polyacrylic acid).

Sonawane et al. [94] have depicted FTIR spectrum of ZnO loaded with PANI (pattern A), ZnO loaded with PANI and then Benzotriazole (pattern B) and ZnO loaded with PANI-

Benzotriazole-Polyacrylic acid (pattern C) in Figure 8. Figure 8 (pattern A) reports the formation of PANI layer on the ZnO nanoparticles. The peaks at 3506, 3408 and 3342 cm^{-1} are the characteristics peaks of polyaniline, indicating that polyaniline has been formed on the surface of ZnO nanoparticles. Further the characteristics peaks at 3506.7 and 2995 cm^{-1} are due to the NH_2 stretching and C-H bonds respectively. The peaks at 1400, 1278, 1203, 1120.68 and 768 cm^{-1} depicts the formation of benzotriazole layer (Figure 8, pattern B). The bands close to 786 cm^{-1} are typical of the benzene ring vibration and the band near to 1400 cm^{-1} is characteristic of the aromatic and the triazole rings stretching vibration [96]. In Figure 8 (pattern C), it is observed that the peaks of the PANI (3408 and 3342 cm^{-1}), as seen in Figure 8 (pattern B), was not observed because of the benzotriazole and PAA interfered with PANI chains.

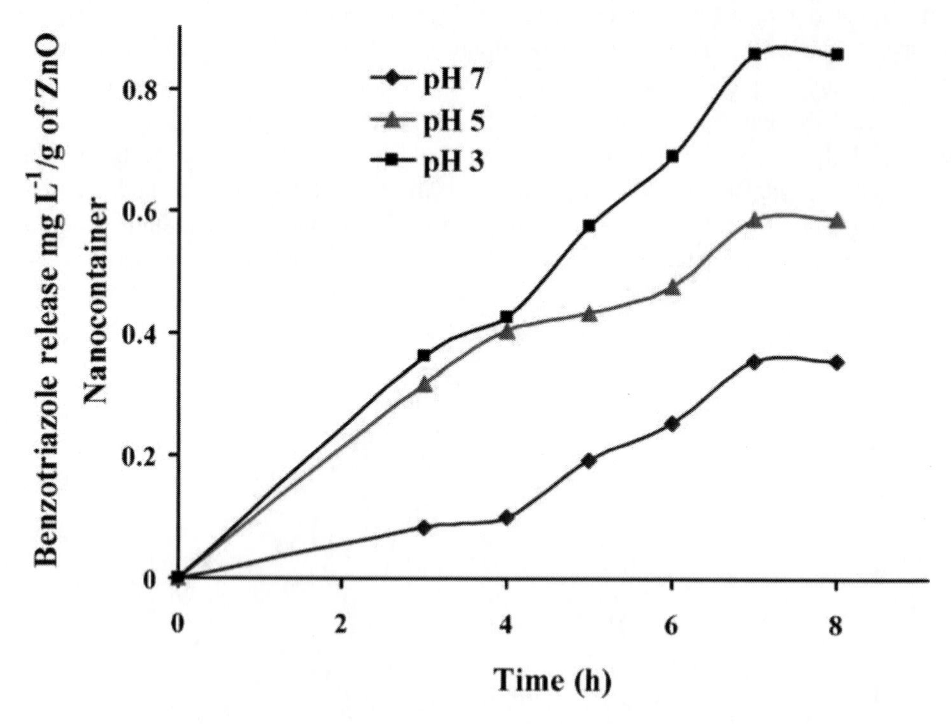

Figure 9. Release of benzotriazole from nanocontainers at three different pH values.

Benzotriazole acts as an inhibitor for ferrous metals under acidic conditions [97-98] as well as under neutral conditions [99-100]. The inhibition of iron in acid solutions results from the adsorption of benzotriazole in its molecular or protonated form with the formation of compact passive layer. Further benzotriazole gives better corrosion inhibition in acidic medium; therefore release performance of the benzotriazole containing nanocontainers has been reported by Sonawane et al. [94] in aqueous media of 3, 5, 7 pH. As depicted in figure 9, at any operating pH, a stable release is observed after 7 hours (all pH values), which indicates that the prepared nanocontainer shows proper encapsulation of the benzotriazole into the polyelectrolyte (PANI) layer. Maximum release at 8 hrs (0.86 mg Benzotriazole release mg L^{-1}/g of ZnO nanocontainer) was reported in the case of operating pH as 3. At neutral pH, the release was 0.36 mg L^{-1}/g of ZnO nanocontainer, which is intermediate between the acidic and alkaline conditions.

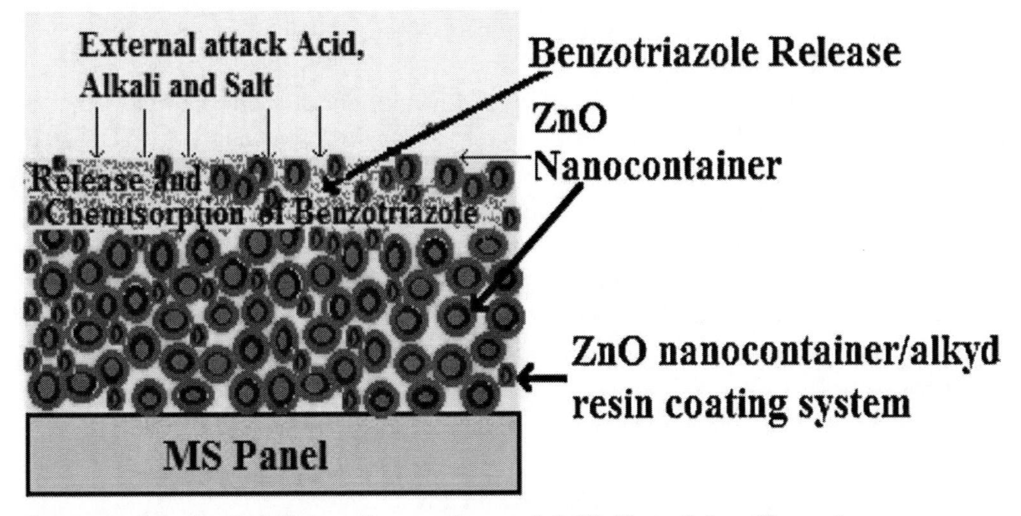

Figure 10. Release mechanism of corrosion inhibitor (benzotriazole) in the polymer.

Finally the release mechanism of benzotriazole as depicted schematically in figure 10 has been reported by Sonawane et al. [94]. The release of benzotriazole takes place in the presence of corrosion medium (pH 3, 5 and 7). As reported earlier, benzotriazole gives better corrosion inhibition in acidic medium and hence the release of benzotriazole has been tested under conditions of pH 3, 5 and 7 by Sonawane et al. [94]. The formation of the chemisorption passive layer, consisting of a complex between metal (in this case mild steel) and benzotriazole, takes place on MS surface when benzotriazole gets released from ZnO nanocontainer in the polymer/alkyd coatings. The passive layer formed on MS panel inhibits the attack of corrosion species, which results in the improvement in the anticorrosion performance of ZnO nanocontainer/alkyd coating.

6. Future Prospects

Rapid progress in the past few years has demonstrated that the potential exists for the development of new technology centered on hybrid functional nanocomposites. The PANI based hybrid nanocomposites which has been covered in this book chapter shows that polyaniline modification of different inorganic nanoparticles can play a vital role in the development of novel materials with unusual electrical, magnetic and optical properties for various applications. It is most exciting to realize that the most exceptional applications could result from the prospective synthesis (like ultrasound) and fundamental studies of these novel materials, which could be a good example of functional or multifunctional hybrid materials advanced to their specific-may be still unknown-future applications.

CONCLUSION

The recent development of PANI based hybrid nanocomposites prepared via ultrasound assisted miniemulsion with the addition of different nanoparticles such as Ag, $CaCO_3$, TiO_2, $ZrBiTiO_4$, ZnO have been mentioned along with their applications. In these processes the ultrasound triggers and accelerates chemical reactions with acoustic energy, which effectively shortens reaction time, improves rates of polymerization, reduces reaction condition and achieves some reactions that cannot be accomplished by conventional methods. Utilization of ultrasonication opens up a way for fabrication of various PANI hybrid nanocomposites via efficient encapsulation which will enhance the synergy between PANI and added nanoparticle.

REFERENCES

[1] Shirakawa, H.; Louis, E. J.; MacDiarmid, A.G.; Shiang, C.K.; Heeger, A. J. *J. Chem. Soc. Chem. Commun.* 1977, *16,* 578-580.

[2] Ziegler, C. Handbook of Organic Conductive Molecules and Polymers; Wiley, 1997.

[3] Sonawane, S; Neppolian, B.; Teo, B.; Grieser, F.; Ashokkumar, M. *J. Phys. Chem. C.* 2010, *114*, 5148-5153.

[4] Barkade, S. S.; Naik, J. B.; Sonawane, S. H. *Colloids Surf. A.* 2011, *378*, 94-98.

[5] Bhanvase, B. A.; Sonawane, S. H. *Chem. Eng. J.* 2010, *156*, 177-183.

[6] Romero, P.; Sanchez, C. Functional Hybrid Materials, Wiley-Vch Verlag GmbH & Co. KGaA, Weinheim, 2004.

[7] Ameen, S.; Akhtar, M.; Husain, M. *Sci. Adv. Mater.* 2010, *2,* 441-462.

[8] Rajesh, T.; Ahujab, D. K. *Sens. Actuators B.* 2009, *136*, 275-286.

[9] Xia, H.; Wang, Q. *Chem. Mater.* 2002, *14*, 2158-2165.

[10] Lu, X.; Mao, H.; Chao, D.; Zhang, W.; Wei, Y. *J. Solid State Chem.* 2006, *179*, 2609–2615.

[11] Raut, B. T.; Chougule, M. A.; Nalage, S. R.; Dalavi, D. S.; Mali, S.; Patil, P. S.; Patil, V. B. *Ceram. Int.* 2012, *38*, 5501–5506.

[12] Sivakumar, M.; Gedanken, A. *Synthetic Met.* 2005, *148*, 301-306.

[13] Macdiarmid, A. G.; Chiang, J. C.; Richter, A. F. *Synth. Met.* 1987, *18*, 285-290.

[14] Gangopadhyay, R.; De, A. *Chem. Mater.* 2000, *12*, 608-622.

[15] Novak, P.; Muller, K.; Santhanam, K. S. V.; Haas, O. *Chem. Rev.* 1997, *97*, 207-281.

[16] Wang, J.; Hu, Y.; Tang, Y. Chen, Z. *Mater. Res. Bull.* 2003, *38*, 1301-1308.

[17] Park, J. E.; Atobe, M.; Fuchigami, T. *Electrochim. Acta* 2005, *51,* 849-854.

[18] Zhang, Y. P.; Lee, S. H.; Reddy, K. R.; Gopalan, A. I.; Lee K. P. *J. Appl. Polym. Sci.* 2007, *104,* 2743–2750.

[19] Haldorai, Y.; Long, P. Q.; Noh, S. K.; Lyoo, W. S.; Shim, J. J. *Polym. Adv. Technol.* 2009, 22, 781-787.

[20] Jia, W.; Su, L.; Lei, Y. *Biosens. Bioelectron* 2011, *30*, 158-164.

[21] Langa, X.; Zhanga, L.; Fujita T.; Dingc, Y.; Chen, M. *J. Power Sources* 2012, *197*, 325-329.

[22] Belaabed, B.; Wojkiewicz, J.; Lamouri S.; Kamchi N.; Lasri, T. *J. Alloys Compd.* 2012, *527*, 137-144.
[23] Deng, J.; Ding, X.; Zhang. W.; Peng, Y.; Wang, J.; Long, X.; Li, P.; Chan, A. S. C. *Eur. Polym. J.* 2002, *38*, 2497-2501.
[24] Divya, V.; Sangaranarayanan, M. V. *Eur. Polym. J.* 2012, *48*, 560-568.
[25] Wang, G.; Xing, X.; Zhuo, S. *Electrochim. Acta* 2012, *66*,151-157.
[26] Islama, R.; Witcomb, M.; Lingen, E., Scurrell, M.; Otterlo, W.; Mallick, K. *J. Organomet. Chem.* 2011, 696, 2206-2210.
[27] Wang, Z.; Suna, S.; Hao, X.; Ma, X.; Guanb, G.; Zhanga, Z.; Liua, S. *Sens. Actuators B* 2012, 171, 1073-1080.
[28] Paulraj, P.; Janaki, N.; Sandhya, S.; Pandian, K. *Colloids Surf. A.* 2011, *377*, 28-34.
[29] Zhang, L. J.; Wan, M. X. *J. Phys. Chem. B*, 2003, *107*, 6748-6753.
[30] Parvatikar, N.; Jain, S.; Khasim, S.; Ravansiddappa, M.; Bhoraskar, S.V., Ambikaprasad, M. V. N. *Thin Solid Films* 2006, *514*, 329–333, 2006.
[31] Tai, H.; Juang, Y.; Xie, G.; Yu, J.; Chen, X. *Sens. Actuators B*, 2007, *125*,664-650.
[32] Tudorache, F.; Grigoras, M. *Optoelectron adv mat* 2010, *4*, 43-47.
[33] Deshpande, N.; Gudage, Y.; Sharma, R.; Vyas, J.; Kim, J.; Lee, Y. *Sens. Actuators B* 2009, *138*, 76-84.
[34] Jaroslav, S.; Miroslava, T.; Jitka, B.; Pete, K.; Svetlana, F.V.; Jan, P.; Josef, Z. *J. Colloid Interface Sci.* 2006, 298, 87-93.
[35] Katoch, A.; Burkhart, M.; Hwang, T.; Kim, S. *Chem. Eng. J.* 2012, *192*, 262–268.
[36] Qiao, Y.; Bao, S. J.; Li, C, M.; Cui, X. Q.; Lu, Z. S.; Guo, J. *ACS Nano.* 2008, 2, 113-119.
[37] Kowsari, E.; Faraghi, G. *Ultrason. Sonochem.* 2010, *17*, 718-725.
[38] DeBerry, D. W. *J. Electrochem. Soc.*1985, *132,* 1022-1026.
[39] Chen, S. A.; Fang, Y. *Synth. Met.* 1993, *60*, 215-222.
[40] Herod, T. E.; Schlenoff, J. B. *Chem. Mater.* 1993, *5*, 951-955.
[41] Osawa, S.; Ito, M.; Tanaka, K.; Kuwano, J. *Synth. Met.* 1987, *18*, 145-150.
[42] Xia, H.; Wang, Q. *J. Nanopart. Res.* 2001, *5-6*, 399–409.
[43] Atobe, M.; Chowdhury, A. N.; Fuchigami, T.; Nonaka, T. *Ultrason. Sonochem.* 2003, *10*, 77-80.
[44] Parra, C.; González, G.; Albano, C. *e-Polymers* 2005, *25*, 1618-7229.
[45] Price, G. J. *Ultrason. Sonochem.* 2003, *10*, 277-283.
[46] Teo, B. M.; Prescott, S. W.; Ashokkumar, M.; Grieser, F. *Ultrason. Sonochem.* 2008, 15, 89-94.
[47] Landfester, K. *Angew Chem Int Ed* 2009, *48,* 4488-507.
[48] Sonawane, S. H.; Teo, B.; Brotchie, A.; Grieser, F.; Ashokkumar, M. *Ind. Eng. Chem. Res.* 2010, *49*, 2200-2205.
[49] Palaniappan, S.; John, A. *Prog. Polym. Sci*. 2008, *33*, 732–758.
[50] Jiang, J.; Ai, L. H. *Appl. Phys. A: Mat. Sc. Eng.* 2008, *92*, 2341- 2344.
[51] Alam, J.; Riaz, U.; Ashraf, S. M.; Ahmad, S. *J. Coat. Technol. Res.* 2008, *5*, 123-128.
[52] Bedre, M.D.; Basavaraja, S.; Salwe, B. D.; Shivakumar, V.; Arunkumar, L.; Venkataraman, A. *Polym. Compos.* 2009, *30*, 1668-1677.
[53] Chowdhari, A. *Sens. Actuat. B: Chem.* 2009, *138*, 318–325.
[54] Khanna, P. K., Singh, N.; Charan, S.; Viswanath, A. K. *Mater. Chem. Phys.* 2005, *92*, 214-219.

[55] Wang, J.; Neoh, K. G.; Kang, E. T. *J. Colloid Interface Sci.* 2001, *239*, 78-86.
[56] DeBarros, R. A.; DeAzevedo, W. M. *Synth. Met.* 2008, *158*, 922-926.
[57] Somani, P.R.; Marimuthu, R.; Mulik U. P.; Sainkar S. R; Amalnerkar D. P. *Synth. Met.* 1999, 106, 45-52.
[58] Hu, Z.; Xie, Y.; Wang, Y.; Mo, L.; Yang, Y.; Zhang, Z. *Mater. Chem. Phys.* 2009, 114, 990-995.
[59] Parvatikar, N.; Jain, S.; Khasim, S.; Ravansiddappa, M.; Bhoraskar, S.V., Ambikaprasad, M. V. N. *Sens. Actuator B* 2006, *114*, 599-603.
[60] Eskizeybeka, V.; Sar, F.; Gulceb, H.; Gulceb, A.; Avc, A. *Appl Catal B-Environ.* 2012, *119*, 197-206.
[61] Wang, H.; Wang, R.; Wang, L.; Tian, X. *Colloids Surf. A.* 2011, *384*, 624-629.
[62] Aleahmad, M.; Taleghani, H.; Eisazadeh, H. *Synth. Met.* 2011, *161*, 990-995.
[63] Li, L.; Jiang, J.; Xu, F. *Materials Letters* 2007, *61*, 1091-1096,
[64] Mathur, R.; Sharma, D. R.; Vadera, S. R.; Kumar N. *Acta mater.* 2001, *49*, 181-187.
[65] Li, L.; Liu, H.; Wang, Y.; Jiang, J.; Xu, F. *J. Colloid Interface Sci.* 2008, *321*, 265-271.
[66] Aphesteguy J. C., Jacobo S. E., *Physica B* 2004, *354*, 224-227.
[67] Jacobo, S. E.; Aphesteguy, J. C.; Anton, R. L., Schegoleva, N. N.; Kurlyandskaya, G. V. *Eur. Polym. J.* 2007, *43*, 1333-1346.
[68] Wan, M.; Fan, J. *Polym. Sci. part A: Polym. Chem.* 1998, *36*, 2749-2755.
[69] Kim, Y.; Kim, D. Lee, C. *Physica* B 2003, 337, 42-51.
[70] Qu, Y.; Yang, H.; Yang, N.; Fan, Y.; Zhu, H.; Zou, G. *Mater. Lett.* 2006, *60*, 3548–3552.
[71] Gonzalez-Sandoval, M.P.; Beesley A. M.; Yoshida, M. M.; Cobas, L. F., Aquino, J. A. M. *J. Alloys Compd.* 2004, *369*, 190-194.
[72] Zhu J.; Wei, S.; Zhang, L.; Mao, Y.; Ryu, J.; Haldolaarachchige, N.; Young, D.; Guo, Z. *J. Mater. Chem.* 2011, ***21***, 3952-3959.
[73] Dey, A.; De, S.; De, A.; De, S. K. *Nanotechnology* 2004, *15* 1277-1283.
[74] DeAraujo, A. C.V.; DeOliveira, R. J.; Junior, S.; Rodrigues, A. R., Machado, F.L.A., Cabral, F.A.O.; DeAzevedo, W.M. *Synth. Met.* 2010, 160, 685-690.
[75] Olad, A.; Naseri, B. *Prog. Org. Coat.* 2010, *67*, 233-238.
[76] Caruso, F. *Adv. Mater.* 2001, *13*, 11–22.
[77] Gong, J.; Li, Y.; Hu, Z.; Zhou, Z.; Deng, Y. *J. Phys. Chem. C* 2010, *114*, 9970–9974.
[78] Kumar, R.; Yavuz, O.; Lahsangah, V.; Aldissi, M. *Sens. Actuat. B: Chem.* 2005, 106, 750–757.
[79] Negi, Y. S.; Adhyapak, P. V., *J. Macromol. Sci. Polym. Rev.* 2002, *42*, 35–53.
[80] Athwale, A. A.; Bhagwat, S. V.; Katre, P. P. *Sens. Actuat. B: Chem.* 2006, *114*, 263–267.
[81] Wang, Y.; Jing, X. *Mater. Sci. Eng., B* 2007, *138*, 95-100.
[82] Jyongsik, J.; Hak, O. *Adv. Funct. Mater.* 2005, *15*, 494-502.
[83] Heng, L.; Zhai, J.; Zhao, Y.; Xu, J.; Sheng, X.; Jiang, L. *Chem. Phys. Chem.* 2006, 7, 2520-2525.
[84] Zarras, P.; Anderson, N.; Webber, C.; Irvin, D. J.; Guenthner, A.; Stengersmith, J. D. *Rad. Phy. Chem. 2003, 68*, 387-394.
[85] Chisholm, N.; Mahfuz, H.; Rangari, V. K.; Ashfaq, A.; Jeelani, S. *Comp. Struct.,* 2005, *67*, 115-124.

[86] Yu, H. J.; Wang, L.; Shi, Q.; Jiang, G.H.; Zhao, Z.R.; Dong, X.C. *Prog. Org. Coat. 2006*, 55, 296-300.

[87] Dispenza, C.; Presti, C. L.; Belfiore, C.; Spadaro, G.; Piazza, S. *Polym.* 2006, 47, 961-971.

[88] Abu, Y. M.; Aoki, K. *J. Elec. Chem.* 2005, 583, 133-139.

[89] Revie, R. W. Corrosion and Corrosion Control, 4th ed., John Wiley & Sons, New Jersey, 2008.

[90] Zheludkevich, M. L.; Serra, R.; Montemor, M. F.; Ferreira, M. G. S. *Electrochem. Commun.* 2005, 7, 836–840.

[91] Evaggelos, M.; Ioannis, K.; George, P.; George, K. *J. Nanopart. Res.* 2011, 13, 541–554.

[92] Tedim, J.; Poznyak, S. K.; Kuznetsova, A.; Raps, D.; Hack, T.; Zheludkevich, M. L.; Ferreira, M. G. S. *Appl. Mater. Inter.* 2010, 2, 1528–1535.

[93] Bell, D. J.; Sun, Y.; Zhang, L. Dong, L. X.; Nelson, B. J.; Grutzmacher, D. *In: Proc. of the 13th Int. Conf. on Solid-State Sensors, Actuators and Microsystems*, Seoul, 2005, 15–18.

[94] Sonawane, S. H.; Bhanvase, B. A.; Jamali, A. A.; Dubey, S. K.; Kale, S. S.; Pinjari, D. V.; Kulkarni, R.D.; Gogate, P. R.; Pandit, A. B. *Chem. Eng. J.* 2012, 189, 464- 472.

[95] Jose, E.; Pereira, S.; Susana, I.; Cordoba, T.; Roberto, M.T. *Corros. Sci.* 2005, *47*, 811–822.

[96] Mennucci, M. M.; Banczek, E. P.; Rodrigues, P. R. P., Costa, I. *Cem. Concr. Compos.* 2009, *31*, 418–424.

[97] Matheswaran, P.; Ramasamy, A. K. *J. Appl. Electrochem.* 2003, *33*, 1175–1182.

[98] Popova, A.; Christov, M. *Corros. Sci.* 2006, 48, 3208–3221.

[99] Cao, P. G.; Yao, J. L.; Zheng, J. W.; Gu, R. A.; Tian, Z. Q. *Langmuir* 2002, *18*, 100-104.

[100] Ramesh, S. Rajeswari, *Electrochim. Acta* 2004, *49*, 811–820.

In: Trends in Polyaniline Research
Editors: T. Ohsaka, Al. Chowdhury, Md. A. Rahman et al.
ISBN: 978-1-62808-424-5

Chapter 4

STRUCTURE CONTROL SYNTHESIS OF POLYANILINE MATERIALS USING ELECTROCHEMICAL AND CHEMICAL POLYMERIZATIONS IN UNIQUE REACTION FIELDS AND MEDIA

Mahito Atobe[*]
Graduate School of Environment and System Sciences,
Yokohama National University, Tokiwadai, Hodogaya-ku, Yokohama, Japan

ABSTRACT

Conducting polymers such as polyanilines exhibit not only the electroconductivity but also unique optical and chemical properties. The diversity of properties exhibited by conducting polymers offers these materials to be used in numerous technological applications.

Generally, the properties of the conducting polymers originate from their chemical (molecular) and physical (morphological) structures. Therefore, it follows that the structures of the polymers should be controlled in order to tailor them to the purposes of their utilization. Their chemical structures can be controlled by changing the molecular structures of the corresponding monomers. On the other hand, the methods for controlling their physical structures have been relatively limited because of poor processability of the conducting polymers, but recently, many studies were focused on applying ionic liquids and supercritical fluids as electrolytic media for this purpose. In addition, the application of mechanical energies such as ultrasound and centrifugal force to polymerization processes also enabled their physical structures to control for the purposes of their utilization.

This chapter provides an introduction and examples of structure control synthesis of polyaniline materials using electrochemical and chemical polymerizations in unique reaction fields and media.

Keywords: Ultrasound; centrifugal field; magnetic field; ionic liquid; supercritical fluid

[*] Tel and Fax: (+81) 45-339-4214. E-mail: atobe@ynu.ac.jp.

INTRODUCTION

Over the last couple of decades, conducting polymers such as polyaniline and their derivatives have attracted considerable attention for a wide variety of potential technological importance. The unique electrical, optical and chemical properties offer these materials to be used in anticorrosion coatings [1, 2], electromagnetic shielding [3], rechargeable polymeric batteries [4], polymer photovoltaics [5], polymer actuators [6, 7], and so on. However, a major problem in their successful utilization is their poor mechanical properties and processability due to their insoluble nature in common organic solvents. Incorporation of polar functional groups or long and flexible alkyl chains in the polymer backbone is a common technique to prepare polyaniline type polymers, which are soluble in water and/or organic solvents. For example, substituted polyanilines like polytoluidines, polyanisidines, poly(N-methylanilines), and poly(N-ethylanilines) are more soluble in common organic solvents than the unsubstituted polyaniline but less conductive [8]. On the other hand, structure control synthesis of polyaniline materials is another approach to overcome their poor processability. Therefore, many methods for controlling their physical structures have been reported so far. Solvent, concentrations, temperature and additives have all been shown to play a role for this purpose [9].

Furtheremore, in recent years, an increase in interest has been observed in controlling polyaniline morphological properties using original techniques such as the use of magnetic field [10], centrifugation [11, 12] or ultrasound [13-17]. In addition, many studies were also focused on applying ionic liquids [18-20] and supercritical fluids [21-23] as electrolytic media for controlling their morphological structures. For this reason, this chapter is dedicated to these innovative techniques and to its contribution to controlling physical structure of polyaniline materials.

CHEMICAL AND ELECTROCHEMICAL POLYMERIZATIONS UNDER ULTRASONICATION

Ultrasonication techniques have proved advantageous in terms of polyaniline morphological properties if they are performed during chemical and electrochemical polymerizations. The work of Atobe and co-workers was probably the first 'modern' example investigating electrochemical polymerization of aniline under ultrasonication [13, 14].

Reactors used in these studies consist mainly of undivided cells equipped with platinum working (anode, 1 x 1 cm^2), platinum counter (cathode, 1 x 1 cm^2) and SCE reference electrodes (Figure 1). An ultrasonic stepped horn (a titanium alloy rod) connected with a PZT oscillator (20 kHz) was inserted into the electrolytic solution. The working electrode (anode) surface was positioned perpendicularly to the propagating direction of the ultrasonic wave. The electrochemical polymerization of aniline (0.1 M) was cyclic-voltammetrically carried out in 4 M HCl aqueous solution.

Surface of the film polymerized with ultrasonication was visibly bright, while that without ultrasonication seemed dull. Figure 2 shows SEM photographs of polyaniline films deposited on the anode surface. A porous structure with grains is observed in the film polymerized without ultrasonication, but there is no grain in that with ultrasonicatiaon.

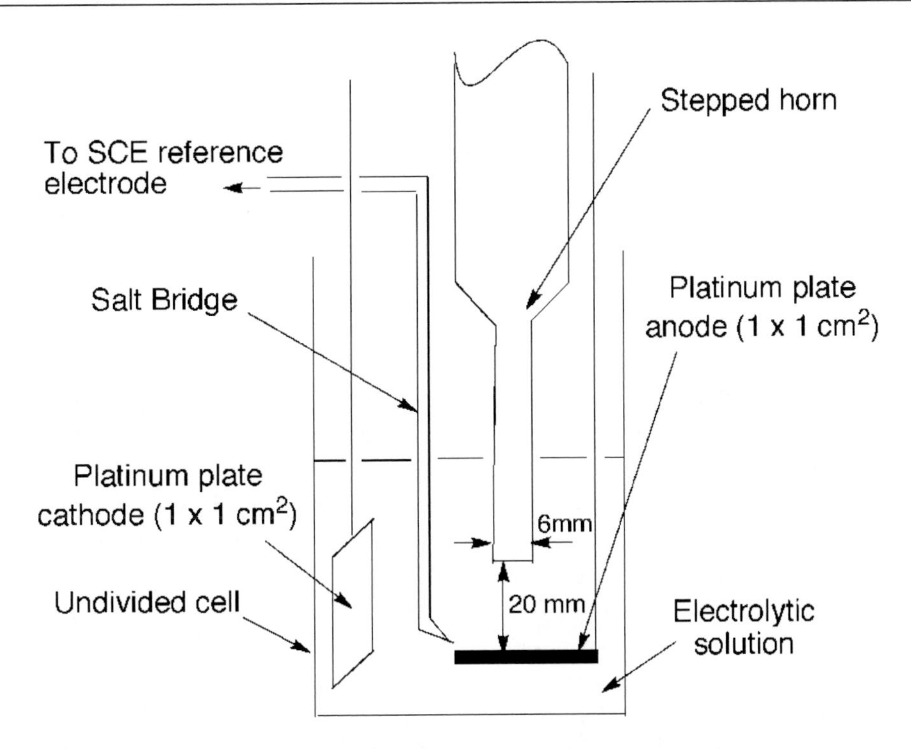

Figure 1. Schematic representation of the electrolytic cell for the electrochemical polymerization of aniline under ultrasonication.

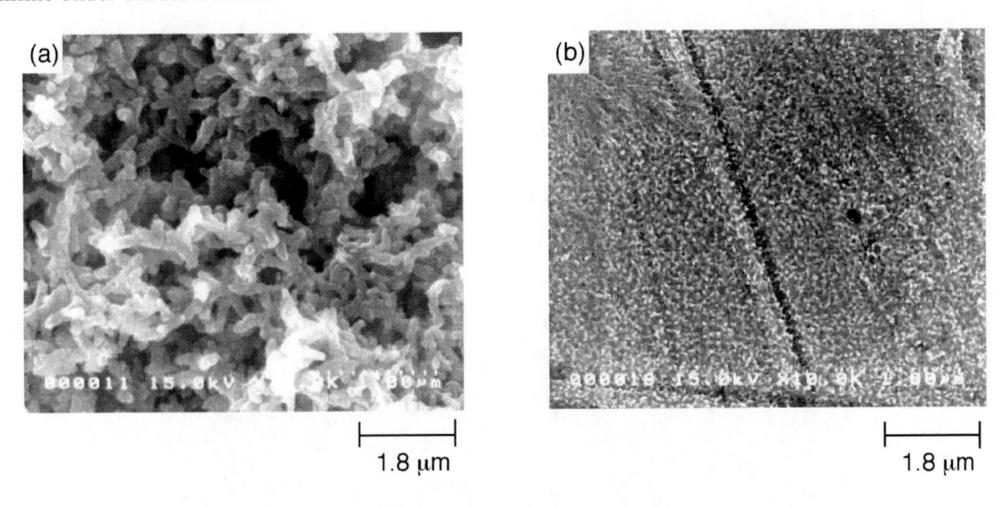

Figure 2. SEM photographs of the polyaniline films electropolymerized (a) without and (b) with ultrasonication.

The film prepared with ultrasonication is so thin that linear traces derived from polishing scars of the Pt substrate appear on the photograph.

Table 1 summarized the thickness and densities of the films prepared with and without ultrasonication. From the data in Table 1, it is stated that the thin, uniform, and dense film of polyaniline can be prepared using an ultrasonic effect. In addition, it was also found that the significant effect of ultrasounds on the film properties occurs at an ultrasonic intensity lager than the cavitation threshold value.

Table 1. Properties of polyaniline films electropolymerized by potential scanning method[a] without and with ultrasonication

Sonication	Potential scanning /time	Surface appearance		Tickness /μm	Density /x10^{-2} g cm^{-3}
		Visibly	SEM		
Without	50	Dull	Porous	120	3.7
Without	15	Dull	Porous	40	1.1
With	50	Bright	Dense	5	10

[a] Potential scan range: 0.0 – 1.0 V vs. SCE, Potential scan rate: 0.1 V s^{-1}.

Ultrasonication technoques also provide a useful method for controlling the structure and properties of polyaniline colloids formed by the chemical polymerization [15, 16]. Figure 3 shows SEM images of the dried-down polyaniline colloids. It can be observed that the morphological structure of the polyaniline colloids prepared in the absence of ultrasonication possesses needle-shaped and finely net-worked structures. On the other hand, the morphology of colloids prepared under the sonication is quite different from the above one, *i.e.* grains are deposited and they are not net-worked. In addition, electroconductivity of polyaniline colloids prepared in the presence of ultrasonication (3.0 x 10^{-2} S cm^{-1}) was found to be ca. 17 times as high as that in the absence. In general, it is known that needle-shaped particles have a conductivity superior to that of grain-shaped ones [24]. However, polyaniline colloids prepared under ultrasonication exhibit relatively high conductivity regardless of its morphological structure.

As shown in Figure 4, the FT-IR spectrum of the vacuum dried powder of the colloids prepared with and without ultrasonication were almost similar to each other in most of the wavelength region, but the band at 1140 cm^{-1} is seem to be very intense and broad in the spectra of the sample with sonication compared to that without sonication. This band is a vibrational mode of B-NH=Q or B-NH-B (B: benzoid unit, Q: quinoid unit) and may be attributed to the existence of the positive charge and the distribution of the dihedral angle between the benzoid and quinoid rings [25].

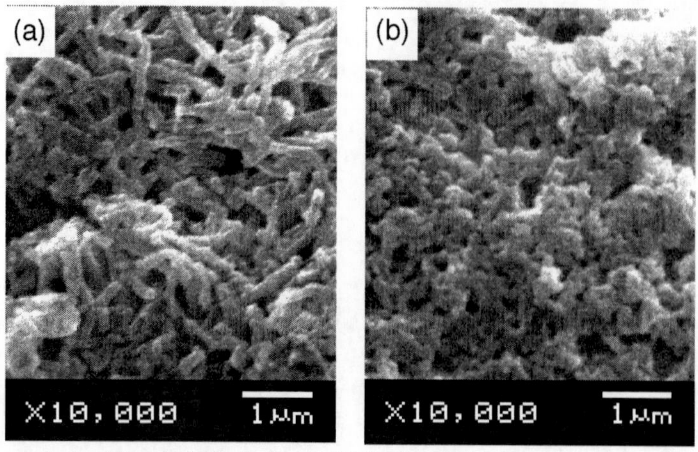

(Reprinted with permission from [15] Copyright (2003) Elsevier Ltd).

Figure 3. SEM photographs of the dried-down polyaniline colloids formed (a) without and (b) with ultrasonication.

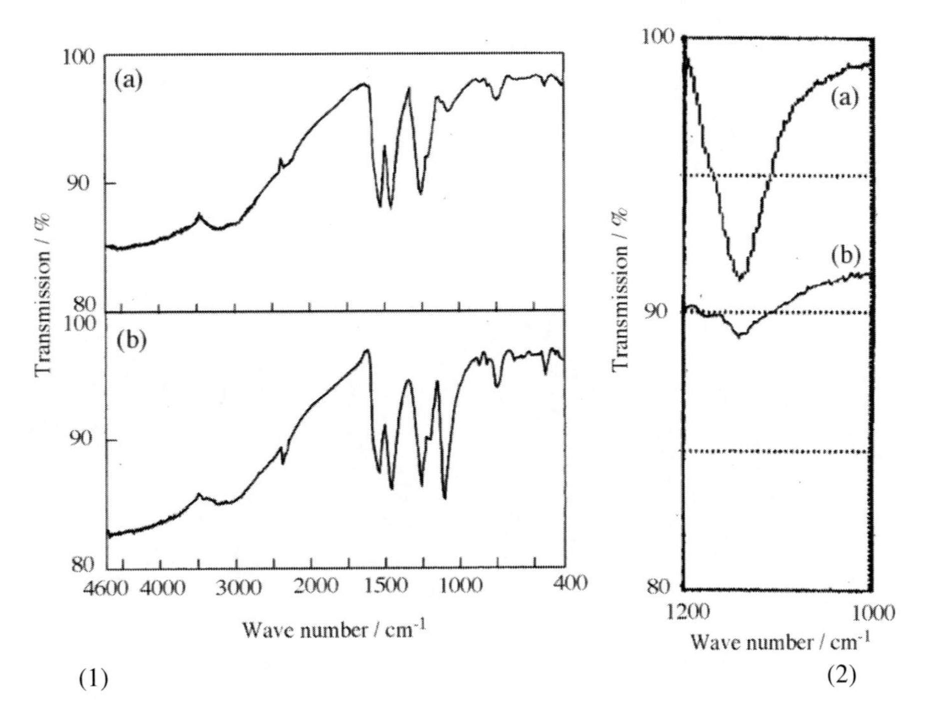

(Reprinted with permission from [15] Copyright (2003) Elsevier Ltd).

Figure 4. FT-IR spectrum of the polyaniline colloids prepared (a) without and (b) with ultrasonication (1) in the region of 4600 – 400 cm^{-1} and (2) in the region of 1200 - 1000 cm^{-1}.

Therefore, it could be considered that the polyaniline colloids prepared with ultrasonication have a higher doping level than those without ultrasonication. This may result in an increase in the electroconductivity of the colloids.

The pulsed sonoelectrochemical synthesis involving alternating sonic and electric pulses has been employed to obtain the shape and size controlled synthesis of the polyaniline particles [17].

The sonoelectrochemical formation of these particles was accomplished by applying an electric pulse to nucleate and grow the electrodeposit, followed by a burst of ultrasonic energy that removes the deposited particles (see Figure 5). Therefore, the particle size can be controlled by adjusting the width of ultrasonic and electric pulses, and moreover large amounts of the particles can be obtained by repeated application of the pulses.

Figures 6a-e show the SEM images of the as-deposited polyaniline microspheres on the anode surface by applying a square wave electric (potential) pulse. It can be seen from the SEM images that the spherical particles are deposited and the average diameter of the microspheres increases with an increase in the pulse width. On the other hand, Figures 6f-j show the SEM images of the polyaniline microspheres obtained by the application of alternating electric and sonic pulses.

Although these samples were removed from the anode surface by a burst of ultrasonic energy, their shapes and sizes were never changed compared with those of the as-prepared polyaniline microspheres on the anode surface by applying only a electric pulse, in other words, the subsequent sonication had the sole role to ablate the microspheres formed by from the anode surface and did not break the spheres.

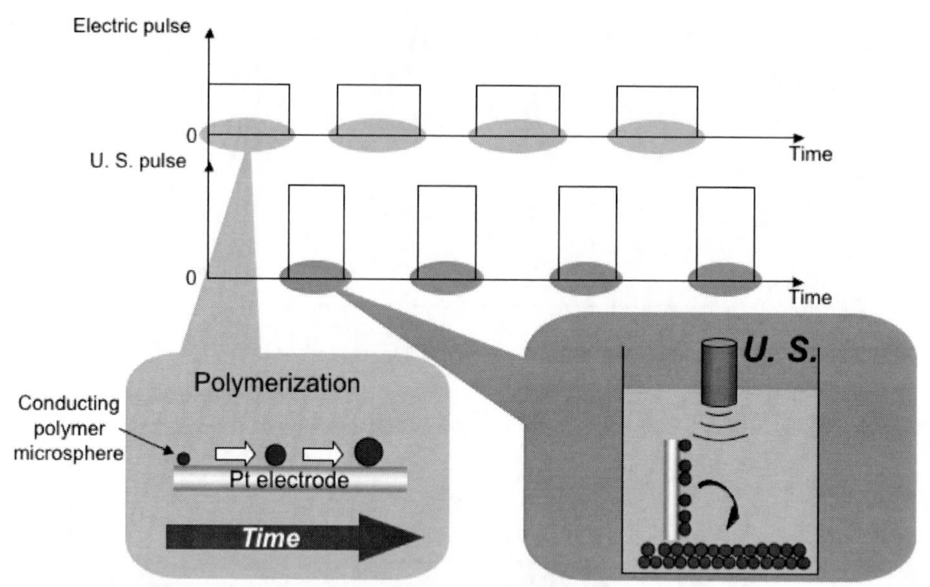

(Reprinted with permission from [17] Copyright (2009) Wiley Ltd).

Figure 5. Schematic representation of the pulsed sonoelectrochemical method for the production and recovery of conducting polymer microspheres.

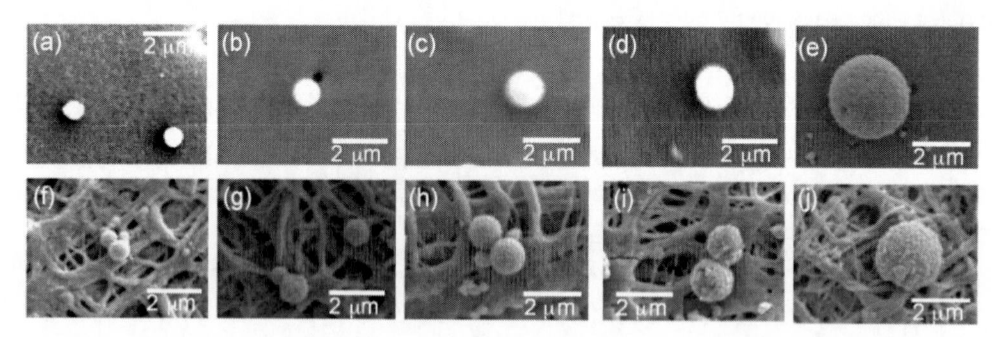

(Reprinted with permission from [17] Copyright (2009) Wiley Ltd).

Figure 6. SEM images of the PNMA microspheres (a-e) as-deposited on the anode surface by applying a potential pulse and (f-j) ablated from the anode surface by applying a subsequent sonic pulse. Experimental conditions: the potential pulse, 0.75 V vs. SCE; the ultrasonic pulse density, 31 W cm^{-2}; the duration of the ultrasonic pulse, 5 s; the duration of the potential pulse, (a, f) 10 s, (b, g) 20 s, (c, h) 30 s, (d, j) 40 s, (e, j) 90 s.

Electrochemical Polymerization under Centrifugal Fields

The use of centrifugation leads to beneficial modifications in structural properties of the polyaniline deposits. High gravity fields have been applied to broad areas in science and technology, particularly to material processing, with the purpose of producing unique and improved materials that cannot be prepared under normal earth conditions. High gravity fields can be generated by means of aircrafts, rockets, and so on [26]. Experiments in such fields are costly, but the use of small but powerful centrifuge equipment offers an excellent alternative [27-30].

Atobe et al. reported a successful fabrication of a centrifuge by improving commercially available centrifuge equipment with a 17 cm arm to generate a stable acceleration of roughly 300 g during the electrochemical deposition of polyaniline [11, 12]. The experimental set-up consisted of a cylindrical electrolytic cell made of polytetrafluoroethylene resin, 7 mm long and 14 mm in diameter. Contacts between classical three-electrode cell and potensiosta/galvanostat were possible through silver rotating rings and carbon brushes. The entire system was suspended from the lid of a centrifuge tube with a polyethylene line, as shown in Figure 7.

Typical cyclic voltammograms were recorded during the electropolymerization of aniline at various centrifugal accelerations (Figure 8). Although their shapes seem to be similar, the oxidation and reduction peak currents are increased at 315 g (electrode A) and decreased at 290 g (electrode B – with a reversed direction force), compared to the electropolymerization performed under 1 g. These results apparently suggest that the polymerization rate of aniline was increased and decreased at 315 g and 290 g of centrifugal acceleration forces on the electrodes A and B, respectively, compared to the polymerization conducted under 1 g.

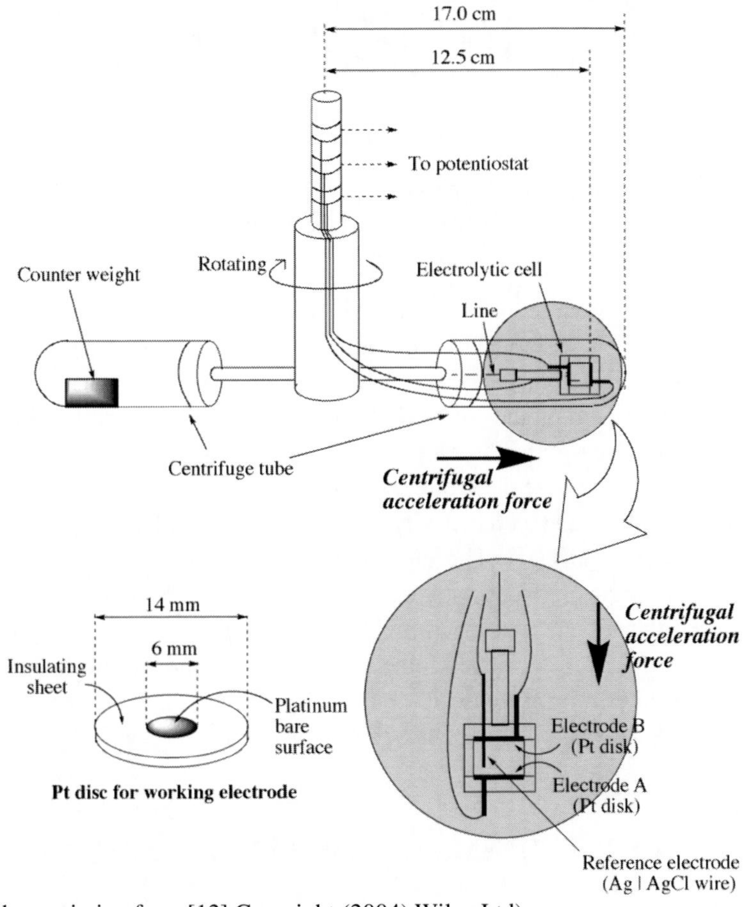

(Reprinted with permission from [12] Copyright (2004) Wiley Ltd).

Figure 7. Centrifuge setup equipped with an electrolytic cell.

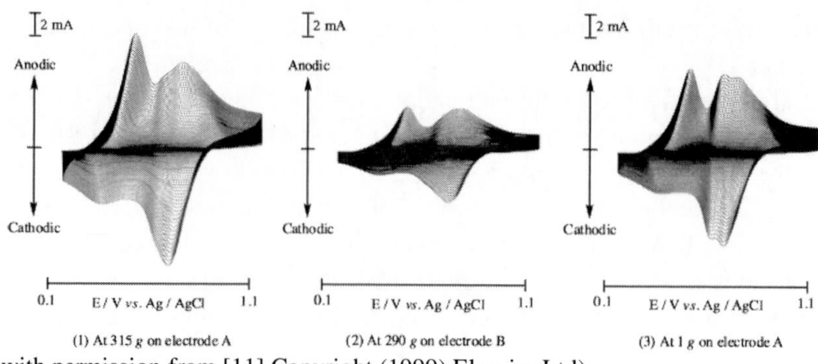

(Reprinted with permission from [11] Copyright (1999) Elsevier Ltd).

Figure 8. Cyclic volutammograms in the course of the electrochemical polymerization of aniline at 50 cycles of potential scanning at various centrifugal acceleration forces.

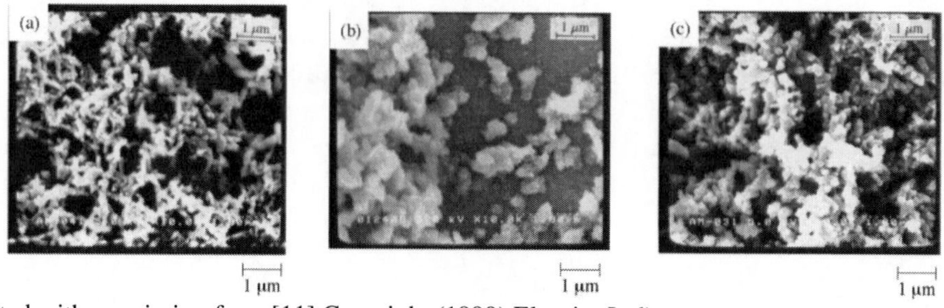

(Reprinted with permission from [11] Copyright (1999) Elsevier Ltd).

Figure 9. SEM photographs of the surface of the polyaniline films prepared at 50 cycles of potential scanning, (1) at 315g on the electrode A, (2) at 290 g on the electrode B and (3) at 1 g on the electrode A.

The morphological structure of polyaniline films was also greatly affected by the centrifugal field (Figure 9). The surface structure of the film on the electrode A at 315 g seems to be uniform and finely networked, compared with that prepared at 1 g. On the other hand, the structure of the film on the electrode B at 290 g is quite different from that of either of the above ones, *i.e.* large grains are deposited, and consequently they are not networked and the surface of a platinum base electrode remained partially bare.

The chemical bonding and conductivity of the obtained polymers were also affected considerably by the gravity fields. The clear anisotropy of the effects led the authors to propose mechanisms for the electropolymerization under centrifugal forces based on the basic differences between the solvent and macromolecules, shown concretely by different inertial behaviors. The accumulation of the oligomers produced at the electrode opposite to the centrifugal force resulted in an increase in nucleation and precipitation at the surface. As a final result, the high density of oligomers closer to the surface led to denser polymer films.

Electrochemical Polymerization under Magnetic Fields

Another specific method used to influence polymerization tested with polyaniline consists in performing electrosynthesis under magnetic fields. It is generally accepted that an applied magnetic field influences on the molecular or/and electronic spin state of materials, and the chemical reactivity of radical pairs in a solution may be influenced by nuclear spins [31], meaning that polymers can be oriented as well as electrochemical reactions can be affected.

In particular, the morphological changes are observed under the magnetic field due to the large anisotropy of their magnetic susceptibility. Moreover, the formation of special orientation of the conjugated polymers would result in preparing chiral surfaces on electrodes during their electrosynthesis and therefore lead to the formation of enantioselective electrode. Mogi et al. have tested polyaniline electrosynthesis on platinum disc under an applied magnetic field of 5 T [10]. Electrochemical polymerization was carried out at a constant potential of 0.9 V (vs. Ag/AgCl). The experimental set-up is shown in Figure 10. The applied magnetic field was parallel (+) or antiparallel (-) to faradaic currents.

The polyaniline films prepared by the electropolymerization under the magnetic field of 5 T parallel (+5 T) or antiparallel (-5 T) were used as a modified electrode and their chiral properties were examined for L-ascorbic acid (L-AA) and D-isoascorbic acid (D-AA) by cyclic voltammetry (Figure 11). The oxidation current of L-AA on the +5 T-film electrode was larger than on the -5 T-film electrode, and the result for D-AA was opposite. Such a chiral recognition implies that the magneto-electropolymerization films possess a molecular-level chiral structure. A molecular model study suggests that polyaniline easily forms a helical structure as shown in Figure 11.

In the absence of a magnetic field, the electropolymerization produces the racemic state of right-handed and left-handed helical chains. The experimental results in Figure 10 indicate that Lorenz force changes the growth probabilities of right-handed and left-handed helical chains, causing chirality.

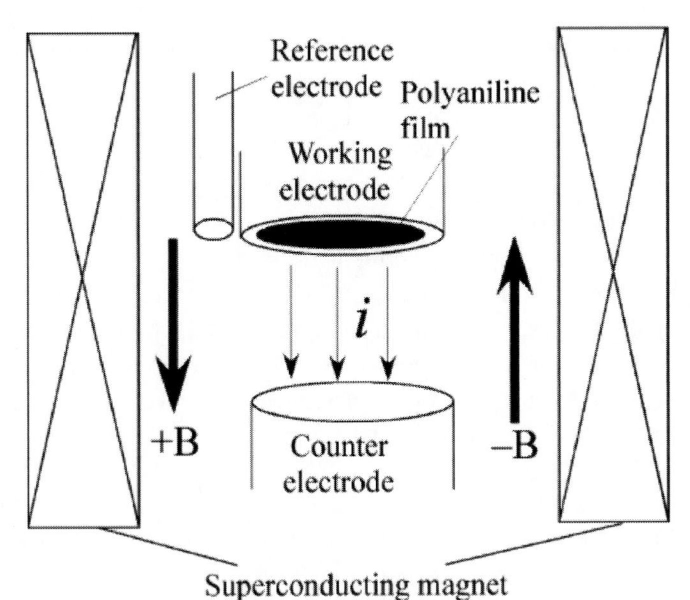

(Reprinted with permission from [10] Copyright (2006) Elsevier Ltd).

Figure 10. Electrode configuration in the magnetoelectropolymerization process in a superconducting magnet. Magnetic fields *B* are applied parallel (+) or antipallarel (-) to faradaic currents, and they are perpendicular to the electrode surface.

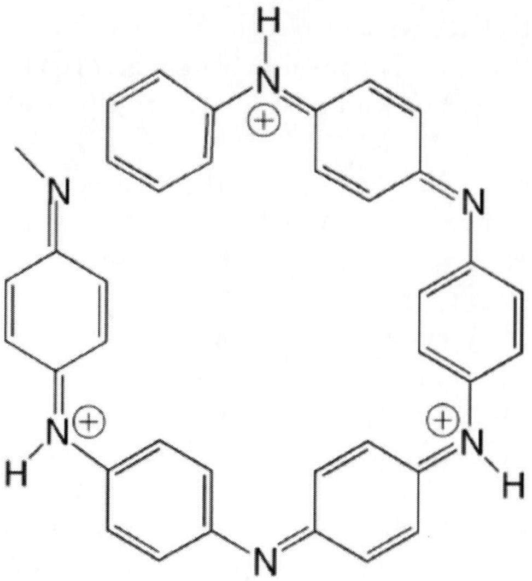

(Reprinted with permission from [10] Copyright (2006) Elsevier Ltd).

Figure 11. A proposed helical structure of polyaniline.

CHEMICAL AND ELECTROCHEMICAL POLYMERIZATIONS IN IONIC LIQUIDS

Ionic liquids exhibit non-flammability and non-volatility, and therefore they are becoming widely recognized as solvents for 'green' organic and polymer syntheses [32]. Moreover, the ionic liquids have been applied to a variety of electrochemical processes since they have also excellent properties as electrolytes such as a high ionic conductivity and wide potential window [33-37].

From the above points of view, the ionic liquids are expected to be peculiar media for electroorganic syntheses and electropolymerizations. Pioneer investigations were performed by Koura et al. at the beginning of 1990s whereby polyaniline films were prepared by electropolymerization [18]. However, the electropolymerization was carried out in moisture-sensitive chloroaluminate ionic liquids, and the hydrolysis of the ionic liquids resulted in the formation of highly corrosive products such as HCl. Therefore, the polymer films were decomposed rapidly by the corrosive product. In addition, the chloroaluminate ionic liquids are intractable materials, and their treatment requires special equipment such as a glove box.

The breakthrough for the above problem should be achieved by using an imidazorium ionic liquid having a stable counter anion. Fuchigami et al. reported that the electrooxidative polymerization of aniline leading to formation of polyaniline films was carried out in the air- and moisture-stable ionic liquid like 1-ethyl-3-methylimidazolium trifluoromethanesulfonate ($EMICF_3SO_3$) [19]. Figure 12 shows SEM photographs of the polyaniline films electrodeposited in an aqueous solution and $EMICF_3SO_3$. The surface morphology of both polyaniline films seems to be smooth. However, the former film contains small globular grains while there is no grain on the latter one. In addition, the electroconductivity of polyaniline prepared in $EMICF_3SO_3$ (4.1 S cm^{-1}) was found to be ca. 8.0 x 10^3 times as high as that in the aqueous solution.

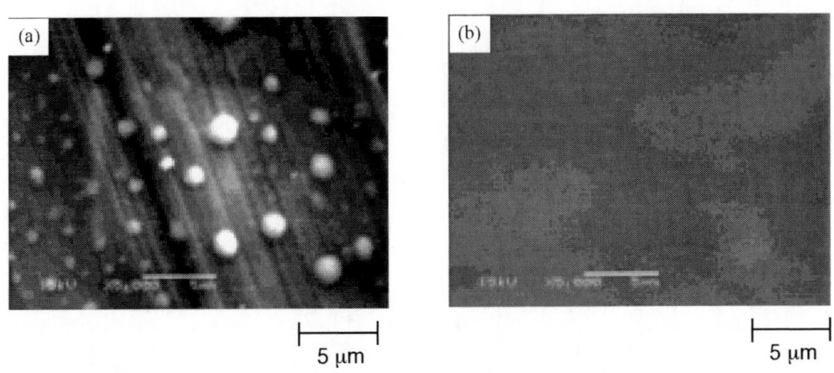

(Reprinted with permission from [19] Copyright (2003) Elsevier Ltd).

Figure 12. SEM photographs of the polyaniline films (a) polymerized in 1.0 M CF_3SO_3H / H_2O and (b) polymerized in 1.0 M CF_3SO_3H / $EMICF_3SO_3$.

On the other hand, it was also found that the ionic liquids have a crucial effect on the morphological properties of polyaniline prepared chemical oxidative polymerization. Liu and co-workers prepared polyaniline nanoparticles by the chemical oxidation of aniline monomer with ammonium persulfate in a pure ionic liquid such as 1-hexadecyl-3-methylimidazolium chloride; they synthesized nanofibers or irregularly shaped agglomerates (or mixtures of both) by changing only the aniline to ionic liquid mole ratio [20].

Electrochemical Polymerization in Supercritical Fluids

Supercritical fluids are practical green alternatives to harsh and environmentally hazardous solvents traditionally used in electrosynthesis of polyaniline such as sulfuric acid or acetonitrile [38]. In addition, the use of supercritical fluids leads to beneficial modifications in morphological structure of the polyaniline electrodeposits. Marbrouk et al. successfully demonstrated that electrochemical polymerization of aniline proceeded in supercritical carbon dioxide ($scCO_2$) containing tetrabuthylammonium hexafluorophosphate as a supporting electrolyte [21]. The polyaniline film prepared in $scCO_2$ was visibly smoother and flatter and exhibited distinctive surface characteristics that could prove advantageous in optical, dielectric, and anticorrosion applications. On the other hand, Jikei and co-workers demonstrated the electrochemical polymerization of aniline in $scCO_2$ in–water emulsion using a high-pressure reactor shown in Figure 13 [22]. A commercially available reactor was modified for the purpose of electrochemical reactions under high pressure [23]. The total volume of the reactor was 50 mL. Platinum wire insulated to the reactor by a PEEK tube was attached to the platinum electrodes and the electrodes were placed at a distance of 2 cm.

Figure 14 shows the effect of the reaction medium used for the electrochemical polymerization on the morphology of the resulting polyaniline films. The film prepared in the emulsion was homogeneously rough and small nodules were observed from the early stage of the polymerization. In contrast, the initial film prepared in aqueous electrolyte showed some spikes and a flat surface, and an irregular rough morphology was observed for the grown films. The authors claim that $scCO_2$ continuously cleans the electrode surface, which allows it to propagate and to aggregate the initial nodules to form the fine uneven texture.

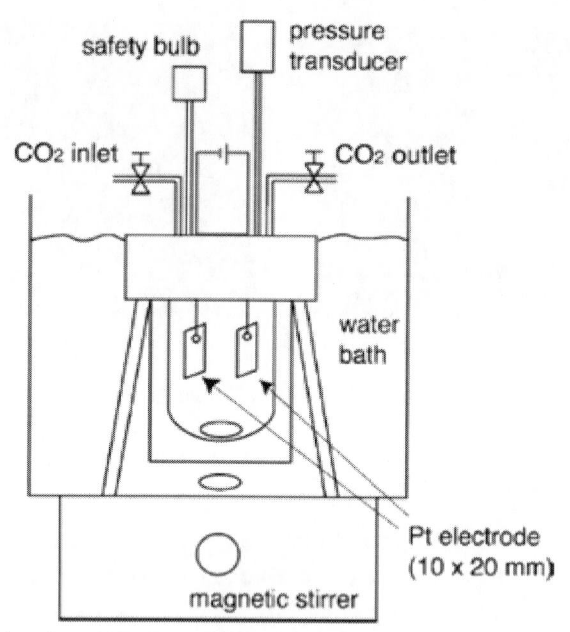

(Reprinted with permission from [23] Copyright (2006) Elsevier Ltd).

Figure 13. High pressure reactor and reaction setting of the electrochemical polymerization in the $scCO_2$ in–water emulsion.

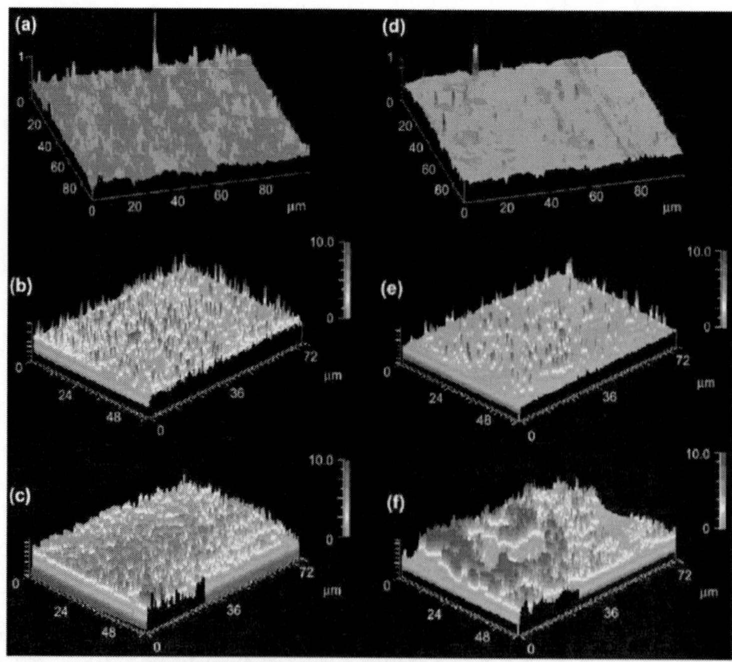

(Reprinted with permission from [22] Copyright (2007) Elsevier Ltd).

Figure 14. AFM (a, d) and confocal scanning microscope (b, c, e, f) images of the polyaniline films prepared on Pt in the $scCO_2$ in–water emulsion or in water by the electrochemical polymerization at 35 oC in the $scCO_2$ in–water emulsion for 100 ms (a); for 10 s (b); for 500 s (c); in water for 100 ms (d); for 10 s (e); for 500 s (f).

CONCLUSION

It is shown in this chapter that the use of unique reaction fields and media for controlling morphological structures of the polyaniline materials has an advantage over many other methods due to the unusual experimental conditions. The author of this chapter expects these techniques to significantly expand and be applied to fabricating the structured polyaniline materials that are difficult and costly to prepare using other conventional and traditional methodologies.

REFERENCES

[1] DeBerry, D. W. *J. Electrochem. Soc.* 1985, *132*, 1022-1026.
[2] Ahmed, N.; MacDiarmid, A. G. *Synth. Met.* 1996, *78*, 103-110.
[3] Joo, J.; Epstein, A. J. *Appl. Phys. Lett.* 1994, *65*, 2278-2280.
[4] MacDiarmid, A. G.; Mu, S. L.; Somatiri, N. L. D.; Wu, M. *Mol. Cryst. Liq. Cryst.* 1985, *121*, 187-190.
[5] Chen, S. A.; Fang, Y. *Synth. Met.* 1993, *60*, 215-222.
[6] Herod, T. E.; Schlenoff, J. B. *Chem. Mater.* 1993, *5*, 951-955.
[7] Kaneto, K.; Kaneko, M.; Min, Y.; MacDiarmid, A. G. *Synth. Met.* 1995, 71, 2211-2212.
[8] Kapli, A.; Taunk, M.; Chand, S. *Synth. Met.* 2009, *159*, 1267-1271.
[9] Yang, M.; Xiang, Z.; Wang, G. *J. Coll. Int. Sci.* 2012, *367*, 49-54.
[10] Mogi, I.; Watanabe, K. *Sci. Tech. Adv. Mater.* 2006, *7*, 342-345.
[11] Atobe, M.; Hitose, S.; Nonaka, T. *Electrochem. Commun.* 1999, *1*, 278-281.
[12] Atobe, M.; Murotani, A.; Hitose, S.; Suda, Y.; Sekido, M.; Fuchigami, T.; Chowdhury, A.-N.; Nonaka, T. *Electrochim. Acta* 2004, *50*, 977-984.
[13] Atobe, M.; Fuwa, S.; Sato, N.; Nonaka, T. *Denki Kagaku* (presently *Electrochemistry*) 1997, *65*, 495-497.
[14] Atobe, M.; Kaburagi, T.; Nonaka, T. *Electrochemistry* 1999, *67*, 1114-1115.
[15] Atobe, M.; Chowdhury, A.-N.; Fuchigami, T.; Nonaka, T. *Ultrason. Sonochem.* 2003, *10*, 77-80.
[16] Chowdhury, A.-N.; Atobe, M.; Nonaka, T. *Ultrason. Sonochem.* 2004, *11*, 77-82.
[17] Atobe, M.; Ishikawa, K.; Asami, R.; Fuchigami, T. *Angew. Chem. Int. Ed.* 2009, *48*, 6069-6072.
[18] Koura, N.; Ejiri, H.; Takeishi, K. *Denki Kagaku* (presently *Electrochemistry*) 1991, *59*, 74-75.
[19] Sekiguchi, K.; Atobe, M.; Fuchigami, T. *J. Electroanal. Chem.* 2003, *557*, 1-7.
[20] Miao, Z.; Wang, Y.; Liu, Z.; Huang, J.; Han, B.; Sun, Z.; Du, J. *J. Nanosci. Nanotechnol.* 2006, *6*, 227-230.
[21] Anderson, P. E.; Badlani, R. N.; Mayer, J.; Mabrouk, P. A. *J. Am. Chem. Soc.* 2002, *124*, 10284-10285.
[22] Jikei, M.; Yasuda, H.; Itoh, H. *Polymer* 2007, *48*, 2843-2852.
[23] Jikei, M.; Saitoh, S.; Yasuda, H.; Itoh, H.; Sone, M.; Kakimoto, M.; Yoshida, H. *Polymer* 2006, *47*, 1547-1554.
[24] Cooper, E. C.; Vincent, B. *J. Phys. D: Appl. Phys.* 1989, *22*, 1580-1585.

[25] Tang, J.; Jing, X.; Wang, B.; Wang, F. *Synth. Met.* 1988, *24*, 231-238.

[26] Briskman, V. A. *Adv. Space Res.* 1999, *24*, 1199-1210.

[27] Kuiken, H. K.; Tijburg, R. P. *J. Electrochem. Soc.* 1983, *130*, 1722-1729.

[28] Roco, M. C. *Corrosion* 1990, *46*, 424-431.

[29] Shin, C. G.; Economou, D. J. *J. Electrochem. Soc.* 1991, *138*, 527-538.

[30] Sato, M.; Aogaki, R. *Materials Sci. Forum* 1998, *289/292*, 459-464.

[31] Maret, G.; Dransfeld, K. *Strong and Ultrastrong Magnetic Fields and Their Applications*; Springer: Berlin, 1985.

[32] Noda, A.; Watanabe, M. *Electrochemistry* 2002, *70*, 140-144.

[33] Hagiwara, R. *Electrochemistry* 2002, *70*, 130-136.

[34] Matsumoto, H.; Matsuda, T.; Tsuda, T.; Hagiwara, R.; Ito, Y.; Miyazaki, Y. *Chem. Lett.* 2001, *30*, 26-27.

[35] Nanjundiah, C.; McDevitt, S. F.; Koch, V. R. *J. Electrochem. Soc.* 1997, *144*, 3392-3397.

[36] Fuller, J.; Carlin, R. T.; Osteryoung, R. A. *J. Electrochem. Soc.* 1997, *144*, 3881-3886.

[37] Takahashi, S.; Koura, N.; Kohara, S.; Sabouugi, M.-L.; Curtiss, L. A. *Plasma and Ions* 1999, *2*, 91-105.

[38] Cooper, A. I. *J. Mater. Chem.* 2000, *10*, 207-234.

In: Trends in Polyaniline Research
Editors: T. Ohsaka, Al. Chowdhury, Md. A. Rahman et al.
ISBN: 978-1-62808-424-5

Chapter 5

ELECTROCHEMICAL DOPING OF POLYANILINE USING BIPOLAR ELECTRODES

Shinsuke Inagi[1], Naoki Shida and Toshio Fuchigami[2]
Department of Electronic Chemistry, Tokyo Institute of Technology, Japan

ABSTRACT

Bipolar electrode, i.e. a wireless electrode having both an anodic pole and a cathodic pole with a potential gradient was used for electrochemical doping of conducting polymers such as polyaniline (PAni). Upon the bipolar electrolysis in the U-type electrolytic cell, the gradual color change of PAni film was observed at the anodic side of the bipolar electrode due to the application of different potential at each position.The multicolor gradation of the PAni was found to be an indicator for the actual potential applied on the bipolar electrode. The combination of PAni and the anthraquinone-based dopant was effective to afford the multicolor electrochromism device due to the electrochemical doping of PAni and the electrochemical conversion of quinone/hydroquinone system. Further development of the bipolar electrode system using the cylinder driving electrodes could realize the spot application of the anodic potential on the bipolar electrode. This helped to successfully carry out spot electrochemical doping of the PAni film. The inverse of polarity of the driving electrodes resulted in complete erasing of the doped spot. This chapter would provide a new insight to the electrochemical doping behavior of the conducting polymer films such as PAni.

Keywords: Electrochemical doping, electrochromism, bipolar electrochemistry, patterning

1. INTRODUCTION

Conjugated polymers in which electron can delocalize through the main chain are defined as conducting polymers. Polyacetylene, polyphenylene, polypyrrole, etc. have π-conjugated

[1] E-mail: inagi@echem.titech.ac.jp.
[2] E-mail: fuchi@echem.titech.ac.jp.

structure, whereas polysilanes have σ-conjugation in the polymer chain. They are readily available by chemical polymerization such as polyaddition and polycondensation of appropriate monomers. Electrochemical polymerization in other words, electropolymerization of aromatic or heteroaromatic compounds also occurs with ease but has proved to be a powerful method to produce conducting polyarene film on electrode [1].

When a conducting polymer on electrode is positively (anodically) charged, polaron (radical cation) or bipolaron (dication) species are generated in the polymer main chain [2]. To compensate the positive charge, counter anions derived from supporting electrolytes are introduced into polymer chain as dopants. This *p*-doped state can be made neutral by cathodic reduction. Conversely, application of cathodic potential to conducting polymers gives *n*-doped polymers.

Cyclic voltammetric analyses show pair of broad current responses, which are usually stable with good cycle properties to provide necessary evidences to confirm these doping and dedoping phenomena. However, the application of much higher potential to conducting polymers often results in over-oxidation or over-reduction to form a non-conducting state (non-conjugated state) and/or to cause degradation of polymer chain by undesirable attack of nucleophiles. The stable color change of conducting polymers can be monitored by spectroelectrochemical measurements, for instance, UV-visible absorption spectral measurement with a constant-potential applied to a polymer thin film on a transparent electrode.

The multiple doping states of Polyaniline (PAni) under acidic conditions contribute to the multicolor chromic appearances depending on the applied potential. The electrochromic behavior of PAni is one of the most promising outputs. This chapter describes the novel electrochemical doping method for conducting polymer films in gradient and local manners based on the bipolar electrochemistry.

2. Bipolar Electrochemistry

In a conventional electrochemical setup, a pair of feeding electrodes (anode and cathode) is equipped in an electrolytic cell system. On both electrode surfaces, homogeneous potential is applied to involve electrochemical reactions, namely oxidation at the anode, and reduction at the cathode. To reduce the resistance of solution (IR-drop), a large amount of supporting electrolyte is necessary. When the concentration of the supporting electrolyte is low, the difference of the applied potential between the anode and the cathode is consumed not only as potential on both electrodes, but also as IR drop of the electrolytic solution. In such case, a conducting material placed between the electrodes is polarized to have a potential distribution on the surface. One edge close to the feeding anode becomes cathodic pole, and the other side becomes anodic pole. When the potential difference on the conducting material surface induced by the IR drop is beyond a threshold, the electrochemical reactions on both edges of the conducting material take place simultaneously. That is to say, the inserted conducting material behaves as an electrode having both anode and cathode, and there is a potential gradient between the poles. Such an electrode system is termed “bipolar electrode”.

Bipolar electrode has been used for industrial-scale electrolysis with a number of stacking to expand the effective electrode surface. Furthermore, when the bipolar electrodes are stacked with extremely narrow gap, the supporting electrolyte is no longer necessary. This offers the capillary gap cell system as a promising tool for a green electrochemical methodology.

Recent progress of bipolar electrode system is quite noticeable [3-10]. The wireless character and the potential distribution on bipolar electrode are useful features for a wide variety of applications. For example, Crooks and coworkers developed a wireless DNA sensor based on the chemoluminescence driven on bipolar electrode [11, 12]. They developed a number of analytical techniques using bipolar electrochemistry [13, 14]. Kuhn and coworkers demonstrated that the bipolar electrode made of metal crystals or particles is propelled by the electrochemical reactions involved at the both edges [15-17]. Björefors and coworkers produced gradient surface of self assembled monolayer of alkanethiol on gold substrate by local reductive elimination of thiol group [18, 19].

3. Bipolar Doping for Conducting Polymers

We propose the hybridization of electrochemical doping of conducting polymers and their bipolar electrochemistry. The gradual doping of the polymer films appears to cause drastic color change in a gradient manner depending on the potential applied on the bipolar electrode. Moreover, the dopant (anion derived from electrolyte) content in the doped polymer is also gradient across the film. Thus the obtained polymer should have both cationic charge and composition gradients.

At first, we prepared poly(3-methylthiophene) (PMT) on indium-tin-oxide (ITO) electrode by potential-sweep electropolymerization of 3-methylthiophene in 0.1 M tetrabutylammonium hexafluorophosphate (Bu_4NPF_6). The originally red-colored PMT film turned to blue when oxidatively doped. The PMT film on ITO substrate was then subjected to the electrolytic cell with a pair of driving electrodes containing 5 mM Bu_4NPF_6/acetonitrile. When a constant current was passed between the driving electrodes, the bipolar electrode worked and the PMT at the anodic pole changed to blue color. The color change gradually proceeded to the center of the film; however, the edge of the film was detached from the ITO substrate because of the application of high potential. The linear potential gradient caused the degradation of polymer on the edge. Upon bipolar doping, the reaction on the cathodic side seemed to be the reduction of contaminating water or oxygen. No *n*-type doping for the PMT could not at all be observed.

Consequently, we designed and fabricated a novel electrolytic cell for bipolar electrolysis having insulating wall to control the electric field around the bipolar electrode (Figure 1) [20]. In this U-type cell system, the potential distribution on the bipolar electrode was rather sigmoidal with a slight slope around both edges. The ITO substrate with PMT film (5 mm × 20 mm) was used as the bipolar electrode. Under the constant current electrolysis (0.5 mA) in 5 mM Bu_4NPF_6/acetonitrile, the bipolar electrolysis was carried out. As the current was passed between the driving electrodes, color changed gradually from the anodic pole to the center of the bipolar electrode. This color change was originated from the anodic doping of the PMT to give quinoid structures in the polymer main chain. When the polarity of the driving electrode was switched inversely, the polarity of the bipolar electrode was totally changed, the color of doped PMT turned to red, together with the simultaneous color change from red to blue on the new anodic side. The reversible switching of gradient electrochromism was confirmed by pulse electrolysis without any noticeable loss in of absorbance.

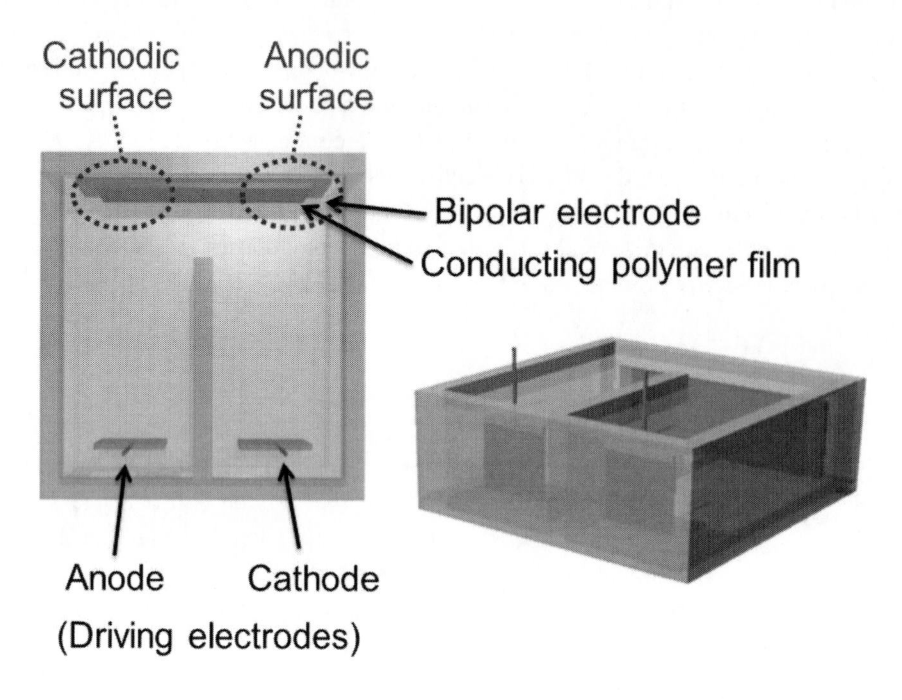

Figure 1. U-type electrolytic cell containing conducting polymer film on bipolar electrode and a pair of driving electrodes.

The anodically doped PMT film was stable and could be kept in its charged state even when the dopant was rinsed with acetonitrile under open-circuit conditions. To quantify the doping, energy dispersive X-ray (EDX) analysis of the doped PMT film on ITO was performed at each position after rinsing with acetonitrile. The amount of the elements in the dopant, gradually increased from the cathodic side to the anodic side.

Poly-(3,4-ethylenedioxythiophene) (PEDOT) is one of the most intensively investigated conducting polymers due to its very highconductivity and excellent environmental stability. Furthermore, transparency of PEDOT in the *p*-doped state is potentially applicable to organic transparent electrodes and electrochromic materials. The localized and gradient doping for PEDOT film by bipolar technique would also be interesting. We prepared a PEDOT film on an ITO working electrode by potential sweep electropolymerization in 0.1 M Bu_4NPF_6/acetonitrile. After further dedoping of the film, the substrate was used as a bipolar electrode in the U-type cell (vide supra). The as-prepared (neutral) PEDOT film was dark blue with low transmittance. Bipolar doping of the PEDOT film with 0.5 mA gave rise to a distinct color change at the anodic surface to transparency depending on the charge passed.

Besides poly(thiophene)s, PAni is well-known for its ease of synthesis, environmental stability, and unique acid/base, doping/dedoping, and oxidation/reduction chemistry [21]. The color of PAni changes diversely from transparent yellow to green, and further to blue depending on the applied potential [22]. The electrochemicalpolymerization and doping of PAni are usually conducted in a water system. This prompted us to explore the scope and limitations of the bipolar technique in water. A PAni film was prepared on an ITO working electrode by potential sweep electropolymerization in 0.1 M H_2SO_4. The obtained PAni film was further dedoped in 5 mM H_2SO_4 to transparency. This was placed in the U-type cell as the bipolar electrode. Then, the bipolar doping of the film with 4 mA in 5 mM H_2SO_4 for 0.05 C resulted in the formation of multicolored film with gradation as shown in Figure 2. This phenomenon is highly important not only for use in novel electrochromic application, but also to estimate the applied potentials on each position of the bipolar electrode.

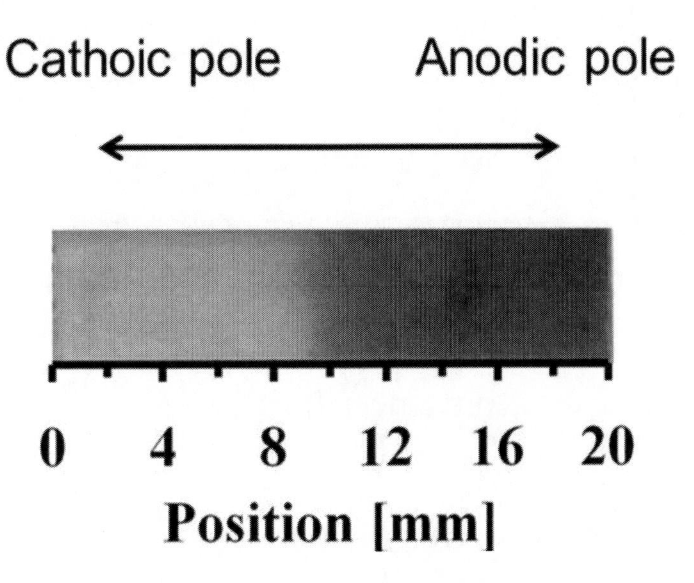

Figure 2. Photograph of the PAni film produced by bipolar doping after charging with 0.05 C in 5 mM H_2SO_4.

Subsequently, we measured absorption spectra at each position of the PAni film. Figure 3 represents the absorption spectra of the film measured at 2mm intervals from the cathodic edge of the bipolar electrode. Early report on the spectroelectrochemical study of PAni film [22] showed that (a) there is almost no absorption around the visible region at –0.2 V; (b) the application of more anodic potential (0.30 V) induced the absorption at the near-IR region with the absorption maximum (λ_{max}) at 810 nm; (3) the absorption band was shifted to the visible region by further application of anodic potential (λ_{max} = 730 nm at 0.55 V, λ_{max} =620 nm at 0.80 V).

The absorption spectra of the PAni film obtained by the bipolar doping at 2 to 8 mm positions suggest that the film has not been yet anodically doped. On the other hand, the spectrum at 10 mm position shows an absorption band the near-IR region and the potential at this position is roughly estimated as ca. 0.3 V. At the more anodic side (16 and 18 mm), the spectra are approximately equal to that measured at 0.8 V; consequently, the potential applied on the anodic side should be >0.8 V. The gradual change of the spectra at 12 and 14 mm positions is attributed to the potential gradient between 0.3 and 0.8 V. Bipolar patterning of conducting polymer film in combination with spectroelectrochemical result reveals the actual potential distribution across the bipolar electrode.

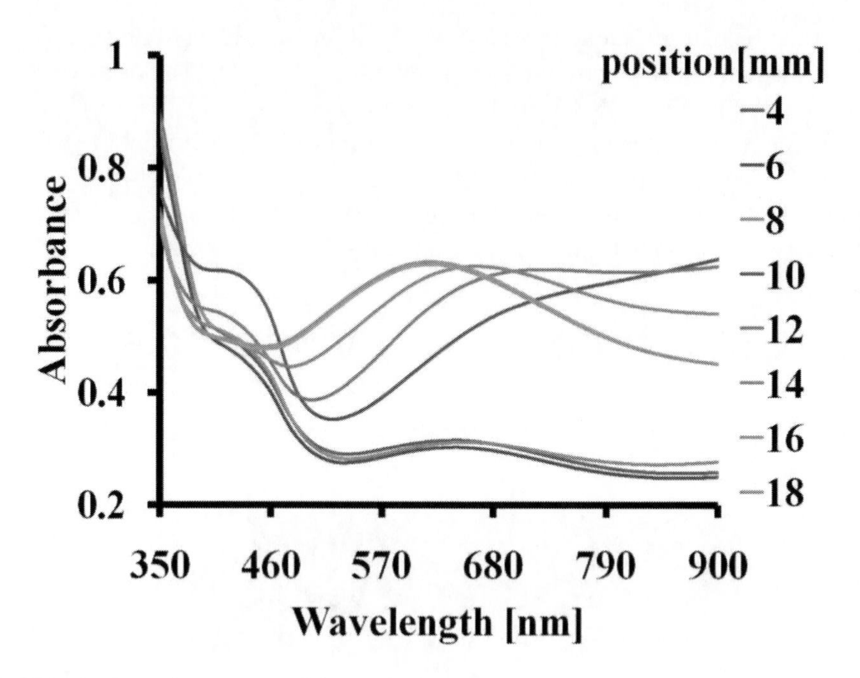

Figure 3. UV–vis absorption spectra of the gradually doped PAni film (for 0.05 C), measured at each distance from the cathodic edge.

4. Bipolar Chlorination of Poly(3-methylthiophene)

If a charged polymer film is unstable in the presence of reactive reagents in an electrolytic cell, a subsequent chemical reaction takes place that will modify the polymer film. In the case of poly(thiophene) derivatives, selective chlorination onto the 3- or 4-positions of the repeating thiophene unit occurred when they were anodically charged in the presence of Et_4NCl as the supporting electrolyte [23-25]. Therefore, we investigated the electrochemical chlorination of the PMT film on a bipolar electrode. Using the electrolytic cell, a PMT film on ITO as the bipolar electrode was placed into an acetonitrile solution containing 5 mM Et_4NCl. A constant current (1 mA) was passed between the anode and cathode for 0.2 C. The bipolar electrode was then removed from the cell and treated by cathodic dedoping at –0.3 V (vs. SCE) to remove chloride anions attached as the dopant from the PMT film.

The chlorine content in the PMT film after the passage of various amount of charge (0.05, 0.1, and 0.2 C) was estimated by EDX analysis. Although the doped area on the film cannot be distinguished by its appearance, the EDX results provided clear evidence of the successful introduction of chlorine atoms into the film on the anodic surface, which reflects the potential gradient of the bipolar electrode. The chlorine atoms must be substituted covalently onto the 4-position of the repeating 3-methylthiophene unit. The degree of chlorination at the anodic surface increased during passage of charge, in keeping with the composition gradient. The highest chlorination area was observed in the sample charged with 0.2 C, and this represents approximately 80% chlorine substitution per repeating thiophene unit. X-ray photoelectron spectroscopy (XPS) analysis of the surface of the chlorinated PMT revealed that chlorine atoms introduced were covalently linked to the thiophene rings.

Chlorine substitution on poly(thiophene) derivatives often imparts a tolerance to oxidation, because it lowers the HOMO energy level. Therefore, the chlorine-gradient PMT film has an intrinsic energy gradient across the film. Anodic doping of the chlorine-gradient film (0.1 C) was also performed in 0.1m Bu_4NPF_6/acetonitrile. When the gradient film was charged at 0.8 V (vs. SCE), only the non-chlorinated area was doped and subsequently a bluecolor around the center of the film could be observed. The interface of the doped and undoped polymer was gradually shifted to a chlorine-rich area during the course of the applied potential. All of the polymer film was doped when charged at 1.4 V (vs. SCE)[26].

5. Bipolar Doping of PAni with a Color-Changeable Dopant

It is known that PAni with 1-amino-bromoanthraquinone-2-sulfonate as a dopant can show multi-electrochromism driven by the doping of PAni main chain and the color change of the quinone unit of the dopant [27]. The quinone derivative is known to show color change from red to colorless for the oxidized and reduced states. The polymer film on ITO substrate was prepared by the potential sweep method in 0.1 M H_2SO_4 containing 5 mM sodium 1-amino-bromoanthraquinone-2-sulfonate. The obtained PAni film was red and transparent. The bipolar electrolysis was then conducted for the film in the U-type cell containing 5 mM H_2SO_4 with a passage of 4 mA current. After the passage of 0.04 C between the driving electrodes, PAni on the anodic side of the bipolar electrode gradually turned purple due to the anodic doping of the PAni (blue) and the quinone structure of the dopant (red). Further passage of charge (0.07 C) shifted the visible color change to the proximity of the center. However, at the cathodic side, no color change was observed during the bipolar electrolysis. Other combinations of conducting polymer and multicolor dopant are expected to give more drastic color change in the single polymer. Moreover, the composite of conducting polymer films also seems to be candidate for multicolor electrochromism applications.

6. Spot Application of Potential on Bipolar Electrode

The development of the U-type cell for bipolar electrolysis made it possible to control the electric field, namely the potential distribution on the bipolar electrode around the insulating wall. We exploited this fascinating phenomemon for the development of a novel device for spot application of electric potential on bipolar electrode. The spot bipolar configuration is shown in Figure 4. In the present study, conducting polymer films were prepared on ITO electrode using potential-sweep method in 0.1 M Bu_4NPF_6/acetonitrile containing aromatic monomers. After further dedoping of the film, the substrate was used as the bipolar electrode, which was placed into a container equipped with an external cathode wire and an anode ring separated by a plastic insulating cylinder. The device was filled with an electrolytic solution containing a low concentration of salt in order to induce bipolar electrode with a small current. When a current flows via the bipolar electrode, a new anodic area appears under the driving cathode wire together with a surrounding cathodic area with its interface gradient. The cylinders employed were 1 mm in thickness with internal diameter (ID)/outside diameter (OD) of 6 mm/8 mm. The distance between the cylinder and the bipolar electrode was kept constant at 1 mm. PEDOT and PMT films were doped with the bipolar device for 0.01 C in 5 mM Bu_4NPF_6/acetonitrile under a constant current. A clear spot appeared in the anodic pattern of the bipolar electrode corresponding well to the cylinder size. This local color change completely recovered when the whole substrate was treated with cathodic electricity (dedoping). Thus the bipolar electrode system was found to be a novel electrochromic device that can supply spot electric potential on the bipolar electrode without any connection to the conducting polymer canvas itself [28].

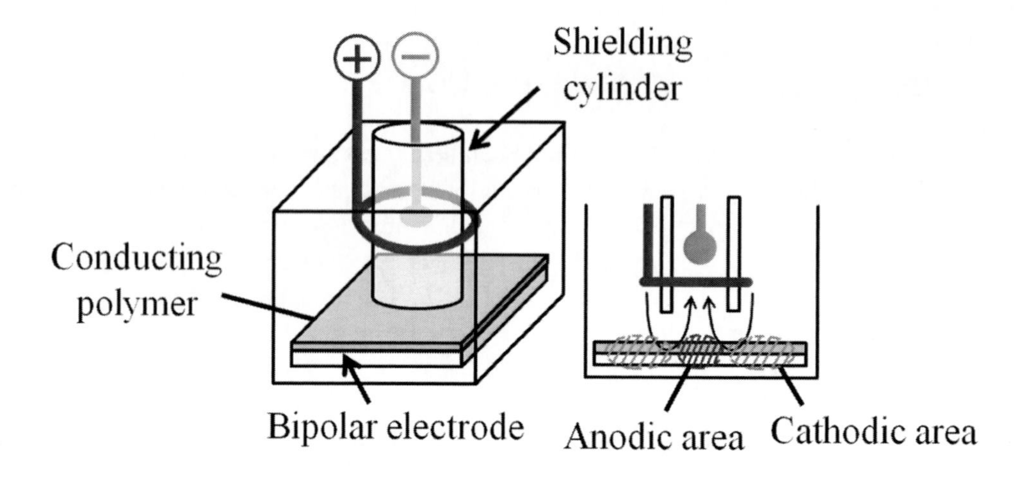

Figure 4. Schematic illustration of the setup used for bipolar patterning.

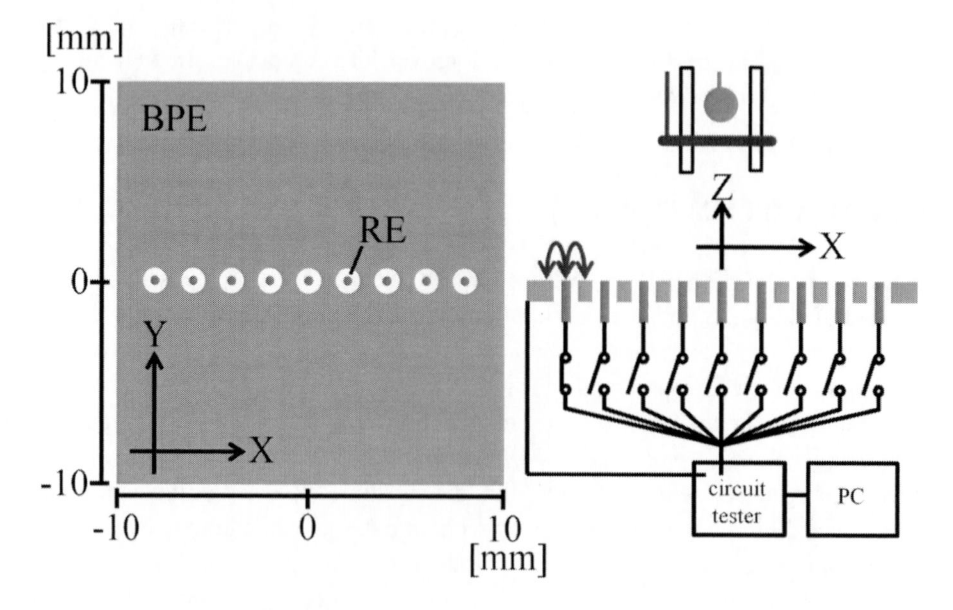

Figure 5. Illustration of the setup used for measuring the potential distribution on the bipolar electrode.

To provide a better understanding of the spot patterning mechanism, we measured the electric potential distribution applied at each position of the bipolar electrode using the setup shown schematically in Figure 5. The voltage between a tip-modified reference electrode (RE) and the surface was measured at different positions. A constant current (1 mA) flowed between the feeder electrodes in 5 mM Et_4NCl/acetonitrile, and the potential differences observed across the X axis of the bipolar electrode are shown in Figure 6. The electric potential profile of the small-diameter cylinder (ID/OD = 4 mm/6 mm) showed a maximum at the center of the cylinder and gradually decreased toward the border. In the case of the large-diameter (ID/OD = 8 mm/10 mm), the electric potential profile was broader. This direct observation of the potential difference on the BPE strongly supports the spot application of potential, which can realize patterned doping and chlorination on conductingpolymer films.

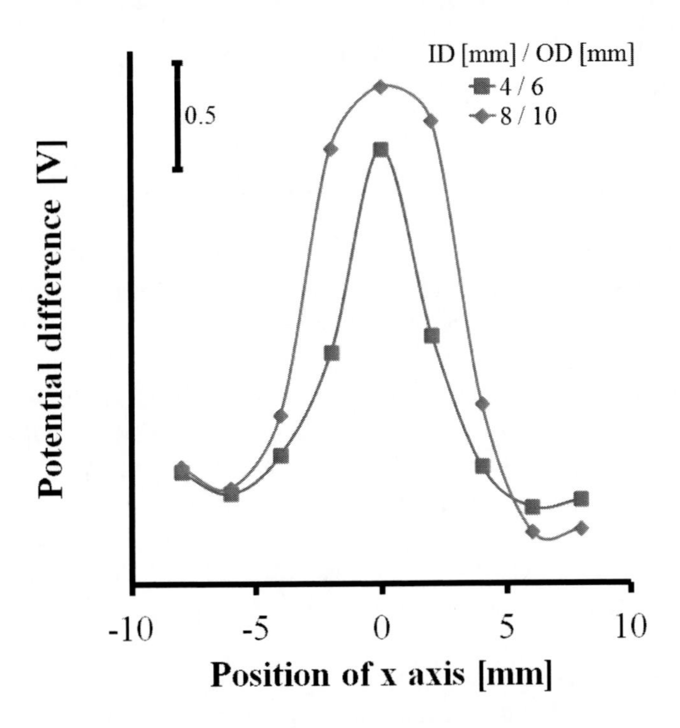

Figure 6. Plot of measured potential differences between the bipolar electrode and reference electrode (Y = 0 mm, Z = 1 mm) using various cylinder sizes (ID/OD = 4 mm/6 mm and 8 mm/10 mm).

Next, the spot patterning of the PAni film using the cylinder electrode system (ID/OD = 6 mm/8 mm) with a passage of constant current (0.5 mA) was successfully carried out to give a blue spot. This blue spot was easily erased by the bipolar electrolysis using the inversed feeding current (Figure 7). Thus the spot writing and erasing were fully reversible. When the shielding cylinder was split into three parts, anodic spots were induced on the corresponding position, resulting in three circle patterns with equal size in spite of the fact that onlya pair of driving electrodes was used. Furthermore, we also demonstrated that the array type device, in which feeder electrodes could be controlled independently, successfully produced the spot patterning on PAni film.

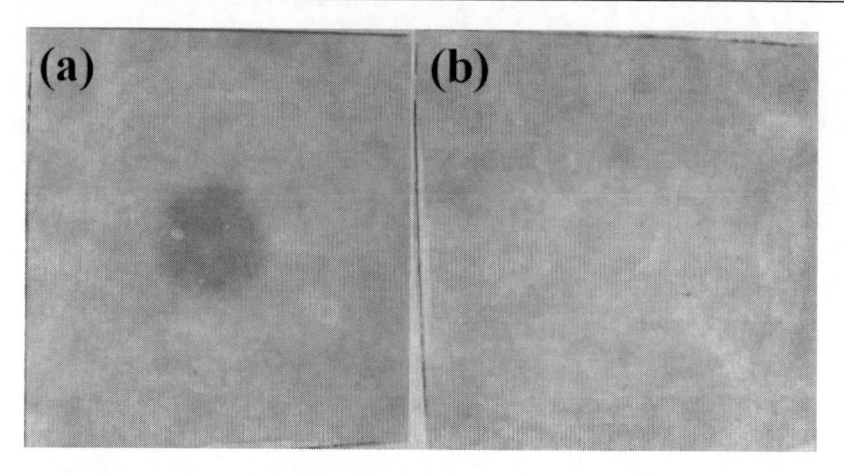

Figure 7. Photograph of the PAni film on the bipolar electrode: (a) as doped with spot anodic application and (b) after the application of a reverse bias.

Conclusion

In conclusion, we have successfully demonstrated the gradient electrochemical doping of conducting polymers such as PMT, PEDOT, and PAni on bipolar electrode using the U-type cell. The multicolor gradation of the PAni film on the bipolar electrode was found to be an indicator for the actual potential applied. The combination of PAni and the anthraquinone-based dopant was effective to afford a multicolor electrochromism device through both electrochemical doping of PAni and the electrochemical conversion of quinone/hydroquinone system. Further development of the bipolar electrode system using cylinder driving electrodes could realize the spot application of the anodic potential on the bipolar electrode. Thus the spot electrochemical doping of the PAni film was successfully demonstrated to show a blue circle spot. The inverse of polarity of the driving electrodes resulted in the complete erase of the doped spot. Both bipolar devices could reflect the potential distribution to the electrochemical doping of PAni.

It is noteworthy that these bipolar techniques are driven without attachment of the polymer to an electrical circuit. This advantage makes this quite new, simple, and powerful method suitable for fabricating composition-gradient materials. Therefore, the method has the potential for novel electrochromic applications involving simultaneous multicolored imaging.

References

[1] Heinze, J.; Frontana-Uribe, B. A.; Ludwigs, S. *Chem. Rev.* 2010, *110*, 4724–4771.
[2] Brédas, J. L.; Street, G. B.*Acc. Chem. Soc.*1985, *18*, 309–315.
[3] Arora, A.; Eijkel, J. C. T.; Morf, W. E.; Manz, A. *Anal. Chem.*2001, *73*, 3282–3288.
[4] Bradley, J.; Ma, Z. *Angew. Chem., Int. Ed.* 1999, *38*, 1663–1666.
[5] Bradley, J.; Babu, S.; Ndungu, P. *Fullerenes, Nanotubes, Carbon Nanostruct.* 2005, *13*, 227–237.

[6] Bradley, J.; Chen, H.; Crawford, J.; Eckert, J.; Ernazarova, K.;Kurzeja, T.; Lin, M.; McGee, M.; Nadler, W.; Stephens, S. *Nature* 1997,*389*, 268–271.

[7] Warakulwit, C.; Nguyen, T.; Majimel, J.; Delville, M.; Lapeyre, V.; Garrigue, P.; Ravaine, V.; Limtrakul, J.; Kuhn, A. *Nano Lett.* 2008,*8*, 500–504.

[8] Ramakrishnan, S.; Shannon, C. *Langmuir* 2010, *26*, 4602–4606.

[9] Ramakrishnan, S.; Shannon, C. *Langmuir* 2011, *27*, 878–881.

[10] Bouchet, A.; Descamps, E.; Mailley, P.; Livache, T.; Chatelain, F.; Haguet, V. *Small* 2009, *5*, 2297–2303.

[11] Zhan, W.; Alvarez, J.; Crooks, R. M. *J. Am. Chem. Soc.* 2002,*124*, 13265–13270.

[12] Chow, K. F.; Mavré, F.; Crooks, R. M. *J. Am. Chem. Soc.* 2008,*130*, 7544–7545.

[13] Mavré, F.; Chow, K. F.; Sheridan, E.; Chang, B. Y.; Crooks, J. A.;Crooks, R. M. *Anal. Chem.* 2009, *81*, 6218–6225.

[14] Fosdick, S. E.; Crooks, J. A.; Chang, B. -Y.; Crooks, R. M. *J. Am.Chem. Soc.* 2010, *132*, 9226–9227.

[15] Loget, G.; Kuhn, A.*J. Am. Chem. Soc.* 2010, *132*, 15918-15919.

[16] Loget, G.; Kuhn, A. *Nat. Commun.* 2011, *2*, 535

[17] Loget, G.; Kuhn, A. *Lab Chip* 2012, *12*, 1967-1971.

[18] Ulrich, C.; Andersson, O.; Nyholm, L.; Björefors, F. *Angew.Chem., Int. Ed.* 2008, *47*, 3034–3036.

[19] Ulrich, C.; Andersson, O.; Nyholm, L.; Björefors, F. *Anal.Chem.* 2009, *81*, 453–459.

[20] Ishiguro, Y.; Inagi, S.; Fuchigami, T. *Langmuir* 2011, *27*, 7158 – 7162.

[21] Li, D.; Huang, J.; Kaner, R. B. *Acc. Chem. Res.* 2009, *42*, 135–145.

[22] Watanabe, A.; Mori, K.; Iwasaki, Y.; Nakamura, Y.; Niizuma, S.*Macromolecules* 1987, *20*, 1793–1796.

[23] Inagi, S.; Hayashi, S.; Hosaka, K.; Fuchigami, T. *Macromolecules* 2009, *42*, 3755 – 3760.

[24] Hayashi, S.; Inagi, S.; Hosaka, K.; Fuchigami, T. *Synth. Met.* 2009, *159*, 1792 – 1795.

[25] Inagi, S.; Hosaka, K.; Hayashi, S.; Fuchigami, T. *J. Electrochem. Soc.* 2010, *157*, E88 – E91.

[26] Inagi, S. Ishiguro, Y.; Atobe, M.; Fuchigami, T. *Angew. Chem. Int. Ed.* 2010, *49*, 10136 – 10139.

[27] Yano, J.; Kitani, A. *Synth. Met.* 1995, *69*, 117–118.

[28] Ishiguro, Y.; Inagi, S.; Fuchigami, T. *J. Am. Chem. Soc.* 2012, *134*, 4034 – 4036.

In: Trends in Polyaniline Research
Editors: T. Ohsaka, Al. Chowdhury, Md. A. Rahman et al.
ISBN: 978-1-62808-424-5

Chapter 6

RECENT ADVANCES IN THE APPROACH OF POLYANILINE AS ELECTRODE MATERIAL FOR SUPERCAPACITORS

Singu Bal Sydulu*[1] *and Srinivasan Palaniappan*[2,3]*
[1]Department of Chemistry, Osmania University, Hyderabad, India
[2]Polymers and Functional Materials Division,
Indian Institute of Chemical Technology, Hyderabad, India
[3]CSIR – Network Institutes for Solar Energy (NISE)

ABSTRACT

Supercapacitors, also called ultracapacitors or electrochemical capacitors, are an ideal electrochemical energy-storage system, suitable for rapid storage and release of energy. They are in between high-power-output conventional capacitors and high-energy-density batteries. In comparison with that of batteries, the energy density of supercapacitors is much lower. The aim of supercapacitor development is improving the energy density without sacrificing the high power density. Supercapacitors can be classified into two major types based on the active electrode material used: an electrochemical double-layer capacitor consists of carbon electrode and pseudocapacitor or redox capacitor with the use of metal oxide or conducting polymers. Among these electroactive materials, conducting polymers are promising materials for supercapacitors due to the advantages of high specific capacitance, reasonable conductivity, redox properties, environmental stability and eco-friendly quality. Among the conducting polymers, polyaniline is mostly being studied for supercapacitor application because of its easy processability, light weight, safe, low cost and eco-friendly nature. This article presents an overview of literature data on the historical background of capacitor, classification of supercapacitors and status of polyaniline based materials (polyaniline and its composites with carbon and/or metal oxide) for supercapacitor work. These issues are summarized and discussed.

* Correspondence: S. Palaniappan: Phone: +91-40-27191474; Fax: +91-40-27193991, Email: palaniappan@iict.res.in, palani74@rediffmail.com.

1. Energy Storage Devices

Storing energy allows humans to balance the supply and demand of energy. Energy storage became a dominant factor in economic development with the widespread introduction of electricity and refined chemical fuels, such as gasoline, kerosene and natural gas in the late 19th century. Unlike other common energy storage fuels in prior use such as wood or coal, electricity must be used as it is being generated, or converted immediately into another form of energy such as potential, kinetic or chemical. Until recently electrical energy has not been converted and stored on a major scale; however new efforts to that effect began in the 21st century.

The worldwide electricity demand is predicted to double by the middle of the century and triple by the end of the century. Energy storage systems in commercial use today can be broadly categorized as mechanical (compressed air energy storage, flywheel energy storage and hydroelectric energy storage), electrical (capacitor, supercapacitor and superconducting magnetic energy storage), chemical (hydrogen, biofuels and liquid nitrogen), biological (starch and glycogen) and thermal (ice storage, hot bricks, cryogenic liquid air and steam accumulator). However, so far no ideal energy storage device and method are available to meet all technical and economical requirements from a growing spectrum of applications.

One of the distinctive characteristics of the electric power sector is that the amount of electricity that can be generated is relatively fixed over short periods of time, although demand for electricity fluctuates throughout the day. Developing technology to store electrical energy so it can be available to meet demands whenever needed would represent a major breakthrough in electricity distribution. Helping to try and meet this goal, electricity storage devices can manage the amount of power required for uninterrupted supply. These devices can also help make renewable energy, whose power output cannot be controlled by grid operators, smooth and dispatchable. An energy storage system is usually based on batteries. Such components store a big amount of energy, but with a limited instantaneous power. Compared to batteries, supercapacitors are emerging as one of the most interesting new developments in the field of energy storage. Supercapacitors can be used for electrical storage devices because of their high energy storage density, even if their energy density is 10 times lower than a battery. But they exhibit a promising set of features such as high power density, fast rates of charge-discharge, reliable cycling life, and safe operation. Such outstanding properties make them promising candidates for energy storage devices in a wide range of applications, such as hybrid electric vehicles, mobile electronic devices, large industrial equipments, memory backup systems, military devices and energy harvesting applications, where high power density and long cycle-life are highly desirable. However, the key factor limiting the widespread deployment of supercapacitors in everyday technology is their moderate energy density. Energy density has long been limited to values well below 10 Wh/kg [1, 2].

The modern supercapacitor is not a battery per seen but crosses the boundary into battery technology by using special electrodes and electrolyte. It is carbon-based, has an organic electrolyte that is easy to manufacture and is the most common system in use today.

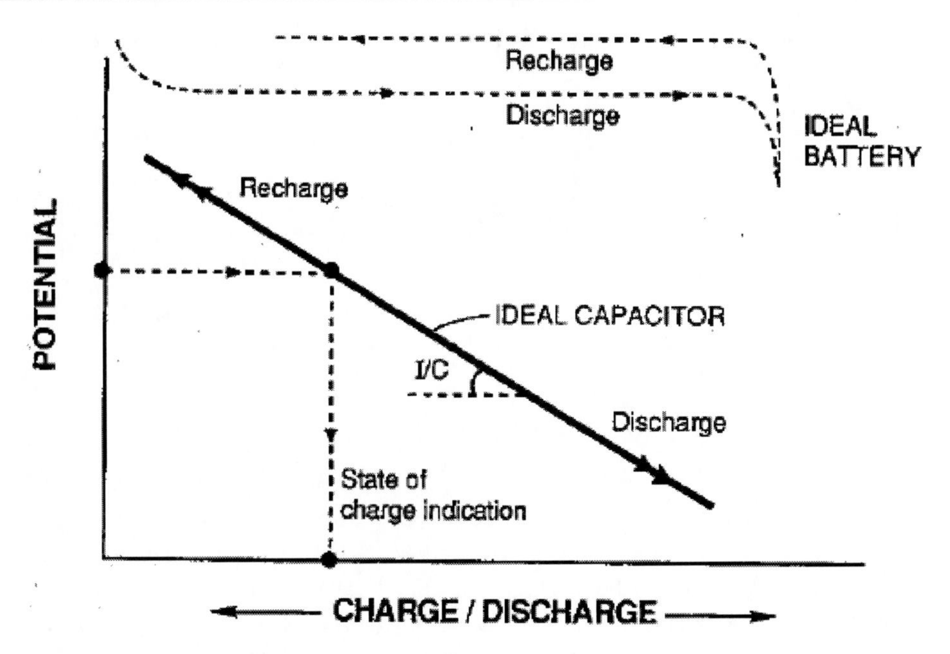

Figure 1. Charge-discharge characteristics of batteries and supercapacitors.

When used in conjunction with batteries and fuel cells, by virtue of their high power densities and rapid charge–discharge characteristics (Figure 1), it finds a range of applications in portable power sources and in electric vehicles.

2. Historical Background of Capacitors

Capacitors are two-terminal electrical elements. Capacitors are essentially two conductors, usually conduction plates - but any two conductors - separated by an insulator - a dielectric - with connection wires connected to the two conducting plates.

Capacitors can be classified into three types, i.e. electrostatic, electrolytic and super capacitors (Figure 2). Electrostatic capacitor consists of two electrodes with a dry separator. This capacitor has a very low capacitance and is used to filter signals and tune radio frequencies. The size ranges from a few picofarad (pF) to low microfarad (μF). The second one is the electrolytic capacitor, which is used for power filtering, buffering and coupling. The electrolytic capacitor has several thousand times the storage capacity of the electrostatic capacitor and uses a moist separator. The third type is the supercapacitor, rated in farads, which is again thousands of times higher than the electrolytic capacitor. The supercapacitor is ideal for energy storage that undergoes frequent charge and discharge cycles at high current and short duration.

Capacitor devices first appeared in Leyden, a city in the Netherlands sometime before 1750. It was discovered by E. G. von Kleist and Pieter van Musschenbroek who found that charge could be stored by connecting a generator by a wire to a volume of water in a hand-held glass jar. Von Kleist's hand and the water acted as conductors and the jar as a dielectric. Benjamin Franklin investigated the Leyden jar, and proved that the charge was stored on the glass, not in the water.

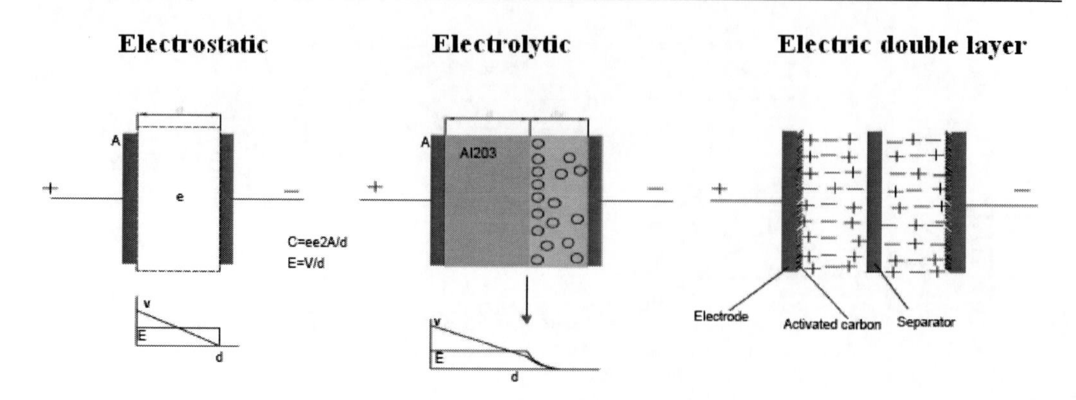

Figure 2. Electrostatic, Electrolytic and Supercapacitors.

Electrolytic capacitors are next generation capacitors, the principle of the electrolytic capacitor was discovered in 1886 by Charles Pollak, as part of his research into anodizing of aluminum and other metals, which are commercialized in full scale. They are similar to batteries in cell construction but the anode and cathode are made up of the same materials. They are aluminium, tantalum and ceramic capacitors where they use solid/liquid electrolytes with a separator between two symmetrical electrodes [3]. Some of the examples under the conductive polymer based electrolytic capacitors, Tetracyanoquinodimethane (TCNQ), polyaniline, polypyrrole (Aluminium capacitors based on PPY are manufactured by Panasonic (SP Cap), Rubycon, and others. Tantalum capacitors based on PPY are manufactured by NEC and Sanyo (POS-CAP).), polyethylenedioxythiophene (PEDT). With introduction of conductive polymer electrolyte in electrolytic capacitor, remarkable improvements like ESR, dielectric strength and long term stability (50,000h), high in capacitance (3.3 to 2700 μF), operating temperature -55 ^{0}C TO 125 ^{0}C, suitable for applications used at low temperature (below 0 ^{0}C). Applications of the Electrolytic capacitors are noise removing, backup and bypass capacitors for digital equipment, cellular phone, personal computers, home appliance, automotive electric equipment, industrial equipment, etc.

Electric/electrochemical double layer capacitor (EDLC) is a unique electrical storage device, which can store much more energy than conventional capacitors and offer much higher power density than batteries. EDLCs fill up the gap between the batteries and the conventional capacitor, allowing applications for various power and energy requirements. The Electric Double Layer Capacitor effect was first noticed in 1957 by General Electric. Standard Oil of Ohio (SOHIO) re-discovered this effect in 1966 [4]. At this time SOHIO acknowledged that the double layer at the interface behaves like a capacitor of relatively high specific capacity. Standard Oil of Ohio gave the licensing to Nippon Electric Company (NEC), which in 1978 marketed the product as a “supercapacitor”, for memory backup applications, finding their way into consumer appliances. Finally, in the last two decades there has been with the introduction of high-voltage decomposition electrolytes (e.g. organic and ionic liquids) [5-7], the improvement of activated carbon properties and the introduction of new configurations (increased combination of supercapacitor and batteries components) and the growth of market demand on specific high power devices, an improvement in the classes and designs of supercapacitors. The supercapacitors are basically working according to the same principles of conventional capacitors, both electrostatic and electrolytic, with much different technical properties, an supercapacitor has specific capacitance and energy 10 times greater than that of an electrolytic capacitor and 100 times that of an electrostatic capacitor. Redox properties of polypyrrole had been first reported in 1963 by Weiss and co-workers [8-10]. Later used in supercapacitors electrode materials in the mid-1990s [11].

3. Classification of Supercapacitors

Depending on the charge-storage mechanism as well as the active materials used, supercapacitors can be divided into two major types: (1) an electrochemical double-layer capacitor (EDLC) that stores the energy non-Faradaically by charging an electrochemical double layer at the interface between the porous electrode (usually carbon-based active materials with high surface area) and the electrolyte, and (2) a redox supercapacitor or pseudocapacitor that stores energy Faradaically using the pseudocapacitive behavior of a redox-active material. Recently, more research has been concentrated on investigating redox supercapacitors because the specific pseudocapacitance, in many cases, exceeds that of carbon materials using double-layer charge storage (depending on the charge-storage mechanism and construction, supercapacitor cells are classified as shown in Figure 3).

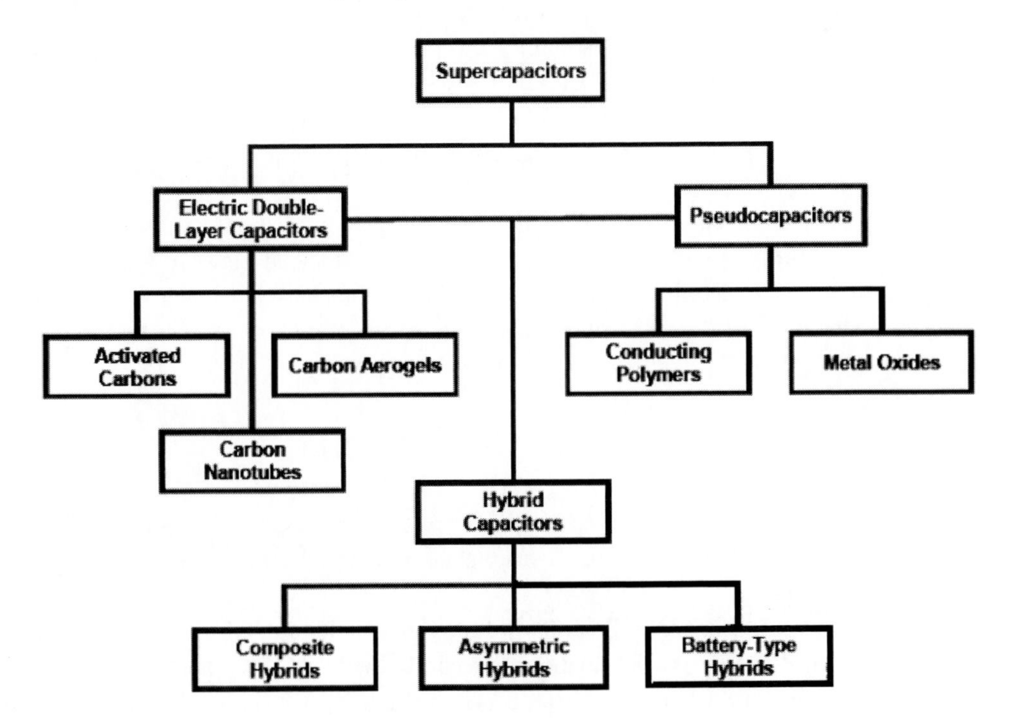

Figure 3. Classification of supercapacitors.

3.1. Electrochemical Double Layer Capacitors

Electrochemical double layer supercapacitors store electrical energy using reversible adsorption of ions from an electrolyte onto two porous electrodes to form an electric double layer at an electrode/electrolyte interface. Electrochemical supercapacitors that only involve physical adsorption of ions in this manner, without any chemical reactions, are called electric double-layer capacitors (EDLCs).

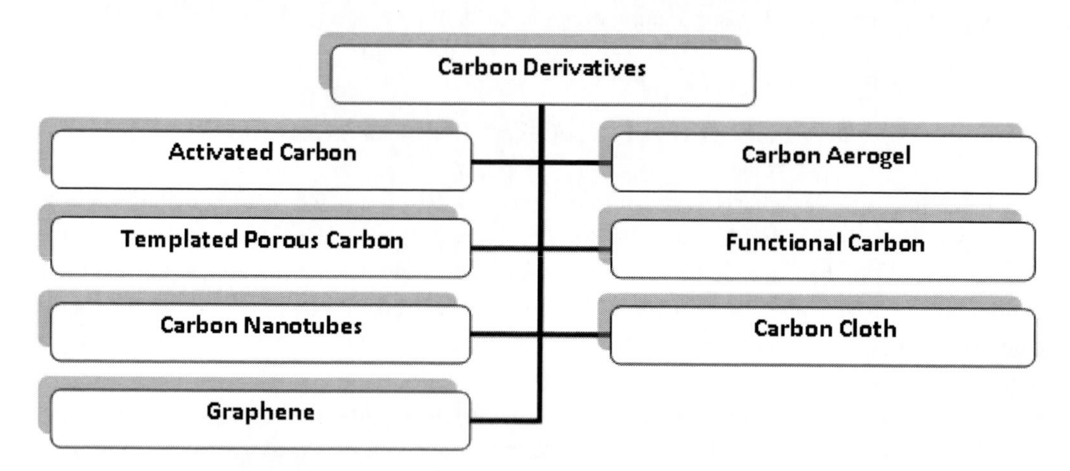

Figure 4. Various forms carbon materials used in supercapacitor.

The EDLC version of the supercapacitor is the most developed form of electrochemical capacitor. Carbon, in its various forms (Figure 4), is currently the most extensively examined and widely utilised electrode material in EDLCs with development focusing on achieving high surface-area on the order of 1000 to 2000 m^2/g, with low matrix resistivity. A number of carbon manufacturers are now targeting supercapacitors as a market for their products [12, 13].

Supercapacitors are constructed much like a battery in that there are two electrodes immersed in an electrolyte, with an ion permeable separator located between the electrodes. In such a device, each electrode–electrolyte interface represents a capacitor so that the complete cell can be considered as two capacitors in series. For a symmetrical capacitor (similar electrodes), the cell capacitance (C_{cell}), will therefore be:

$$\frac{1}{C_{cell}} = \frac{1}{C_1} + \frac{1}{C_2}$$

where C_1 and C_2 were represents the capacitance of the first and second electrodes, respectively.

The double layer capacitance is generally a function of potential, which is different from a standard capacitor. Several models, such as Helmholtz model [14], Gouy-Chapman model [15], and Gouy-Chapman-Stern model [16], and Grahame model [17] have been proposed to explain the behavior observed for electrodes under potentiostatic control in solution.

EDLCs have much higher power density with speed at which the energy can be delivered to the load. Batteries, which are based on the movement of charge carriers in a liquid electrolyte, have relatively slow charge and discharge times. Capacitors, on the other hand, can be charged or discharged at a rate that is typically limited by current heating of the electrodes. Existing EDLCs have energy densities that are perhaps 1/10 that of a conventional battery, their power density is generally 10 to 100 times greater. This makes them most suited to an intermediary role between electrochemical batteries and electrostatic capacitors, where neither sustained energy release nor immediate power demands dominate one another.

3.2. Pseudo-Capacitors

The capacitance of an electrode can be expressed as: $C = C_{dl} + C_{\Phi}$ where C_{dl} is the double layer capacitance and C_{Φ} denotes the amount of charge stored due to pseudocapacitance. This pseudocapacitance part of the total capacitance comes from a redox process for the general reaction $Ox + e^{-} \rightarrow Red$. The species are adsorbed on the surface and thus the coverage of the surface is an important factor in determining the capability of charge storage in a pseudocapacitor cell. The amount of charge stored on the surface of an electrode due to this reaction depends on the potential.

In principle, any materials with fast and reversible redox properties can be used for fabricating pseudocapacitors. For practical applications, however, they should be cheap and insoluble in electrolyte. Up-to-date, the most attractive materials for pseudo-capacitors are conducting polymers (e.g. polyaniline) and cheap transition metal oxides.

One of the drawbacks of carbon-based EDLCs is their low specific energy density compared with that of modern secondary batteries. One approach to enhance the specific energy is the addition of a pseudo-capacitor component into EDLC electrodes.

Several types of Faradaic processes occur in the pseudocapacitive electrodes: (1) reversible surface adsorption of proton or metal ions from the electrolyte; (2) redox reactions involving ions from the electrolyte; and (3) reversible doping dedoping processes in conducting polymers [18, 19]. The first two processes are primarily surface reactions and hence are highly dependent on the surface area of the electrodes. The third process is more of a bulk process and the specific capacitance of the electrode materials is much less dependent on surface area, although a relatively high surface area with micropores is desirable to distribute efficiently the ions to and from the electrodes.

In principle, pseudo-capacitors can provide a higher energy density than EDLCs, especially in systems where multiple oxidation states can be accessed, but because the electrodes undergo physical changes during charge/discharge they have relatively poor durability compared with EDLCs.

The main difference between pseudocapacitance and EDL capacitance lies in the fact that pseudocapacitance is Faradaic in origin, involving fast and reversible redox reactions between the electrolyte and electro active species on the electrode surface. For ECs based on pseudocapacitance, the essential process can be Faradaic, similar to that in a battery. However, an essential fundamental difference from battery behavior arises because, in such systems, the chemical and associated electrode potentials are a continuous function of the degree of charge, unlike the thermodynamic behavior of single-phase battery reactants. ''Mirror-image'' relations between anodic and cathode current–response profiles in cyclic voltammetry are important characteristic of reversible pseudocapacitance charge/discharge processes. In contrast, battery charge/discharge processes are usually entirely irreversible, in this sense, i.e., the anodic and cathodic current profiles are not mirror images.

Pseudocapacitance is faradaic in origin, and quite different from the classical electrostatic capacitance observed in the double layer. In the case of the pseudocapacitance, charge will transfer across the double layer, similar to discharging and charging in a battery, thus the capacitance can be calculated by using the extent of charge stored (Δq) and the change of the potential (ΔV). The relationship between them can be described by the following equation.

$$C = \Delta q/\Delta V$$

Generally in a double layer carbon capacitor, there is about 1-5% pseudocapacitance due to the functional groups on the surface, and there is also about 5-10% double layer capacitance in a battery. Pseudocapacitance can be caused by electrosorption of H^+ or metal atoms and redox reactions of electroactive species, which strongly rely on the chemical affinity of the surface to ions in the electrolyte. Pseudocapacitance can remarkably enhance the capacitance of supercapacitors, and on the one hand, it can also deteriorate other properties, such as life cycle.

Transition metal oxides as well as electrically conductive polymers are typical examples of pseudocapacitive active materials because they can be readily charged and discharged by converting between their different redox states. This conversion process is associated with ion diffusion into/out of the conductive polymer or metal oxide films to keep their electroneutrality.

3.2.1. Metal Oxide

As reported by Conway et al. the capacitance of a redox supercapacitor can be 10–100 times higher than that of EDLCs. However, a redox supercapacitor usually suffers from relatively lower power density than an EDLCs because faradaic processes are normally slower than non-faradaic processes. Moreover, because redox reactions occur at the electrode, a redox supercapacitor often lacks stability during cycling, similar to batteries. The pioneering work on the pseudocapacitive behavior of manganese oxide in an aqueous solution was published in 1999 by Lee and Goodenough [20, 21]. This was followed by several studies to establish the charge storage mechanism in manganese oxide electrodes, due to their low cost and high specific capacitance, abundance and environmentally friendly nature, attracted interest as active electrode materials for electrochemical supercapacitors.

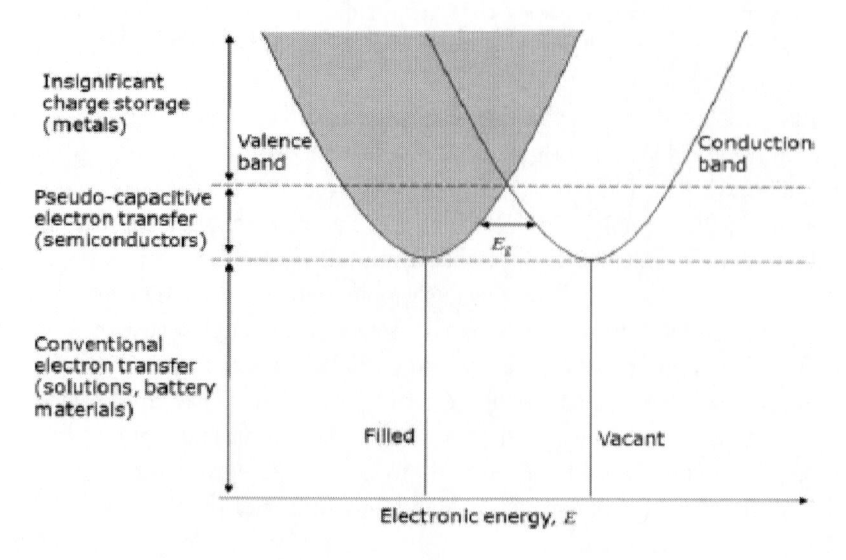

Figure 5. Electron transfer in pseudocapacitive materials.

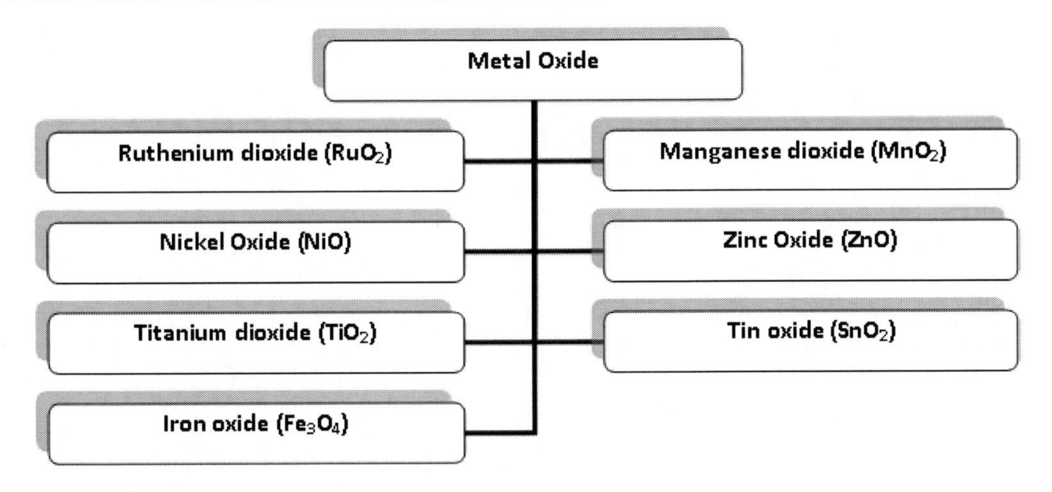

Figure 6. Representive examples metal oxides used in supercapacitors.

The electrons involved in the non-faradaic electrical double-layer charging are the itinerant conduction-band electrons of the metal or carbon electrode, whereas the electrons involved in the faradaic processes are transferred to or from valence-electron states (orbitals) of the redox cathode or anode reagent [22]. (Figure 5).

The electrons may, however, arrive in or depart from the conduction-band states of the electronically conducting support material, depending on whether the Fermi level in the electronically conducting support lies below the highest occupied state of the reductant or above the lowest unoccupied state of the oxidant.

In pseudocapacitors, the non-faradaic double-layer charging process is usually accompanied by a faradaic charge transfer [23]. Recently other metal oxides for supercapacitors have also been investigated, such as RuO_2 [24], IrO_2 [25], MnO_2 [26], NiO [27], SnO_2 [28], and Fe_3O_4 [29]. (Figure 6).

3.2.2. Conducting Polymers

Conducting polymers (CPs) possess many advantages that make them suitable materials for supercapacitor [30], such as low cost, low environmental impact, high conductivity in a doped state, high voltage window, high storage capacity/porosity/reversibility, and adjustable redox activity through chemical modification. CPs offer capacitance behavior through the redox process.

When oxidation takes place, ions are transferred to the polymer backbone, and when reduction occurs, the ions are released from this backbone into the electrolyte. These redox reactions in the conducting polymer come about throughout its entire bulk, not just on the surface. Because the charging and discharging reactions do not involve any structural alterations such as phase changes, the processes are highly reversible and the corresponding mechanism as follows (Figure 7).

The π-conjugated backbone of the polymer provides a perfect environment for the incorporation of a number of organic and inorganic moieties to yield functional hybrid materials. Some of the conducting polymers used in supercapacitor study are showed in Figure 8.

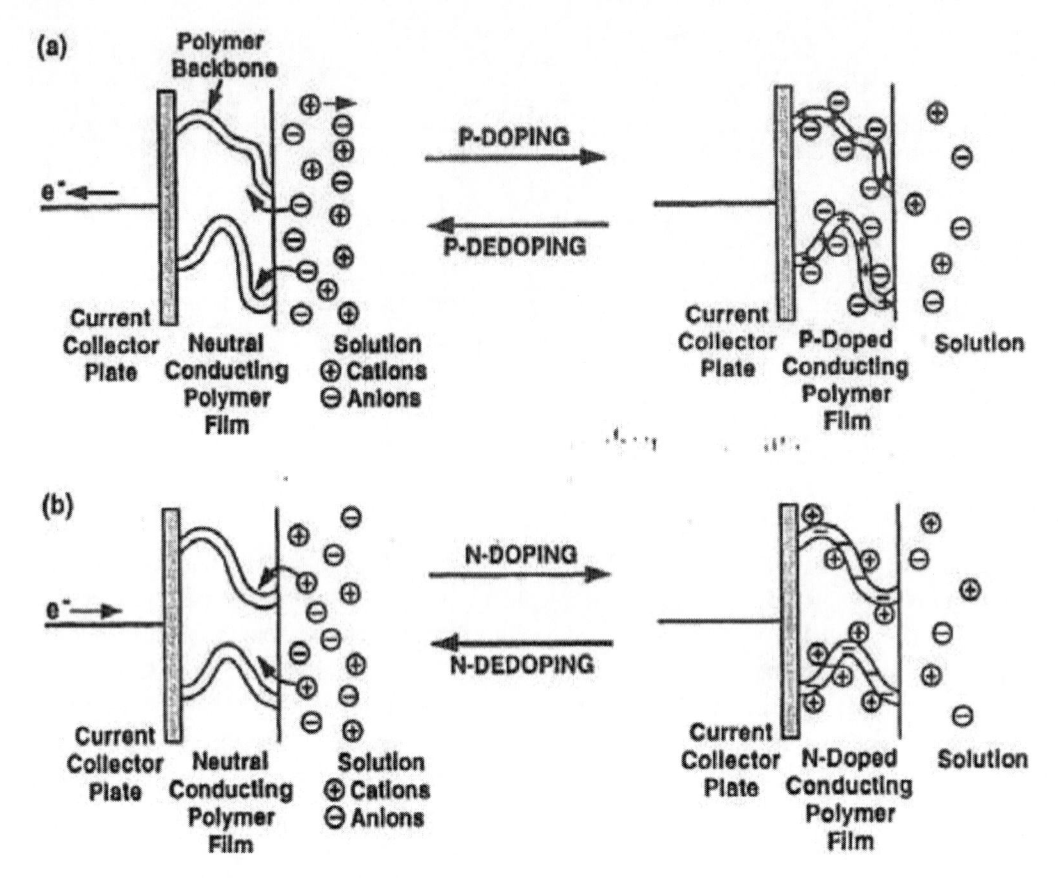

Figure 7. Charge storage mechanism in conducting polymers.

Conducting polymers can be doped with (counter) anions when oxidized and n-doped with (counter) cations when reduced. The simplified equations for these two charging processes are as follows:

$C_P \rightarrow C_P^{n+} + ne-$ (p-doping)
$C_P + ne- \rightarrow (C^+)n\ CP^{n-}$ (n-doping)

The discharge reactions are, of course the reverse of the above equations, as illustrated in Figure 7. Capacitor systems utilizing electroactive conducting polymers were classified into three types [31-35].

Type I capacitor (symmetric): both the electrodes are polymers of the p-dopable type, with the oxidation leading to positively charged polymer chains. In the fully charged state, one electrode will be in the fully p-doped (positive) state and other in the uncharged state.

Type II capacitor (asymmetric): two different p-dopable polymers are used that have different ranges of potentials for oxidation and reduction.

Type III capacitor (asymmetric): using the same polymer for both electrodes with the p-doped form used as the positive electrode and the n-doped form used as the negative electrode. Example for Type III capacitor is functionalized type of polythiophene.

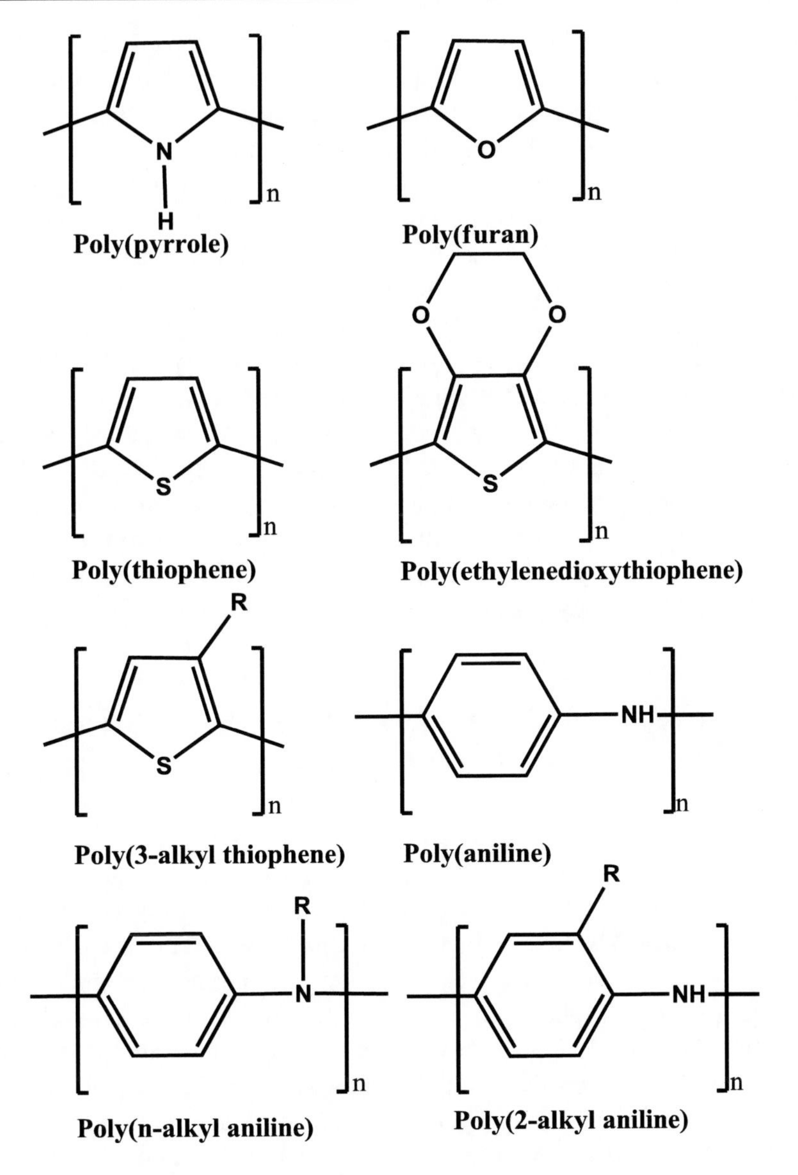

Figure 8. Representative candidates of simplified conducting polymer structures.

3.3. Hybrid Systems

Organic-inorganic hybrid materials represent the natural interface between two worlds of chemistry (organic and inorganic) each with very significant contributions to the field of material science, and each with characteristic properties that result in diverse advantages and limitations. The main idea when developing hybrid materials is to take advantage of the best properties of each component that forms the hybrid, trying to decrease or eliminate their drawbacks getting in an ideal way a synergic effect, which results in the development of new materials with new properties.

In the Hybrid supercapacitors attempt to exploit the relative advantages and mitigate the relative disadvantages of EDLCs and pseudocapacitors to realize better performance characteristics like in developing high cycle life, high-energy supercapacitors, the tremendous flexibility in tuning the design and performance of hybrid capacitors. This leads them to surpass EDLCs as the most promising class of supercapacitors. Utilizing both Faradaic and non-Faradaic processes to store charge, hybrid capacitors have achieved energy and power densities greater than EDLCs without the sacrifices in cycling stability and affordability that have limited the success of pseudocapacitors. Hybrid supercapacitors combine these two charge storage mechanisms (Faradaic and non-Faradaic), resulting in improved device characteristics [36-39]. Typically, carbon based materials are good electrodes for EDLCs, while transition metal oxides and electrically conducting polymers are good candidates for pseudocapacitors. Research has focused on three different types of hybrid capacitors, distinguished by their electrode configuration: composite, asymmetric, and battery-type respectively. Recently, asymmetric or hybrid supercapacitor, regarded as the trend in electrochemical capacitors, has been reported. Asymmetric supercapacitor can be fabricated with one electrode being of a double-layer carbon material and the other electrode being of a pseudo-capacitance material. It is possible to reach the high working voltage and high energy density by choosing a proper electrode material, contributed to a significant increase in the overall energy density of the supercapacitor devices.

4. Polyaniline Based Materials as Electrode in Supercapacitor

4.1. Polyaniline

Recently, many attempts have been made to use electronically conducting polymers (CPs) as electrode materials in electrochemical capacitors, often called supercapacitors. Among the various CPs, polyaniline (PANI) is a unique and promising candidate for practical applications because of its good processability, environmental stability, low cost, and reversible control of electrical properties by simple acid-base doping-dedoping chemistry.

Polyaniline (PANI) is a conducting polymer of the semi-flexible rod polymer family. Although the compound itself was discovered over 150 years ago, only since the early 1980s has polyaniline captured the intense attention of the scientific community. This interest is due to the rediscovery of high electrical conductivity. Amongst the family of conducting polymers and organic semiconductors, polyaniline has many attractive processing properties. Because of its rich chemistry, polyaniline is one of the most studied conducting polymers of the past 50 years.

PANI synthesized by chemical and electrochemical method is being investigated through the performance of electrode material in supercapacitor. PANI synthesized by chemical method has advantages when compared to electrochemical method. In the electrochemical method, mass production of composite electrodes is not possible and also it is not suitable for preparing controlled polymer films with thickness above 100μm.

General Structure of Polyaniline: The base form of polyaniline, in principle, can be described by the following general formula (Figure 10a):

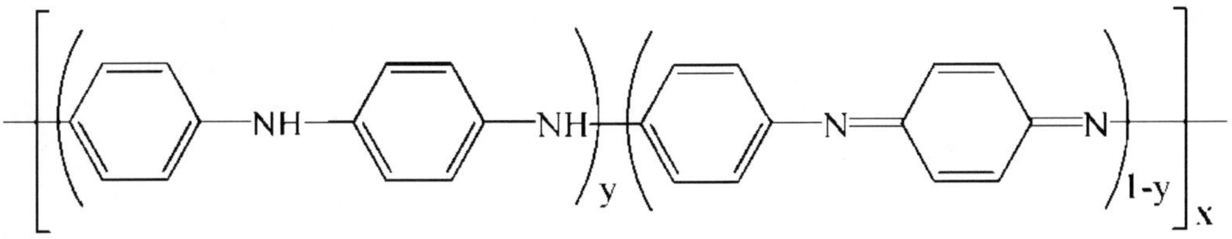

Figure 10a. General formula for polyaniline base.

In the generalized base form, (1- y) measures the function of oxidized units. When (1- y) = 0, the polymer has no such oxidized groups and is commonly known as a leucoemeraldine base. The fully oxidized form, (1 - y) = 1 is referred to as a pernigraniline base. The half oxidized polymer, where the number of reduced units and oxidized units are equal, i.e., (1- y) = 0.5, is of special importance and is termed the emeraldine oxidation state or the emeraldine base. The corresponding emeraldine salt can be represented by the following formula (Figure 10b).

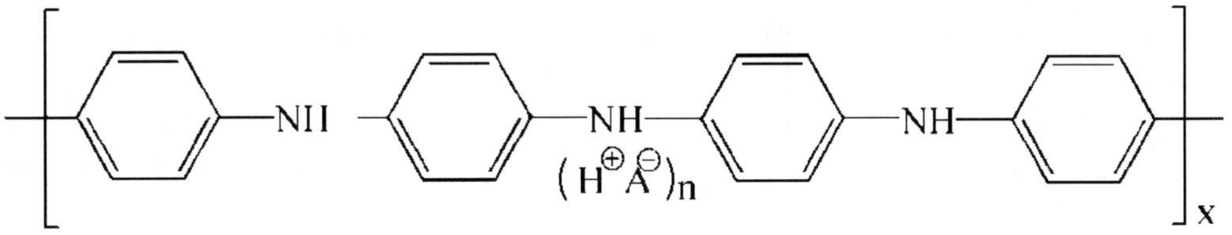

Figure 10b. General formula for polyaniline salt.

General Procedure for Electrochemical Synthesis of PANI Electrode

Polyaniline was obtained by an electrochemical oxidation by a galvanostatic/potentiostatic procedure in a one-compartment cell with a three-electrode configuration. Typically, Au, SS, C, Ni, Pt, Ti, ITO, etc., are being used as the working electrode. Saturated calomel electrode (SCE) or Ag/AgCl are being used as reference electrodes. Electrolyte solution was composed of different concentration of aniline and different concentration of dopants like H_2SO_4, $NaClO_4$, PTSA, $HClO_4$+ $NaClO_4$, etc. Polymerization was carried out at the constant current/voltage. The deposited material is being washed with methanol, acetone and distilled water to removed unreacted monomers.

General Chemical Synthesis Route to Prepare PANI Electrode

Generally, aqueous solution of aniline is added to aqueous solution of protonic acid. Polymerization is started by the addition of oxidizing agent such as aqueous solution of ammonium persulfate to the above solution for a particular temperature. The mixture is constantly stirred for a particular period of time.

The formed precipitate is filtered, washed with water and acetone several times until the filtrate become colourless and drying is carried out in vacuum or in an oven. Fabrication of working electrode is generally carried out by mixing the electro active PANI powder, carbon black (additive) and poly(tetrafluoroethylene) binder and ethanol/n-methylpyrrolidone solvent and coating or pressing the mixture on metal substrates such as. Au, SS, C, Ni, Pt and Ti) followed by drying.

General Chemical Synthesis Route to Prepare PANI by Interfacial Polymerization

Typically the interfacial reaction was performed in a 100 mL beaker. A required molar amount of aniline was dissolved in the organic phase (50 mL) such as hexane, benzene, toluene, xylene, diethyl ether, carbon disulfide, carbon tetrachloride, chloroform, *o*-dichlorobenzene, or methylene chloride. Calculated amount of Ammonium peroxydisulfate was dissolved in 50 mL of dopant acid solution. A great variety of dopant acids can be used, including hydrochloric, sulfuric, nitric, phosphoric, perchloric, acetic, formic, tartaric, camphorsulfonic, methylsulfonic, ethylsulfonic, or 4-toluenesulfonic acid, among others. The oxidant solution is added to the organic solution without disturbance of the 100 ml beaker. After completion of the reaction the aqueous layer is filtered, washed with water and acetone for several times until the filtrate becomes colourless and drying is carried out in vacuum or in an oven.

General Chemical Synthesis Route to Prepare PANI by Emulsion Polymerization

In the emulsion polymerization pathway, calculated amount of sodium lauryl sulfate was dissolved in 150 ml of 1M aq.H_2SO_4 solution. The resulting solution was stirred for the 1hr. 1 ml of aniline was added to the above reaction mixture and stirred for 1 hr. Benzoyl peroxide (appropriate amount) was dissolved in 50 ml of chloroform and then added to the above reaction mixture. The reaction mixture was stirred for 24 hr at ambient temperature and then the resulting reaction mixture was poured in 250 ml acetone. The precipitate was filtered, washed with an ample amount of water and finally with 250 ml of acetone. The powder sample was dried at 50^0 C till a constant weight was recorded. PANI samples are generally characterized by Fourier transform infrared spectroscopy (FT-IR), X-ray diffraction study (XRD), scanning electron microscopy (SEM), transmission electron microscope (TEM) and electrical conductivity measurements.

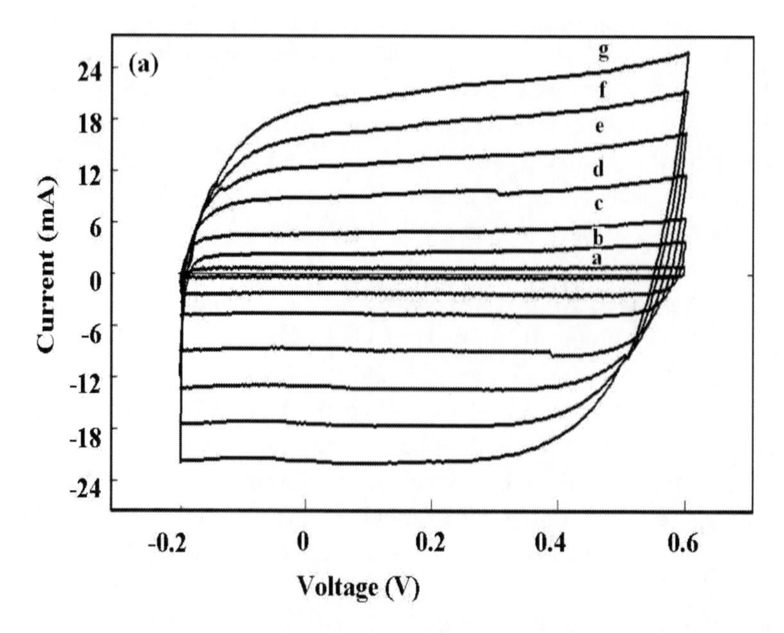

Figure 11. Cyclic voltammogram of PANI salt at different scan rates (a) 1, (b) 5, (c) 10, (d) 20, (e) 30, (f) 40 and (g) 50 mV/s.

Electrochemical Performance of Supercapacitor Cell Such as Cyclic Voltammetry

Electrochemical performance of PANI electrode is generally carried out by cyclic voltammogram with a configuration of PANI as working electrode, metal as counter electrode, standard electrode and electrolyte. Electrochemical performance of PANI is carried out using symmetric or asymmetric configuration of electro active PANI separated by separator containing electrolyte in a particular voltage window at various scan rates. As a representative system, salt is recorded in the voltage range of -0.2 to 0.6 V using two electrode systems, the electrodes are separated by cellulose cloth in 1.0 M H_2SO_4 electrolyte [40]. Corresponding cyclic voltammograms shown in Figure 11.

Charge-Discharge Study

In order to better understand the behavior of the PANI supercapacitor cell (two electrode system), galvanostatic charge/discharge study carried out at current density of 5 mA was performed on PANI salt and corresponding charge-discharge curves are shown in above Figure 12 [40]. Discharge specific capacitance (Cd), energy density (E), and power density (P) were calculated from the formula reported in literature [41].

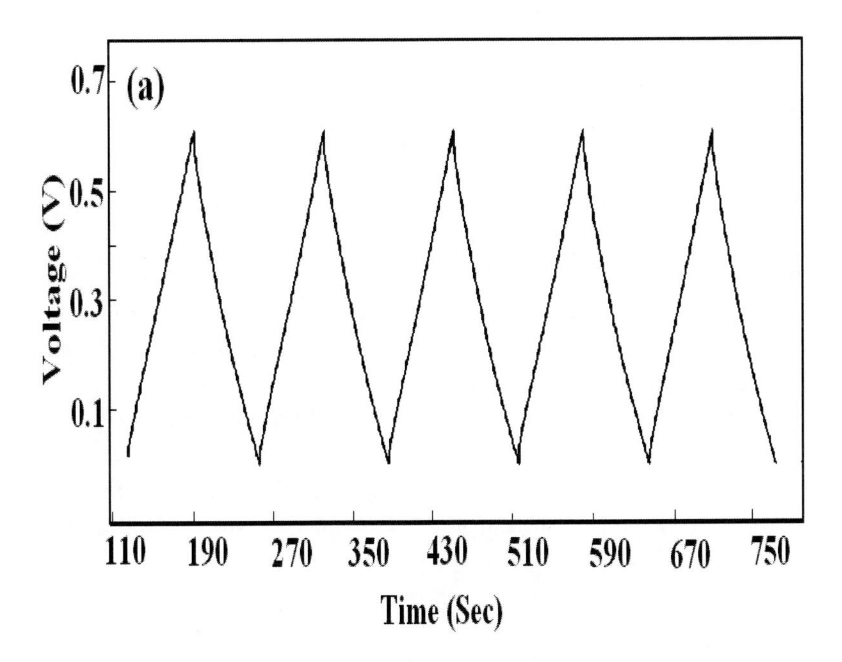

Figure 12. Galvanostatic charge-discharge curves.

Electrochemical Impedance Spectroscopy

Electrochemical impedance spectroscopy (EIS) measurements were carried out with IM6ex zahner-Elektrik, Germany) by applying an AC voltage of 5 mV amplitude in the 40 kHz to 10 mHz frequency range at various voltages using three electrode cell configuration i.e., polyaniline salt as working electrode, platinum counter electrode and Ag/AgCl reference electrode (containing saturated KCl solution) in 1M H_2SO_4 electrolyte. Corresponding nyquist plot is shown in Figure 13. The capacitance can be calculated from the following formula: $C = -(2\pi f z_{im})^{-1}$.

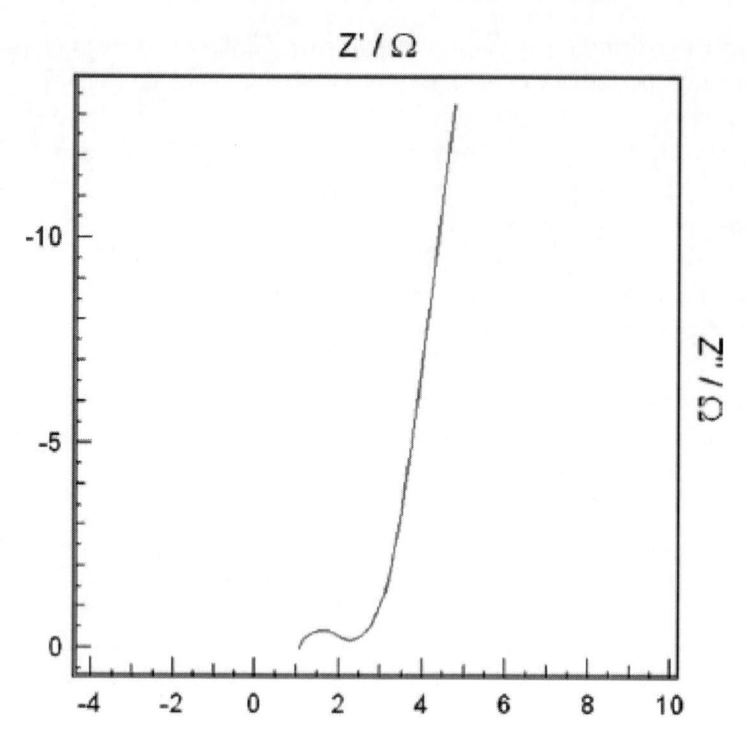

Figure 13. Nyquist plot of PANI salt in the frequency range of 40 kHz to 10 mHz.

Literature work on polyaniline based electrode materials in supercapacitor is given in Table 1 for ready reference.

Munichandraiah et al. synthesized PANI by electrochemical method on stainless steel electrode of 24 cm^2 area. The amount of PANI deposited on the electrode was found to be 2.74 g. Supercapacitor was assembled in a polypropylene container by stacking seven PANI electrodes separated by polypropylene separator with acid electrolyte and among the seven electrodes, three electrodes used as the positive terminal and the remaining four electrodes was used as the negative terminal. The specific capacitance value was calculated by discharge-charge of the supercapacitor at a constant current density of 750 mA. Initial capacitance was reported as 450 F, the capacitance remained same up to 100 cycles and then decreased to 440 F at the end of the 1000 cycles [42].

PANI-HCl and PANI-$LiPF_6$ salts are prepared by chemical oxidative polymerization technique and used as the active material for symmetric redox supercapacitors. The supercapacitor property was studied with two different types electrolytes namely $Et_4NCF_3SO_3$ and Et_4NBF_4. Lower internal resistance and larger discharge specific capacitance were observed in the case of Et_4NBF_4 electrolyte [43]. PANI was electrochemically deposited on high surface area (1778 m^2/g) carbons and used as electrode in supercapacitor cell by Chen et al. [46]. The chemical composition of the PANI-deposited carbon electrode was determined by XPS and it revealed the presence of C, O, S and N. CV and EIS measurements were used to investigate the electrochemical properties of electrodes.

Cyclic voltammogram recorded in the range -0.2 to 0.6 V indicated the electrode stability in acid solution. An equivalent circuit was proposed to successfully fit the EIS data, and the significant contribution of pseudocapacitance from PANI was thus identified.

Table 1.

Ref. No/ Year	System	Ch/ Ech	Sub	Electrolyte	Conf	Mor	Vol.Ran (V)	CV/ CD	SC (F/g)	ED	PD
42/2002	P	Ech	SS		SC dev	---	0-0.75	CD	1300	110	900
43/2002	P-HCl	Ch	Al foil	1M Et_4NBF_4	2 elect	---	1-0.01	CD	107	---	---
44/2003	P-H_2SO_4	Ech	Carbon paper	1M H_2SO_4	2 elect	Fibers	0-0.60	CV	230	4.25	1200
45/2003	P-$LiPF_6$	Ch	Al mesh	1M H_2SO_4	3 elect	Grain	1-0.01	CD	115	---	---
46/2003	P-H_2SO_4	Ech	Carbon	1M H_2SO_4	3 elect	---	0-0.60	CD	180	---	---
47/2004	P-HNO_3	Ech	Carbon	1M $NaNO_3$	2 elect	Fibers	0-0.60	CV	300	18.3	340
48/2005	P-H_2SO_4	Ech	SS	1M H_2SO_4	3 elect	Nano Wires	0-0.70	CD	742	68.0	16000
49/2005	P-H_2SO_4	Ech	SS	1M $NaClO_4$	2 elect	Fibers, Granular	0-0.75	CD	609	26.8	---
50/2006	P-PMo_{12}	Ch	---	---	---	Needle	0-1.0	CD	168	---	---
51/2006	P-PTS	Ech	Ni	0.5M PTS	3 elect	Particles	0-0.70	CD	405	---	---
52/2006	P-$HClO_4$ -$NaClO_4$	Ech	Ni	0.1M $HClO_4$ 3.0M $NaClO_4$	3 elect	Particles	0-0.70	CD	2300	---	---
53/2006	P-PTS	Ech	SS	0.5M PTS	3 elect	Fiber, Tubular	0-0.70	CD	805	---	---
54/2006	P-H_2SO_4	Ech	SS	1M H_2SO_4	3 elect	Nano Wires	0-0.70	CV	775	---	---
55/2007	P-H_2SO_4	Ech	Pt, SS, Carbon	0.1M $HClO_4$ 3.0M $NaClO_4$	3 elect	Fiber	0-0.75	CD	1600		
56/2007	P-H_2SO_4	Ech	Carbon	1M H_2SO_4	2 elect	---	0-0.80	CD	70	---	---
57/2007	P-DMS	Ch	Al foil	1M $LiPF_6$	2 elect	Granular	0-1.00	CV	115	---	---
58/2008	P-H_2SO_4	Ech	Gold foil	1M H_2SO_4	2 elect	Nano Wires	0-0.70	CD	700	2.7	1000
59/2008	P-MO	Ch	Graphite	1M H_2SO_4	2 elect	Fibers	-0.2-0.80	CD	428	---	---
60/2008	P-CA	Ch	SS	1M H_2SO_4	3 elect	Fibers	-0.2-1.0	CV	298	---	---
61/2008	P-H_2SO_4	Ech	AAO/Ti/Si	1M H_2SO_4	3 elect	Nano Wires	0-0.60	CD	1142	---	---

Table 1. (Continued)

Ref. No/ Year	System	Ch/ Ech	Sub	Electrolyte	Conf	Mor	Vol.Ran (V)	CV/ CD	SC (F/g)	ED	PD
62/2009	P	Ech	ITO	0.1M H_2SO_4	3 elect	Nanorods spheres	0-1.0	CV	592	---	---
63/2009	P-H_2SO_4	Ech	ITO	0.5M H_2SO_4	3 elect	Fiber	0-1.0	CV	606	---	---
64/2010	P-$HClO_4$	Ech	Pt, Au, SS	1M $HClO_4$	3 elect	Nano Wires	0-0.70	CD	950	130	700
65/2010	P-H_2SO_4	Ech	Ti	1M H_2SO_4	3 elect	Nanobelts	-0.25-0.75	CV	873	---	---
66/2010	P	Ch	Carbon	1M H_2SO_4	3 elect	Mat	-0.6-0.6	CV	222	---	---
67/2010	P-HCl	Ch	SS	1M H_2SO_4	3 elect	Fibers	0-0.70	CD	548	36.0	127
68/2010	PZn^{+2} -HCl	Ch	Pt	1M H_2SO_4	3 elect	Granules	0-0.70	CD	369	---	---
69/2010	P-SDS	Ch	Graphite	1M H_2SO_4	3 elect	2D-Lamellar	0-0.75	CD	560	---	---
70/2010	P-H_2SO_4	Ch	SS	1M H_2SO_4	3 elect	---	0-0.65	CD	837	---	---
71/2011	P	Ech	ITO	---	---	Fiber	-0.05-0.65	CV	3407	---	---
72/2011	P-HCl	Ch	Graphite	1M H_2SO_4	3 elect	Mat	-0.2-0.80	CD	232	32	200
73/2011	P-SA	Ch	Carbon	1M H_2SO_4	3 elect	Fibers	0-0.80	CD	2093	---	---

Ref. No: Reference number. CD: Charge-Discharge. Ch: Chemical synthesis. SC: Specific capacitance. Ech: Electrochemical synthesis. ED: Energy density (Wh/Kg). Sub: Substrate. PD: Power density (W/Kg). Conf: Configuration. P: Polyaniline. Mor: Morphology. Vol.Ran: Voltage range. CV: Cyclic voltammetry.

Table 2.

Ref. No/ Year	System	Ch/ Ech	Sub	Electrolyte	Conf	Mor	Vol.Ran (V)	CV/ CD	SC (F/g)	ED	PD
74/2002	P-AC	Ch	Ni	6M KOH	2 elect	---	0.8-1.60	CV	380	18	1250
75/2004	P-$LiPF_6$	Ch	Al	$1MEt_4NBF_4$	2 elect	---	0.0-3.0	CD	58	---	---
76/2004	P-ST	Ch	Ni	1M $NaNO_3$	3 elect	---	-0.2-0.80	CD	190	---	---
77/2005	P-MT	Ech	Pt	0.5M H_2SO_4	3 elect	---	0.0-0.80	CD	---	---	---
78/2005	P-MT	Ech	Carbon	1 M HCl–0.5 M LiCl	3 elect	Globular	0.0-0.70	CV	3.5 F/cm^2	---	---
79/2005	P-CNT	Ch	---	1M $NaNO_3$	3 elect	---	0.0-1.0	CD	183	6.34	---
80/2005	P-CNF	Ch	SS	1 M H_2SO_4	3 elect	---	0.0-0.80	CV	264	---	---
81/2006	P-MC	Ch	SS	1 M H_2SO_4	3 elect	Throns	-0.2-0.65	CD	940	---	---
82/2006	P-ST	Ech	SS	1 M H_2SO_4	3 elect	---	0.0-0.70	CD	485	228	2250
83/2006	P-ST	Ech	SS	1 M H_2SO_4	3 elect	---	0.0-0.70	CD	463	---	---
84/2006	P-C	Ch	---	2 M H_2SO_4	3 elect	Sphere	0.0-0.70	CD	437		
85/2007	P-HPCM	Ech	Carbon	1 M H_2SO_4	3 elect	Porous	0.0-0.70	CD	2200	300	470
86/2007	P-MT	Ech	Graphite	1 M H_2SO_4	2 elect	---	-0.2-0.80	CD	322	22	83
87/2007	P-PC	Ch	SS	1 M H_2SO_4	3 elect	---	-0.2-0.80	CV	148	---	---
88/2007	P-MT	Ch	Ti	1 M H_2SO_4	2 elect	Fibers	0.0-0.60	CD	554	---	---
89/2008	P-AC	Ech	SS	0.5 MH_2SO_4	3 elect	Gravel, porous, Fibers	-0.2-0.70	CV	587	---	---
90/2008	P-AC	Ech	Graphite	1 M HCl-0.5 M KCl	3 elect	---	0.0-0.60	CD	200	---	---
91/2008	P-MT	Ch	Ni	30.0 wt % KOH	3 elect	---	0.0-0.50	CD	102	---	---
92/2008	P-MT	Ch	Ni	1M $NaNO_3$	3elect	Porous	-0.2-0.80	CD	224	---	---
93/2009	P-MC	Ch	Pt	10 wt% H_2SO_4	3 elect	Porous	0.0-0.80	CD	96	---	---
94/2009	P-OMC	Ch	Graphite	30 wt % KOH	30 wt % KOH	Fasciculi	0.0-0.90	CD	747	---	---

Table 2. (Continued)

Ref. No/ Year	System	Ch/ Ech	Sub	Electrolyte	Conf	Mor	Vol.Ran (V)	CV/ CD	SC (F/g)	ED	PD
95/2009	P-GO	Ch	SS	1 M H_2SO_4	3 elect	Fibers	0.0-0.80	CD	216	---	---
96/2009	P-G	Ech	Graphene	1 M H_2SO_4	3 elect	Sheets, Particles	-0.2-0.80	CV	233	---	---
97/2009	P-GR	Ch	---	1 M KCl	3 elect	Sheets	-0.2-1.00	CD	---	---	---
98/2009	P-DOM	Ch	Ti	1 M $LiClO_4$	3 elect	Macropore	2.5-4.00	CD	111 mAh/g	---	---
99/2009	P-Li	Ch	Al	1 M $LiPF_6$	2 elect	---	0.0-1.0	CD	120	---	---
100/2009	P-MT	Ech		0.5MH_2SO_4	3 elect				500	---	---
101/2009	P-CNT	Ch	Ni	1 M$NaNO_3$	2 elect	Porous	0.0-0.60	CD	>35	---	---
102/2009	P-CNT	Ch	Pt	1 M $LiClO_4$	2 elect	Globular	2.9-4.20	CV	289	---	---
103/2010	P-G	Ch	Graphite	1 M HCl	3 elect	Rod, wire, fiber	0.20-0.70	CV	490	4.86	8750
104/2010	P-NG	Ch	SS	1 M H_2SO_4	3 elect	Sheet	-0.2-0.60	CD	1126	37.9	141
105/2010	P-ST	Ech	Paper	0.5M H_2SO_4	3 elect	Porous	0.0-0.70	CV	501	---	---
106/2010	P-CNT	Ech	Pt	0.1M H_2SO_4	3elect	Porous	0.0-0.70	CV	516	---	---
107/2010	P-ND	Ch	Au	1M H_2SO_4	2 elect	Particles	-0.6-0.60	CV	640	---	---
108/2010	P-GO	Ch	Au	1 M H_2SO_4	2 elect	Sheets, Wires	0.0-0.70	CV	555	---	---
109/2010	P-GO	Ch	SS	1 M H_2SO_4	2 elect	Fiber, Granular	0.0-0.80	CV	746	---	---
110/2010	P-MC	Ch	Carbon	2 M H_2SO_4	3 elect	Fiber	0.0-0.80	CD	1490	49	---
111/2010	P-AC	Ech	Carbon	0.5M H_2SO_4	3 elect	Fiber,	0.0-0.80	CD	958	---	---
112/2010	P-CA	Ch	SS	1 M H_2SO_4	2 elect	---	0.0-0.60	CD	226	---	---
113/2010	P-HCS	Ch	SS	2 M H_2SO_4	2 elect	Spheres	-0.2-0.60	CV	525	17.2	---
114/2010	P-CA	Ch	SS	1 M H_2SO_4	3 elect	Sphere	-0.2-0.80	CD	710	58.2	---
115/2010	P-CNT	Ch	Carbon	1 M H_2SO_4	3 elect	---	-0.6-0.40	CV	305	---	---
116/2010	P-G	Ch	Carbon	2 M H_2SO_4	3 elect	Sheets, Fibers	-0.2-0.80	CD	480	---	---
117/2010	P-GN	Ch	Ni	6M KOH	3 elect	Sheets, Particles	-0.6-0.40	CD	1046	145	522
118/2010	P-G	Ch	Pt	1 M H_2SO_4	2 elect	Sheets, Fibers	0.0-0.80	CD	210	---	---

Ref. No/ Year	System	Ch/ Ech	Sub	Electrolyte	Conf	Mor	Vol.Ran (V)	CV/ CD	SC (F/g)	ED	PD
119/2010	P-NG	Ch	SS	1 M H_2SO_4	3 elect	Sheets	-0.2-0.60	CD	1126	34.8	136
120/2010	P-CNT	Ch	---	0.5M H_2SO_4	3 elect	Tubes, Fibers	0.0-0.80	CD	350	7	2189
121/2011	P-G	Ch	Si	2 M H_2SO_4	3 elect	Flakes	0.4-0.80	CD	300-500	---	---
122/2011	P-CF	Ch	Pt	1 M H_2SO_4	---	Fibers, Granular	0.0-0.80	CD	188	---	---
123/2011	P-AC	Ch	Pt	0.5M H_2SO_4	3 elect	Particle	0.0-1.00	CV	115	---	---
124/2011	PCNF	Ch	Sample	1M H_2SO_4	3 elect	Fiber	0.0-0.80	CD	638	---	---
125/2011	P-CNT	Ch	SS	1M H_2SO_4	3 elect	Tubes	-0.2-0.80	CD	837	---	---
126/2011	P-G	Ech	ITO	1M H_2SO_4	3 elect	Sheets	-0.5-1.20	CD	640	---	---
127/2011	P-G	Ch	SS	1M H_2SO_4	3 elect	Sheets, Rods	-0.2-0.60	CV	931	---	---
128/2011	P-GN	Ch	Carbon	1M H_2SO_4	3 elect	Sheets, Fibers	0.0-0.80	CD	1130	---	---
129/2011	P-G	Ch	ITO	1M H_2SO_4	3 elect	Sheets, Fibers	0.0-0.80	CD	301	---	---
130/2011	P-G	Ch	Pt	1M H_2SO_4	3 elect	Sheets	0.0-0.80	CD	372	---	---
131/2011	P-G	Ch	Carbon	2M H_2SO_4	3 elect	Sheets	-0.2-0.80	CD	526	---	---
132/2011	P-MC	Ch	Pt	1M H_2SO_4	3 elect	Porous	0.0-0.90	CV	470	76.4	112
133/2011	P-MT	Ch	Pt	1M H_2SO_4	2 elect	Rods, Fiber	-0.2-0.80	CV	515	---	---
134/2011	P-CNT	Ch	Pt	1M H_2SO_4	3 elect	Sheets, Layers	-0.2-0.80	CV	450	15.6	1125
135/2011	P-MPC	Ch	Ni	6M KOH	3 elect	Irregular	0.0--0.75	CD	400	---	---
136/2011	P-OMC	Ch	Ni	30%MKOH	3 elect	Porous, Granular	0.0-0.90	CD	409	---	---
137/2011	P-ST	Ch	Wiper cloth	1M H_2SO_4	2 elect	Nano wires	0.0-0.70	CD	410	26.6	7000
138/2011	P-ST	Ch	Carbon	0.5MH_2SO_4	3 elect	---	-0.3-0.90	CV	1000	---	---
139/2012	P	Ch	---	1M H_2SO_4	2 elect	Tubes	---	CV	846	---	---
140/2012	P-AU	Ch	Au	H_2SO_4-PVA gel	2 elect	Porous	0.0-0.80	CD	1500 F/cm^3	0.078Wh/ cm^3	190W/cm 3
141/2012	P-rGO	Ech	rGO	0.5MH_2SO_4	3 elect	Nanorods	0.0-0.90	CD	970	---	---

MT: MWNT. ST: SWNT.

Table 3.

Ref. No/ Year	System	Ch/ Ech	Sub	Electrolyte	Conf	Mor	Vol.Ran (V)	CV/ CD	SC (F/g)	ED	PD
142/2002	P-Pt	Ech	Graphite	0.5M H_2SO_4	3 elect	Grains	-0.2-0.6	CD	300 mF/cm^2	-	-
143/2003	P-POM	Ch-Ech	Carbon	0.5M H_2SO_4	2elect	Porous	0.0-0.50	CD	195 mF/cm^2	24.4 mJ/cm^2	-
144/2007	P-MnO_2	Ch	Ni	0.1MNa_2SO_4	3 elect	Plate	0.0-0.85	CD	330	---	---
145/2007	P-RuO_2	Ch	Pt	0.5M H_2SO_4	3 elect	Globular	-0.2-0.60	CV	475	---	---
146/2008	P-SiO_2	Ech	Carbon	0.5M H_2SO_4	3 elect	Globular	0.15-0.60	CD	558 mF/cm^2	---	---
147/2009	P-TiO_2	Ch	Carbon	0.5M H_2SO_4	3 elect	Fiber	0.05-0.55	CD	330	---	---
148/2009	P-SnO_2	Ch	Graphite	1.0M H_2SO_4	3 elect	Pores	-0.20-0.80	CD	305	42.4	666.6
149/2010	P-TiO_2	Ch	Carbon	1.0M H_2SO_4	3 elect	Granular	-0.20-0.70	CD	495	---	---
150/2010	P-ND -MnO_2	Ch	Carbon	1.0M$NaNO_3$	2 elect	Rod like	0.0-0.65	CD	80	---	---
151/2010	P-MnO_2	Ch	Ni	1.0M H_2SO_4	3 elect	Porous, Particles	0.0-0.65	CD	510	---	---
152/2010	P-SiO_2	Ech	Graphite	0.5M H_2SO_4	3 elect	Fiber	0.0-0.80	CD	470	30.0	220
153/2011	P-DBSA -Fe_3O_4	Ch	SS	1.0M H_2SO_4	3 elect	Particles	0.0-0.75	CD	228	6.33	407
154/2011	P-MnO_2	Ch	---	1.0M H_2SO_4	---	1D Nano	0.0-0.70	CD	626	17.8	600
155/2011	P-WO_3	Ech	Carbon	1.0M H_2SO_4	3 elect	Fibers, Particles	-0.5-0.70	CD	168	33.6	---

Table 4.

Ref. No/ Year	System	Ch/ Ech	Sub	Electrolyte	Conf	Mor	Vol.Ran (V)	CV/ CD	SC (F/g)	ED	PD
156/2010	P-GNS -CNT	Ch	Ni	6M KOH	3 elect	Particles Rod	-0.70-0.30	CD	1035	---	---
157/2011	P-CNT-MnO_2	Ch	Graphite	0.5M Na_2SO_4	3 elect	Flakes	-0.20-0.80	CV	330	---	---
158/2011	P-GN-CNT	Ch	Carbon	1 M/HCl	---	Sheets, Fibers	-0.20-0.80	CV	569	---	---
159/2011	P-AC-MnO_2	Ch	Carbon	1M $LiClO_4$	2 elect	Fiber	0.00-1.20	CD	413	61	172
160/2011	P-GO-CNT	Ch	Pt	1 M H_2SO_4	3 elect	Sheets, Particles	-0.20-0.60	CD	464	---	---
161/2012	P-GO-Mo_3O_{10}	Ch	SS	1 M H_2SO_4	3 elect	Sheets, Particles	-0.40-0.60	CD	553	76.8	276.3

A comparative analysis of the electrochemical properties of bare-carbon electrodes was also conducted under similar conditions. The performance of the capacitors equipped with the resulting electrodes in 1 M H_2SO_4 (sandwich type) was evaluated by constant current charge-discharge cycling within a potential range from 0 to 0.6 V. The PANI-deposited electrode exhibited high specific capacitance of 180 F/g, in comparison with a value of 92 F/g for the bare-carbon electrode; after completion of the 1000 cycles the capacitance value decreased to 90 and 163 F/g respectively.

Hu et al. had deposited polyaniline on polyacrylonitrile (PAN)-activated carbon fabrics (ACFs) via electrochemical polymerization. The surface morphology of ACFs-PANI analyzed by scanning electron microscope showed that activated carbon was present in the nano fiber form and PANI uniformly coated on the activated carbon nano fibers. The capacitance value calculated from sandwich type supercapacitor device using 1M $NaNO_3$ was around 300 F/g [47].

Long nanofibers form of PANI had been deposited electrochemically on stainless steel substrate by Gupta et al. The capacitance value of PANI nano fibers were calculated from the CD studies, which showed 742 F/g. The cycle life of PANI nanofibers capacitor cell was performed at 3 mA/cm^2 for 1500 cycles. The decrease in capacitance value for the first 500 cycles was almost 7% and subsequent 1000 cycles, the decrease in capacitance value was around 1% [48].

Zhou et al. had deposited polyaniline on stainless steel substrate by pulse galvanostatic and galvanostatic methods from the aqueous solution containing 0.5 M H_2SO_4 and 0.2 M aniline [49]. The resultant PANI materials synthesized by pulse galvanostatic and galvanostatic methods showed nanofibrous and granular morphology respectively. The capacitive performance for the two samples was carried out by CD study at a constant current density of 1.5 mA/cm^2. Among the two samples, PANI nanofibers form showed higher values of capacitance of 609 F/g and energy density of 26.8 Wh/Kg.

Girija and Sangaranarayanan had reported the synthesis of polyaniline on Ni foil by potentiodynamic oxidation of aniline in the presence of p-toluene sulfonic acid and Triton X-100 , in the voltage range of -0.2 to 1.2 V. The capacitance of polyaniline coated Ni foil observed was 405 F/g. Interestingly, the same group subsequently synthesized polyaniline on Ni foil using Triton X-100 instead of using p-toluene sulfonic acid and observed very high capacitance of 2300 F/g. The increasing in supercapacitor was attributed to the alteration of molecular and supramolecular structure, conductivity, and stability of the polymers by Triton X-100 [51, 52]. The same group had deposited polyaniline on SS foil potentiodynamically on a using p-toluene sulfonic acid. The resultant PANI micrograph showed mixed fibers and tubular morphology. The capacitance value was calculated from CD study and obtained as 805 F/g for the first cycle, which decreased to783.0 F/g at the end of 1000 cycles. They had demonstrated that PANI coated on SS is protecting it from the corrosion when the electrode is in contact with the electrolyte [53].

PANI in nanowires has been synthesized by electrochemical on SS by the oxidation of 0.05 M aniline in 1M H_2SO_4 by Gupta and Miura. The deposited PANI nanowires on SS is used for the calculation of capacitance using cyclic voltammetry, which showed 775 F/g at the sweep rate of 10 mV/sec, the capacitance value decreased to 562 F/g at scan rate of 200 mV/sec. The life time of PANI nanowires was studied by carrying out 1500 cycles at a sweep rate of 100 mV/sec. Capacitance value was constant for the first 100 cycles and after that 8 % decrease in the next 500 cycles and 1% in the remaining cycles [54].

Polyaniline has been deposited electrochemically on platinum, SS, and carbon substrates using 0.2 M aniline and 2M H_2SO_4, with applying voltage -0.2 to 1.2 V and sweep rate 100 mV/sec by Munichandraiah et al. The capacitance of polyaniline coated Pt, SS and C was calculated from CD study, which showed 74, 240 and 1100 F/g respectively. PANI/C electrode was further subjected to cycle life test. The initial capacitance value of around 1200 F/g changed to 800 F/g in about 300 cycles, and thereafter the capacitance value was almost constant up to 1000 cycles [55].

PANI nanowires have been synthesized by Yanyan and Thomas by the potentiostatic method. Preparation of nanowires with tubular ends and then to open nanowires with increase in concentration of H_2SO_4 have been demonstrated. Maximum capacitance was found to be 710 F/g at a current density of 5 A/g [58].

Polyaniline nanowires of 30 nm size have been synthesized electrochemically using template (anodic aluminium oxide) method by Zhao et al. CD studies of the PANI nanowires cell carried out at different current densities of 1, 2.5, 5.0 A/g showed capacitance of 988, 1100, 1142 F/g respectively. Further, cycle life test was carried out at 2.5 A/g current density for 500 cycles and showed 5 % capacitance loss [61].

Montilla et al. have carried out electrochemical deposition of PANI on ITO and ITO coated silica substrate. The synthesized PANI showed nano fibers of 200-300 nm size in both cases. The value of specific capacitance of the silica-PANI composites [63] is higher than that of the equivalent ITO/PANI electrodes with the maximum capacitance (607 F/g).

Amarnath et al. deposited PANI on ITO substrate by chemical polymerization method using 3-(triethoxysilyl)-propyl isocyanate as self assembled mono layer. PANI was obtained in nanorods and nano spheres form. Capacitance value of PANI nanorods (592 F/g) was found to be higher than that of PANI nano spheres (214 F/g) [62].

Wang. et al. have prepared PANI nanowires on different substrates such as Au, Pt, SS and Graphite by galvanostatic deposition process [64]. The capacitance value was calculated from galvanostatic CD study at two current density of 1 and 40 A/g and it is reported as 950 and 780 F/g respectively.

Li et al. have synthesized PANI in the form of nano belts with diameter of 50 nm and length of 20 μm by galvanostastically using a current density 1.0 mA/cm^2 for 12h at 70 ^0C. From FE-SEM study for PANI present in the nanobelt form, the diameter of the PANI is around 50 nm and length of the PANI nanobelts is 20 μm. Capacitance value from CD measurements was calculated at scan rates of 10 and 200 mV/sec and it was found to be 873 and 622 F/g respectively [65]. Interestingly, the capacitance value was reported as almost constant for 1000 cycles.

Porous and non porous form of polyaniline has been synthesized by soft template method using ammonium persulphate oxidant. The surface area of porous PANI (211 m^2/g) was found to be higher than that of non porous PANI (6 m^2/g). Correspondingly, higher capacitance was obtained from CD carried out in the voltage range 0 to 0.7 V for the porous PANI (837 F/g) in comparison with that of non porous PANI (519 F/g) [70].

Polyaniline-poly(styrene sulfonate) (PANI-PSS) in the form of hydrogels and colloids have been chemically synthesized via supramolecular self assembly between positively-charged PANI chains and negatively-charged PSS chains by Dai and Jia. Capacitance of PANI-PSS hydrogel and PANI-PSS colloid have been calculated from CD at a current density of 0.2 A/g and are reported as 232 and 212 F/g, respectively. PANI-PSS hydrogel showed specific energy density of 32 Wh/kg and specific power density of 200 W/kg [72].

Li et al. recently synthesized polyaniline-sodium alginate (PANI-SA) by chemical polymerization method, showed mat-like network with a diameter of 50-100 nm. The capacitance value of PANI-SA from the galvanostatic charge-discharge study at a current density of 1 A/g was reported as 2093 F/g [73]. The cycle life was performed for 1000 cycles and the retention of capacitance value was 74 %.

4.2. Polyaniline-Carbon

Literature work on polyaniline-carbon based electrode materials in supercapacitor is given in Table 2 for ready reference.

PANI-carbon composite was prepared by *in situ* chemical polymerization method by Park and Park [74]. Hybrid electrochemical capacitor tested with PANI-D as positive electrode and activated carbon as negative electrode. The specific capacitance, energy and power densities of the cell showed 380.0 F/g, 18 Wh/kg and 1200 W/kg respectively. Initially capacitance value reduced itself to 20 % and then remained constant up to 4000 cycles.

PANI-$LiPF_6$ was prepared by the chemical polymerization method. The supercapacitor was constructed in a symmetric type using PANI-$LiPF_6$ (redox type) and in an asymmetric type with PANI-$LiPF_6$-activated carbon (hybrid type) using 1M Et_4NBF_4. The capacitance value was found to be higher in the case of the hybrid supercapacitor system [75].

Polyaniline-single walled carbon nanotubes (PANI-SWNTs) has been prepared by chemical polymerization method. The capacitance properties of PANI-SWNTs were compared with PANI and SWNTs. PANI-SWNTs (190.6 F/g) system showed higher capacitance compared to that of PANI (169.6 F/g) and SWNTs (40 F/g). Also, PANI-SWNTs (0.386 ms) showed lower time constant than that of PANI (0.135 ms), indicating the fast charge-discharge property by PANI-SWNTs [76].

Polyaniline-Multi walled carbon nanotubes (PANI-MWNTs) had been prepared by electrochemical polymerization method. Higher capacitance value was reported for PANI-MWNTs (3.5 F/cm^2) and PANI (2.3 F/cm^2) [78].

Polyaniline-Carbon nanofibers (PANI-CNFs) with conductivity of 30 S/cm had been prepared by vapour deposition polymerization technique. Increase in capacitance value was observed with thickness and further increase in thickness resulted in decrease in capacitance value [80].

Polyaniline had been deposited on single walled carbon nanotubes (PANI-SWNTs) by potentiodynamically using aniline by Gupta and Miura. The supercapacitor cell showed capacitance - 485.0 F/g, energy density - 228.0 Wh/kg and power density - 2250 W/kg [82]. The same group had also [83] prepared polyaniline-single walled carbon nano tubes (PANI-SWNTs) composite by potentiostatic deposition. The dependence of capacitance on the microstructure and the amount of PANI deposited on the SWNTs has been demonstrated. The maximum capacitance value obtained when 73 wt % of PANI deposited on the SWNTs was around 463 F/g at current density of 10 mA/cm^2 and the capacitance value decreased 5 % for first 500 cycles and just 1 % during the next 1000 cycles.

High porous carbon monolith (HPCM) had been prepared by nano casting technique using silica monolith as hard template and mesophase pitch was used as the carbon precursor for HPCM. Polyaniline have been deposited on HPCM electrochemically and used in supercapacitor study. Very high capacitance was reported for this system 2200 F/g with energy and power density 0.47 kW/kg and 300 Wh/kg respectively [85].

Polyaniline-multiwalled carbon nanotubes (PANI-MWNTs) was synthesized by microwave-assisted polymerization method, using MWNTs of 20-50 nm size. After deposition of polyaniline on MWNTs, the diameter of the composite increased to 150-180 nm indicated that PANI acts as a shell and MWNTs as a core. The capacitance value of this electrode system calculated from CD was around 322 F/g with energy density 22 Wh/kg, which was about 12 times higher than that of MWNTs [86].

Sivakumar et al. [88]. had reported the synthesis of PANI by interfacial method using $HClO_4$ and compared this with that of its composite of MWNTs. Morphological analysis was carried out of PANI, MWNTS and they were found to be 90, 40 and 85 nm respectively. PANI-MWNTs showed better capacitance and cycle life compared to PANI.

PANI-MWNTs have been synthesized by chemical oxidative polymerization method with increasing amount of MWNTs. The effect of the amount of MWNTs on capacitance of supercapacitor has been studied by Kong et al. The capacitance increased with increase in the amount of MWNTs i.e., the capacitance of PANI-MWNTs with 0.2, 0.4, 0.8 wt % of MWNTs showed 200.5, 205.0 and 224.0 F/g respectively [92].

Polyaniline loaded ordered mesoporous carbon (PANI-OMC) composites were prepared by *in situ* chemical oxidative polymerization method using OMC and physical mixing of PANI and OMC. The electrochemical performances of the prepared electrodes were characterized by CD, CV and EIS measurements [94]. PANI-OMC showed specific capacitance of 747 F/g at a current density of 0.1 A/g from the charge-discharge study.

Wang et al. have demonstrated the advantage of using GO along with PANI by comparing the capacitance value of chemically synthesized PANI and its composite with GO. Higher capacitance was observed for PANI-GO (531 F/g) compared to that of PANI (216 F/g) [95].

Wang et al. have used graphene film electrode and deposited polyaniline by anodic polymerization and compared the capacitance value of graphene and PANI-graphene. Higher capacitance was observed for graphene (233 F/g) compared to that of PANI-graphene (147 F/g) [96].

PANI-MWNTs and polyaniline-poly(2,5-dimercapto-1,3,4-thiadiazole)-multi-walled carbon nanotubes (PANI-PDMcT-MWNTs) have been prepared by chemical oxidative polymerization using APS as the oxidizing agent. The electrical conductivity of the PANI-MWNTs and PANI-PDMcT-MWNTs was found to be 49.0 and 96.8 S/cm respectively. The introduction of poly(2, 5-dimercapto-1, 3, 4-thiadiazole) showed in PANI-MWNTs (154.4 F/g) showed higher capacitance (289.4 F/g) [102].

An asymmetric supercapacitor was made using PANI nanofibers as anode and graphene as cathode electrode material. A PANI nanofiber was synthesized by chemical oxidative polymerization using pottasium persulfate as the oxidizing agent [103]. The cyclic voltammogram has been recorded in the voltage window of 0.2-0.7 V, showing specific capacitance of 490 F/g. A symmetric supercapacitor shows energy and power density of 4.86 Wh/Kg and 8.75 kW/kg respectively.

A nanostructured PANI-G has been prepared by in-situ chemical polymerization of aniline with G. This electrode showed reasonably good capacitance (1126.0 F/g) with retention life of 84 % after 1000 cycles [104].

Flexible PANI-single-walled carbon nanotube (PANI-SWNT) composite films have been synthesized through an *in situ* electrochemical polymerization of aniline in 1M H_2SO_4 solution on SWNT bucky paper attached on Ni thin film working electrode This electrode material showed initial capacitance of 501.8 F/g and this electrode was subjected to 90 polymerization cycles and showed higher capacitance value of 706.7 F/g [105].

Xu et al. [108]. have also prepared PANI-GO by *in situ* dilute chemical polymerization method using APS oxidant in 1M $HClO_4$ solution. Morphology analysis showed that PANI nanowire arrays are aligned vertically on GO substrate. The capacitance value of the PANI-GO electrode was calculated at two different current densities (0.2 and 2 A/g), which showed capacitance value of 555.0 and 227.0 F/g respectively. It showed retention capacitance of 92 % after completion of 2000 consecutive cycle.

The effect of mesh sizes of GO in PANI-GO have been brought out by Wang et al. *In situ* chemical polymerization of aniline was carried out with two different sizes of GO (12,500 and 500 mesh). PANI-GO composite prepared with 12,500 mesh size showed higher capacitance value 746 F/g compared with that of 500 mesh (627 F/g) [109].

Polyaniline was deposited on three-dimensional ordered mesoporous carbon (3DOM) by the electrochemical polymerization of aniline. The electrochemical properties supercapacitors were analyzed using three electrode cell configuration in 2M H_2SO_4 electrolyte. The electrochemical results showed high capacitance value (1490 F/g) [110].

Carbon aerogel (CA) was prepared by the micro emulsion templated sol gel polymerization method. PANI-CA was prepared by *in situ* chemical oxidative polymerization method using aniline, APS, HCl and CA. The capacitance calculated at a current density of the 50 mA/cm^2 from charge-discharge study showed that PANI-CA exhibited higher capacitance (226 F/g) than that of PANI (89.0 F/g) [112].

Hollow carbon spheres (HCS) were prepared by chemical vapor deposition with ferrocene as the carbon precursor and colloidal silica spheres as the template, with specific surface area 2239 m^2/g. PANI-HCS composites were prepared by in situ chemical oxidative polymerization method by Lei et al. [113]. Higher capacitance value was obtained for PANI-HCS composite (525 F/g) than its HCS (268 F/g).

Carbon aerogel (CA) was prepared from pyrolysis of a resorcinol formaldehyde gel. The PANI coated CA was prepared by adsorption of aniline on CA followed by chemical oxidative polymerization method. Higher capacitance was reported for PANI-CA (710.7 F/g) than its carbon aerogel electrode (143.8 F/g) [114].

Conductivity of 143 S/cm with capacitance of 480 F/g has been obtained for PANI-G composite by Zhang et al. [116]. PANI-G was synthesized by chemical method. Graphene oxide was synthesized from graphite by modified hummors method. Polyaniline-graphene oxide (PANI-GO) nanofiber composites were prepared by in-situ polymerization in the presence of graphene oxide under acidic conditions, later modified to polyaniline-graphene (PANI-G) using hydrazine followed by reoxidation and reprotanation of the reduced PANI to give the graphene/PANI nanocomposites.

Graphene nanosheets (GNS) were prepared by reduction of graphite oxide with hydrazine hydrate. The PANI-GNS composite was synthesized using in situ polymerization method [117]. The capacitance of supercapacitor consisting of GNS, PANI and PANI-GNS was calculated. The result showed that high capacitance value was observed for PANI-GNS (1046 F/g) than its pure PANI material (115 F/g). PANI-GNS showed energy density of 39 Wh/kg and power density of 7 KW/kg.

PANI-G, wherein, PANI nanofibers are sandwiched between the graphene layers, has been prepared by Wu et al. [118]. by interfacial polymerization method. Higher conductivity value was obtained for PANI-G (550 S/cm) than that of the pure PANI nanofibers (50 S/cm). However, low capacitance value was observed for PANI-G (210.0 F/g at current density of 0.3 A/g).

4.3. Polyaniline-Metal Oxides

Literature work on polyaniline-metal oxide based electrode materials in supercapacitor is given in Table 3 for ready reference.

The capacitive behavior of electrochemically deposited PANI deposited on Pt showed very low capacitance of 300 mF/cm^2 [142].

Organic and inorganic composite, phospomolybidic acid (POM) and PANI (PANI-POM) had been prepared by electrochemical and chemical-electrochemical polymerization methods. PANI-POM prepared by electrochemical polymerization showed itself to be more microporous than PANI-POM prepared by chemical-elctrochemical polymerization method. This hybrid composite shows a specific capacitance value of 195 mF/cm^2 and energy density of the 24.4 mJ/cm^2 at a current density of the 0.125 mA/cm^2. The capacitance value decreased by 10-15 % at the end of 1000 cycles [143].

A novel synthesis of polyaniline-intercalated layered manganese oxide nano composite through ion-exchange reaction between *n*-octadecyltrimethylammonium-intercalated precursors and polyaniline in an organic solution by Xiong et al. PANI-MnO_2 showed nano form and used as electrode in supercapacitor system. The capacitive behaviour of PANI-MnO_2 was compared with PANI and MnO_2. The specific capacitance of PANI-MnO_2, PANI, and MnO2 were observed as 330, 187 and 208 F/g respectively from the galvanostatic charge-discharge study at a current density of 1 A/g. The cycle life of PANI-MnO_2, PANI and MnO_2 analysed using 1000 cycles of operation, the electrodes showed retention of capacitance of 94, 90 and 75% respectively [144].

Song et al. [145]. have prepared PANI/Nafion/RuO_2 composite in globular form by chemical polymerization method followed by electrochemical modification process. The specific capacitance value calculated from cyclic voltammetry using three electrode cell configuration showed an initial specific capacitance value of 325 F/g and after 10000 cycles it showed 260 F/g (80% capacitance retention). Globular 3D nano structure of hybrid film of PANI-SiO_2 was prepared by electrochemical polymerization method using aniline and silica nanoparticles. The capacitor consisting of the hybrid film showed capacitance value of 558 mF/cm^2 in 0.5 M H_2SO_4 electrolyte using current density 1.5 mA/cm^2 [146].

Fiber form of polyaniline containing 20 wt % of nano TiO_2 (PANI-TiO_2) have been prepared by *in situ* chemical oxidative polymerization method. The electrical conductivity of PANI-TiO_2 is 2.45 S/cm. This PANI-TiO_2 composite showed a capacitance value of 330 F/g at 1.5 A/g current density in the voltage range of 0.05-0.55V. PANI-TiO_2 and after 10000 cycles, the specific capacitance value decreased to 305 F/g [147].

The PANI-SnO_2 nano composite have been synthesized by chemical oxidative polymerization method, where the diameter of SnO2 nano particles was 20-60 nm size. FE-SEM shows that micropores morphology under high magnification, SnO_2 nanoparticles are embedded within the netlike structure built by PANI chains. PANI-SnO_2 showed a specific capacitance of 305.3 F/g, specific energy density of 42.4 Wh/kg with a coulombic efficiency of 96%. Furthermore, there was 4.5% decay in the available capacity over 500 cycles [148].

Li et al. have prepared PANI-TiO_2 composite by chemical oxidative polymerization method using aniline and $Ti(SO_4)_2$ with ammonium persulfate as reaction initiator, the resulting product was immersed in ammonia solution and the product was filtered. Lastly it was treated at 180 ^{0}C under inert atmosphere. The specific capacitance of PANI-TiO_2 was calculated from the charge-discharge study at a constant current density of 2.5 A/g, which showed a capacitance of 495 F/g and after 3000 cycles, the capacitance decreased to 245.0 F/g (the retention in capacitance value is 70%) [149].

PANI-MnO_2 composite film was obtained via controlled electro co-polymerization of aniline and N-substituted aniline grafted on surfaces of MnO_2 nanoparticles on a carbon cloth. Nano composite consists of PANI nanorods and MnO_2 nanoparticles. In the PANI-MnO_2 composite, it was observed that MnO_2 nano particles are present in the α and γ phase. Supercapacitor cell consists of PANI-ND-MnO_2 composite film showed capacitance value of 80 F/g and after 1000 cycles the capacitance value retained was 85 % with 98 % coulombic efficiency [150].

MnO_2 nano particles are prepared by using and $KMnO_4$. PANI-MnO_2 nano composite have been synthesized by *in situ* polymerization reaction of aniline with nano MnO_2 in the presence of non-ionic surfactant (Triton X-100). Morphology study of PANI-MnO_2 showed a uniform porous structure, wherein, MnO_2 (5-10 nm) nano particles were coated on PANI. The results showed that the materials had specific capacitance as large as 510 F/g in the potential range from 0.0 to 0.65V at a charge–discharge current density of 1.0A/g [151].

Polyaniline-silica (PANI-SiO_2) composite have been prepared by electrochemical polymerization method using ultra small size silica. PANI-SiO_2 was obtained in fibre form of 600 nm size; however, PANI showed fibre form with 150 nm size. Electrochemical studies of PANI-SiO_2 showed high power (220 Kw/kg) and energy (30 Wh/Kg) densities [152].

Recently, PANI-Fe_3O_4 composite with 3.8 wt % of Fe_3O_4 have been prepared by in-situ chemical polymerization method. An supercapacitor consisting of PANI-Fe_3O_4 composite showed the values of capacitance, energy and power densities as 180 F/g, 6.33 Wh/kg and 0.407 Kw/kg [153]. Jaidev et al. [154]. have prepared a novel binary nanocomposite based on polyaniline and α- MnO_2 nanotubes. PANI-MnO_2 binary nano composite showed a capacitance of 626.0 F/g, corresponding energy density of 17.8 Wh/Kg and power density of 600 W/Kg. Cycle life of the capacitance was also reported and the capacitance value decreased rapidly for first 600 cycles, and got stabilized after 400 cycles. PANI-MnO_2 binary nanocomposite showed equivalent series resistance (ESR) of 0.65Ω.

Polyaniline-tungsten oxide (PANI-WO_3) has been electro deposited by cyclic voltammetry from a mixed solution of aniline and tungsten oxide. PANI-WO_3 showed good electrochemical performance over wide potential range of -0.5 to 0.7 V, due to electro activities of WO_3 in negative potential range. PANI-WO_3 showed capacitance value around 168 F/g with energy density of 33.6 Wh/Kg [155].

4.4. Polyaniline-Ternary Composite

Literature work on polyaniline-ternary composite based electrode materials in supercapacitor is given in Table 4 for ready reference.

Graphene nanosheet/carbon nanotube/polyaniline composite (GNS/CNT/PANI) has been synthesized by in-situ chemical polymerization pathway and a comparison of GNS/CNT/PANI with GNS/PANI, CNT/PANI and PANI electrode materials in supercapacitor application has been made by Yan et al. Capacitance of GNS/CNT/PANI calculated from cyclic voltammetry (1035 F/g) carried out at 1mV/s scan rate in 6M of KOH is a little lower than that of GNS/PANI composite (1046 F/g), but much higher than pure PANI (115 F/g) and CNT/PANI composite (780 F/g). Though a small amount of CNTs (1 wt. %) is added into GNS, the cycle stability of GNS/CNT/PANI composite is greatly improved due to the maintenance of highly conductive path as well as mechanical strength of the electrode during doping/dedoping processes. After 1000 cycles, the capacitance decreases only 6% of the initial capacitance compared to 52% and 67% for GNS/PANI and CNT/PANI composites. Therefore, the intriguing GNS/CNT/PANI composite is quite a suitable and promising electrode material for supercapacitors. Impedance of the GNS/CNT/PANI composite after the 1st and 1000th cycle was measured in the frequency range of 100 kHz to 0.1 Hz. After 1000th cycle, the internal resistance increases from 1.48 to 1.66Ω and the diffusion limitation is also enhanced, which is probably attributable to the composite crack during charge/discharge and then part of the PANI deposits loses contact with each other due to the swelling and shrinkage [156].

MWCNT/PANI/MnO_2 ternary composite can be synthesized chemically in two steps. The structure of MWCNT/PANI/MnO_2 is analyzed by using XRD, XPS, TEM and FE-SEM. The SC calculated from cyclic voltammetry of MWCNT/PANI/MnO_2 is analyzed using three electrode cell configuration in a potential window from -0.2 to 0.8V in 0.5 M Na_2SO_4 solutions. The maximum SC showing by MWCNT/PANI/MnO_2 is 330 F/g [157].

GN/PANI/CNT is prepared by reduction of GO/PANI/CNT followed by re-oxidation of GN/PANI/CNT ternary composite due to restoring conducting properties of polyaniline. Form the SEM study, authors have shown that PANI/CNT nano composite is uniformly sandwiched in between graphene nanosheets. The ternary composite is shows a specific capacitance value of 569 and 341 F/g at a current density of 0.1A/g and 10 A/g [158].

PANI/MnO_2/AC composite have been prepared on activated carbon coated on stainless steel by electrochemical polymerization pathway using 0.5M H_2SO_4 solution which contains 0.4M aniline and 0.1M $MnSO_4$. Asymmetric supercapacitor cell were carried out with the use of PANI/MnO_2/AC as anode and AC electrode as cathode and PVDF-HFP as separator and 1 M $LiClO_4$ in acetonitrile as electrolyte. Energy and power density were found to be 61 Wh/Kg and 172W/Kg respectively [159].

Recently, GO-CNT-PANI has been prepared by *in situ* chemical polymerization method by the oxidation of aniline in presence of GO and carbon nanotubes. In the second step, RGO-CNT-PANI was prepared by reducing GO-CNT-PANI using hydrazine hydrate as reducing agent. The resulted RGO-CNT-PANI ternary composite showed higher capacitance value (464 F/g) compared to that of its RGO-PANI binary composite (453 F/g) [160].

Reduced graphene oxide-molybdenum oxide-polyaniline ternary composite (RGO-MoO_3-PANI) was synthesized by chemical method, initially graphene oxide (GO) reduced by Mo_3O_{10}, the resultant product is used in the synthesis of RGO-MoO_3-PANI ternary composite. RGO-MoO_3-PANI ternary composite showed maximum specific capacitance of 553.0 F/g in 1M H_2SO_4 and 363.0 F/g in 1M Na_2SO_4 electrolytes. RGO-MoO_3-PANI ternary composite showed energy density of 76.8 Wh/kg at a power density of 276.3 W/Kg in 1M H_2SO_4 [161].

5. Applications of Supercapacitors

1) Heavy and public transport: Some of the earliest uses were motor startup capacitors for large engines in tanks and submarines, and as the cost has fallen they have started to appear on diesel trucks and railroad locomotives. In the 2000s they attracted attention in the electric car industry, where their ability to charge much faster than batteries makes them particularly suitable for regenerative braking applications. New technology in development could potentially make EDLCs with high enough energy density to be an attractive replacement for batteries in all-electric cars and plug-in hybrids, as EDLCs charge quickly and are stable with respect to temperature. Some countries like China, Germany and France started running of buses with supercapacitors.
2) Solar energy: smaller applications like home solar energy systems where extremely fast charging is a valuable feature.
3) Home applications: Home appliances such as TVs, microwave ovens, dishwashers, and refrigerators; data routers, personal computers, energy management controls, thermostats, point of sale terminals, process controllers, mobile phones and radio tuners.
4) Other applications where supercapacitors are used: Electric and water utility meters, vehicle tracking systems, starters, ignitors, actuators, disc drives, coin metering devices, security systems, toys, hand tools or flashlights, radars and torpedoes in the military domain, elevators, cranes or pallet trucks in the electric transport domain, memory supplies in phones or computers.

Key Features of Supercapacitors

1) Safer Than Batteries: These capacitors will not explode or be damaged if short circuited. They can be installed with wave soldering equipment or surface mount package with IR or vapor phase reflow equipment.
2) Cycle life: long cycle life of more than 500,000 cycles.
3) Calendar life: Long calendar life 10 to 20 years
4) Charging method: Simple charging methods. No special charging or voltage detection circuits required. Very fast charge and discharge. Can be charged and discharged in seconds. Much faster than batteries.
5) Charge/discharge efficiency: 0.90-0.95
6) Energy density: 1-10 Wh/kg
7) Power density: 1000-2000 W/kg [limited by IR drop or equivalent series resistance (esr)]
8) Voltage: very high cell voltages possible
9) Cost: cheap materials compare to batteries

Construction: simple principle and mode of construction can employ battery construction technology.

Results of Observation from the Reported Work

- Excellent market potential is available for supercapacitors
- Among the supercapacitor based systems of polyaniline, the work on polyaniline with carbon derivatives are concentrated more.
- Electrochemically polymerized polyaniline system give higher capacitance compared to that of chemically synthesized one. However, larger quantities of polyaniline can not be synthesized by electrochemical polymerization method.
- Capacitance calculated from single electrode are higher than that of supercapacitor cell configuration
- Supercapacitor studies are carried using various parameters i.e., scan rates, charge-discharge voltage range, charge-discharge rate, concentration of electrolyte solution, nature of substrates, electrodes and electrolytes, cycle life. Hence it is difficult to compare the results.
- Capacitance value of supercapacitor in cell configuration gives better idea about performance. Polyaniline systems synthesized by chemical polymerization method generally showed capacitance value less than 300 F/g and in some cases it reached to 700 F/g
- Device fabrication studies are very few

CONCLUSION

Carbon, metal oxide, conducting polymer and their combination are being used in supercapacitor. Carbon materials only possess double layer capacitance (non-Faradaic), while metal oxide or conducting polymer possess Faradaic capacitance, and the Faradaic pseudocapacitance is almost 10–100 times higher than that of double layer capacitance. In the case of metal oxides, extensive fundamental and development work on the ruthenium oxide type of electrochemical capacitor have been carried out because of its high specific capacitance of 700 F/g.

However, due to the high cost and toxic nature, its use in practical applications is limited. The other types of electrochemically prepared metal oxides have shown good specific capacitance of 400–500 F/g, but the deposited mass is generally very low, giving low specific capacitance per unit area.

Among the conducting polymers, polyaniline has emerged as the one of the most promising class of active materials for electrochemical capacitor, due to its easy processing, high capacitive characteristics (capacitance, power density, energy density), low cost and environmental friendliness. However, the development of supercapacitor technologies for their energy management is still inferior.

Carbon based supercapacitor devices are commercially available and however, conducting polymer based devices are still under development stage. It is necessary to improve the performance, such as energy density, self life, cycle life and device fabrication with performance evaluation, for commercial exploitation.

ACKNOWLEDGMENTS

The authors thank CSIR, New Delhi under the TAPSUN programme (NWP-0056) for funding. Dr. Vijayamohanan K Pillai, Director and Dr. S. Gopukumar, Scientist, Central Electrochemical Research Institute, Karaikudi for their helpful inputs are greatly acknowledged. Authors also thank Dr. K.V.S.N. Raju, Head, PFM, IICT, Hyderabad for his encouragement. One of the author SBS is thankful to UGC for providing research fellowship.

REFERENCES

[1] Simon, P.; Gogotsi, Y. *Nat. Mater*. 2008, *7*, 845–854.

[2] Zhang, L. L.; Zhao, X. S. *Chem. Soc. Rev*. 2009, *38*, 2520–2531.

[3] Niwa, S.; Taketani, Y., *J. Power Sources*. 1996, *60*, 165-171.

[4] Boos, D.L. "*Electrolytic capacitor having carbon paste electrodes*", U.S. Patent 3536963, 27 Oct 1970.

[5] Alessandrini, F.; G. B. Appetecchi, G. B.; Conte, M. ; Passerini, S. *ECS Trans*. 2006, *1*, 67-71.

[6] Laforgue, *A*.; Simon, P.; Fauvarque, J. F.; Mastragostino, M.; Soavi, F.; Sarrau, J. F.; Lailler, P.; Conte, M.; Rossi, E.; Saguatti, S. *J. Electrochem. Soc*. 2003, *150*, A645-A651.

[7] Supercapacitors: World Markets, Technologies and Opportunities: 1999–2004; Paumanok Publications, Inc.: USA, 2000, p 7.

[8] McNeill, R.; Weiss, D. E.; Wardlaw, J. H.; Siudak, R. *Aust. J. Chem*. 1963, *16*, 1056-1075.

[9] Bolto, B. A.; Weiss, D. E. *Aust. J. Chem*. 1963, *16*, 1076-1089.

[10] Bolto, B. A.; McNeill, R.; Weiss, D. E. *Aust. J. Chem*. 1963, 16, 1090-1103.

[11] Mastragostino, M.; Soavi, F.; Arbizzani, C. *Advances in Li-ion Batteries*. 2002, 69-80.

[12] Reimerink, M. Advanced Capacitor World Summit: Building the Technology Applications and New Business Opportunities for High Performance Electrochemical Capacitors (ECs), Washington, DC, 2003.

[13] Otsuka, K.; Segal, C. L. Advanced Capacitor World Summit: Building the Technology Applications and New Business Opportunities for High Performance Electrochemical Capacitors (ECs), Washington, DC, 2003.

[14] Von Helmhlotz, H. *Ann. Phys*. 1853, *89*, 211-233.

[15] Chapman. D. L. *Phil. Mag*. 1913, *25*, 475-481.

[16] Stein, O. *Zeit. Elektrochem*. 1924, *30*, 508-516.

[17] Grahame, D. C. *Chem. Rev*. 1947, *41*, 441-501.

[18] Burke, A. *J. Power Sources*, 2000, *91*, 37-50.

[19] Vol'fkovich, Y. M.; Serdyuk, T. M. *Russ. J. Electrochem*. 2002, *38(9)*, 935-958.

[20] Lee, H. Y.; Goodenough, J. B. *J. Solid State Chem*. 1999, *144*, 220–223.

[21] Lee, H. Y.; Manivannan, V.; Goodenough, J. B. *C. R. Acad. Sci., Ser. IIc: Chim*. 1999, *2*, 565–577.

[22] Bard, A. J.; Faulkner, L. R. Electrochemical methods: fundamentals and applications; John Wiley and Sons: New York, 2001.

[23] Shukla, A. K.; Sampath, S.; Vijayamohanan, K. *Curr. Sci.* 2000, *79*, 1656-1661.
[24] Trasattu, S.; Buzzanca. G. *J. Electroanal. Chem. Interfacial Electrochem.* 1971, *29*, 1-2.
[25] Liu, T. C.; Pell, W. G.; Conway, B. E. *Electrochim. Acta* 1997, *42*, 3541-3552.
[26] Mitchell, D.; Rand, D. A. J.; Woods. R. *J. Electroanal. Chem.* 1973, *43*, 9-36.
[27] Srinivasan, V.; Weidner. J. W. *J. Electrochem. Soc.*1997, *144*, L210-L213.
[28] Wu. N. L. *Mater. Chem. Phys.* 2002, *75*, 6-11.
[29] Wu, N. L.; Lan, Y. P.; Han, C. Y.; Wang, S. Y.; Shiue, L. R. *Proceedings-Electrochemical Society* 2002, 2002-7(Electrochemical Capacitor and Hybrid Power Sources), 95-106.
[30] Rudge, A.; Iraistrick, Gottesfeld, S. *J. power sources* 1994, *47*, 89-107.
[31] Ryu, K. S.; Kim, K. M.; Park, Y. J.; Park, N. G.; Kang, M. G.; Chang, S. H. *Solid State Ionics* 2002,*152*, 861-866.
[32] Villers, D.; Jobin, D.; Soucy, C.; Cossement, D.; Chahine, R.; Breau, L.; Belanger, D. *J. Electrochem. Soc.* 2003, *150*, A747-A752.
[33] Hashmi, S. A.; Upadhyaya, H. M. *Solid State Ionics* 2002, *152*, 883-889.
[34] Vol'fkovich, Y. M.; Serdyuk, T. M. *Russ. J. Electrochem.* 2002, *38*, 935-958.
[35] Arbizzani, C.; Mastragostino, M.; Meneghello, L.; Paraventi, R. Adv. Mater. 1996, *8*, 331-334.
[36] Simon, P.; Gogotsi, Y. *Nat. Mater.* 2008, *7*, 845-854.
[37] Zhang, L. L.; Zhao, X. S. *Chem. Soc.* Rev. 2009, *38*, 2520-2531.
[38] Chen, Z.; Augustyn, V.; Wen, J.; Zhang, Y. W.; Shen, M. Q.; Dunn, B.; Lu, Y. F. *Adv. Mater.* 2011, *23*, 791-795.
[39] Liu, J. W.; Essner, J.; Li, J. *Chem. Mater.* 2010, *22*, 5022-5030.
[40] Bal, S. S.; Palaniappan, S.; Srinivas, P. *J. Electrochem. Soc.* 2012, *159*, A6-A13.
[41] Palaniappan, S.; Bal, S. S.; Prasanna, T. L.; Srinivas, P. *J. App. Poly. Sci.* 2011, *120*, 780–788.
[42] Prasad, K.; Munichandraiah, N. *J. Power Sources.* 2002, *112*, 443-551.
[43] Ryu, K. S.; Jeong, S. K.; Joo, J.; Kim, K. M. *J. Phys. Chem. B.* 2002, *111*, 731-739.
[44] Talbi, H.; Just, P.; Dao, L. *J. App. Electrochemistry* 2003, *33*, 465-473.
[45] Ryu, K. S.; Wu, X. L.; Lee, Y. G.; Chang, S. H. *J. App. Polym. Sci.* 2003, *89*, 1300-1304.
[46] Chen, W. C.; Wen, T. C.; Teng, H. S. *Electrochim. Acta.* 2003, *48*, 641-649.
[47] Hu, C. C.; Li, W. Y.; Lin, J. Y. *J. Power Sources.* 2004, *137*, 152-157.
[48] Gupta, V.; Miura, N. *Electrochem. Solid St.* 2005, *8*, A630-A632.
[49] Zhou, H. H.; Chen, H.; Luo, S. L.; Lu, G. W.; Wei, W. Z.; Kuang, Y. F. *J. Solid State Electr.* 2005, *9*, 574-580.
[50] Vaillant, J.; Lira-Cantu, M.; Cuentas-Gallegos, K.; Casan-Pastor, N.; Gomez-Romero, P. *Prog. Solid State Chem.* 2006, *34*, 147-159.
[51] Girija, T. C.; Sangaranarayanan, M. V. *J. Power Sources.* 2006, *156*, 705-711.
[52] Girija, T. C.; Sangaranarayanan, M. V. *J. Power Sources.* 2006, *159*, 1519-1526.
[53] Girija, T. C.; Sangaranarayanan, M. V. *Synth. Met.* 2006, *156*, 244-250.
[54] Gupta, V.; Miura, N. *Mater. Lett.* 2006, *60*, 1466-1469.
[55] Mondal, S. K.; Barai, K.; Munichandraiah, N. *Electrochim. Acta.* 2007, *52*, 3258-3264.
[56] Tamai, H.; Hakoda, M.; Shiono, T.; Yasuda, H. *J. Mater. Sci.* 2007, *42*, 1293-1298.
[57] Ryu, K. S.; Jeong, S. K.; Joo, J.; Kim, K. M. *J. Phys. Chem. B.* 2007, *111*, 731-739.

[58] Cao, Y.; Mallouk, T. E. *Chem. Mater*. 2008, *20*, 5260-5265.
[59] Mi, H.; Zhang, X.; Yang, S.; Ye, X.; Luo, *J. Materials Chemistry and Physics*. 2008, *112*, 127-131.
[60] Subramania, A.; Devi, S. L. *Poly. Adv. Techno*. 2008, *19*, 725-727.
[61] Zhao, G. Y.; Li, H. L. *Micropor. Mesopor. Mater*. 2008, *110*, 590-594.
[62] Amarnath, C. A.; Chang, J. H.; Kim, D.; Mane, R. S.; Han, S. H.; Sohn, D. *Mater. Chem. Phys*. 2009, *113*, 14-17.
[63] Montilla, F.; Cotarelo, M. A.; Morallon, E. J. *Mater. Chem*. 2009, *19*, 305–310.
[64] Wang, K.; Huang, J. Y.; Wei, Z. X. *J. Phys. Chem. C*. 2010, *114*, 8062-8067.
[65] Li, G. R.; Feng, Z. P.; Zhong, J. H.; Wang, Z. L.; Tong, Y. X. *Macromolecules*. 2010, *43*, 2178-2183.
[66] Wilson, G. J.; Looney, M. G.; Pandolfo, A. G. *Synth. Met*. 2010, *160*, 655 -663.
[67] Guan, H.; Fan, L. Z.; Zhang, H. C.; Qu, X. H. *Electrochim. Acta*. 2010, *56*, 964-968.
[68] Li, J.; Cui, M.; Lai, Y. Q.; Zhang, Z. A.; Lu, H.; Fang, J.; Liu, Y. X. *Synthetic Met*. 2010, *160*, 1228-1233.
[69] Shi, L.; Wu, X.; Lu, L.; Yang, X.; Wang, X. *Synth. Me*. 2010, *160*, 989-995.
[70] Liu, J. L.; Zhou, M. Q.; Fan, L. Z.; Li, P.; Qu, X. H. *Electrochim. Acta*. 2010, *55*, 5819-5822.
[71] Peng, C.; Hu, D.; Chen, G. Z. *Chem Commu*. 2011, *47*, 4105-4107.
[72] Dai, T.; Jia, Y. *Polym*. 2011, *52*, 2550-2558.
[73] Li, Y. Z.; Zhao, X.; Xu, Q.; Zhang, Q. H.; Chen, D. J. *Langmuir*. 2011, *27*, 6458-6462.
[74] Park, J. H.; Park, O. O. *J. Power Sources*. 2002, *111*, 185-190.
[75] Ryu, K. S.; Lee, Y. G.; Han, K. S.; Park, Y. J.; Kang, M. G.; Park, N. G.; Chang, S. H. *Solid State Ionics*. 2004, *175*, 765-768.
[76] Zhou, Y. K.; He, B. L.; Zhou, W. J.; Huang, J.; Li, X. H.; Wu, B.; Li, H. L. *Electrochim. Acta*. 2004, *49*, 257-262.
[77] Guo, D. J.; Li, H. L. *J. Solid State Electrochem*. 2005, *9*, 445-449.
[78] Wu, M.; Snook, G. A.; Gupta, V.; Shaffer, M.; Fray D. J.; Chen, G. Z. *J. Mater. Chem*. 2005, *15*, 2297-2303.
[79] Meigen, D.; Bangchao, Y.; Yongda, H. *J. Mater. Sci*. 2005, 5021-5023.
[80] Jang, J.; Bae, J.; Choi, M.; Yoon, S. H. *Carbon*. 2005, *43*, 2730-2736.
[81] Wang, Y. G.; Li, H. Q.; Xia, Y. Y. *Adv. Mater*. 2006, *18*, 2619-2623.
[82] Gupta, V.; Miura, N. *Electrochim. Acta*. 2006, *52*, 1721-1726.
[83] Gupta, V.; Miura, N. *J. Power Sources*. 2006, *157*, 616-620.
[84] Lai, Y. Q.; Li, J.; Lu, H.; Zhang, Z. A.; Liu, Y. X. *J. Cent. South Univ. Techno*. 2006, *13*, 353-359.
[85] Fan, L. Z.; Hu, Y. S.; Maier, J.; Adelhelm, P.; Smarsly, B.; Antonietti, M. *Adv. Funct. Mater*. 2007, *17*, 3083-3087.
[86] Mi, H. Y.; Zhang, X. G.; An, S. Y.; Ye, X. G.; Yang, S. D. *Electrochem. Commu*. 2007, *9*, 2859-2862.
[87] Bleda-Martinez, M. J.; Morallon, E.; Cazorla-Amoros, D. *Electrochim. Acta*. 2007, *52*, 4962-4968.
[88] Sivakumar, S. R.; Kim, W. J.; Choi, J. A.; MacFarlane, D. R.; Forsyth, M.; Kim, D. W. *J. Power Sources*. 2007, *171*, 1062-1068.
[89] Wang, Q.; Li, J. L.; Gao, F.; Li, W. S.; Wu, K. Z.; Wang, X. D. *New Carbon Materials*. 2008, *23*, 275-280.

[90] Bleda-Martinez, M. J.; Peng, C.; Zhang, S. G.; Chen, G. Z.; Morallon, E.; Cazorla-Amoros, D. *J. Electrochem. Soc*. 2008, *155*, A672-A678.

[91] Xing, W.; Zhuo, S.; Cui, H.; Si, W.; Gao, X.; Yan, Z. *J. Porous Materials*. 2008, *15*, 647-651.

[92] Kong, L. B.; Zhang, J.; An, J. J.; Luo, Y. C.; Kang, L. *J. Mater. Sci*. 2008, *43*, 3664-3669.

[93] Xing, W.; Yuan, X.; Zhuo, S. P.; Huang, C. C. *Polym. Adv. Techno*. 2009, *20*, 1179-1182.

[94] Li, L. X.; Song, H. H.; Zhang, Q. C.; Yao, J. Y.; Chen, X. H. *J. Power Sources*. 2009, *187*, 268-274.

[95] Wang, H. L.; Hao, Q. L.; Yang, X. J.; Lu, L. D.; Wang, X. *Electrochem. Commu*. 2009, *11*, 1158-1161.

[96] Wang, D. W.; Li, F.; Zhao, J. P.; Ren, W. C.; Chen, Z. G.; Tan, J.; Wu, Z. S.; Gentle, I.; Lu, G. Q.; Cheng, H. M. *Acs Nano*. 2009, *3*, 1745-1752.

[97] Nikzad, L.; Alibeigi, S.; Vaezi, M. R.; Yazdani, B.; Rahimipour, M. R. *Chem. Eng. Techno*. 2009, *32*, 861-866.

[98] 98) Wooa, S. W.; Dokkob, K.; Nakanoa, H.; Kanamura, K. *J. Power Sources*. 2009, *190*, 596-600.

[99] Fang, J.; Cui, M.; Lu, H.; Zhang, Z. A.; Lai, Y. Q.; Li, J. *J. Cent. South Univ. Techno*. 2009, *16*, 434-439.

[100] Zhang, J.; Kong, L. B.; Wang, B.; Luo, Y. C.; Kang, L. *Synth. Met*. 2009, *159*, 260-266.

[101] Koysuren, O.; Du, C. S.; Pan, N.; Bayram, G. *J. Appl. Polym. Sci*. 2009, *113*, 1070-1081.

[102] Sheila C. C.; Dalva, A. L. A.; Carla, P. F.; Silmara, N. *Electrochim Acta*. 2009, *54*, 6383-6388.

[103] Hung, P. J.; Chang, K. H.; Lee, Y. F.; Hu, C. C.; Lin, K. M. *Electrochim. Acta*. 2010, *55*, 6015-6021.

[104] Wang, H. L.; Hao, Q. L.; Yang, X. J.; Lu, L. D.; Wang, X. *Nanoscale*. 2010, *2*, 2164-2170.

[105] Liu, J. L.; Sun, J.; Gao, L. A. *J. Phys. Chem. C*. 2010, *114*, 19614-19620.

[106] Viorel, B.; Florina, B.; Luisa, P. *Surface Interface Analysis*. 2010, *42*, 1266-1270.

[107] Kovalenko, I.; Bucknall, D. G.; Yushin, G. *Adv. Fun. Mater*. 2010, *20*, 3979-3986.

[108] Xu, J. J.; Wang, K.; Zu, S. Z.; Han, B. H.; Wei, Z. X. *Acs Nano*. 2010, *4*, 5019-5026.

[109] Wang, H. L.; Hao, Q. L.; Yang, X. J.; Lu, L. D.; Wang, X. *Acs Applied Materials and Interfaces*. 2010, *2*, 821-828.

[110] Zhang, L. L.; Li, S.; Zhang, J. T.; Guo, P. Z.; Zheng, J. T.; Zhao, X. S. *Chem. Mater*. 2010, *22*, 1195-1202.

[111] Lan, L.; Wei, W.; Wu-yuan, Z.; Ben-lin, H.; Ming-liang, S.; Min, W.; Xue-fei, X. *J. Solid State Electrochem*. 2010, *14*, 2219-2224.

[112] Xu, F.; Zheng, G.; Wu, D.; Liang, Y.; Li, Z.; Fu, R. *Phys. Chem. Chem. Phy*. 2010, *12*, 3270-3275.

[113] Lei, Z. B.; Chen, Z. W.; Zhao, X. S. *J. Phys. Chem. C*. 2010, *114*, 19867-19874.

[114] An, H. F.; Wang, Y.; Wang, X. Y.; Li, N.; Zheng, L. P. *J. Solid State Electrochem*. 2010, *14*, 651-657.

[115] Li, F.; Shi, J. J.; Qin, X. *Chinese Science Bulletin*. 2010, *55*, 1100-1106.

[116] Zhang, K.; Zhang, L. L.; Zhao, X. S.; Wu, J. S. *Chem. Mater*. 2010, *22*, 1392-1401.

[117] Yan, J.; Wei, T.; Shao, B.; Fan, Z.; Qian, W.; Zhang, M.; Wei, F. *Carbon*. 2010, *48*, 487-493.

[118] Wu, Q.; Xu, Y. X.; Yao, Z. Y.; Liu, A. R.; Shi, G. Q. *Acs Nano*. 2010, *4*, 1963-1970.

[119] Wang, H.; Hao, Q.; Yang, X.; Lu, L.; Wang, X. *Nanoscale*. 2010, *2*, 2164-2170.

[120] Meng, C.; Liu, C.; Chen, L.; Hu, C.; Fan, S. *Nano Lett*. 2010, *10*, 4025–4031.

[121] Humberto, G.; Manoj K. R.; Farah, A.; Villalba, P.; Elias, Stefanakosc.; Ashok, K. *J. Power Sources*. 2011, *196*, 4102-4108.

[122] Fonseca, C. P.; Almeida, D. A. L.; Baldan, M. R.; Ferreira, N. G. *Chem. Phys. Lett*. 2011, *511*, 73-76.

[123] Misoon, O.; Seok, K. *Electrochim. Acta*. 2012, *59*, 196-201.

[124] Xingbin Yan, Zhixin Tai, Jiangtao Chena and Qunji Xue. *Nanoscale*. 2011, *3*, 212-216.

[125] Liping, Z.; Xianyou, W.; Hongfang, A.; Xingyan, W.; Lanhuan, Y.; Li, B. *J. Solid State Electrochem*. 2011, *15*, 675–681.

[126] Feng, X. M.; Li, R. M.; Ma, Y. W.; Chen, R. F.; Shi, N. E.; Fan, Q. L.; Huang, W. *Adv. Fun. Mater*. 2011, *21*, 2989-2996.

[127] Hao, Q. L.; Wang, H. L.; Yang, X. J.; Lu, L. D.; Wang, X. *Nano Research*. 2011, *4*, 323-333.

[128] Li, J.; Xie, H. Q.; Li, Y.; Liu, J.; Li, Z. X. *J. Power Sources*. 2011, *196*, 10775-10781.

[129] Sheng, L.; Xiaohong, L.; Zhangpeng, L.; Shengrong, Y.; Jinqing, W. *New J. Chem*. 2011, *35*, 369-374.

[130] Zhao, Y.; Hua, B.; Yue, H.; Yan, L.; Liangti, Q.; Shaowen, Z.; Gaoquan, S. *J. Mater. Chem*. 2011, *21*, 13978-13983.

[131] Mao, L.; Zhang, K.; Chan, H. S. O.; Wu, J. S. *J. Mater. Chem*. 2012, *22*, 80-85.

[132] Zhou, S. L.; Mo, S. S.; Zou, W. J.; Jiang, F. P.; Zhou, T. X.; Yuan, D. S. *Synth. Met*. 2011, *161*, 1623-1628.

[133] Zhu, Z. Z.; Wang, G. C.; Sun, M. Q.; Li, X. W.; Li, C. Z. *Electrochim. Acta*. 2011, *56*, 1366-1372.

[134] Zhou, G. M.; Wang, D. W.; Li, F.; Zhang, L. L.; Weng, Z.; Cheng, H. M. *New Carbon Mater*. 2011, 26, 180-186.

[135] Dou, Y. Q.; Zhai, Y. P.; Liu, H. J.; Xia, Y. Y.; Tu, B.; Zhao, D. Y.; Liu, X. X. *J. Power Sources*. 2011, *196*, 1608-1614.

[136] Liu, W. X.; Liu, N.; Song, H. H.; Chen, X, H. *New Carbon Mater*. 2011, *26*, 217-223.

[137] Wang, K.; Zhao, P.; Zhou, X. M.; Wu, H. P.; Wei, Z. X. *J. Mater. Chem*. 2011, *21*, 16373-16378.

[138] Mikhaylovaa, A. A.; Tusseevaa, E. K.; Mayorovaa, N. A.; Rychagova, A. Y.; Volfkovicha, Y. M.; Krestininb, A. V.; Khazova, O. A. *Electrochim. Acta*. 2011, *56*, 3656-3665.

[139] Wang, Z. L.; Guo, R.; Li, G. R.; Lu, H. L.; Liu, Z. Q.; Xiao, F. M.; Zhang, M. Q.; Tong, Y. X. *J. Mater. Chem*. 2012, *22*, 2401-2404.

[140] Lang, X. Y.; Zhang, L.; Fujita, T.; Ding, Y.; Chen, M. W. *J. Power Sources*. 2012, *197*, 325-329.

[141] Xue, M. A. Q.; Li, F. W.; Zhu, J.; Song, H.; Zhang, M. N.; Cao, T. B. *Adv. Fun. Mater*. 2012, *22*, 1284-1290.

[142] Chi, C. H.; Eve, C.; Jeng, Y. L. *Electrochim. Acta*. 2002, *47*, 2741-2749.

[143] Gomez-Romero, P.; Chojak, M.; Cuentas-Gallegos, K.; Asensio, J. A.; Kulesza, P. J.; Casan-Pastor, N.; Lira-Cantu, M. *Electrochem. Commu*. 2003, *5*, 149-153.

[144] Xiong, Z.; Liyan, J.; Shichao, Z.; Wensheng, Y. *J. Power Sources*. 2007, *173*, 1017-1023.

[145] Song, R. Y.; Park, J. H.; Sivakkumar, S. R.; Kim, S. H.; Ko, J. M.; Park, D. Y.; Jo, S. M.; Kim, D. Y. *J. Power Sources*. 2007, *166*, 297-301.

[146] Xiao, X. L.; Ya, B. L.; Li, J. B.; Yu, Q. D.; Yu, Q. H. *J. Solid State Electrochem*. 2008, *12*, 909-912.

[147] Chaoqing, B.; Aishui, Y.; Haoqing, W. *Electrochem. Commu*. 2009, *11*, 266-269.

[148] Hu, Z. A.; Xie, Y. L.; Wang, Y. X.; Mo, L. P.; Yang, Y. Y.; Zhang, Z. Y. *Mater. Chem.* Phys. 2009, *114*, 990-995.

[149] Xingwei, L.; Han, Z.; Gengchao, W.; Zhihui, J. *J. Mater. Chem*. 2010, *20*, 10598-10601.

[150] Chen, L.; Sun, L. J.; Luan, F.; Liang, Y.; Li, Y.; Liu, X. X. *J. Power Sources*. 2010, *195*, 3742-3747.

[151] Ni, W. B.; Wang, D. C.; Huang, Z. J.; Zhao, J. W.; Cui, G. E. *Mater. Chem. Phys*. 2010, *124*, 1151-1154.

[152] Liu, Q.; Nayfeh, M. H.; Yau, S. T. *J. Power Sources*. 2010, *195*, 3956-3959.

[153] Radhakrishnan, S.; Rao, C. R. K.; Vijayan, M. *J. Appl. Poly. Sci*. 2011, *122*, 1510-1518.

[154] Jaidev; Jafri, R. I.; Mishra, A. K.; Ramaprabhu, S. *J. Mater. Chem*. 2011, *21*, 17601-17605.

[155] Ben, X. Z.; Ying, L.; Xiao, X. L.; Dermot, D.; King, T. L. *J. Power Sources*. 2011, *196*, 4842-4848.

[156] Yan, J.; Wei, T.; Fan, Z. J.; Qian, W. Z.; Zhang, M. L.; Shen, X. D.; Wei, F. *J. Power Sources*. 2010, *195*, 3041-3045.

[157] Li, Q. A.; Liu, J. H.; Zou, J. H.; Chunder, A.; Chen, Y. Q.; Zhai, L. *J. Power Sources*. 2011, *196*, 565-572.

[158] Lu, X. J.; Dou, H.; Yang, S. D.; Hao, L.; Zhang, L. J.; Shen, L. F.; Zhang, F.; Zhang, X. G. *Electrochim. Acta*. 2011, *56*, 9224-9232.

[159] Wu, Y. Z.; Wei, W.; Ben, L. H.; Ming, L. S.; Yan, S. Y. *J. Power Sources*. 2010, *195*, 7489-7493.

[160] Sathish, M.; Mitani, S.; Tomai, T.; Honma, I. *J. Mater. Chem*. 2011, *21*, 16216-16222.

[161] Xifeng, X.; Qingli, H.; Wu, L.; Wenjuan, W.; Hualan, W.; Xin, W. *J. Mate. Chem*. 2012, *22*, 8314-8320.

In: Trends in Polyaniline Research
Editors: T. Ohsaka, Al. Chowdhury, Md. A. Rahman et al.
ISBN: 978-1-62808-424-5

Chapter 7

POLYANILINE: A FASCINATING MATRIX FOR COMPOSITE MATERIALS

Md. Abu Bin Hasan Susan[1,*], Md. Sirajul Islam[1], M. Muhibur Rahman[2] and M. Yousuf A. Mollah[1]

[1]Department of Chemistry, University of Dhaka, Dhaka, Bangladesh
[2]University Grants Commission of Bangladesh, Agargaon, Dhaka, Bangladesh

ABSTRACT

Polyaniline (PAni) is one of the most technologically important and versatile conducting polymers because of its flexibile chemistry, facile processibility, environmental stability, and relatively low cost. PAni can be combined with ease with wide range of metals, metal oxides and even with other polymers to prepare composite materials for diverse applications. PAni serves as a leading matrix for silver, gold and platinum to yield PAni-metal composites for stunning applications. Inorganic metal oxides and salts of Ti, Si, Nb, Ba, Mo, Zr, Zn, Cu, Fe when incorporated in PAni show significantly improved physicochemical properties. Polystyrene, polyethylene, nylon, polyurethane, poly (methyl methacrylate), etc. may also be conveniently used to prepare composites with PAni. Depending on the nature of composite materials and their applications, different polymerization methods need to be carefully chosen for the preparation of such composites, which *inter alia, include:* oxidative polymerization, electrochemical deposition polymerization, template synthesis, self assembly, mechanical blending and interfacial polymerization. Techniques commonly used for characterization of composite materials based on PAni are inverse gas chromatography (IGC), thermogravimetric analysis (TGA), differential thermal analysis (DTA), scanning electron microscopy (SEM), atomic force microscopy (AFM), X-ray diffraction (XRD), X-ray photoelectron spectroscopy (XPS), Fourier transform infrared spectroscopy (FT-IR), inductance capacitance and resistance (LCR) meter, electrometer, and UV-visible spectroscopy. Characterization techniques depend on the properties to be characterized, for instances, IGC is used to study the thermodynamic properties, SEM and AFM are

* Corresponding author: Md. Abu Bin Hasan Susan. Department of Chemistry, University of Dhaka, Dhaka 1000, Bangladesh. Tel: 88029661920-73/7162; Fax: 88028615583; E-mail: susan@du.ac.bd.

employed to reveal morphological behavior, TGA is used to investigate the thermal stability of the composites, FT-IR is used to confirm the presence of functional groups of the component materials, LCR meter and electrometer are used to examine the electric conductivity. The ease of preparation and tunable properties of the composites have catered PAni as a fascinating matrix for composite materials. This chapter provides a comprehensive review of research works centred on PAni based composite materials and addresses the perspective of the use of PAni in composites with variety of materials, characterization techniques and multi-facet applications in multidisciplinary areas.

1. INTRODUCTION

Polyaniline (PAni), because of its high electrical conductivity, easy producibility, environmental stability, ease of preparation and relatively low cost, is known as one of the most technologically important conducting polymers. It was discovered as "aniline black", a part of an organic polymer, melanin in 1934. Although PAni already has a long and productive history, only recent findings demonstrate that PAni is a flexible polymer and is highly useful for manifold applications, such as, electrochemical displays, sensors, catalysis, redox capacitors, electromagnetic shielding, computer chips as well as in secondary batteries [1-5].

The prospect of PAni for manifold applications in multidisciplinary fields relies on its chemical structure and ability to combine with ease with wide range of metals, metal oxides, metal salts and even with other polymers to make composites. The mutual influence of the individual constituents and synergism of their properties render composites based on PAni so promising and PAni-based composite materials have been harvesting several intriguing properties within themselves.

Nowadays composites based on PAni have multi-facet applications. PAni blends are being used as corrosion protectors on many metals. PAni incorporated with carbon black enhances its thermal stability [6], while PAni and Bi, Sb, Te composites show improved thermoelectric property [7]. In addition, invention of nanoscaled composites of PAni has opened a new world of surprise to the chemists and researchers. A hallmark of PAni nanocomposites and nanoparticles is its interdisciplinary nature - its practice requires researchers to cross the traditional boundaries between the experimental and theoretical fields of chemistry and physics, materials science and engineering, biology and medicine, to work together in close-knit teams on challenging problems of global importance. In the wake of increasing pollution and depleting oil reserves, the need for the supercapacitors as alternative sources of high power has been gaining momentum for the last decade. PAni composites with different materials have been considered as the most promising materials for electrode material in the redox supercapacitors because of its distinctive characteristics of conducting pathways, surface interactions, and nanoscale dimensions and ease of synthesis. PAni and gold nanoparticles are an excellent nonvolatile plastic digital memory device [8]. PAni and CZO (Cu, Zn and O) nanocomposites exhibit an outstanding performance in antibacterial activity [9]. The incorporation of fly ash into the polymeric network introduces uniform porosity and is expected to be advantageous for gas sensing and biosensing applications [10]. PAni nanofiber is used in biocompatible chemosensors.

Even in biological systems, PAni is used as biosensors due to its capability to act as a biomolecule entrapment matrix [11]. PAni, due to thousands of such applications and tunable properties, has emerged as a fascinating matrix for composite materials.

2. Chemical Structure of PAni

PAni, polymerized from the aniline monomer, can exist in one of three well-defined oxidation states: leucoemeraldine, emeraldine and pernigraniline [12] (Scheme 1).

Leucoemeraldine and pernigraniline are the fully reduced (all the nitrogen atoms are amine) and the fully oxidized (all the nitrogen atoms are imine) forms, respectively, and the ratio of -N- and -N= is 0.5 in emeraldine.

Typical chemical or electrochemical oxidation converts electrically insulating leucoemeraldine to electrically conducting emeraldine which upon further oxidation through a second redox process yields a new insulating material, pernigraniline. A decrease of conductivity by ten orders of magnitude is obtained just by treatment of the conducting emeraldine in neutral or alkaline media. Though the number of π-electrons in the chain remains constant, protonation induces an insulator-to-conductor transition.

The emeraldine form of PAni, frequently referred to as emeraldine base (EB) is neutral, but by protonating imine nitrogens with an acid, EB can easily be converted into emeraldine salt (ES) (Scheme 2). Protonation helps to delocalize the otherwise trapped diiminoquinone-diaminobenzene state.

Emeraldine base is regarded as the most useful form of PAni due to its high stability at room temperature and the fact that, upon doping with acid, the resulting emeraldine salt form of PAni is highly electrically conducting [13]. Leucoemeraldine and pernigraniline are poor conductors, even when doped with an acid.

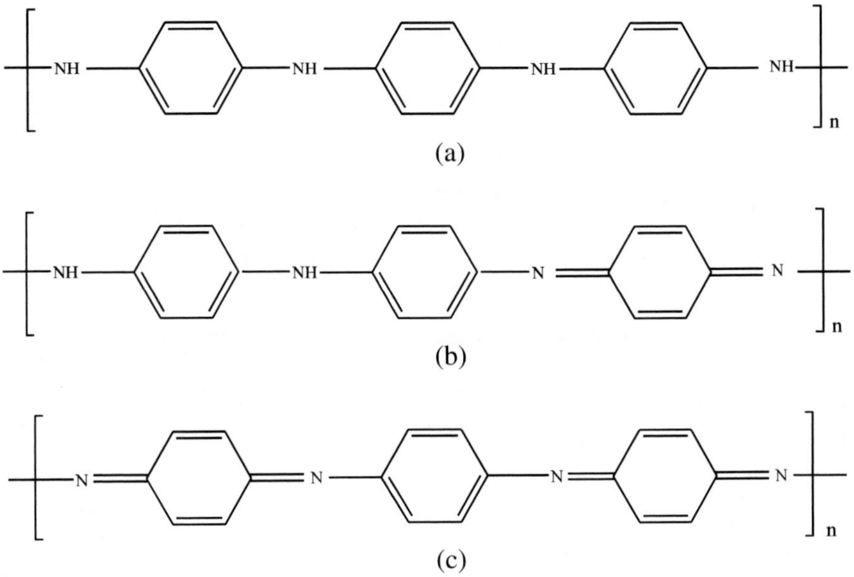

NH NH NH NH n

(a)

NH NH N N n

(b)

N N N N n

(c)

Scheme 1. Chemical structure of (a) Leucoemeraldine, (b) Emeraldine, and (c) Pernigraniline.

3. Basic Pre-Requisites for Composite Materials

The preparation, characterization and applications of composite materials have received an upsurge of interest from many researchers in diversified areas, such as, chemistry, physics, material science, biology and the corresponding engineering.

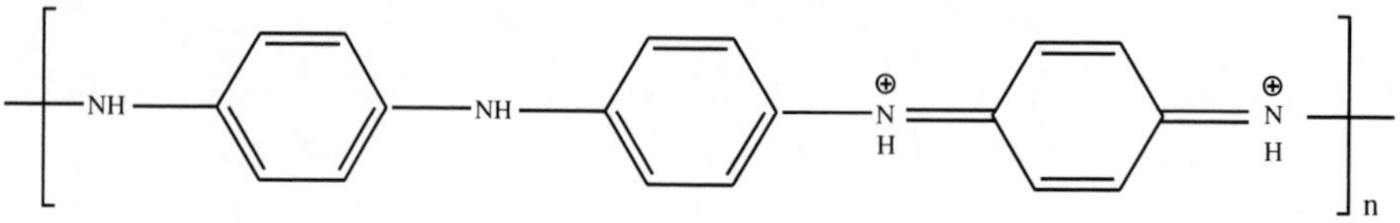

Scheme 2. Chemical structure of emeraldine salt.

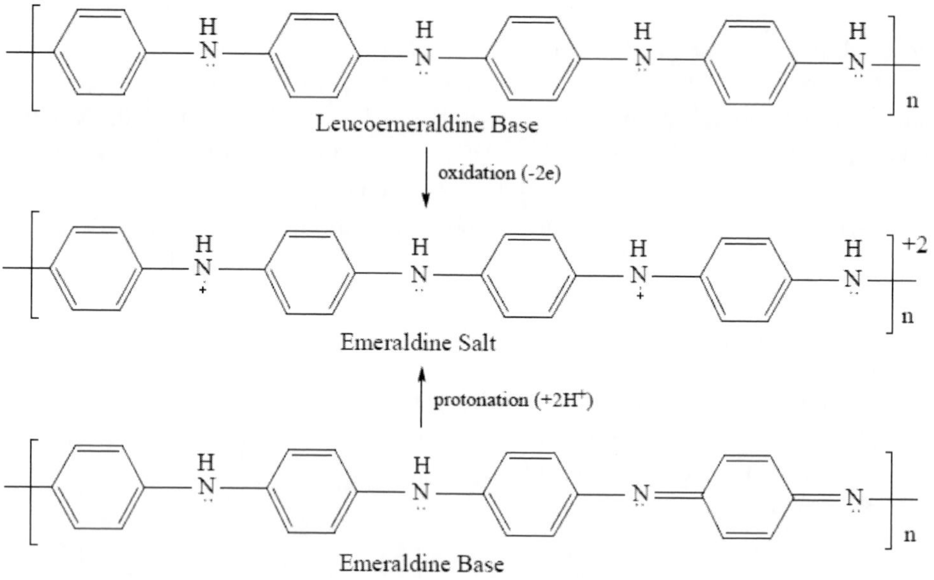

Leucoemeraldine Base

oxidation (-2e)

Emeraldine Salt

protonation ($+2H^+$)

Emeraldine Base

Scheme 3. Formation of a PAni salt following oxidation and protonation of bases.

The conducting emeraldine salt form can be obtained by oxidative doping of leucoemeraldine base or by protonation of emeraldine base (Scheme 3).

Different oxidation states of PAni give rise to different colors and the associated color change can be used in sensors and electrochromic devices [14] (Table 1).

Table 1. Color and electronic conductivity of different forms of PAni

Polymer	*Color*	*Electronic conductivity*
Leucoemeraldine	White/clear	Insulator
Emeraldine base	Blue	Insulator
Emeraldine salt	Green	Conductor (10^{-4} - 10^{2} Scm^{-1})
Pernigraniline	Transparent yellow/yellow	Insulator

Although the change in color can be successfully exploited for the fabrication of PAni based biosensors, the best method is arguably to take advantage of the dramatic changes in electrical conductivity between the different oxidation states or doping levels [15].

With technological advances, we have been increasingly relying on the unusual combinations of properties that cannot be met by the conventional metal alloys, ceramics, and polymeric materials.

According to the principle of combined action, better property combinations are fashioned by judicious combination of two or more distinct materials [16]. Combination of strength and stiffness with lightness is a pre-requisite for the use of composite materials.

By choosing an appropriate combination of reinforcing and matrix material, it is possible to have properties that exactly fit the requirements for a particular structure for a particular purpose. Particles of different nature and size can be combined with the conducting polymers, giving rise to a host of composites with interesting physical properties and potential for important applications.

4. Versatility of PAni As Matrix in Composites

PAni is a conductive polymer and has properties similar to some metals. The unique properties of PAni have led it to an interest in numerous potential applications. It can easily be composited with metal, metal oxide and more readily with other polymeric materials.

Moreover, PAni undergoes oxidation or reduction reaction through electron donation or withdrawal. The pristine PAni is sensitive to many chemical species and the application is limited for indiscriminate interactions with many different species. The redox reaction of the polymer and electron donor or acceptor interactions often occurs spontaneously without the ability to control the processes.

Therefore, to improve both selectivity and sensitivity for the targeted chemical interactions, incorporation of secondary components into PAni, such as metals, metal oxides or organic molecules into PAni is often considered to be an ideal means.

4.1. Composites of PAni with Metals

For preparation of composites with metals, the number of choices is limited since the number of metal elements is small and limited. However, PAni can be funcntionalized or the aniline derivatives may be used for polymerization to obtain vast number of molecules for use as polymer matrix for composite materials. Quite reasonably, combination of these two parent families can enrich the property library and widen the scope of applications offered by each of them individually. Incorporating metals in conducting polymers enhances electron transfer through a direct or mediated mechanism with improved conductivity and enhanced stability [17]. PAni, a conducting organic polymer, has been used for a number of particular application oriented composite materials due to the fact that the electronic interactions between metallic particles and PAni [18] are strong and it is also easy to control the shape and size of these composites for defining their properties, such as the electronic band gap [19], conductivity, and light-emission efficiency [20].

For instance, nanoparticles or clusters of noble metals, including gold, palladium, and platinum provides the polymer with high surface area metal templates for chemical catalysis and sensing interactions.

The combination of PAni and metal can also produce composites with novel and synergistic properties that are not originally present in each component [21]. They are capable of providing selective surfaces for the withdrawal or donation of electron density through controlled interactions and chemically distinct reaction sites. The efficiency and versatility of PAni-metal composites is reflected in the literature.

Athawale et al. [22] reported that a PAni-Pd nanocomposite is highly sensitive and selective sensor for methanol vapors and give a stable response for a sufficiently long time. This can be accounted in terms of the enhanced degree of interactions between the nanocomposite and the vapors due to the larger surface area provided by the Pd nanoparticles present in the nanocomposite near the imine nitrogens [23]. The coinage metal group, Ag, Au, Cu and Pt has been most extensively studied for polymer-metal nanocomposites. The excellent and unusual properties of gold nanoparticles coupled with their ability to tune the properties of conducting polymers leading to the fabrication of materials with varied applications have continued to attract interest from researchers. It was demonstrated by Yang et al. [24] that PAni/gold nanocomposite has a good promotion toward the reduction/ oxidation of immobilized microperoxidase-11 (MP-11).

The high conductivity, large specific surface area, and excellent electroactivity of the PAni/gold nanocomposite make it an excellent matrix for enzyme immobilization and electrocatalysis. When combined with MP-11, the nanocomposite electrode displays a potential application for the detection of H_2O_2 with a low detection limit and rapid response time. Moreover, the nanocomposite-modified electrode exhibits high stability. Platinum has catalyzed many processes that are important for technologies of energy production and environmental protection.

Grzeszczuk and Poks [25] reported the incorporation of platinum nanoparticles into PAni matrix. The PAni-platinum nanocomposite electrodes grown potentiostatically were used to study electrochemical process of hydrogen evolution. Liu et al. [26] prepared PAni-poly (styrene sulfonic acid)-Pt nanocomposite through an interfacial polymerization route. The composites consisting of Pt nanoparticles loaded into conducting PAni-PSS (poly(styrenesulfonate) polymers was reported to find applications in direct methanol fuel cell. PAni-Ag composite nanofibres prepared by Pillalamarri et al. [27] were reported to have 50 times the electrical conductivity of pure PAni nanofibres. Shu-hong et al. [28] also reported composites of Bi, Sb, Te with PAni and suggested that these composites reduce the contact resistance between multiphase and improve the power factor of the composites.

4.2. Composites of PAni with Metal Oxide

Composites containing PAni and different metal oxides have been widely studied, especially nanostructured composites, such as nano-tubes, nano-rods, nano-fibres or core-shell nanostructures. Conducting polymer systems with nanoparticles of metal oxides can be tailored to obtain desired mechanical as well as novel electrical, magnetic or optical properties. Metal oxides exhibit electrical behavior that can vary from electrically insulating

(MgO and Al_2O_3), wide-band semiconducting (TiO_2, SnO_2, ZnO, Ti_2O_3), to metal-like conducting (V_2O_3, ReO_3, RuO_2) behavior.

Some oxides have several stable oxidation states that are very important in surface chemistry. Among metal oxide nanoparticles, iron oxides have technological importance due to their magnetic, catalytic and biological properties. Tudorache and Grigoras [29] reported that PAni-Fe_2O_3 composites exhibit a better sensitivity to vapors compared with simple iron oxides and can be used as a sensor. Conn et al. [30] developed a PAni-PtO_2 based selective H_2 sensor and reported that the conductivity of PAni increases with H_2 exposure, due to the formation of water. It is known that water present in the polymer takes part in charge transfer leading to an increase in the conductivity of PAni which is reversible. Wang et al. [31] developed PAni intercalated MoO_3 thin film sensors and reported that the conductivity change is due to the reversible absorption of analyte. Parvatikar et al. [32] developed PAni-WO_3 composite based sensors and reported that conductivity of the film increases with increasing humidity. Ram *et al.* [33] developed PAni-SnO_2 and TiO_2 nanocomposite thin film based sensors and reported that conductivity of the film increases with NO_2 exposure. Geng et al. [34] synthesized PAni-SnO2 hybrid materials by a hydrothermal process for gas sensing applications. They found that hybrid materials are sensitive to ethanol and acetone vapor at 60 to 90 °C. Jumali et al. [35] have demonstrated that ZnO incorporated PAni composite have an excellent methanol sensing properties and this sensitivity is due to the influence of PANi on the microstructure of composite films, which gives rise to variation of the sensor response properties. The presences of porous fibrous structure with small grain sizes in the composite sensor enables the sensor to detect methanol vapor at room temperature with superior sensing properties compared to ZnO alone. Mho et al. [36] have showed that composites of PAni-V_2O_5 exhibit definite synergistic effect of inorganic layer structure with improved lithium ion intercalation and accessibility of polymer with flexibility and conductivity. Rechargeable lithium batteries assembled with the PAni-V_2O_5 composite cathode show better cyclability with high specific capacities. Ahn et al. [37] illustrated that controlled PAni nanocoating of Li_2MnO_3-based cathode material is an effective alternative for improving the cycle life of a cell.

4.3. Composites of PAni with Metal Salt

The incorporation of metal salts into PAni matrices results in materials with modified properties which are of interest for their electronics properties. Metal salts, due to their ability to influence the polymer conformations, play an important role in the electrical property of the polymer. The PAni-metal salts composites are widely used materials as gas sensors, electrodes in many electronic devices. Virji et al. [38] synthesized a number of composites of PAni with $CuCl_2$, $CuBr_2$, CuF_2, $Cu(O_2CCH_3)_2$, $Cu(NO_3)_2$, $EuCl_2$, $NiCl_2$, $FeCl_3$, and $CoCl_2$ and reported that the copper(II)bromide/PAni nanofiber composite yields the best response with greater than an order of magnitude change in resistance upon exposure to arsine. Virji et al. [39] also studied the composites of PAni with transition metal salts for hydrogen sulfide gas sensing and demonstrated that the composites have remarkable responses to hydrogen sulfide that are up to four orders of magnitude greater than the corresponding unmodified nanofibers.

In supercapacitors, PAni-metal salts composites are frequently employed as promising materials. Cui et al. [40] prepared a series of composites of PAni with $ZnCl_2$, $CuCl_2$ and

$FeCl_3$ and concluded that Zn^{2+} doped PAni is suitable for aqueous redox supercapacitors and has the potential to be used as electrode material for non-aqueous redox supercapacitors with lithium salt.

4.4. Composites of PAni with Other Polymers

PAni can be easily combined with other polymers to form composites with desired shapes. However, one of the key problems is the difficulty in processability of PAni, either from melt or from solution. With a rigid π-conjugated backbone molecular structure, PAni undergoes decomposition without melting transition when heat is applied resulting in hindered processability from melt.

PAni is not also soluble in many common solvents and cannot be processed from solution either. There have been numerous attempts to overcome this disadvantage through efficient increase in its processability. Notables are: substitution in the aniline monomer prior to polymerization, copolymerization with any other suitable polymers, making blends or composites with other conventional synthetic polymers, modifying the preformed Langmuir–Blodgett film [41] and preparation of dispersed colloidal particles [42, 43]. Incorporating a conducting polymer in a plastic or rubber matrix is a promising method to introduce flexibility and toughness into that polymer. In fact, PAni blends or composites with other polymers have widely been studied. For example, poly(vinyl alcohol) (PVA), a water soluble well-studied conventional polymer used in several industrial applications [44], can be combined with PAni.

Blends of PAni have their own importance. Special interest has been focused on PAni doped with *dodecyl benzene sulfonic acid* (DBSA) due to excellent thermal and environment stability combined with relatively high level of electroconductivity. Ray et al. [45] and Morgan and Foot [46] reported blends of poly(methyl methacrylate) (PMMA)/PAni. DBSA showed low conductivity and high percolation threshold. PMMA was reported to be more miscible in the presence of hydroquinone and the conductivity of the blend enhanced in presence of hydroquinone [47, 48]. Manuel [49] synthesized blends of PAni and polystyrene (PS) and polyethylene oxide (PSO) to obtain thinner diameter fibers of PAni blends.

4.5. Composites of PAni in Nano Form

In recent years, there has been growing interest in research on conducting polymer nanostructures (i.e., nano-rods, -tubes, -wires, and -fibers) since they combine the advantages of organic conductors with low-dimensional systems [50-55]. Combination of PAni with metals, metal oxides, metal salts or other polymers may have nano-form composites, but the composites of intrinsically nanostructured PAni is still a challenge. Kulkarni et al. [56] synthesized a PAni-multiwalled carbon nanotube (PAni-MWCNT) nanocomposite for optical pH sensor. The synthesized nanocomposite was successfully used for the determination of pH of the solution (pH 1-12) and found to be a potential candidate for optical pH sensing. Graphene, an allotrope of carbon, is another promising material for designing nanostructured PAni composites since it can produce a dramatic improvement in properties at very low filler content. Zhao et al. [57] prepared a PAni/graphene nanosheet/carbon nanotube (PAni/ GNS/

CNT) composite and reported that specific capacitance of the PAni/GNS/CNT composites is much higher than that of pure PAni and PAni/CNT composites. The enhanced specific capacitance was attributed to the synergistic effect between GNS and PAni [58]. In fact, the discovery of graphene as nanofiller has opened a new dimension for the production of light weight, low cost, and high performance composite materials for a range of applications.

5. Synthesis of PAni and Its Composites

The changes in physicochemical properties of PAni in response to various external stimuli are exploited in various applications [59-60], specifically, in organic electrodes, sensors, and actuators [61-63].

Other uses are based on the combination of electrical properties typical of semiconductors with parameters characteristic of polymers, like the development of "plastic" microelectronics [63-64], electrochromic devices [65], tailor-made composite systems [66-67], and "smart" fabrics [68]. It is therefore very crucial to establish well-defined methodology for preparation of PAni and its composite materials to have physical properties properly known.

In practice, depending on the nature of composite materials and their applications, different polymerization methods need to be carefully chosen, which *inter alia,* include: chemical oxidative polymerization, electrochemical deposition polymerization, template synthesis, self assembly, mechanical blending, interfacial polymerization and electrospinning.

5.1. Chemical Oxidative Polymerization

The most commonly used method to prepare PAni and/or PAni based composites is chemical oxidative polymerization. The efficient polymerization of aniline is achieved only in an acidic medium, where aniline exists as an anilinium cation. A variety of inorganic and organic acids of different concentration have been used in the syntheses of PAni and its composites; the resulting PAni and/or PAni based composites, protonated with various acids, differs in solubility, conductivity, and stability [69]. Peroxydisulfate is the most commonly used oxidant, and its ammonium salt was preferred to the potassium counterpart because of its better solubility in water. Scheme 4 shows a typical course of oxidative polymerization.

4n C_6H_5–$NH_2.HA$ + 5n $(NH_4)_2S_2O_8$

↓

$-[\overset{\oplus}{N}H(A^{\ominus})=C_6H_4=\overset{\oplus}{N}H(A^{\ominus})-C_6H_4-NH-C_6H_4-NH-C_6H_4-NH]-$

2n HA + 3n H_2SO_4 + 5n $(NH_4)_2SO_4$

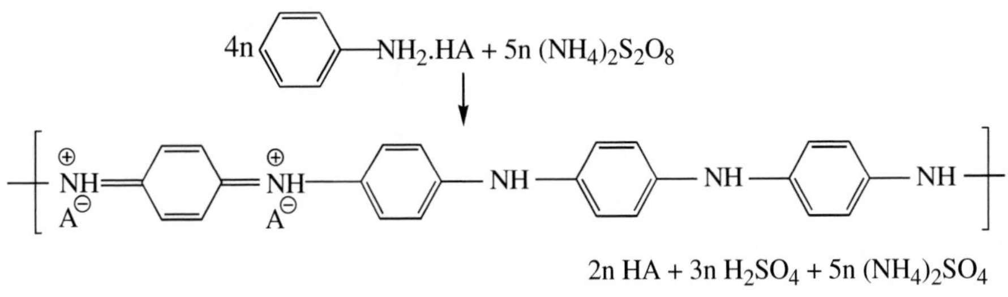

Scheme 4. Chemical reactions in chemical oxidative polymerization of PAni and/or PAni based composites. A- is an arbitrary anion, e.g., chloride.

5.2. Electrochemical Deposition Polymerization

Electrochemical synthesis of PAni and/or PAni based composites is generally carried out in aqueous protonic acid medium and this can be achieved by any of the following methods:

I *Galvanostatic*: constant current in the range of 1-10 mA.
II *Potentiostatic*: at constant potentials −0.7 to 1.1 V versus SCE.
III *Sweeping the potential*: between two potential limits −0.2 V to +1.0 V vs. SCE.

Electropolymerization of PAni and/or PAni based composites is a radical combination reaction and is diffusion-controlled. Oxidation of the aniline monomer generates the radical cation on the electrode surface. The process of formation of radical cation is considered to be the rate-determining step. The formation of paraformed dimers occurs mainly by radical coupling and elimination of two protons. Chain propagation proceeds with oxidation of both the dimeric and monomeric aniline on the electrode surface. The radical cation of the oligomer couples with a radical cation of aniline monomer in this step.

Finally, PAni is doped by the acid present in solution. The polymers are formed at higher rate as the more monomers are deposited onto the polymer surface. It involves the adsorption of the anilinium ion onto the oxidized form of PAni, followed by electron transfer to form the radical cation and subsequent reoxidation of the polymer to its most oxidized state. The growth of PAni is thus considered to be self catalyzed.

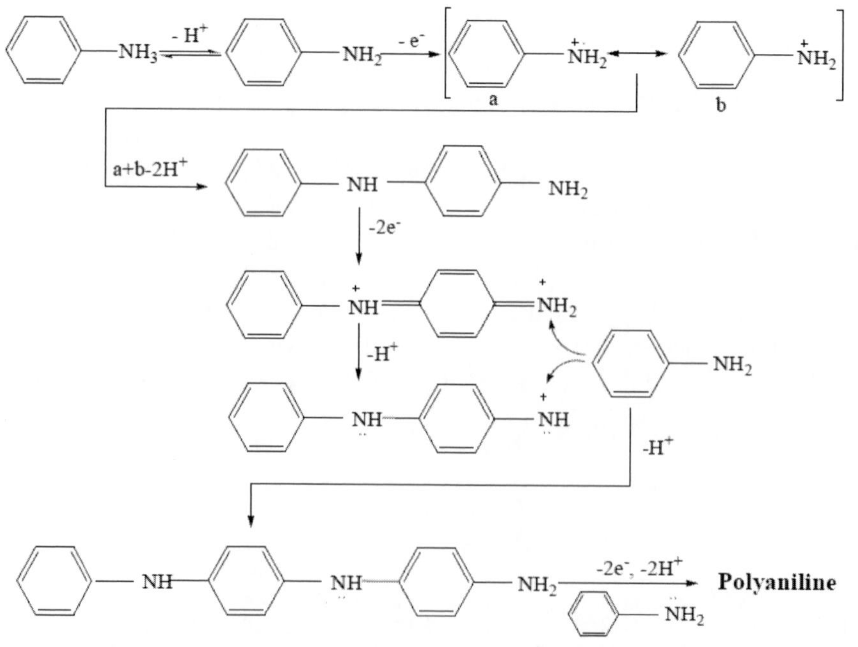

Scheme 5. Mechanism of the polymerization of aniline, proposed by Wei et al. [70-73].

The most generalized mechanism of aniline polymerization suggested by Wei et al. (Scheme 5) [70-73], is based mainly on kinetic studies of the electrochemical polymerization of aniline. The slowest step in the polymerization of aniline is the oxidation of aniline monomer to form dimeric species (i.e. *p*-aminodiphenylamine, PADPA, *N-N*-diphenylhydrazine and benzidine), because the oxidation potential of aniline is higher than those of dimers, subsequently formed oligomers and polymer.

The dimers formed are immediately oxidized and react with an aniline monomer via an electrophilic aromatic substitution, followed by further oxidation and deprotonation to afford the trimers. This process continues repeatedly leading eventually to the formation of PAni and/or PAni based composites.

5.3. Template Polymerization

Template polymerization is a process of synthesis of polymers by which a preformed macromolecule (template) assists in the organization of monomer units and refers to a single phase system wherein the monomer and template are soluble in the same solvent. The template method of polymerization first proposed by Martin *et al.* [74-82] is an effective technique to synthesize arrays of aligned polymer micro-/nanotubes and wires with controllable length and diameter. The template synthesizing route of conducting polymer includes hard template and soft template methods.

The former replicates existing nanostructure by physical or chemical interactions while the latter relies on molecular self-assembly to form nanostructures.

5.3.1. Hard Template

The growth of conducting polymers relies on the use of a physical template as a scaffold. Colloidal particles and some templates with a nanosized channel, such as anodized alumina oxide (AAO) and mesoporous silica/carbon templates are usually used as hard template scaffold [83, 84].

For the synthesis using micro/nanoparticles as templates, the target material is precipitated or polymerized on the surface of the template [85], which results in a core-shell structure [86, 87]. After removal of the template, hollow nanocapsules or nanotubes can be obtained [88-90].

The most commonly used hard templates include monodispersed inorganic oxide nanoparticles [91, 92] and polymer microspheres [93, 94]. Templates of this variety are advantageous for several reasons: narrow size distribution, ready availability in relatively large amounts, and availability in a wide range of sizes from commercial sources, and simplicity of synthesis using well known formulations. However, the removal of the template often affects the hollow structures. Furthermore, the post-processing for template removal is tedious. Wan [95] developed a template self-removing process to produce a PAni hollow structure with octahedral cuprous oxide as template, which was spontaneously removed by reaction with an oxidative initiator, ammonium peroxydisulfate [96]. The method simplified the process to produce PAni hollow structures in a quantitative way. The only potential drawback of the method is that a reduced emeraldine form of PAni and/or PAni based composites was produced because of the reducibility of the cuprous oxide template.

The AAO template can be used to fabricate conducting polymer composite with well-tuned nanostructures by controlled electrochemical deposition [97-105]. The first reported transmission electron microscopy (TEM) image of conductive polymer nanotubes using the AAO template was reported by Martin et al. [84, 100]. One of the most attractive advantages of this route is that the ordered array of conducting polymers nanotubes can be produced using the AAO template. Lahav et al. [106] fabricated core-shell and segmented polymer-metal composite nanostructures by sequentially depositing PAni and gold via an electrochemical route. Some mesoporous materials with open nanochannels can be used as template to produce conducting polymer nanofiber or its composites.

Wu and Bein [107] prepared conducting filaments of PAni in the 3 nm wide hexagonal channel of the aluminosilicate MCM (Mobile Crystalline Material)-41. Aniline vapor was adsorbed onto the dehydrated host. This was followed by a reaction with peroxydisulfate, leading to encapsulated PAni filaments. The materials showed good low-field conductivity, which demonstrated for the first time that conjugated polymers, can be encapsulated in nanometer channels and still support mobile charge carriers.

5.3.2. Soft Template

The soft template synthesis, also named self-assembly method, employs micelles formed by surfactants to confine the polymerization of conducting polymers into low dimensional nanomaterials. Typical synthesis includes microemulsion polymerization [108] and reversed microemulsion polymerization in which surfactants are involved, and the non-template (or self-template) synthesis in which the monomer or its salt forms micelles by itself.

Microemulsion (oil-in-water) polymerization produces conducting polymer nanoparticles with good control over the size of nanoparticles. The structure and concentration of surfactants and monomers are critical factors for controlling the morphological parameters of products. Zhou et al. [109-111] used sodium dodecyl sulfate (SDS) and HCl solution to control the morphology of PAni. They found that the pH value of the solution dramatically influenced the self-assembly morphology of the products. PAni in the forms of granules, nanofibers, nanosheets, rectangular submicrotubes, and fanlike/flowerlike aggregates were obtained by using different SDS and HCl concentrations.

Reversed microemulsion (water-in-oil) polymerization generates conducting polymer nanostructures such as monodispersed nanoparticles and nanotubes/rods, with morphology controlled by introducing the interaction between ions and surfactant. Surfactant gel is one kind of soft template that can guide the growth of conducting polymers. PAni nanobelts were synthesized by a self-assembly process using the chemical oxidative polymerization of aniline in surfactant gel.

In this process, cetyltrimethylammonium bromide (CTAB) and aniline self-assembled into belt-like structures, which acted as templates for the formation of PAni nanobelts. The subsequent *in situ* oriented oxidative polymerization of aniline resulted in the formation of PAni nanobelts because of the confinement of the surfactant gel.

Some nanostructural morphology of PAni could be prepared by the template-free or surfactant-free method (self-template method) [95, 112]. In this synthesis, the monomer of conducting polymers or its salts form micelles by themselves, which act as templates for the formation of nanostructures. Wan et al. [95] conducted a thorough study regarding its universality, controllability, and self-assembly mechanism by changing the polymeric chain length, polymerization method, dopant structure, and reaction conditions. They synthesized a variety of micro/nanotubes [113-115], nanofibers, nanotube junctions [116], and hollow microspheres [117] by the template-free method.

The structural parameters of PAni nanostructures were tunable by changing dopant structure, the redox potential of the oxidant, and reaction conditions. By varying the reaction temperature or the molar ratio of dopant to aniline, the PAni 3D hollow spheres, nanotubes, and dendrites with nanotube junctions could be selectively produced. The investigation of these groups shows that the template-free method, which is essentially a kind of a soft-template and self-assembly process, can be a simple and universal approach to synthesizing PAni micro/nanostructures [118].

5.4. Interfacial Polymerization

Interfacial polymerization is an alternative to bulk polymerization of condensation polymers. The process requires two immiscible solvents, with monomer in one solvent reacting with the oxidant species in the other solvent. The interfacial polymerization method provides higher molecular weights of polymer because the monomer is more likely to encounter a growing chain than the opposing monomer leading to termination of the polymerization.

Water-soluble PAni can be prepared via the interfacial polymerization route in the presence of poly (styrene sulfonated, sodium salt) (PSS) dispersed in the water phase. In the setup proposed by Detsri and Dubas [119], the aniline monomers and the PSS are dissolved in $CHCl_3$ organic phase and the aqueous phases, respectively, with the reaction taking place at the interface. PAni is formed at the interface where the aniline monomer and PSS are found within 10 min in a process by which aniline is protonated by the PSS in the presence of the oxidant, ammonium persulfate.

During this oxidative polymerization of aniline, the PSS acts as a proton source for the doping agent and provide the necessary counter ions to the developing charged PAni. The excess sulfonic groups can improve water-soluble properties and the resultant PAni easily disperse in aqueous solution. Figure 1 depicts the progress of PAni synthesized via chemical oxidative interfacial polymerization.

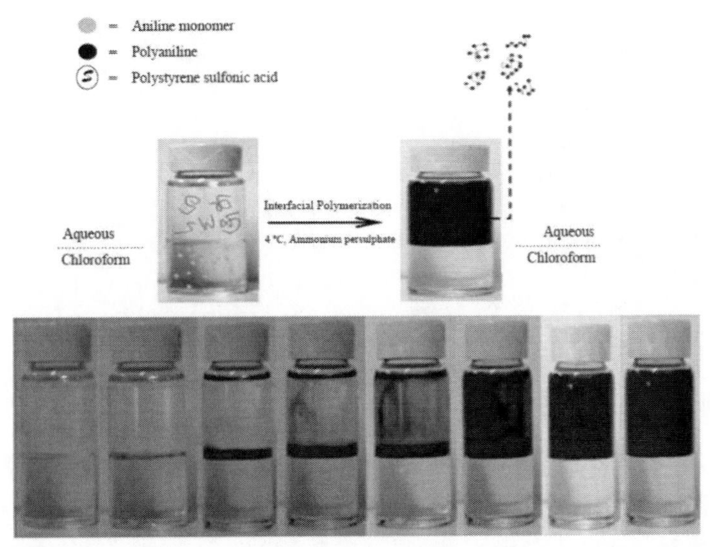

Reprinted from Detsri, E.; Dubas, S. T. *J. Met. Mater. Miner.* 2009, *19*, 39-44, with permission from Journal of Metals, Materials and Minerals.

Figure 1. Interfacial polymerization through the $CHCl_3$/water interface.

At the bottom layer, 10 mM of aniline monomer is dissolved in $CHCl_3$ organic solvent and in the top layer 10 mM of PSS, which is an anionic polymer is mixed with 5 mM ammonium persulfate as the oxidizing agent. The reaction takes place at 4° C without any stirring. The PAni is formed in the interface between organic and aqueous phase within 15 minutes and the reaction is completed within 4 hours. The green solution appears in the top aqueous layer.

5.5. Electrospinning

Electrospinning is a straightforward non-mechanical method to produce (meters) long polymer fibers in air [120]. In this process, the polymer is dissolved into a low boiling point solvent like chloroform, $CHCl_3$, to prepare a viscous solution. The solution is then placed into a pipette and tilted a few degrees below the baseplane depending on the viscosity thus creating a slow flow rate of about one drop per thirty seconds through the pipette. A high voltage (15 kV-25 kV) is then applied to a copper wire inserted into the polymer solution in order to generate an electrical field between the solution and a metallic screen (aluminum foil). As the voltage is increased, the electric field forces overcome the surface tension of the suspended polymer at the end of the pipette tip.

Once the voltage reaches a critical value, a single jet is produced. As the jet travels toward the metal cathode the solvent evaporates and the fibers are collected on the metal screen [121]. A schematic drawing of the electrospinning process is shown in Figure 2.

Through an improved or modified electrospinning device, nanofibers with part or even good orientation could be fabricated.

By now, micro- and nano-scale fibers of PAni/poly(ethylene oxide) (PEO), polypyrrole/ PEO, pure PAni and polypyrrole, poly (3-hexyl-thiophene)/PEO, and PAni/PEO/carbon nanotubes have been prepared by this technique.

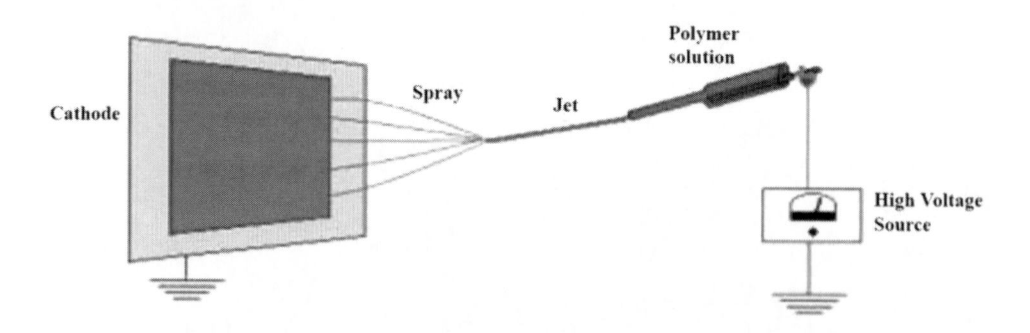

Figure 2. Schematic of the electrospinning process [49].

6. CHARACTERIZATION

For characterization of PAni and its composites, a wide number of techniques can be employed. For instances, IGC is used to study the thermodynamic properties; SEM, TEM and AFM are employed to reveal morphological behavior; TGA is used to investigate the thermal stability of the polymer and its composites; FT-IR is used to confirm the presence of functional groups of the component materials, and LCR meter and electrometer are used to examine the electric conductivity.

Also electroanalytical technique like cyclic voltammetry (CV) is performed to determine the redox behavior of conducting polymers and their composites.

6.1. Thermodynamic Characterization

Inverse gas chromatography (IGC), often known as molecular probe technique, is a promising method for characterization of polymers and polymer blends [122] and the method can be useful in obtaining thermodynamic data on polymeric systems even when the morphology is complex [123]. The term inverse refers to the polymer that can be studied in the solid-phase as the stationary phase, unlike conventional GC where the separation of solvents (solutes) is the prime interest. PAni has a limited solubility in solvents; however, IGC is a method of choice for the characterization of PAni because all small organic molecules will have measurable solubility in solid organic polymers.

The range of interactions which can be probed by the IGC technique is unlimited. IGC has been shown to be valuable for the identification of several types of interaction in molecular and macromolecular systems and for the characterization of the bulk and surface of finely divided materials. IGC has also been recognized for the study of the surface energetic of several systems and their response to actual conditions [124].

The attraction of the IGC method lies in its ability to generate fast and accurate physiochemical data on polymeric systems. It is a relatively rapid, convenient method and has the flexibility of selecting the desired temperature ranges.

Although IGC was successfully applied for the determination of the surface energy of several polymeric systems and fillers [125-139], only a few applications of IGC to conducting polymers were reported [140-142]. Conducting and insulating forms of polypyrrole and chemically synthesized polypyrrole doped by chloride (PPyCl) and its related dedoped form (PPyD) have received much of the attention using IGC [140-142]. The effect of dopant on the dispersive and specific properties, and the Lewis acid–base properties of polypyrrole were investigated [140-142]. Most IGC studies were directed to the composition, conductivity, and adhesion properties of these materials.

IGC was not widely used to characterize PAni and/or PAni based composites until 1999 owing to the difficulty of the solubility of PAni and/or PAni based composites and the preparation of a chromatographic column containing PAni and/or PAni based composites as a stationary phase. However, IGC has currently been used as a convenient and easy characterization method to study the thermodynamic properties of PAni and its innumerable composite materials.

Islam et al. [143] prepared a series of PAni-silica composite materials and characterized them by IGC and successfully unveiled the mechanism for change in surface thermodynamics upon incorporation of silica in the PAni matrix. IGC studies were made for PAni-silica composites using a mixture of several hydrocarbons in the range $C_5 - C_9$ and thermodynamic properties have been evaluated. The incorporation of silica in PAni increased the specific retention volumes of the probes, and more silica in polymer generally caused larger specific retention volumes, suggesting that introducing silica in polymers enhance the interactions between solvents and polymers (Figure 3).

They also demonstrated that incorporation of silica to composite materials could increase the dispersive surface free energy of polymer; which is ascribable to higher surface energy of inorganic silica content (Figure 4).

However, surface energy of PAni and its composites with silica of varying amount decreases with increasing temperature, which probably is due to the expansion of the surface with increasing temperature allowing the microstructure to commence reorganization [144].

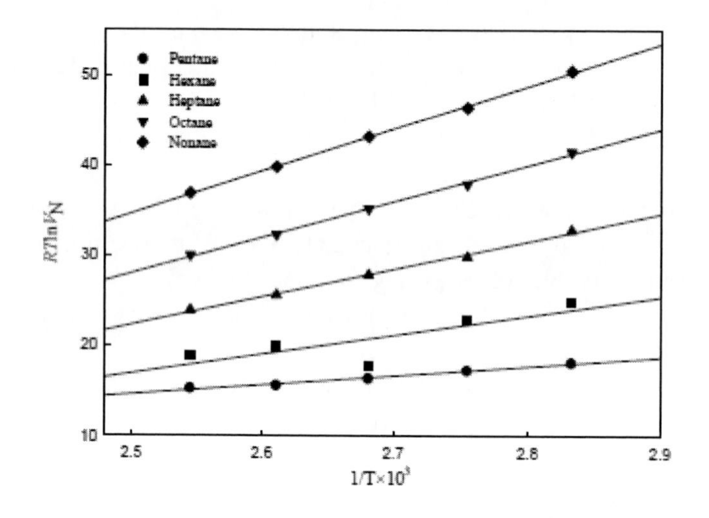

Figure 3. RTlnV_N versus $1/T$ plot for determining ΔH_a of various hydrocarbons (C_5-C_9) on PAni at different temperatures.

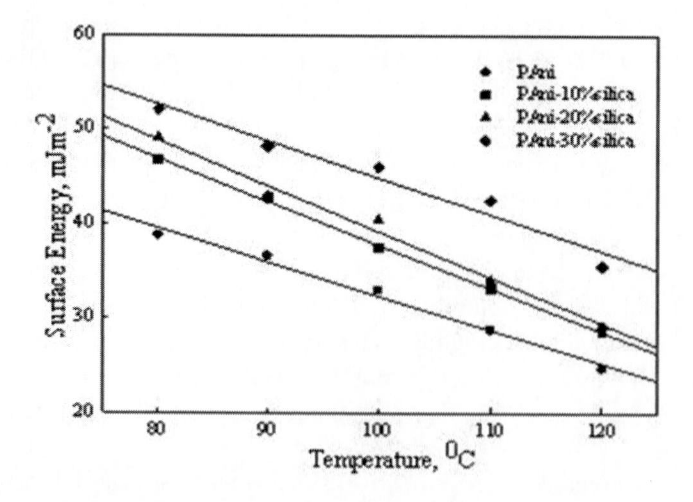

Figure 4. Surface energy versus temperature for n-alkanes (C_5-C_9).

It is also a useful tool to understand morphological change, crystallinity and surface energy of any polymeric systems. Wu et al. [144] used IGC to characterize both doped and undoped PAni, and its blend with nylon-6 using 27 solutes.

They reported that pure polymers were endothermic and exothermic for the blend. Blending nylon-6 with PAni lowered the dispersive surface energy of PAni while increasing the surface energy of nylon-6.

6.2. Thermal Characterization

Thermal analysis is a true workhorse for PAni characterization. Differential scanning calorimetry (DSC) is most pioneering technique for this purpose. The melting or glass transition of PAni is influenced with change in the composition and nature and structure of the additive and these in turn can be linked to the performances of PAni based composite materials. Moreover, to utilize PAni composites for intermediate or high temperature applications, it is necessary to examine its thermal stability. TGA is commonly used to determine polymer degradation temperatures, residual solvent levels, absorbed moisture content, and the amount of inorganic (noncombustible) filler in polymers or composite materials. TGA makes a continuous measurement of weight of a small sample in a controlled atmosphere (e.g., air or nitrogen) as the temperature is increased at a programmed linear rate.

A sample is placed into a sample pan which is attached to a sensitive microbalance assembly. The sample holder portion of the TGA balance assembly is subsequently placed into a high temperature furnace. The balance assembly measures the initial sample weight at room temperature and then continuously monitors changes in sample weight (losses or gains) as heat is applied to the sample. TGA tests may be run in a heating mode at some controlled heating rate, or isothermally. Typical weight loss profiles are analyzed for the amount or percent of noncombusted residue at some final temperature, and the temperatures of various sample degradation process. Islam et al. [143] have demonstrated that incorporation of silica into PAni matrix increases the thermal stability of the PAni-silica composite (Figure 5).

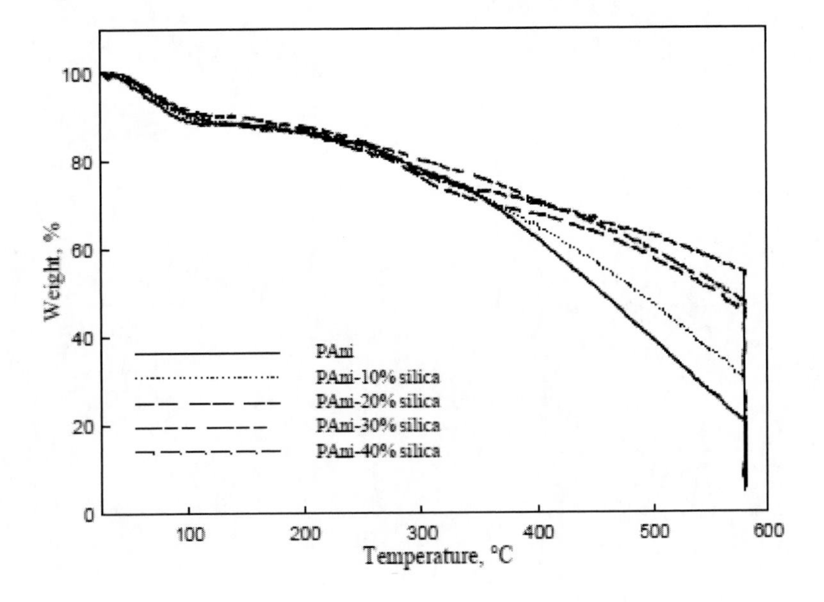

Figure 5. TGA curve of PAni and PAni-silica composites under nitrogen atmosphere.

6.3. Morphological Characterization

Morphological parameters, particularly on a mesoscale (nanometers to micrometers) provide valuable information regarding mechanical properties of many materials. Transmission electron microscopy (TEM) in combination with staining techniques has been the most useful tool for this purpose. In addition, scanning electron microscopy (SEM), scanning probe microscopy (SPM) and other forms of microscopy like atomic force microscopy (AFM) are important tools to optimize the morphology of materials like PAni and its many others composites and blends.

Akhtar et al. [145] worked on PAni-silica composite and reported that TEM image of PAni-40% silica composite they prepared showed an inhomogeneous pore distribution of the composite, although PAni had no significant pores (Figure 6). A portion of the composite was too dark in the TEM image, whereas other part contained bubble like structure to indicate the presence of few pores. Zhang et al. [146] synthesized PAni-ionic liquid-carbon nanofiber (PAni-IL-CNF) nanocomposite for use in amperometric biosensors for phenols and characterized the composite by SEM along with many other techniques (Figure 7). They reported that the composite have a fibrillar morphology with the diameter of around 95 nm, which is larger than 30–50 nm of the pristine CNF [147].

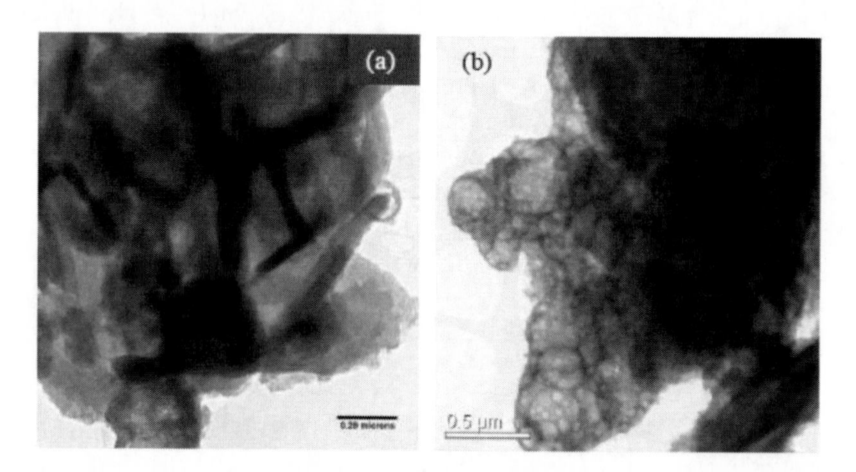

Figure 6. TEM images of (a) PAni and (b) PAni-silica composite.

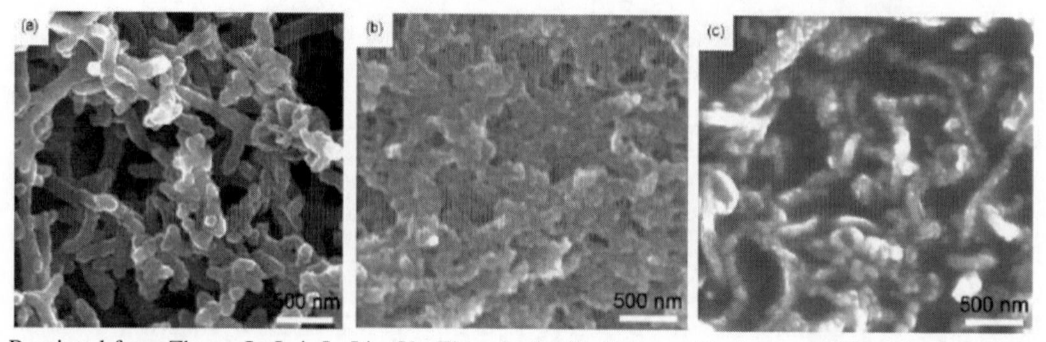

Reprinted from Zhang, J.; Lei, J.; Liu, Y.; Zhao, J.; Ju, H. *Biosens. Bioelectron.* 2009, *24*, 1858–1863 with permission from Elsevier.

Figure 7. SEM images of (a) PAni, (b) PAni–IL and (c) PAni–IL–CNF.

6.4. Spectroscopic Characterization

The spectroscopic characterization of PAni and its composites is usually carried out by Fourier transform infra-red (FT-IR) and ultraviolet-visible (UV-vis) spectroscopic techniques. UV-vis spectroscopy can be used to estimate the band gap energy and defect states, while FT-IR spectroscopy identifies and confirms the structure and presence of various linkages in polymer. UV-vis absorption spectroscopy measures the attenuation of the beam of light after it passes through a sample or after reflection from a sample surface. UV-Vis technique includes transmittance, absorption and reflection measurements in UV, visible and near infra red region.

On the other hand, FT-IR is one of the most powerful analytical technique, which offers the possibility over the other usual method of structural analysis (X-ray diffraction, electron spin resonance, etc.) in the sense that it provides useful information about the structure of the molecules and bonding quickly, without tiresome evaluation method.

Moreover, FT-IR helps rapid identification of chemical structures especially those of the organic ones. FT-IR and UV-vis techniques are frequently used to characterize PAni and its composite materials. Nabid et al. [148] prepared a PAni-TiO_2 nanocomposite and successfully confirmed the presence of PAni and TiO_2 nanoparticles in nanocomposites by FT-IR spectroscopy (Figure 8). Kim et al. [149] synthesized bio-compatible composites of PAni nanofibers and collagen and characterized these by UV-vis spectroscopy to show the different energy band gaps (Figure 9).

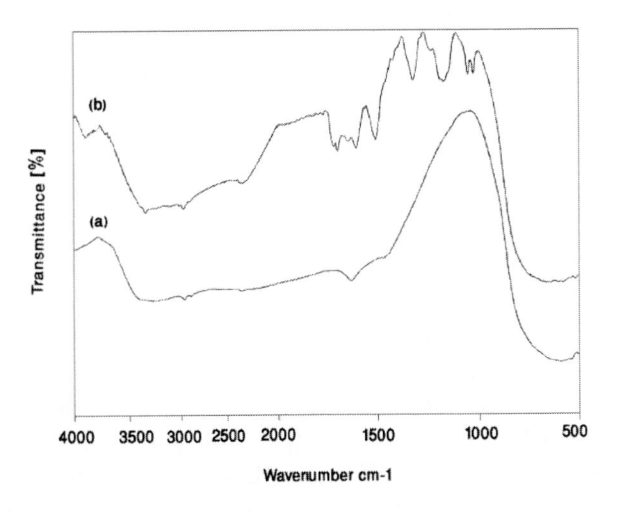

Reprinted from Nabid, M. R.; Golbabaee, M.; Moghaddam, A. B.; Dinarvand, R.; Sedghi, R. *Int. J. Electrochem. Sci.* 2008, *3*, 1117 – 1126 with permission from International journal of Electrochemical Science.

Figure 8. FT-IR spectra of (a) TiO_2 and (b) PAni/TiO_2 nanocomposite.

X-ray diffraction (XRD) and X-ray photoelectron spectroscopy (XPS) are also often used as spectroscopic techniques to study of the structure of PAni and its composites.

XPS is a very useful technique to understand different electronic structures of the conducting polymer samples. Lee and Char [150] worked on the thermal degradation behavior of PAni in PAni/Na^+-montmorillonite ((Na^+-MMT)) nanocomposites and reported that XRD data shows slight decrease of the basal spacing of the nanocomposites with a PAni content of 74.7 wt.% by about 0.1 nm, indicating the presence of PANI backbone chains with slightly deformed structure in the silicate nanolayers (Figure 10).

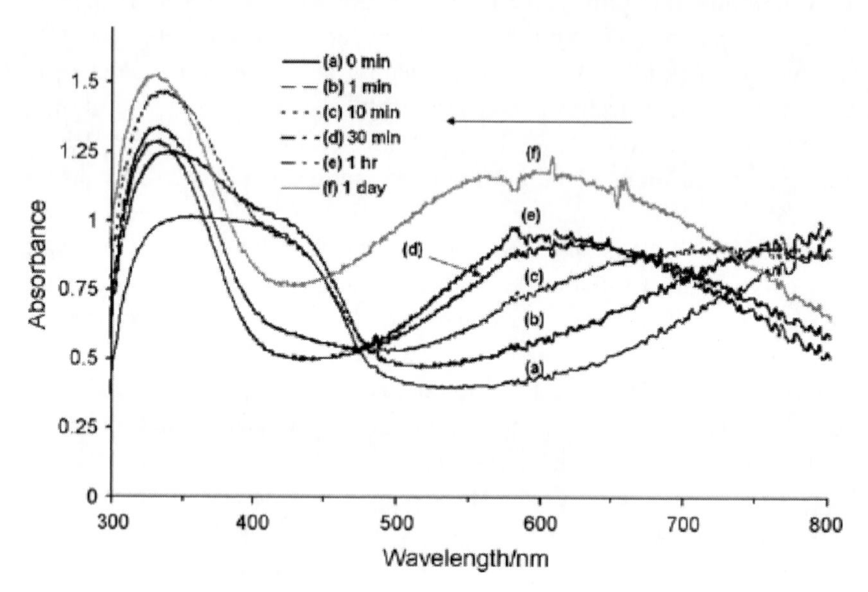

Reprinted from Kim, H. S.; Hobbsb, H. L.; Wanga, L.; Ruttena, M. J.; Wamserc, C. C. *Synth. Met.* 2009, *159*, 1313–1318 with permission from Elsevier.

Figure 9. UV–vis spectra of PAni–collagen composite film (3:1 ratio) dedoped in distilled water after 0, 1 min, 10 min, 30 min, 1 h, and 1 day.

6.5. Surface Area Characterization

Surface area is a measure of how much exposed area an object has. Brunauer-Emmett-Teller (BET) analysis provides precise specific surface area evaluation of materials by nitrogen multilayer adsorption measured as a function of relative pressure using a fully automated analyzer.

The technique encompasses external area and pore area evaluations to determine the total specific surface area in m^2/g yielding important information in studying the effects of surface porosity and particle size in many applications. Barrett-Joyner-Halenda (BJH) analysis can also be employed to determine pore area and specific pore volume using adsorption and desorption techniques.

This technique characterizes pore size distribution independent of external area due to particle size of the sample. Hwang et al. [151] prepared a PAni-silica composite from a water-in-oil microemulsion solution and reported BET surface area analysis results to conclude that the surface area of the porous silica decreases as PAni is incorporated in the pores of silica.

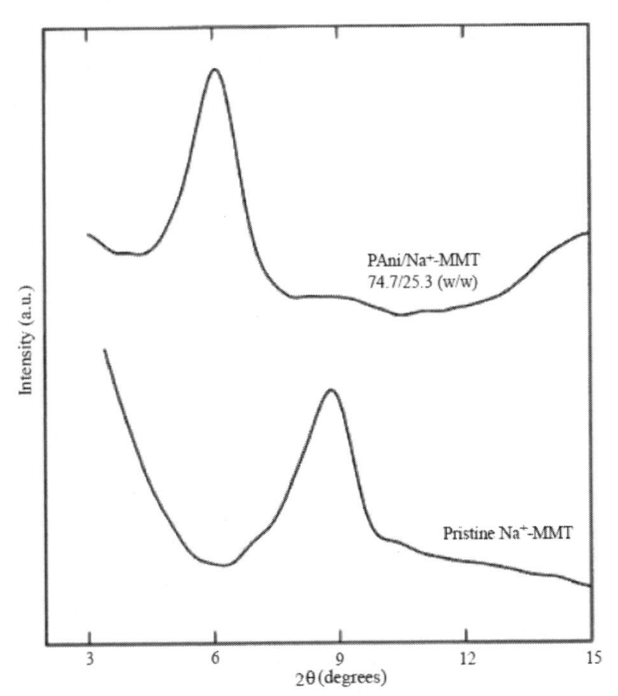

Reprinted from Lee, D.; Char, K. *Polym. Degrad. Stab.* 2002, *75*, 555–560, with permission from Elsevier.

Figure 10. XRD patterns of pristine Na^+-MMT and PAni/Na^+-MMT nanocomposites containing 74.7 wt.% of PAni.

Akhtar et al. [145] also characterized a composite of PAni-silica composite by BJH method. Figure 11(a) shows the adsorption isotherm of nitrogen at liquid nitrogen temperature on PAni-silica composite. It should be noted that PAni-silica underwent evaporation while heating during pre-treatment; therefore, it was processed without application of heat. The desorption of nitrogen at this temperature shows a hysterics branch. This indicates the presence of pores in the sample.

Desorption branch was analysed according to BJH method. Figure 11(b) shows the pore size distribution for PAni-silica composite. The distribution is rather inhomogeneous and particles with pore diameter of about 30 nm appear to be most abundant. Average pore size of PAni-silica was 280 and 175 Å during adsorption and desorption of nitrogen, respectively.

The BET surface area calculated by pore size shows that the surface area increased from 41.5 to 78.3 m^2g^{-1} due to the formation of PAni-silica composite.

6.6. Electroanalytical Characterization

The electroanalytical characterization has been used for determining the polymerization and redox behavior of PAni and its composite materials. Cyclic voltammetry (CV) has been employed for electroanalytical characterization. CV is the most versatile electroanalytical technique for the mechanistic study of redox behavour, number of electrons involved in the redox reaction, electrochemical studies, degradation studies and study of reversibility of redox couples in the conducting polymer systems. By varying the potential between working electrode and a counter electrode in an electrochemical cell, change of color (according to their oxidation level) from one state to another can be investigated. Redox couples can be characterized from the potentials of the peaks on the cyclic voltammogram (CV) and from the changes caused by variation of the scan rate.

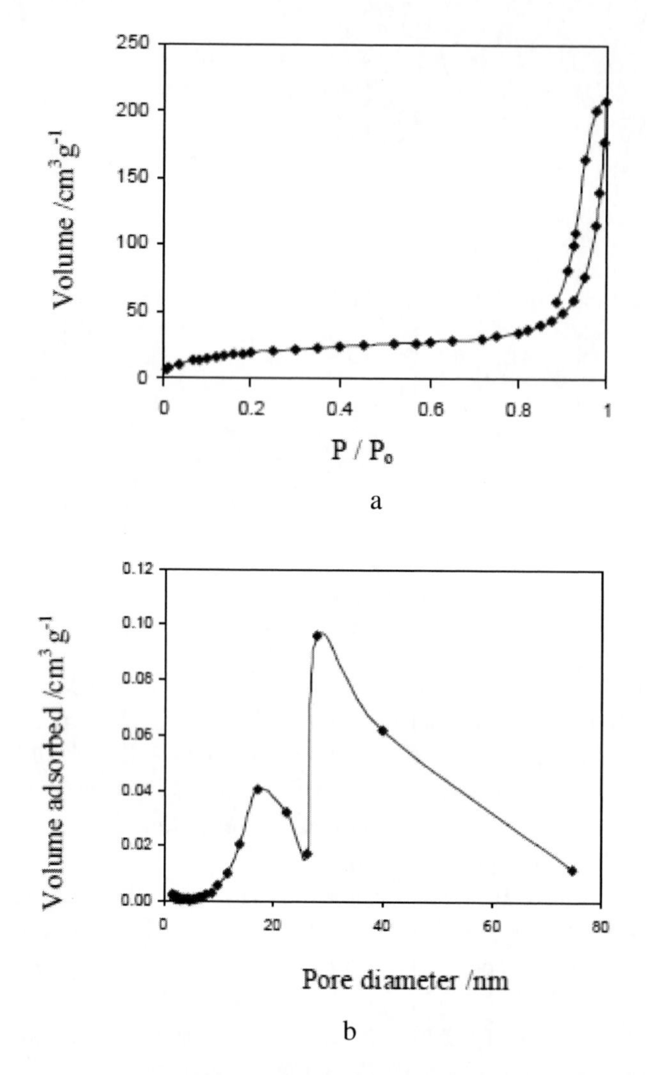

Figure 11. (a) N_2 adsorption–desorption isotherm and (b) pore size distribution for PAni-silica composite.

7. Advantages to Use PAni As a Matrix in Composites

Widely Tuning Conductive Ability

The electrical conductivity of PAni based compositions can simply be controlled over a wide range. Conductivity levels as high as 100 Scm^{-1} can be achieved when the structural units of PAni composites are highly ordered. The full range of conductivity levels from less than 10^{-10} to 10^{-1} Scm^{-1} (melt processing) and 10 Scm^{-1} (solution processing) can be achieved for polymer blends containing PAni compositions.

Highly Melt and Solution Processable Material

PAni based compositions can minimally be processed using conventional techniques such as blow and injection molding, extrusion, calendaring, film casting, and fiber spinning.

Without significant change in electrical properties, the temperature withstanding ability of PAni composite materials is as high as 230-240 °C for short periods of time (5-10 minutes), and can be blended with many commodity polymers.

Conductive Composites and Blends with Many Commodity Polymers

Simple and commonly used solution and melt processing methods can easily produce electrically conductive PAni based composites and blends with many commodity polymers.

Some of the commonly used commodity polymers are polyethylene, polypropylene, polystyrene, poly(vinyl chloride) (PVC), phenol-formaldehyde resins, and different types of thermoplastic elastomers. Unlike conventional filled materials, the mechanical properties of these composites and blends are close to those of the insulating matrix polymer.

Controllable Tinted and Transparent Electrically Conductive Products

As a result of having different oxidation state of PAni, it is easy to get a wide number of electrically conductive, colored and transparent thin films and coatings using PAni based compositions, which would otherwise be difficult to achieve with conventional filled materials.

8. Startling Applications of PAni and Its Composites

PAni has attracted significant attention since last few decades, however with the development of PAni nanofibers and dispersions only in the recent years it has turned to a processable material. The nanofibrillar morphology allows PAni to be used easily in multi-facet fields.

The processability and high surface area allow efficient use in applications such as sensors, artificial muscles, metal nanoparticle scaffold bistable memory devices, electrochromics, water purification, capacitors and many other areas. Improved control of the synthesis of PAni and its derivatives promise an even greater enhancement of the application range for these lightweight and inexpensive materials.

As Sensors in Chemical, Optical and Biotechnology

The electrical and optical properties of PAni can be reversibly changed by a doping/dedoping process. This has been exploited for use of nanostructured PAni and its composite materials for chemical sensors, optical sensors and biosensors due to greater exposure area of the nanostructured polymers than the conventional bulk polymers. Various reports demonstrate that the gas (HCl, NH_3, NO, H_2S, ethanol, liquefied petroleum gas [152], etc.) sensors based on PAni nanofibers usually show a higher sensitivity and shorter response time because of higher surface areas [153-163]. For example, it was found that the response times of the PAni nanofiber sensors exposed to chloroform, toluene and triethylamine were about a factor of 2 faster, with the current variations up to 4 times larger than those of the bulk PAni sensors [156]. Single PAni nanowire chemical sensors showed a rapid and reversible resistance change upon exposure to NH_3 gas at concentrations as low as 0.5 ppm. Pringsheim et al. [164] reported that fluorescent beads coated with PAni can be used as a novel optical pH sensor.

The fluorescence emitted by dye-doped polymer nanobeads is modulated by a thin conductive PAni coating, whose visible-light absorption varies with pH at values around 7. Gu et al. [165, 166] demonstrated a single waveguiding PAni/polystyrene nanowire for highly selective optical detection of gas. In addition, the photosensitivity and photoresponse of a bundle of PAni nanowires were also investigated by Wang et al. [167, 168], which showed that the photocurrent enhanced by 4–5 times under irradiation of an incandescence lamp (12 V, 10 W). These results indicate that PAni nanofibers might be useful in the fabrication of photosensor and photoswitch nanodevices.

In addition, Zhu et al. [169, 170] reported a pH sensor of PAni-perfluorosebacic acid (PFSEA)-coated fabric guided by reversible switching wettability through a PFSEA doping/NH_3 dedoping process. The PAni nanostructure-based sensors showed a fast response, stability, and good reproducibility by changing wettabilities from superhydrophobic (doping) to superhydrophilic (dedoping).

Biosensors based on PAni nanotubes/wires were also explored intensively in recent years [171-180]. For instance, Zhang et al. [172] reported poly(methyl vinyl ether-*alt*-maleic acid) doped PAni nanotubes for oligonucleotide sensors. In addition, Langer et al. [174] reported bacteria nanobiodetector based on a limited number of PAni nanofibers. The device works like an "ON-OFF" switch with nearly linear response above a threshold number of cells in the suspension examined, and may be useful in bio-alarm systems, environmental monitoring and medical applications.

As Transistors and Diodes

Transistors based on PAni nanofibers have been reported extensively [181-183]. Pinto et al. [181] reported an electrospun PAni/poly(ethylene oxide) nanofiber field-effect transistor. In addition, Alam et al. [182] reported electrolyte-gated transistors based on conducting PAni, polypyrrole and poly(3,4-ethylenedioxythiophene) (PEDOT) nanowire junction arrays. In the presence of a positive gate bias, the PAni nanowire-based transistors exhibit a large on/off current ratio of 978, which can vary according to the acidity of the gate medium. Lee et al. [183] also reported electrolyte-gated conducting PAni nanowire field-effect transistors.

Besides nanofiber field effect transitors, nanostructured polymer/ semiconductor (metal) diodes were also reported. Pinto et al. [184, 185] reported a Schottky diode using an *n*-doped Si/SiO_2 substrate and an electrospun fully doped PAni nanofiber.

As Light-Emitting Diodes, Field-Emission and Electrochromic Displays

Highly organized and well-aligned PAni has been explored for polymer light-emitting diodes [186-189], field emission [190-193] and electrochromic [194-198] displays. Boroumand et al. [174] fabricated arrays of nanoscale conjugated-polymer light-emitting diodes, each of which has a hole-injecting contact limited to 100 nm in diameter. Concerning field-emission applications, it was found that PAni nanofibers within hard template membrane showed stable field emission behavior with low threshold voltage of 5–6 Vm^{-1} and high emission current density of 5 $mAcm^{-2}$. This result suggests that PAni nanofibers/tubes can be used as nanotips for field emission displays. In addition, due to the ability to change color under an applied potential, PAni nanostructures have been investigated as the active layer in electrochromic devices.

As Energy Storage Material

Electrochemical studies on PAni nanostructures demonstrate that they usually have higher specific capacitance values and can be beneficial in the development of the next generation energy storage devices [199-204]. Zhang et al. reported that PAni nanofibers showed higher capacitance values and more symmetrical charge/discharge cycles due to their increased available surface area [201]. The highly dense arrays of ordered and aligned nanorods of PAni with 10 nm diameter were found to exhibit excellent electrochemical properties with an electrochemical capacitance value of 3407 Fg^{-1}. Fuel and photovoltaic cells also deal with polymeric nanofibers.

As Actuators

Potential applications of PAni bulk films and nanostructures on actuators or artificial muscles have also drawn much attention [205-212]. Recently, it was reported that flash-welded PAni nanofiber actuators demonstrate unprecedented reversible, rapid actuation upon doping [205]. An asymmetric nanofiber film with a thin dense layer on top of a thick porous nanofibrillar layer is formed by an intense flash of light; the layer can curl more than 72° on exposure to camphor sulfonic acid [205]. Actuation responses (or reversible volume changes with electrode potential) in PAni nanostructures were also investigated by electrochemical AFM [209-212].

Other Applications

Yang et al. [213] reported that PAni nanofibers have better corrosion protection for mild steel than conventional aggregated PAni. PAni microtubes/nanofibers [214, 215] and PAni-multiwalled carbon nanotubes nanocomposites [216] may be used as microwave absorbers and electromagnetic interference shielding materials. Conducting PAni nanofibers were used as nanofillers to improve the electrical properties of a ferroelectric copolymer [217]. In addition, PAni nanofibers decorated with gold nanoparticles exhibited very interesting bi-stable electrical behavior, and can be switched electrically between two states, which may be used in the fabrication of plastic digital non-volatile memory devices [218].

Conclusion

The ease of preparation and characterization associated with tunable properties and potential for diversified applications have catered PAni as a fascinating matrix for composite materials. A fundamental knowledge-base is already established, which may be successfully exploited to underpin further development. In fact, more versatile use of PAni based composite materials is not far away and promising applications of the fascinating materials for intelligent windows, transparent electrodes, EMI shielding, membranes for gas separation, metallization of printed circuit boards, solar cells and fuel cells are likely to be realized in due course of time.

Acknowledgments

The authors gratefully acknowledge financial support for a sub-project (CP-231) from the Higher Education Quality Enhancement Project of the University Grants Commission of Bangladesh financed by World Bank and the Government of Bangladesh. The authors also acknowledge partial financial support for a Research Project on *Syntheis and Characterization of Nanoparticles and their Polymer Composites for Biomedical Applications* from University Grants Commission of Bangladesh.

References

[1] Kang, E. T.; Neon, K. G.; Tan, K. L. *Prog. Polym. Sci.* 1998, 23, 277-324.

[2] Patil, A. O.; Heeger, A. J.; Wudl, F. *Chem. Rev.* 1988, 88, 183-200.

[3] Nakajima T., Kawagoe, T. *Synth. Met.* 1989, 28, 629-638.

[4] Tahir, Z. M.; Alocilia, E. C.; Grooms, D. L. *Biosens. Bioelectron.* 2005, 20, 1690-1695.

[5] Yang, C. H.; Chih, Y. K.; Wu, W. C.; Chen, C. H. *Electrochem. Solid-State Lett.* 2006, 9, 5-8.

[6] Reddy, K. R.; Sin, B. C.; Ryu, K. S.; Noh, J.; Lee, Y. *Synth. Met.* 2009, 159, 1934–1939.

[7] Zhang, B.; Sun, J.; Katz, H. E.; Fang, F.; Opila, R. L. *J. Am. Chem. Soc.* 2010, 2, 3170-3178.
[8] Tseng, R. J.; Huang, J.; Ouyang, J.; Kaner, R. B; Yang, Y. *Nano Lett.*, 2005, 5, 1077-1080.
[9] Liang, X.; Sun, M.; Li, L.; Qiao, R.; Chen, K.; Xiao, Q.; Xu, F. *Dalton Trans.* 2012, 41, 2804–2811.
[10] Khan, R.; Khare, P.; Baruah, B. P; Hazarika, A. K; Dey, N. C. *Adv. Chem. Eng. Sci.* 2011, 1, 37-44.
[11] Wei, D.; Ivaska, A. *Chem. Anal. (Warsaw)* 2006, 51, 839-852.
[12] *http://en.wikipedia.org/wiki/Polyaniline.*
[13] Alan, G.; MacDiarmid, A. G. *Synth. Met.* 2001, 40, 2581-2590.
[14] Huanga, L. M; Chena, C. H; Wen, T. C. *Electrochim. Acta* 2006, 51, 5858–5863.
[15] Virji, S.; Huang, J.; Kaner, R. B.; Weiller, B. H. *Nano Lett.* 2004, 4, 491–496.
[16] Callister, Jr. W. D. *Fundamentals of Materials Science and Engineering*, John Wiley and Sons Inc., New York, 2001.
[17] Muraviev, D. N.; Macanas, J.; Farre, M.; Munoz, M.; Alegret, S. *Sens. Actuators B* 2006, 118, 408-417.
[18] Tian, S.; Liu, J.; Zhu, T.; Knoll, W. *Chem. Mater.* 2004, 16, 4103-4108.
[19] Alivisatos, A. P. *Science* 1996, 271, 933-937.
[20] Huang, J.; Kaner, R. B. *J. Am. Chem. Soc.* 2004, 126, 851-855.
[21] Dan, L. I.; Huang, J.; Kaner, R. B. *Acc. Chem. Res.* 2009, 42, 135-145.
[22] Athawale, A. A.; Bhagwat, S. V.; Katre, P. P. *Sens. Actuators B* 2006, 114, 263-267.
[23] Dimitrriev, O. P. *Polym. Bull.* 2003, 50, 83–90.
[24] Yang, W.; Liu, J.; Zheng, R.; Liu, Z.; Dai, Y.; Chen, G.; Ringer, S.; Braet, F. *Nanoscale Res. Lett.* 2008, 3, 468–472.
[25] Grzeszczuk, M.; Poks, P. *Electrochim. Acta.* 2000, 45, 4171-4177.
[26] Liu, F. J.; Huang, L. M.; Wen, T. C.; Gopalan, A.; Hung, J. S. *Mater. Lett.* 2007, 61, 4400-4405.
[27] Pillalamarri, S. K.; Blum, F. D.; Tokuriho, A. T.; Bertino, M. F. *Chem. Mater.* 2005, 17, 5941-5944.
[28] Shu-hong, H. U.; Hao-dong, P. E. I.; Xin-bing, Z. H. A. O. *Tran. Nonferrous Met. Soc.* 2001, 11, 876-878.
[29] Tudorache, F.; Grigoras, F. *Optoelectron. Adv. Mat.* 2010, 4, 43-47.
[30] Conn, C.; Sestak, S.; Baker, A. T.; Unsworth, J. *Electroanalysis* 1998, 10, 1137-1141.
[31] Wang, J.; Matsubara, I.; Murayama, N.; Woosuck, S.; Izu, N. *Thin Solid Films* 2006, 514, 329-333.
[32] Parvatikar, N.; Jain, S.; Khasim, S.; Revansiddappa, M.; Bhoraskar, S. V.; Prasad, M. A. *Sens. Actuators B* 2006, 114, 599-603.
[33] Ram, M. K.; Yavuz, O.; Aldissi, M. *Synth. Met.* 2005, 151, 77-84.
[34] Geng, L.; Zhao, Y.; Huang, X.; Wang, S.; Zhang, S.; Wu, S. *Sens. Actuators B* 2007, 120, 568-572.
[35] Jumali, M. H. H.; Ramli, N.; Izzuddin, I.; Salleh, M.; Yahaya, M. *Sains. Malays.* 2011, 40, 203-208.
[36] Mho, S. I.; Thieu, M. T.; Kim, Y.; Yeo, I. H. "Preparation and characterization of size-controlled V_2O_5/polyaniline composite electrodes for li battery", *Presented on the*

Seventeenth Annual International Conference on Composites/Nano Engineering, Hawaii, US, 2009.

[37] Ahn, D.; Koo, Y. M.; Kim, M. G.; Shin, N.; Park, J.; Eom, J.; Cho, J.; Shin, T. J. *J. Phys. Chem. C* 2010, 114, 3675–3680.

[38] Virji, S.; Kojima, R.; Fowler, J. D.; Kaner, R. B.; Weiller, B. H. *Chem. Mater.* 2009, 21, 3056–3061.

[39] Virji, S.; Fowler, J. D.; Baker, C. O.; Huang, J.; Kaner, R. B.; Weiller, B. H. J. *Small* 2005, 1, 624-627.

[40] Li, J.; Cui, M.; Fang, J.; Zhang, Z.; Lai Y. *J. Cent. South Univ. Technol.* 2011, 18, 78−82.

[41] Sarkar, D.; Misra, T. N.; Paul, A. *Thin Solid Films* 1993, 30, 255-259.

[42] Armes, S. P.; Aldossi, M.; Agnew, S. F.; Gottesfeld, S. *Langmuir* 1990, 6, 1745-1749.

[43] Han, M. G.; Cho, S. K.; Oh, S. G.; Im, S. S. *Synth. Met.* 2002, 126, 53-60.

[44] Marten, F. L.; Kroschwitz, J. I. *Concise Encyclopedia of Polymer Science and Engineering,* John Wiley and Sons Inc., New York, 1990, 1233-1239.

[45] Ray, S. S.; Pouliot, S.; Bousmin, M.; Utracki, L. A. *Polym.* 2004, 45, 8403-8413.

[46] Morgan, H.; Foot, P. J. S. *J. Mat. Sci.* 2001, 36, 5369 – 5377.

[47] Makeiff, D. A.; Foster, T.; Foster, K. *Technical Memorandum DRDC Atlantic TM,* 2005, 204-301.

[48] Rudin, A. *The Elements of Polymer Science and Engineering,* Academic Press, Massachusetts, US, 1998.

[49] Manuel, J. D. L. *Proceeding of the National Conference on Undergraduate Research,* University of Kentucky, 2001.

[50] Sailor, M. J; Curtis, C. L. *Adv. Mater.* 1994, 6, 688-692.

[51] Neves, S.; Gazotti, W. A.; De Paoli, M. A. *Encyclopedia of Nanoscience and Nanotechnology,* H. S. Nalwa (Ed.), American Scientific Publishers, Los Angeles, 2004, 2, 133-152.

[52] Gangopadhyay, R. *Encyclopedia of Nanoscience and Nanotechnology,* H. S. Nalwa (Ed.), American Scientific Publishers, Los Angeles, 2004, 2, 105-131.

[53] Wallace, G. G.; Innis, P. C.; Kane-Maguire, L. A. P. *Encyclopedia of Nanoscience and Nanotechnology,* H. S. Nalwa (Ed.), American Scientific Publishers, Los Angeles, 2004, 4, 113-130.

[54] Epstein, A. J. *Organic Electronic Materials: Conjugated Polymers and Low Molecular Weight Organic Solids,* R. Farchioni and G. Grosso (Eds.), Springer, Amsterdam, 2001, 41, 3.

[55] Martin, C. R. *Acc. Chem. Res.* 1995, 28, 61-68.

[56] Kulkarni, M. V.; Charhate, N. A.; Bhavsar, K. V.; Tathe, M. A.; Kale, B. B. Development of polyaniline-multiwalled carbon nanotube (PAni-MWCNT) nanocomposite for optical pH sensor, 14th International Meeting on Chemical Sensors, Nuremberg, Germany, 2012, 934-937.

[57] Zhao, L.; Zhao, L.; Xu, Y.; Qiu, T.; Zhi, L.; Shi, G. *Electrochim. Acta* 2009, 55, 491-497.

[58] Yan, J.; Wei, T.; Fan, Z.; Qian, W.; Zhang, M.; Shen, X. *J. Power Sources* 2010, 195, 3041–3045.

[59] Levi, B. G. *Phys. Today* 2000, 53, 19-22.

[60] MacDiarmid, A. G. *Angew. Chem. Int. Ed.* 2001, 40, 2581-2590.

[61] Jin, Z.; Su, Y.; Duan, Y. *Sens. Actuators B* 2001, 72, 75-79.
[62] Sotomayor, P. T.; Raimundo, I. M.; Zarbin, A. J. G.; Rohwedder, J. J. R.; Netto, G. O.; Alves, O. L. *Sens. Actuators B* 2001, 74, 157-162.
[63] Kane-Maguire, L. A. P.; Wallace, G. G. *Synth. Met.* 2001, 119, 39-42.
[64] Hamers, R. J. *Nature* 2001, 412, 489-490.
[65] Rosseinsky, D. R.; Mortimer, R. J. *Adv. Mater.* 2001, 13, 783-793.
[66] Prokes, J.; Krivka, I.; Tobolkova, E.; Stejskal, J. *Polym. Degrad. Stab.* 2000, 68, 261-269.
[67] Elyashevich, G. K.; Terlemezyan, L.; Kuryndin, I. S; Lavrentyev, V. K; Mokreva, P.; Rosova, E. Y.; Sazanov, Y. N. *Thermochim. Acta* 2001, 374, 23-30.
[68] El-Sherif, M. A; Yuan, J.; MacDiarmid, A. J. *J. Intelligent Mater. Syst. Struct.* 2000, 11, 407-414.
[69] Trivedi, D. C. *In: Handbook of Organic Conductive Molecules and Polymers*, Wiley, New York, 1997, 2, 505-572.
[70] Wei, Y.; Jang, G. W.; Chan, C. C.; Hsuen, K. F.; Hariharan, R.; Patel, S. A.; Whitecar, C. K. *J. Phys. Chem.* 1990, 94, 7716–7721.
[71] Wei, Y.; Tang, X.; Sun, Y.; Focke, W. W. *J. Polym. Sci.* 1989, 27, 2385–2396.
[72] Wei, Y.; Hariharan, R.; Patel, S. A. *Macromolecules* 1990, 23, 758–764.
[73] Wei, Y.; Hsueh, K. F.; Jang, G. W. *Polymer* 1994, 35, 3572–3575.
[74] Martin, C. R. *Science* 1994, 266, 1961–1966.
[75] Martin, C. R. *Acc. Chem. Res.* 1995, 28, 61–68.
[76] Granstrom, M.; Inganas, O. *Polymer* 1995, 36, 2867–2872.
[77] Cai, Z. H. L.; Lei, J. T.; Liang, W. B.; Menon, V.; Martin, C. R. *Chem. Mater.* 1991, 3, 960–967.
[78] Duchet, J.; Legras, R.; Demoustier-Champagne, S. *Synth. Met.* 1998, 98, 113–122.
[79] Duvail, J. L.; Retho, P.; Fernandez, V.; Louarn, G.; Molinie, P.; Chauvet, O. *J. Phys. Chem. B* 2004, 108, 18552–185556.
[80] Mativetsky, J. M.; Datars, W. R. *Physica B* 2002, 324, 191–204.
[81] Kim, B. H.; Park, D. H.; Joo, J.; Yu, S. G.; Lee, S. H. *Synth. Met.* 2005, 150, 279–284.
[82] Massuyeau, F.; Duvail, J. L.; Athalin, H.; Lorcy, J. M.; Lefrant, S.; Wery, J.; Faulques, E. *Nanotechnology* 2009, 20, 155701-155708.
[83] Cai, Z. H.; Martin, C. R. *J. Am. Chem. Soc.* 1989, 111, 4138-4139.
[84] Martin, C. R.; Vandyke, L. S.; Cai, Z. H.; Liang, W. B. *J. Am. Chem. Soc.* 1990, 112, 8976-8977.
[85] Beadle, P.; Armes, S. P.; Gottesfeld, S.; Mombourquette, C.; Houlton, R.; Andrews, W. D.; Agnew, S. F. *Macromolecules* 1992, 25, 2526-2530.
[86] Niu, Z.; Liu, J.; Lee, L. A.; Bruckman, M. A.; Zhao, D.; Koley, G.; Wang, Q. *Nano Lett.* 2007, 7, 3729-3733.
[87] Niu, Z. W.; Bruckman, M. A.; Li, S. Q.; Lee, L. A.; Lee, B.; Pingali, S. V.; Thiyagarajan, P.; Wang, Q. *Langmuir* 2007, 23, 6719-6724.
[88] Shyh-Chyang, L.; Hsiao-hua, Y.; Wan, A.; Yu, H.; Ying, J. Y. *Small* 2008, 4, 2051-2058.
[89] Luo, S. C.; Yu, H. H.; Wan, A. C. A.; Han, Y.; Ying, J. Y. *Small* 2008, 4, 2051-2058.
[90] Fu, G. D.; Zhao, J. P.; Sun, Y. M.; Kang, E. T.; Neoh, K. G. *Macromolecules* 2007, 40, 2271-2275.

[91] Yang, M.; Ma, J.; Zhang, C. L.; Yang, Z. Z.; Lu, Y. F. *Angew. Chem.: Int. Edit.* 2005, 44, 6727-6730.
[92] Niu, Z. W.; Yang, Z. H.; Hu, Z. B.; Lu, Y. F.; Han, C. C. *Adv. Funct. Mat.* 2003, 13, 949-954.
[93] Feng, X. M.; Mao, C. J.; Yang, G.; Hou, W. H.; Zhu, J. J. *Langmuir* 2006, 22, 4384-4389.
[94] Wu, Q.; Wang, Z. Q.; Xue, G. *Adv. Funct. Mat.* 2007, 17, 1784-1789.
[95] Wan, M. X. *Adv. Mater.* 2008, 20, 2926-2932.
[96] Zhang, Z. M.; Sui, J.; Zhang, L. J.; Wan, M. X.; Wei, Y.; Yu, L. M. *Adv. Mater.* 2005, 17, 2854-2857.
[97] Penner, R. M.; Martin, C. R. *J. Electrochem. Soc.* 1987, 134, C504-C504.
[98] Cai, Z. H.; Lei, J. T.; Liang, W. B.; Menon, V.; Martin, C. R. *Chem. Mat.* 1991, 3, 960-967.
[99] Liu, C.; Martin, C. R. *Nature* 1991, 352, 50-52.
[100] Penner, R. M.; Martin, C. R. *J. Electrochem. Soc.* 1986, 133, 2206-2207.
[101] Martin, C. R. *Science* 1994, 266, 1961-1966.
[102] Martin, C. R.; Parthasarathy, R.; Menon, V. *Electrochim. Acta* 1994, 39, 1309-1313.
[103] Martin, C. R. *Accounts Chem. Res.* 1995, 28, 61-68.
[104] Mohammadi, A.; Lundstrom, I.; Inganas, O. *Synth. Met.* 1991, 41, 381-384.
[105] Parthasarathy, R. V.; Martin, C. R. *Nature* 1994, 369, 298-301.
[106] Lahav, M.; Weiss, E. A.; Xu, Q. B.; Whitesides, G. M. *Nano Lett.* 2006, 6, 2166-2171.
[107] Wu, C. G.; Bein, T. *Science* 1994, 264, 1757-1759.
[108] Wessling, B. *Adv. Mater.* 1993, 5, 300-305.
[109] Zhou, C. Q.; Han, J.; Guo, R. *Macromol. Rapid Commun.* 2009, 30, 182-187.
[110] Zhou, C. Q.; Han, J.; Guo, R. *Macromolecules* 2009, 42, 1252-1257.
[111] Zhou, C. Q.; Han, J.; Guo, R. *J. Phys. Chem. B* 2008, 112, 5014-5019.
[112] Wan, M. X. *Macromol. Rapid Commun.* 2009, 30, 963-975.
[113] Zhang, Z. M.; Wan, M. X.; Wei, Y. *Adv. Funct. Mater.* 2006, 16, 1100-1104.
[114] Zhang, L. J.; Wan, M. X. *Adv. Funct. Mater.* 2003, 13, 815-820.
[115] Qiu, H. J.; Wan, M. X.; Matthews, B.; Dai, L. M. *Macromolecules* 2001, 34, 675-677.
[116] Wei, Z. X.; Zhang, L. J.; Yu, M.; Yang, Y. S.; Wan, M. X. *Adv. Mater.* 2003, 15, 1382-1385.
[117] Zhu, Y.; Hu, D.; Wan, M. X.; Jiang, L.; Wei, Y. *Adv. Mater.* 2007, 19, 2092-2096.
[118] Zhang, L. J.; Zujovic, Z. A.; Peng, H.; Bowmaker, G. A.; Kilmartin, P. A.; Travas-Sejdic, J. M*acromolecules* 2008, 41, 8877-8884.
[119] Detsri, E.; Dubas, S. T. *J. Met. Mater. Miner.* 2009, 19, 39-44.
[120] Formhals, A. *US Patent 1*, 1934, 975, 504.
[121] Fong, H.; Chun, I.; Reneker, D. H. *Polymer* 1999, 40, 4585-4592.
[122] Al-Saigh, Z. Y. *Int. J. Polym. Charac. Anal.* 1997, 3, 249-291.
[123] Al-Saigh, Z. Y.; Chen, P. *Macromolecules* 1991, 24, 3788-3795.
[124] Guillet, J.; Al-Saigh, Z. Y. *Encyclopedia of Analytical Chemistry: Instrumentation and Applications, Wiley, Chichester,* 2000, 9, 7759-7792.
[125] Balard, H.; Papirer, E. *Progr. Org. Coat.* 1993, 22, 1-17.
[126] Papirer, E.; Schultz, J.; Turchi, C. *J. Eur. Polym.* 1984, 20, 1155-1158.
[127] Papirer, E.; Schultz, J.; Jagiello, J.; Baeza, R.; Clauss, R. F.; Mottola, H. A.; Steinmetz (Eds.), J. R. *Chemically Modified Sur faces, Elsevier, Amsterdam,* 1992, 8, 334-352.

[128] Kuczinski, J.; Papirer, E. *J. Eur. Polym.* 1991, 27, 653-655.
[129] Papirer, E.; Roland, P.; Nardin, M.; Balard, H. *J. Colloid Interface Sci.* 1985, 113, 62-65.
[130] Papirer, E.; Balard, H.; Vidal, A. *J. Eur. Polym.* 1988, 24, 783-790.
[131] Ligner, G.; Vidal, A.; Balard, H.; Papirer, E. *J. Colloid Interface Sci.* 1989, 133, 200-210.
[132] Papirer, E.; Eckhardt, A.; Muller, F.; Yvon, J. *J. Mater. Sci.* 1990, 25, 5109-5117.
[133] Chehimi, M. M.; Abel, M. L.; Landureau, E. P.; Delamar, M. *J. Synth. Met.* 1993, 60, 183-194.
[134] Schmitt, P.; Koerper, E.; Schultz, J.; Papirer, E. *J. Chromato Graphia.* 1988, 25, 786-790.
[135] Flour, C. S.; Papirer, E. *J. Colloid Interface Sci.* 1983, 91, 69-75.
[136] Chen, F. *Macromolecules* 1988, 21, 1640-1643.
[137] Rubio, F.; Rubio, J.; Oteo, J. L. *J. Sol–Gel Sci. Technol.* 2000, 18, 115-130.
[138] Al-Saigh, Z. Y. *Polym. J.* 1999, 40, 3479-3485.
[139] Al-Saigh, Z. Y. *Polym. Int.* 1996, 40, 25-32.
[140] Chehimi, M. M.; Pigois-Landureau, E.; Delamar, M. M. *J. Chim. Phys.* 1992, 89, 1173-1178.
[141] Landureau, E.; Chehimi, M. M. *J. Appl. Polym. Sci.* 1993, 49, 183-194.
[142] Chehimi, M. M; Abel, M. L; Perruchot, C.; Delamar, M.; Lascelles, S. F.; Armes, S. P. *Synth. Met.* 1999, 104, 51-59.
[143] Islam, M. S.; Miran, M. S.; Mollah, M. Y. A.M.; Rahman, M. M.; Susan, M. A. B. H. *J. Nanostructured Polym. Nanocomposites*, 2012, in press.
[144] Wu, R.; Danni, Q.; Al-Saigh, Z. Y. *J. Chromatogr. A* 2007, 1146, 93–102.
[145] Akhtar, U. S.; Miran, M. S.; Susan, M. A. B. H.; Mollah, M. Y. A. M.; Rahman, M. M. *Bangladesh J. Sci. Ind. Res.* 2012, 47, 249-256.
[146] Zhang, J.; Lei, J.; Liu, Y.; Zhao, J.; Ju, H. *Biosens. Bioelectron.* 2009, 24, 1858–1863.
[147] Wu, L. N.; Zhang, X. J.; Ju, H. X. *Anal. Chem.* 2007, 79, 453–458.
[148] Nabid, M. R.; Golbabaee, M.; Moghaddam, A. B.; Dinarvand, R.; Sedghi, R. *Int. J. Electrochem. Sci.* 2008, 3, 1117 – 1126.
[149] Kim, H. S.; Hobbsb, H. L.; Wanga, L.; Ruttena, M. J.; Wamserc, C. C. *Synth. Met.* 2009, 159, 1313–1318.
[150] Lee, D.; Char, K. *Polym. Degrad. Stab.* 2002, 75, 555–560.
[151] Hwang, T.; Lee, H. Y; Kim, H; Kim, G. *Adv.Control, Chem. Eng., Civil Eng., Mech. Eng.*, 2010, 236-240.
[152] Rong, J. H.; Oberbeck, F.; Wang, X. N.; Li, X. D.; Oxsher, J.; Niu, Z. W.; Wang, Q. *J. Mater. Chem.* 2009, 19, 2841-2845.
[153] Huang, J. X.; Virji, S.; Weiller, B. H.; Kaner, R. B. *J. Am. Chem. Soc.* 2003, 125, 314–315.
[154] Virji, S.; Huang, J. X.; Kaner, R. B.; Weiller, B. H. *Nano Lett.* 2004, 4, 491–496.
[155] Sutar, D. S.; Padma, N.; Aswal, D. K.; Deshpande, S. K.; Gupta, S. K.; Yakhmi, J. V. *Sens. Actuat. B-Chem.* 2007, 128, 286–292.
[156] Li, Z. F.; Blum, F. D.; Bertino, M. F.; Kim, C.S.; Pillalamarri, S. K. *Sens. Actuat. B-Chem.* 2008, 134, 31–35.
[157] Yoon, H.; Chang, M.; Jang, J. *Adv. Funct. Mater.* 2007, 17, 431–436.
[158] Chen, Y. X.; Luo, Y. *Adv. Mater.* 2009, 21, 2040–2044.

[159] Rajesh; Ahuja, T.; Kumar, D. *Sens. Actuat. B-Chem.* 2009, 136, 275–286.

[160] Antohe, V. A.; Radu, A.; Matefi-Tempfli, M.; Attout, A.; Yunus, S.; Bertrand, P.; Dutu, C.A.; Vlad, A.; Melinte, S.; Matefi-Tempfli, S.; Piraux, L. *Appl. Phys. Lett.* 2009, 94, 1-3.

[161] Shirsat, M. D.; Bangar, M. A.; Deshusses, M. A.; Myung, N. V.; Mulchandani, A. *Appl. Phys. Lett.* 2009, 94, 083502/1-3.

[162] Bai, H.; Zhao, L.; Lu, C. H.; Li, C.; Shi, G. Q. *Polymer* 2009, 50, 3292–3301.

[163] Yu, X. F.; Li, Y. X.; Kalantar-zadeh, K. *Sens. Actuat. B-Chem.* 2009, 136, 1–7.

[164] Pringsheim, E.; Zimin, D.; Wolfbeis, O. S. *Adv. Mater.* 2001, 13, 819–822.

[165] Gu, F. X.; Zhang, L.; Yin, X. F.; Tong, L. M. *Nano Lett.* 2008, 8, 2757–2761.

[166] Gu, F. X.; Yin, X. F.; Yu, H. K.; Wang, P.; Tong, L. M. *Opt. Express* 2009, 17, 11230–11235.

[167] Wang, X. H.; Shao, M. W.; Shao, G.; Fu, Y.; Wang, S. W. *Synth. Met.* 2009, 159, 273–276.

[168] Wang, X. H.; Shao, M. W.; Shao, G.; Wu, Z. C.; Wang, S. W. *J. Colloid. Inerface Sci.* 2009, 332, 74–77.

[169] Zhu, Y.; Feng, L.; Xia, F.; Zhai, J.; Wan, M. X.; Jiang, L. *Macromol. Rapid Commun.* 2007, 28, 1135–1141.

[170] Zhu, Y.; Li, J. M.; Wan, M. X.; Jiang, L. *Macromol. Rapid Commun.* 2007, 28, 2230–2236.

[171] Gao, M.; Dai, L.; Wallace, G. G. *Electroanalysis* 2003, 15, 1089–1094.

[172] Zhang, L. J.; Peng, H.; Kilmartin, P. A.; Soeller, C.; Travas-Sejdic, J. *Electroanalysis* 2007, 19, 870–875.

[173] Peng, H.; Zhang, L. J.; Spires, J.; Soeller, C.; Travas-Sejdic, J. *Polymer* 2007, 48, 3413–3419.

[174] Langer, J. J.; Langer, K.; Barczynski, P.; Warcho, J.; Bartkowiak, K. H. *Biosens. Bioelectron.* 2009, 24, 2947–2949.

[175] Horng, Y. Y.; Hsu, Y. K.; Ganguly, A.; Chen, C. C.; Chen, L. C.; Chen, K. H. *Electrochem. Commun.* 2009, 11, 850–853.

[176] Xie, H.; Luo, S. C.; Yu, H. H. *Small* 2009, 5, 2611–2617.

[177] Huang, J.; Wei, Z. X.; Chen, J. C. *Sens. Actuat. B-Chem.* 2008, 134, 573–578.

[178] Yang, J. Y.; Martin, D. C. *Sens. Actuat. B-Chem.* 2004, 101, 133–142.

[179] Herland, A.; Inganas, O. *Macromol. Rapid Commun.* 2007, 28, 1703–1713

[180] Xia, L.; Wei, Z. X.; Wan, M. X. *J. Colloid Interface Sci.* 2010, 341, 1–11.

[181] Pinto, N. J.; Johnson Jr, A. T.; Mueller, C. H.; Theofylaktos, N.; Robinson, D. C.; Miranda, F. A. *Appl. Phys. Lett.* 2003, 83, 4244–4246.

[182] Alam, M. M.; Wang, J.; Guo, Y. Y.; Lee, S. P.; Tseng, H. R. *J. Phys. Chem. B* 2005, 109, 12777–12784.

[183] Lee, S. Y.; Choi, G. R.; Lim, H.; Lee, K. M.; Lee, S. K. *Appl. Phys. Lett.* 2009, 95, 013113/1-3.

[184] Pinto, N. J.; Gonzalez, R.; Johnson Jr, A. T.; MacDiarmid, A. G. *Appl. Phys. Lett.* 2006, 89, 033505/1-3.

[185] Perez, R.; Pinto, N. J.; Johnson Jr, A. T. *Synth. Met.* 2007, 157, 231–234.

[186] Granstrom, M.; Berggren, M.; Inganas, O. *Science* 1995, 267, 1479–1481.

[187] Granstrom, M.; Berggren, M.; Inganas, O. *Synth. Met.* 1996, 76, 141–143.

[188] Boroumand, F. A.; Fry, P. W.; Lidzey, D. G. *Nano Lett.* 2005, 5, 67–71.

[189] Grimsdale, A. C.; Chan, K. L.; Martin, R. E.; Jokisz, P. G.; Holmes, A. B. *Chem. Rev.* 2009, 109, 897–1091.
[190] Kim, B. H.; Park, D. H.; Joo, J.; Yu, S. G.; Lee, S. H. *Synth. Met.* 2005, 150, 279–284.
[191] Wang, C. W.; Wang, Z.; Li, M. K.; Li, H. L. *Chem. Phys. Lett.* 2001, 341, 431–434.
[192] Kim, B. H.; Kim, M. S.; Park, K. T.; Lee, J. K.; Park, D. H.; Joo, J.; Yu, S. G.; Lee, S. H. *Appl. Phys. Lett.* 2003, 83, 539–541.
[193] Yan, H. L.; Zhang, L.; Shen, J. Y.; Chen, Z. J.; Shi, G. Q.; Zhang, B. L. *Nanotechnology* 2006, 17, 3446–3450.
[194] Cho, S. I.; Kwon, W. J.; Choi, S. J.; Kim, P.; Park, S. A.; Kim, J.; Son, S. J.; Xiao, R.; Kim, S. H.; Lee, S. B. *Adv. Mater.* 2005, 17, 171-175.
[195] Kim, B. K.; Kim, Y. H.; Won, K.; Chang, H.; Choi, Y.; Kong, K. J.; Rhyu, B. W.; Kim, J. J.; Lee, J. O. *Nanotechnology* 2005, 16, 1177–1181.
[196] Cho, S. I.; Choi, D. H.; Kim, S. H.; Lee, S. B. *Chem. Mater.* 2005, 17, 4564–4566.
[197] Cho, S. I.; Xiao, R.; Lee, S. B. *Nanotechnology* 2007, 18, 405705/1-5.
[198] Kim, Y.; Baek, J.; Kim, M. H.; Choi, H. J.; Kim, E. *Ultramicroscopy* 2008, 108, 1224–1227.
[199] Cao, Y.; Mallouk, T. E. *Chem. Mater.* 2008, 20, 5260–5265
[200] Bian, C. Q.; Yu, A. H.; Wu, H. Q. *Electrochem. Commun.* 2009, 11, 266–269.
[201] Zhang, X. Y.; Manohar, S. K. *Chem. Commun.* 2004, 20, 2360–2361.
[202] Kuila, B. K.; Nandan, B.; Bohme, M.; Janke, A.; Stamm, M.; *Chem. Commun.* 2009, 18, 5749–5751.
[203] Wang, K.; Huang, J. Y.; Wei, Z. X. *J. Phys. Chem. C* 2010, 114, 8062–8067.
[204] Liu, R.; Cho, S. I.; Lee, S. B. *Nanotechnology* 2008, 19, 215710/1-8.
[205] Baker, C. O.; Shedd, B.; Innis, P. C.; Whitten, P. G.; Spinks, G. M.; Wallace, G. G.; Kaner, R. B. *Adv. Mater.* 2008, 20, 155–158.
[206] Jager, E. W. H.; Smela, E.; Inganas, O. *Science* 2000, 290, 1540–1545.
[207] Okamoto, T.; Kato, Y.; Tada, K.; Onoda, M. *Thin Solid Films* 2001, 393, 383-387.
[208] Otero, T. F.; Cortes, M. T. *Chem. Commun.* 2004, 3, 284–285.
[209] Lu, W.; Smela, E.; Adams, P.; Zuccarello, G.; Mattes, B. R.; *Chem. Mater.* 2004, 16, 1615–1621.
[210] He, X. M.; Li, C.; Chen, F. E.; Shi, G. Q. *Adv. Funct. Mater.* 2007, 17, 2911–2917.
[211] Lee, A. S.; Peteu, S. F.; Ly, J. V.; Requicha, A. A. G.; Thompson, M. E.; Zhou, C. W. *Nanotechnology* 2008, 19, 165501/1-8.
[212] Singh, P. R.; Mahajan, S.; Rajwade, S.; Contractor, A. Q. *J. Electroanal. Chem.* 2009, 625, 16–26.
[213] Yang, X. G.; Li, B.; Wang, H. Z.; Hou, B. R. *Prog. Org. Coat.* 2010, 69, 267–271.
[214] Wan, M. X.; Li, J. C.; Li, S. Z. *Polym. Adv. Technol.* 2001, 12, 651–657
[215] Liu, C. Y.; Jiao, Y. C.; Zhang, L. X.; Xue, M. B.; Zhang, F. S. *Acta Metall. Sinica.* 2007, 43, 409–412.
[216] Saini, P.; Choudhary, V.; Singh, B. P.; Mathur, R. B.; Dhawan, S. K. *Mater. Chem. Phys.* 2009, 113, 919–926.
[217] Wang, C. C.; Song, J. F.; Bao, H. M.; Shen, Q. D.; Yang, C. Z. *Adv. Funct. Mater.* 2008, 18, 1299–1306.
[218] Tseng, R. J.; Huang, J.; Ouyang, J.; Kaner, R. B.; Yang, Y. *Nano Lett.* 2005, 5, 1077–1080.

In: Trends in Polyaniline Research
Editors: T. Ohsaka, Al. Chowdhury, Md. A. Rahman et al.
ISBN: 978-1-62808-424-5

Chapter 8

ELECTROSPUN NANOFIBERS OF POLYANILINE-CARBON BLACK COMPOSITE FOR CONDUCTIVE ELECTRODE APPLICATIONS

***Sujith Kalluri*[1]*, A. M. Asha*[1]*, S. Parvathy*[1]*, Taik Nam Kim*[2]*, N. Sivakumar*[1]*, K. R. V. Subramanian*[1]*, Shantikumar V. Nair*[1] *and A. Balakrishnan*[*1]**

[1]Nano Solar Division, Amrita Center for Nanosciences, Kochi, India

[2]Department of Materials Engineering, Paichai University, Daejeon, S. Korea

ABSTRACT

Polyaniline is known for its good thermal stability, high electrical conductivity and corrosion resistance. Incorporating fillers like carbon black as secondary phases enhances these properties, making it available for electrical and electronics applications. Introducing these composites as nanofibers on an electrode overlay can be beneficial from electron mobility standpoint. Electrospinning is one of the commonly pursued methods for synthesizing nanofibers. However, it is difficult to electrospin polyaniline alone as it is insoluble in organic/inorganic solvents. Inorder to overcome this problem, polyaniline is blended with binder solutions like polyvinyl alcohol (PVA). But, the presence of an insulating carrier like PVA introduces a percolation threshold (threshold voltage beyond which a material starts behaving as a conductor) which can affect applications where high conductivity is required. The problem adds up when carbon black is introduced into the polyaniline matrix. Carbon black tends to create a solid gel when mixed with PVA resulting in a high viscosity solution which makes this blend not suitable for electrospinning. In the present chapter, highly conductive porous (~70%) polyaniline-carbon black composite nanofiber mats were fabricated via electrospinning. The fiber mat was electrospun using polyvinyl alcohol as carrier solution which was later decomposed at ~230 °C to get a complete conducting nanofiber network and did not

[*] Contact Details: avinash.balakrishnan@gmail.com (Dr. Avinash Balakrishnan).

result in any structural collapse. This heat treatment reduced the fiber diameter from ~240 nm to ~170 nm, increased surface pore size from 0.4±0.08 μm to 1.3±0.35 μm and the porosity of the mat increased from 40±1.2% to 75±2%. The removal of the carrier phase in the composite was confirmed by Fourier transform infrared spectroscopy. The spatial specific conductance measurements using scanning electrochemical microscopy showed that the presence of polyvinyl alcohol could introduce percolation threshold and removal of the same by heat treatment substantially reduced the percolation threshold and increased the fiber mat conductance. The heat-treated fibers showed four times increase in specific conductance values on removal of carrier phase from the fiber structure. The present chapter discusses the role of carbon black in polyaniline matrix, which can be beneficial as conductive electrode applications in electronic and photovoltaic storage devices.

Keywords: Polyaniline; Carbon black; Electrospinning; Nanofibers; Electrical conductance

INTRODUCTION

Polyaniline (PANI) is one of the widely used electro-active polymers because of ease in synthesis, doping/de-doping chemistry and thermal stability which makes it attractive to be used in solar cells, battery electrodes and sensors [1]. Polymerized from the aniline monomer, PANI exists in distinct oxidation forms such as fully reduced leucoemeraldine base (LEB), the half oxidized emeraldine base (EB) and the fully oxidized pernigraniline base (PNB), which vary in physio-chemical properties [2 ,3].

All these forms show similar back bone chemical structure but with variations in its side chains (Figure 1). Among the three, EB differs substantially from LEB and PNB in the conductivity. EB is considered to be the widely used form of PANI due to the fact that, upon doping with acid, the resulting emeraldine salt (ES) form of PANI is electrically conductive [4, 5].

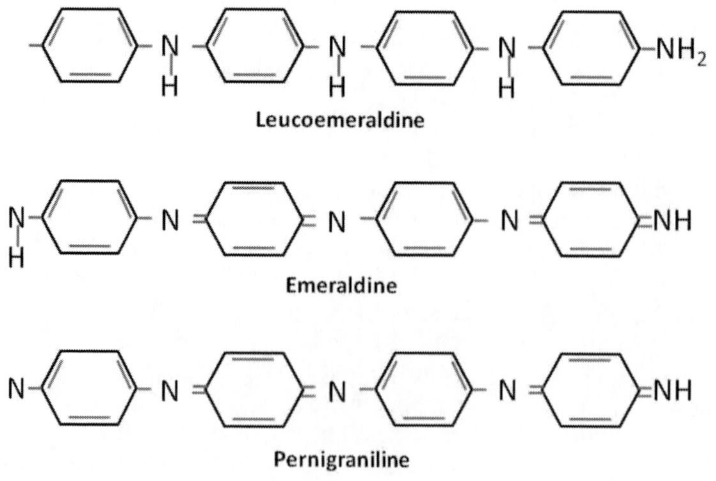

Leucoemeraldine

Emeraldine

Pernigraniline

Figure 1. Chemical structures of different forms of polyaniline [2, 3].

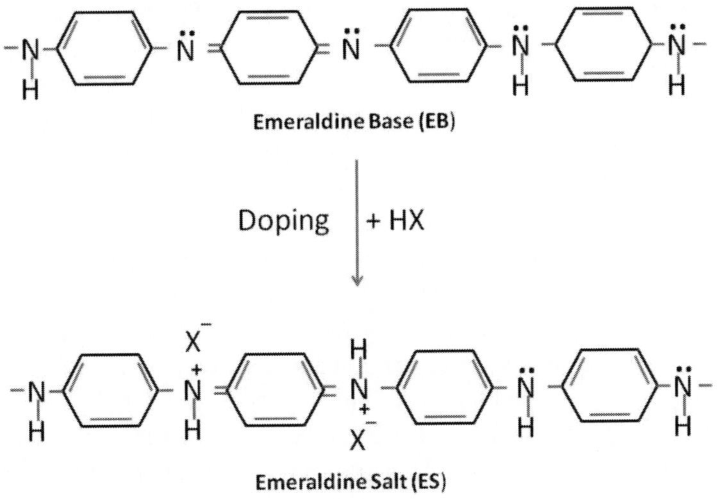

Emeraldine Base (EB)

Doping + HX

Emeraldine Salt (ES)

Figure 2. Protonation of polyaniline emeraldine base [6].

Table 1. Physical Properties of emeraldine salt polyaniline [9, 10]

S.No	Parameter	Value
1	Appearance	powder (infusible)
2	Molecular weight	~Mw >15000
3	Particle size	3-100μm
4	Melting point	~320^{0}C
5	Density	1.36 g/ml
6	Colour	dark green
7	Chemical reaction	acid type
8	Solubility insoluble in common organic solvents	

Table 2. Electrical Properties of emeraldine salt polyaniline [9-11]

S.No	Parameter	Value
1	Electrical conductivity	~5 S/cm
2	Electrical resistivity	~0.2 Ohm-cm
3	Dielectric constant	-4.5 x 10^{-4}

Table 3. Thermal properties of emeraldine salt polyaniline [12-14]

S.No	Parameter	Value
1	Thermal conductivity	~0.5 W/m-k (30^0 - 140^{0}C)
2	Thermal diffusivity	~0.15 mm^2/S (30^0 - 140^{0}C)
3	Glass transition temperature	30^0 - 40^{0}C
4	Deflection temperature	~270^{0}C

ES resulting from doping of EB with aqueous protonic or functionalized acids causes protons to be added to the -N= sites while retaining the number of electrons in the polymer chain constant (Figure 2) [6]. The doping introduces states within the forbidden gap [7]. As the doping level is increased, these states collapse into a closely packed band and finally band gap disappears, thereby making PANI metallic in nature. As pristine (undoped) PANI has normal electrical conductivities, the doping significantly enhances the electrical conductivity [8]. However such a configuration is not achievable in LEB and PNB even after doping with an acid and thereby exhibiting low conductivity. Hereby ES polyaniline will be referred as polyaniline (PANI) in the text. Table 1-3 summarizes the physical, electrical and thermal properties of PANI. Although PANI has a good conductivity, there are issues when it is employed as conductive coatings, such as high resistance and accumulation of static charge carriers on the coating surfaces [15], which limits its application in various fields such as protection from electric shocks in medical operations, explosion in mine environment, dust proofing in integrated circuits and fiber accumulation in spinning industry. The solution to this problem came through the addition of carbon and its derivatives such as graphite, carbon black as a filler material. The addition of graphite to PANI matrix as filler material showed significant improvement (~50 % increase) in conductivity. However, introduction of graphite compromises the physical and mechanical properties of PANI-graphite composite and also processing of such composite into different textures and morphology was difficult [16]. Alternative choice to graphite came in the form of carbon black (CB). Introducing CB as a filler material in PANI, promoted several properties of the composite. There were several advantages of CB that could be exploited, mainly its particular ordered nanostructure, low density, large surface area, high thermal stability, good electrical conductivity, lower cost, corrosion resistance, less material requirement for percolation, and ease of processing into host polymers [17]. The invention of this particular PANI-CB composite [18] had found applications in areas where high microwave absorption, high electrical conductivity and thermal stability were required [19].

BACKGROUND OF INVENTION

Polyaniline/carbon black (PANI-CB) core-shell structure composite was first invented by a group led by Dr. Cheng-Chien Yang, Chemicals System Research Division, Chung-Shan Institute of Science and Technology, Taiwan in 2009. The composite was synthesized by polymerizing emeraldine salt form of PANI in the presence of CB (10~30 wt %). Figure 3 shows the microstructural SEM images of PANI-CB nanocomposites at different proportions of CB.

It was observed that with increase in the proportion of CB, the conductivity of resultant PANI-CB composite also increased. This increase in conductivity was attributed to the bridging effect of CB, which is entanglement of the conducting back chains of PANI to the surface of CB [20].

This entanglement compensates the reduced conducting ability caused by transformation of PANI from emeraldine salt form (ES) to emeraldine base form (EB) during the possible deprotonation chemical reaction involved in the preparation of PANI-CB composite from aniline, which is evident from electron paramagnetic resonance (EPR) analysis [18,21]. Ultraviolet-visible (UV-VIS) spectroscopy studies (Figure 4) showed strong chemical interactions between the quinoid rings of PANI and CB [21].

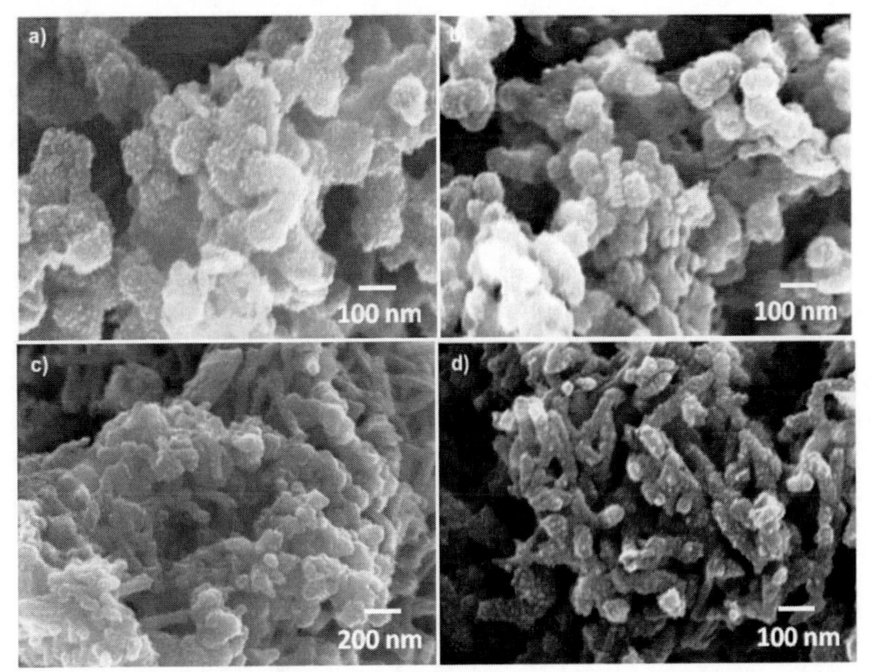

(Courtesy: Dr. Youngil Lee (Associate Professor), Dept. of Chemistry, University of Ulsan, South Korea).

Figure 3. SEM images of the PANI-CB nanocomposites with (a) 5, (b) 9, (c) 13 and (d) 18% CB.

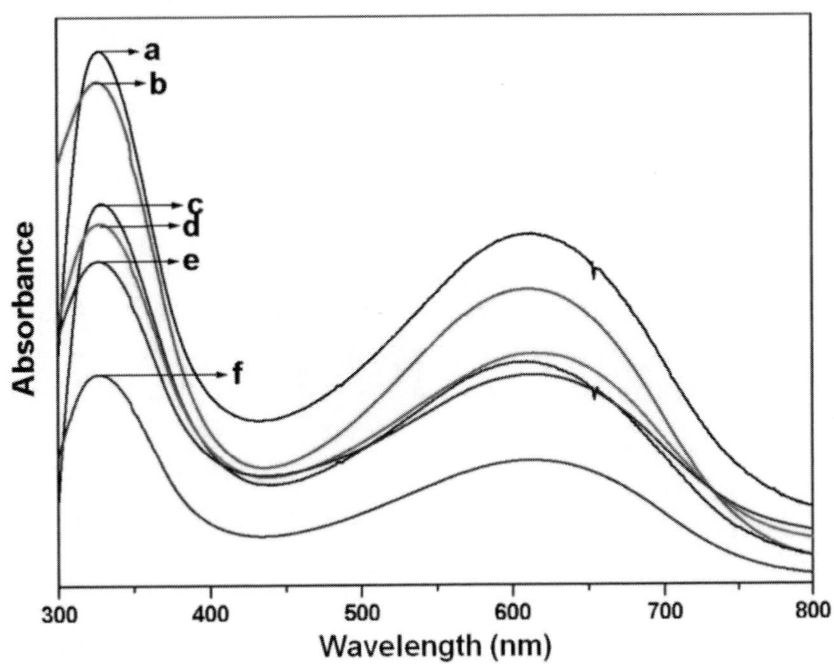

(Courtesy: Dr. Youngil Lee (Associate Professor), Dept. of Chemistry, University of Ulsan, South Korea).

Figure 4. UV-vis spectra of the (a) PANI and PANI-CB nanocomposites with (b) 5, (c) 9, (d) 13, (e) 18 and (f) 23% CB.

Microstructural analysis of the PANI-CB matrix showed PANI encapsulating the CB uniformly. As the amount of CB being added increased, CB exposed outside PANI was more.

Exposure of this CB was found to limit or degrade the mechanical properties and processibility [18]. This saturation limit was found when CB was added in excess of 10 wt % at which it has adequate conductivity without decreasing mechanical property and processibility. Thus, commercially 10 wt% of CB in polyaniline was considered as suitable for most of the applications. At this weight percentage of CB in PANI-CB, comparison of the thermal transitions with pristine PANI revealed that PANI-CB is more thermally stable than PANI alone. This additional thermal stability of the composites can be attributed to the interactions between nanoscale CB particles and the conducting polymer i.e., π-π interactions between the PANI and the aromatic rings of the CB [21].

In addition, percolation threshold phenomenon takes place at critical stage involved in loading of CB, where the three-dimensional continuous conducting network is built throughout the polymer matrix. The percolation threshold of such composites depends on the CB physical structure such as particle size, aggregate shape and structure, on the polymer chemical structure and on the synthesis methodology. In general, larger surface area and smaller particle size of CB results in higher conductivity of the PANI-CB composites [22].

Processing of PANI-CB

The processing method of PANI-CB core-shell composite is well explained by Yang et.al. as shown in figure [18]. Briefly, the processing method involves preparation of two solutions (see Figure 5) i.e., first a solution comprising of a dispersion of nanoscale CB in ethanol solution is prepared. To this solution distilled aniline is added. Second solution is prepared by mixing 2M HCl solution with ammonium persulfate, which acts as an oxidizer responsible for polymerization of aniline.

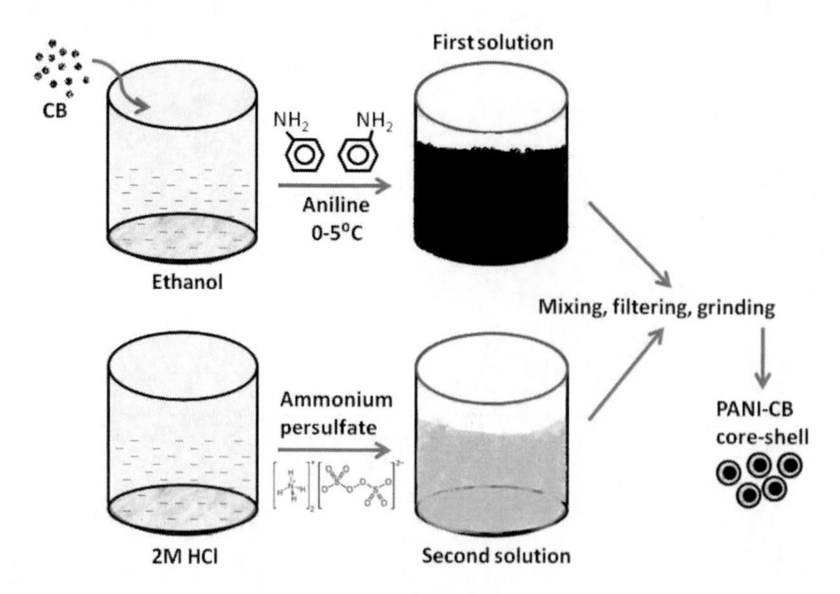

Figure 5. Processing method of PANI-CB core-shell composite.

Upon adding the second solution with first solution, polymerization takes place and results in the formation of the stable PANI-CB core-shell composite.

Modification of PANI-CB Morphology via Electrospinning

These functionalized particles are employed in applications like electronic and electric appliance industry, printed circuit boards, switches, electro thermal materials, electromagnetic wave shielding, and surface protection. However attempts have been made to modulate these materials into one dimensions by nanostructuring them so that their existing properties can be enhanced. One dimensional nanostructures, including nanotubes, nanowires, nanorods and nanofibers have grabbed the attention of scientific community because of their novel and interesting properties such as size dependant excitation or emission [23], quantized conductance [24], coulomb blockade [25] and metal insulator transition [26] for advanced applications in the fields of electronics and optoelectronic devices [27, 28]. Fabrication of nano-fibers [29] can indeed provide good platform for electron transport pathway by increasing the specific surface area and thereby improving the electrical conductivity [30]. Processing nanofibers [31, 32] via electrospinning technique is well known. This technique offers the advantages of simplicity, inexpensiveness and good scalability to form high surface area fibrous mats [33]. The process of electrospinning, namely utilizing electrostatic forces to generate polymer fibers, traces its roots back to the process of electrospraying, in which solid polymer droplets are formed rather than fibers. In fact, a number of processing parameters must be optimized in order to generate fibers as opposed to droplets, and a typical electrospinning apparatus can be used to form fibers, droplets, or beaded structures, depending on various processing parameters, such as distance between source and collector [34]. A typical electrospinning set up consists of a capillary through which the liquid to be electrospun is forced; a high voltage source with positive or negative polarity, which injects charge into the liquid; and a grounded collector [35]. A syringe pump, gravitational forces, or pressurized gas are typically used to force the liquid through a small-diameter capillary forming a pendant drop at the tip. An electrode from the high voltage source is then immersed in the liquid or can be directly attached to the capillary if a metal needle is used. The voltage source is then turned on and charge is injected into the polymer solution. Increasing the electric field strength causes repulsive interactions between like charges in the liquid and attractive forces between the oppositely charged liquid and collector to begin to exert tensile forces on the liquid, elongating the pendant drop at the tip of the capillary [36]. As the electric field strength is further increased a point will be reached at which the electrostatic forces balance out the surface tension of the liquid leading to the development of the Taylor cone [37]. If the applied voltage is increased beyond this point a fiber jet will be ejected from the apex of the cone and be accelerated toward the ground collector. Figure 6 shows the actual and schematic representation of an electrospinning unit.

Processing Parameters

Despite electrospinning's relative ease of use, there are a number of processing parameters that can greatly affect fiber formation and structure. Grouped in order of relative impact to the electrospinning process, these parameters are applied voltage, polymer flow rate, concentration, distance between the capillary and collection screen, ambient parameters (temperature, humidity and air velocity in the chamber), and motion of target screen.

Solution Parameters

In addition to the processing parameters, a number of solution parameters play an important role in fiber formation and structure. In relative order of their impact on the electrospinning process these include polymer concentration, solvent volatility and solvent conductivity. In addition system parameters include molecular weight, molecular weight distribution and architecture (branched, linear etc.) of the polymer.

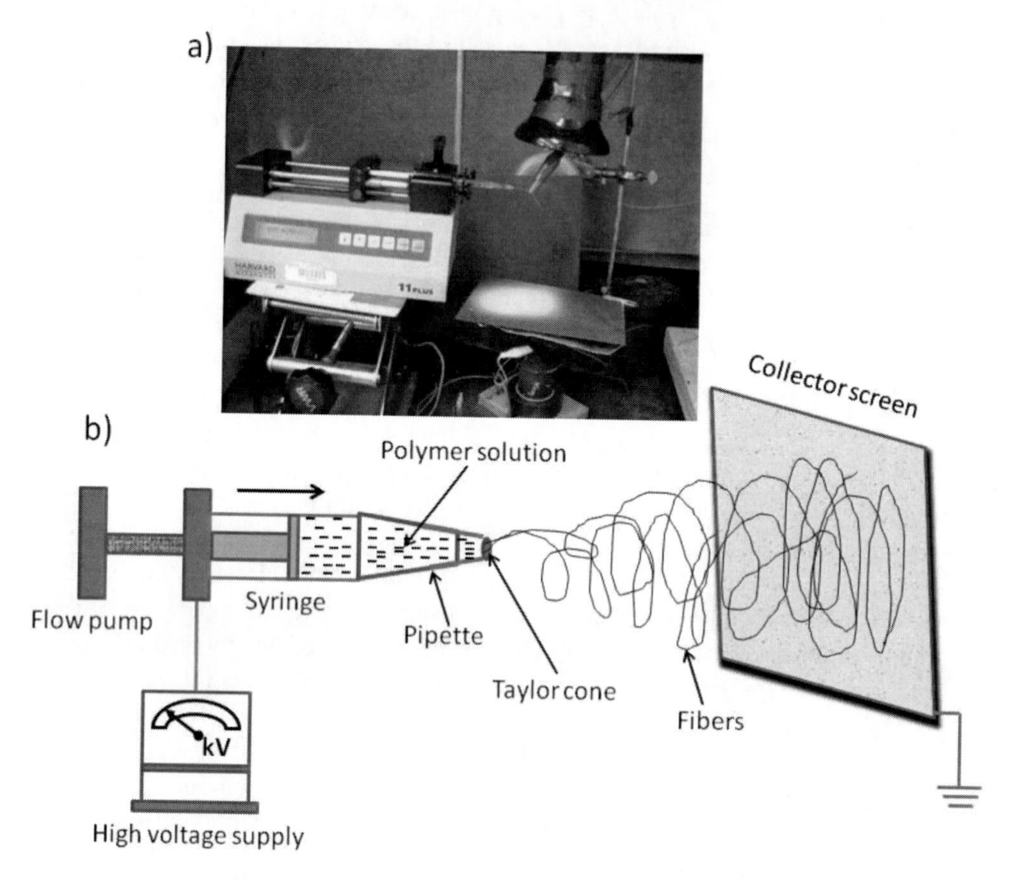

Figure 6. (a) Actual and (b) schematic representation of electrospinning set up.

The formula that predicts the terminal diameter of the fiber, which is controlled by flow rate, electric current and the surface tension of the fluid [38], is:

$$ht = \left(\gamma\epsilon\frac{Q^2}{I^2}\frac{2}{\pi(2ln\varphi - 3)}\right)^{1/3}$$

Typical electrospinning conditions are:

Voltage: 0-35 kV (gamma high voltage research)
Flow rate Q: 0.1-25 ml/hr (Kd scientific infusion pump)

Tip to target distance: 2 cm-25 cm

I: Electric current,

φ: Dimensionless parameters of instability,

γ: Surface tension.

In addition to adjusting solution or processing parameters, the type of electrospinning process can greatly influence the resulting product [39]. This can include choices in nozzle configuration (single, single with emulsion, side-by-side or coaxial nozzles) or solution vs. melt spinning.

Electrospinning of PANI

There have been reports of electrospinning PANI alone by blending it with different carrier solutions. For instance, Kahol and Pinto [40] studied different concentrations of polyethylene oxide (PEO), while Hui et.al. [41] used different concentrations of polyvinyl alcohol (PVA) as carrier solutions for electrospinning PANI. However, electrospinning of PANI and carbon composites fibers to yield nanofiber composites has not been reported to date. This could be attributed to two reasons: PANI by itself is insoluble in many common organic solvents [42], thus requiring carrier solutions. Secondly, under non-optimized conditions, presence of CB causes rapid gelation of the polymeric blend making the solution difficult to be electrospun.

First reports of electrospinning PANI-CB was reported by K. Sujith et al., [43] using an optimized polymeric blend of PANI-CB composite and polyvinyl alcohol (PVA) solution (carrier solution) The PVA was later sublimated to get stand alone PANI-CB conductive nano-fibrous mats.

The basic processing step involved in fabrication PANI-CB composite nanofibers is the use of polyvinyl alcohol (PVA) as a binder material. 8 wt% of PVA solution was prepared. 3-10 wt% of polyaniline-carbon black (PANI-CB) composite solutions (530565-25G, 20 wt% carbon black, Sigma Aldrich Grade, Conductivity: 30 S/cm) were prepared using chloroform as solvent, which was then mixed with PVA binder solution. 3 ml of the processed polymeric blend solution was filled in a 20 ml syringe and fixed with to an electrospinning unit (Zeonics System, India). Electrospinning was performed for 4 hrs under room temperature at an applied potential of 15 kV maintaining a tip to target distance of 10 cm and a pump flow rate of 0.5 ml/hr. The relative humidity (RH) over the entire experiment was maintained at 55 %. The electrospinning was performed over Al foils and Ti foils with dimensions 8 cm X 8 cm X 0.2 mm. After the electrospinning process, these nanofibrous mats were calcined at 230 ºC to sublimate the PVA binder to obtain all conducting PANI-CB electrospun webs.

It was observed that the formation of the PVA/PANI-CB composite fiber depended on the concentration ratio of PVA: PANI-CB in the polymeric blend solution. Blending of PVA with PANI-CB under optimized concentrations (i.e. ratio PVA: PANI-CB=1:1) resulted in a viscous solution (viscosity η= 3.1 Pa.s) which was suitable for electrospinning morphologically stable nanofibers. For concentrations ratio of PVA: PANI-CB < 1, rapid gelation occurred which made the solution unsuitable for electrospinning. For concentrations ratio of PVA: PANI-CB > 1 the resultant polymeric blend solution was less viscous.

Electrospinning this less viscous polymeric blend solution resulted in the formation of thin fibers with beads as shown in Figure 7a. In addition, experimental results (see Figure 9a), indicated that as the fiber diameter increased proportionally with concentration increasing upto 12 wt% of the composite beyond which the polymeric solution starts splitting without spinning. This was attributed to the gelation of polymeric blend solution. Similar observations have been reported in studies by Fong and Reneker [46], who proposed that electrospinning polymer solution with low viscosity results in the formation of the beads caused by the capillary breakup of the jet due to surface tension.

Figure 8 shows the photograph of the physical appearance of electrospun fibers of as-prepared (dark) and heat treated (yellowish) coatings over the Al foil. SEM image of electrospun fibers (Figure 7 bandc) of as-prepared and heat treated samples showed that the fibers were randomly oriented and structurally stable, with their lengths extending up to several hundred nanometers. SEM image did not reveal any phase separation images between CB and PANI matrix possibly due to the poor contrast between the two phases of the composite. It was interesting to note that the fibers did not collapse even as the binder, PVA was sublimated. Porosity measurements of the prepared samples, performed by using refractive index obtained from UV analysis (P_{UV}) and image analysis model (P_{Model}), [44] showed an increase in porosity values for heat treated samples (Table 4). Image analysis model is a simulation model which measures the porosity of the given sample from the high resolution SEM image. Ultraviolet visible (UV-VIS) spectrometer was used to measure the porosity from the refractive index values [45] using the following equation (1):

$$1 - \mathrm{P} = \frac{n^2 - 1}{n_t^2 - 1} \frac{n_t^2 + 2}{n^2 + 2} \tag{1}$$

where P is porosity, n is refractive index of a composite material and n_i is refractive index of the individual materials in the composite.

Figure 9b shows the influence of electric potential on fiber diameter, which illustrates that increase in electric potential results in the decrease of fiber diameter, when all other parameters were kept constant. But beyond a threshold electric potential, the increase in electric potential results in high bead density [46-48]. Because of high electric potential, the beads will combine together to form a thick diameter fibers [49]. The flow rate parameter in electrospinning technique determines the speed at which the feed solution is being electrospun. When the flow rate was increased, fiber diameter was found to increase as shown in Figure 9c. This could be due to the fact that as there is a large amount of solution that is drawn from syringe needle tip, the jet will take longer time to dry and hence cause the fibers to fuse together [48,50]. Based on the properties of polymeric blend solution, the influence of varying distance between tip to target may or may not have a significant role in the resultant fiber morphology. But in the present context PANI-CB composite showed a peculiar property as shown in Figure 9d. With increase in tip to target distance, the fiber diameter decreases till the optimized point target distance is 18 cm; beyond this point, the solution started splitting which results in bead formation. This bead formation can be attributed to the increased electric field strength between the needle tip and the fiber collector [51]. Decreasing the tip to target distance has similar effects as increasing the electric potential supplied (see Figure 9b) and this will result in an increase of electric field strength. The possible reason for decreasing fiber diameter with increasing tip to target distance can be due to the longer flight time for the polymeric solution to be stretched before it is deposited on the collector [52, 53]. However, there are a few reports [54] where the fiber diameter increases with increase in tip to target distance.

This can be due to less stretching of the fibers because of the reduction in electric field strength. When the tip to target distance is very large, no fibers are deposited on the collector [52]. Therefore, it is evident that there is an optimal point of electric field strength below which the stretching of the solution will decrease resulting in increased fiber diameter.

In addition, the exposure of polymeric solution with surrounding parameters such as humidity, temperature and pressure results in the variation of electrospun fiber morphology. For example high humidity was found to close the pores on the surface of the nanofibers [55]. When electrospinning is carried out at normal atmosphere and high humidity, there could be a chance of water condensing on the surface of fibers and hence may have effect on its morphology by forming circular pores on fiber surface [56]. But in the case of PANI-CB composite, the relative humidity was maintained at 55%, beyond which we are unable to get continuous electrospun fibers.

For further characterization two sets of samples were prepared 1) electrospun PVA/PANI-CB and 2) heat treated electrospun PVA/PANI-CB, hereby designated as as-prepared and heat treated samples respectively.

Heat treatment of PVA/PANI-CB fibers at 230^{0}C resulted in the sublimation of PVA and resulted in standalone PANI-CB nano-fiber mats. The heat treated nanofibers showed high discrepancies in the diameter which was seen as large scattering in the diameter data. Diameter size distribution (DSD) analysis based on Figure 7 bandc showed a skewed distribution and bimodal distribution for as prepared and heat treated samples respectively (Figure10 aand b).

Table 4. Physical and optical properties of as-prepared and heat treated fiber mats

	Diameter(nm)	Refractive index	P_{UV}(%)	*P_{Model}(%)	Thickness (μm)	Surface pore size(μm)
As-prepared	240±20	1.4±0.005	40±1.2	43±2.8	65±10	0.4±0.08
Heat treated	170±40	1.25±0.01	75±2	70±4	35±9	1.3±0.35

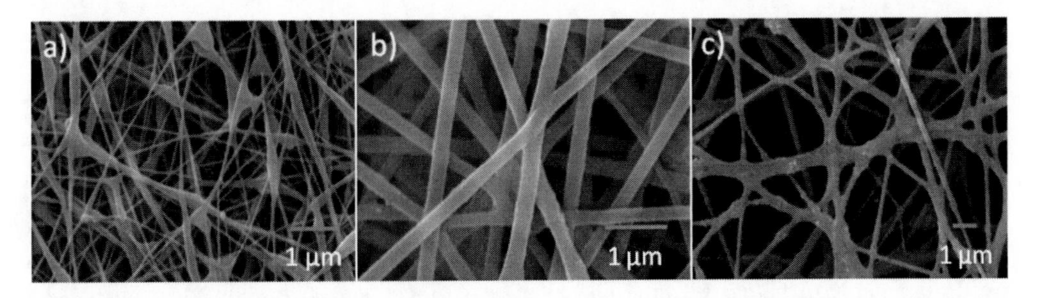

Figure 7. SEM images of electrospun PANI-CB composite (a) thin fibers with beads (PVA: PANI-CB = 4:1), (b) as-prepared fibers and (c) heat treated fibers.

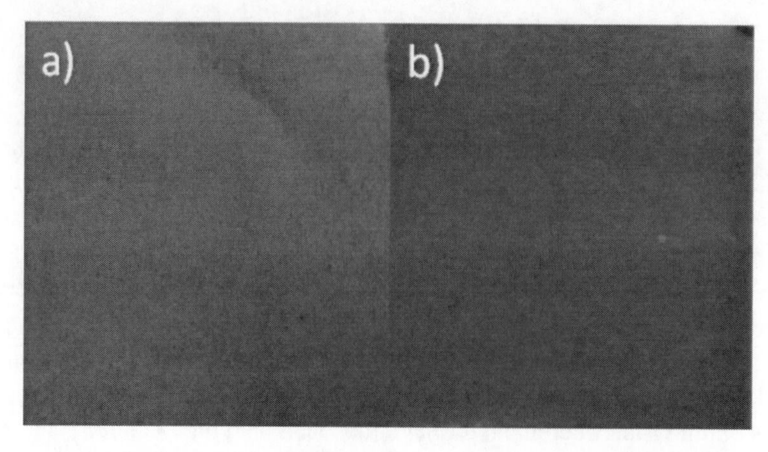

Figure 8. Photograph of electrospun PANI-CB (a) as-prepared fiber mats and (b) heattreated fiber mats coating over the Al foil.

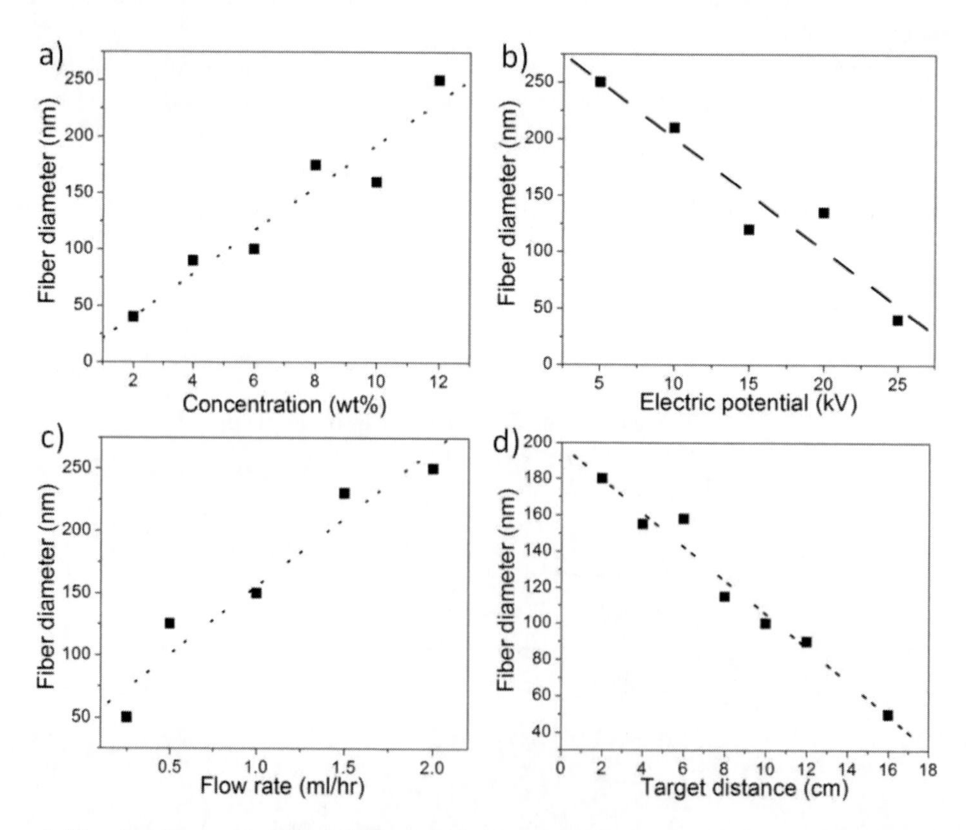

Figure 9. Plot showing trend of fiber diameter with (a) concentration, (b) electric potential, (c) flow rate and (d) target distance.

The fiber diameter for as-prepared sample was centered at ~250 nm. While for heat treated samples the average fiber diameter was found ~160 nm of which 60% of the fibers showed diameter size below 90 nm. The average surface pore size showed an increase in surface pore size values for heat treated samples when compared to as-prepared samples (see Table 4 and Figure 11 aand b).

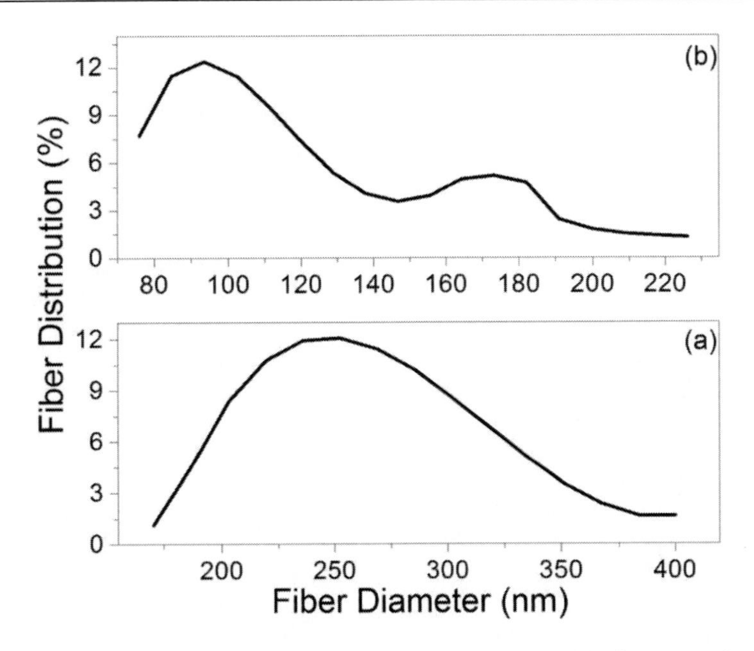

Figure 10. Plot of fiber diameter distribution (a) as-prepared sample and (b) heat treated sample.

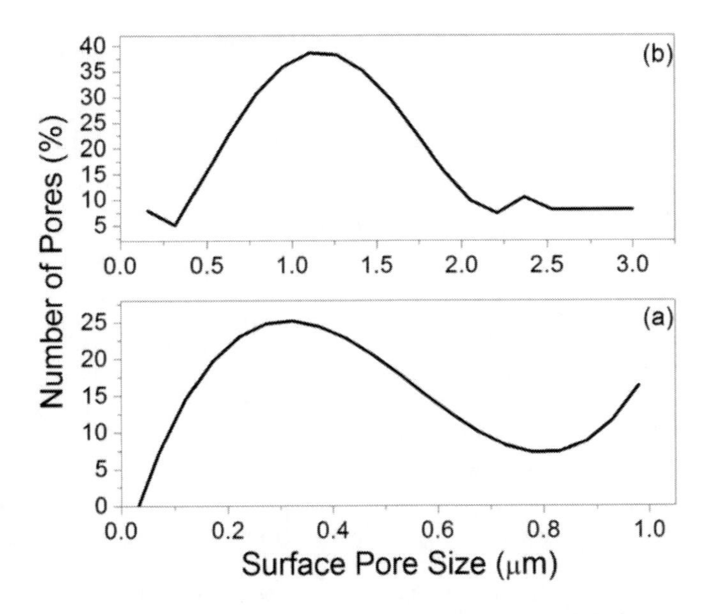

Figure 11. Plot of surface pore size distribution (a) as-prepared sample and (b) heat treated sample.

The average thickness measured (Table 4) for heat treated mats was found to be lower in value (which can be attributed due to shrinkage in fibers) when compared to as prepared samples (see Figure 12 a and b).

The presence of PVA in the as-prepared (non-heat treated) samples was seen as strong infra-red (IR) absorption peaks (see Figure 13) at a wave number of 3450 cm^{-1}, with small indentation in the peak (shown with arrow in the figure) which could be attributed to some cross-linking phenomena that interferes in C-H and O-H stretching of PANI-CB and PVA respectively, indicating some degree interaction between the two phases [57, 58].

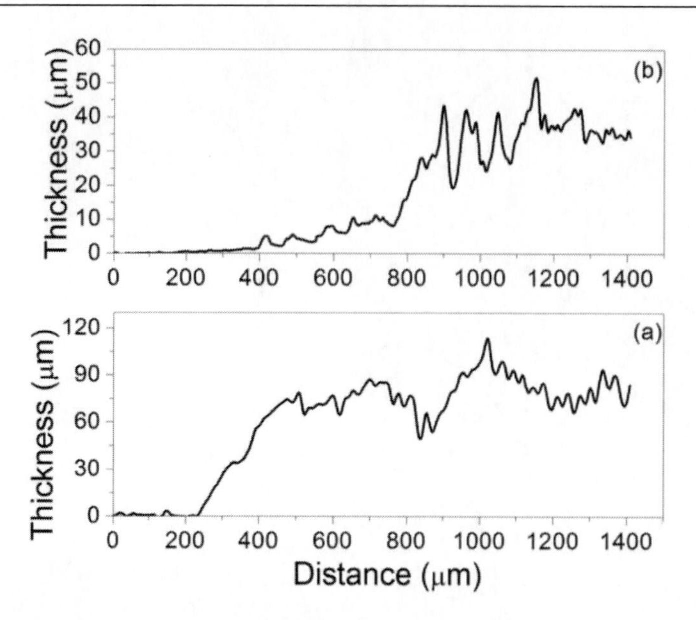

Figure 12. Surface profilometry showing thickness of electrospun PANI-CB (a) as-prepared fiber mats and (b) heat treated fiber mats.

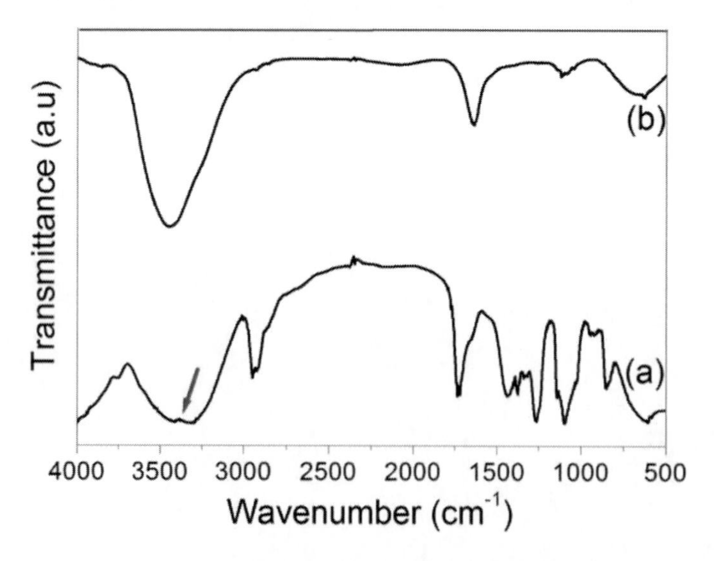

Figure 13. FTIR spectra of electrospun PANI-CB fiber mats (a) as-prepared and (b) heat treated [43].

This indentation was absent in heat treated samples, which meant that PVA was getting completely sublimated from the PANI-CB composite matrix. Characteristic bands of PVA, for C-C stretching of benzenoid rings at ~1435 cm^{-1} [57] was observed in as-prepared samples. Further, an absorption band in as-prepared samples was also seen at ~1400 cm^{-1} which could be attributed to the C-O stretch [59]. C-N stretching mode at 1219 cm^{-1} was observed in both as-prepared and heat treated samples [60]. The peaks at 1092 cm^{-1} and 1250 cm^{-1} for as-prepared samples were attributed to the Ph-CH bend and the symmetric component of the C-C (or C-N) stretching modes in composite [61] and also peak at 1140 cm^{-1} for heated treated samples was attributed to C-H bending mode in PANI-CB [57].

Scanning electrochemical microscope (SECM) is an electrochemical characterization technique capable of probing surface reactivity of materials at microscopic scales and used for measuring the current - voltage characteristics and subsequently the specific conductance (G, S/cm^2) [62]. SECM is based on the extent of precisely adjusting a microelectrode probe close to the sample for characterization. In SECM, the probe is a microelectrode, which is an electrode of micrometer dimension. This microelectrode and sample are immersed into an electrolyte solution containing either oxidizing or reducing agents. The microelectrode is electrically connected so that a redox current and tip current are generated. When this micro tipped electrode is brought near to the sample; the variation in the tip current between electrode and sample surface is recorded as spatial mapping (indicative of the surface topography) based on the current distribution over an area of the sample [63, 64].

Figure 14 a and b shows the spatial mapping of the surface current by keeping the distance between the probe and the conducting surface constant. It was seen that for a scanned area of 100 μm^2, the topography established by the current peaks in heat treated samples was more pronounced on a given current scale, than the as-prepared samples. This was depicted as an I-V curve (Figure 14c) which showed a distinct percolation threshold voltage for as-prepared samples. It was observed that as-prepared samples exhibited diode like characteristics at a break over voltage range of ~1.8 V.

Studies have shown [65] that in a mixture consisting of an insulator and a conducting material, the conductivity and the dielectric constant will show a critical behavior near to the break over voltage region (percolation threshold region) exhibiting smooth transition from insulating region to the conducting region after which a significant increase in electrical conductance values was recorded. Interestingly this feature was also present but less prominent (~0.9 V) in case of heat treated samples. However it should be noted that the SECM gives a localized representation (scan area: 100 μm^2) of the electrical conductance characteristics of the sample surface. Thus I-V values at numerous points were recorded on different samples (4 samples X 5 measurements) to get a wider perspective of the electrical behaviour of these nanofiber mats in both as-prepared and heat treated samples. The specific conductance G_0 measured from Weibull plots (Figure 15) showed higher values for heat treated samples ($(0.14 \pm 0.05) \times 10^{-6}$ S/cm^2) compared to as-prepared samples ($(0.03 \pm 0.005) \times 10^{-6}$ S/cm^2).

These experimental data were analyzed with the conventional ranking method [66] to yield the Weibull parameters, knowledge of which would lead to complete characterization of the statistical properties of the measured specific conductance. This was done by ordering the results measured at each point from the lowest to the highest. The fth result in the set of n=20 data was assigned a cumulative probability of occurrence, P_f, which was calculated with

$$P_f = \frac{i}{n+1} \tag{2}$$

The measured specific conductance (S/cm^2, (G), and the cumulative probability, P_f, were then analyzed, by using a simple, least-square regression method, according to the alternative form of the well-known two-parameter Weibull distribution equation.

$$\ln \ln \left(\frac{1}{(1-P_f)} \right) = m \ln G - m \ln G_0 \tag{3}$$

where m and G_0 are the Weibull modulus and the mean specific conductance, respectively.

It was interesting to note that the scattering in data indicated by the Weibull modulus m was higher (i.e. lesser in scattering) in case of heat treated samples indicating that the sublimation of PVA from the polymer matrix was uniform.

Generally, electron hopping transport mechanism is very prominent in doped materials with localized defect states. In such transport mechanism, phonons play an important role in electrical conductivity. In accordance, because of hopping mechanism the electrical conductivity always increases with rise in temperature, and results in more photons [67].

When polymers are in the non-conducting state, this hopping mechanism predominates and determines the temperature dependencies of electron transport characteristics. In conducting state of conducting polymers free electric charge carriers appear for transportation, and strongly contributes to the electrical conductivity.

During motion of free electric charge carriers, because of phonons and impurities in the material, the free charge carriers undergo some scattering phenomena. This could be possible reason for the decrease in electrical conductivity.

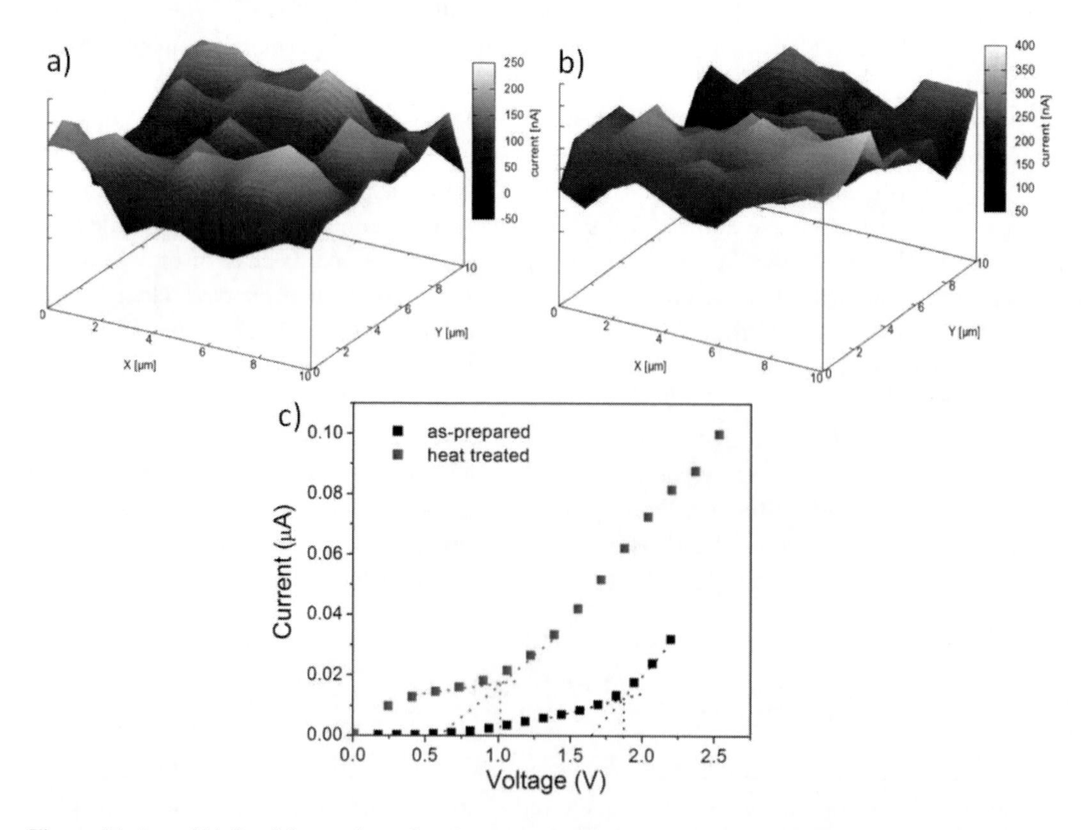

Figure 14. (a and b) Spatial mapping of surface current of electrospun PANI-CB as-prepared and heat treated fiber mats respectively and (c) Comparison of I-V characteristics of electrospun PANI-CB as-prepared and heat treated fiber mats.

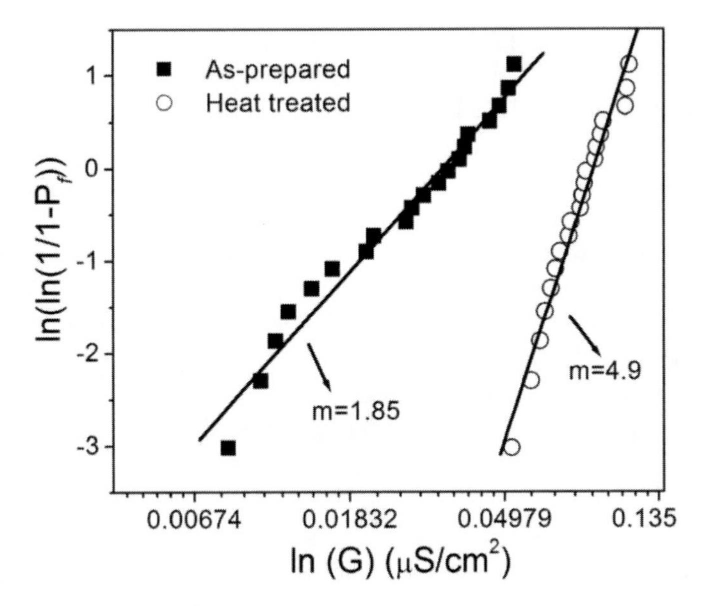

Figure 15. Weibull plot comparison of the specific conductance of electrospun PANI-CB as-prepared and heat treated fiber mats.

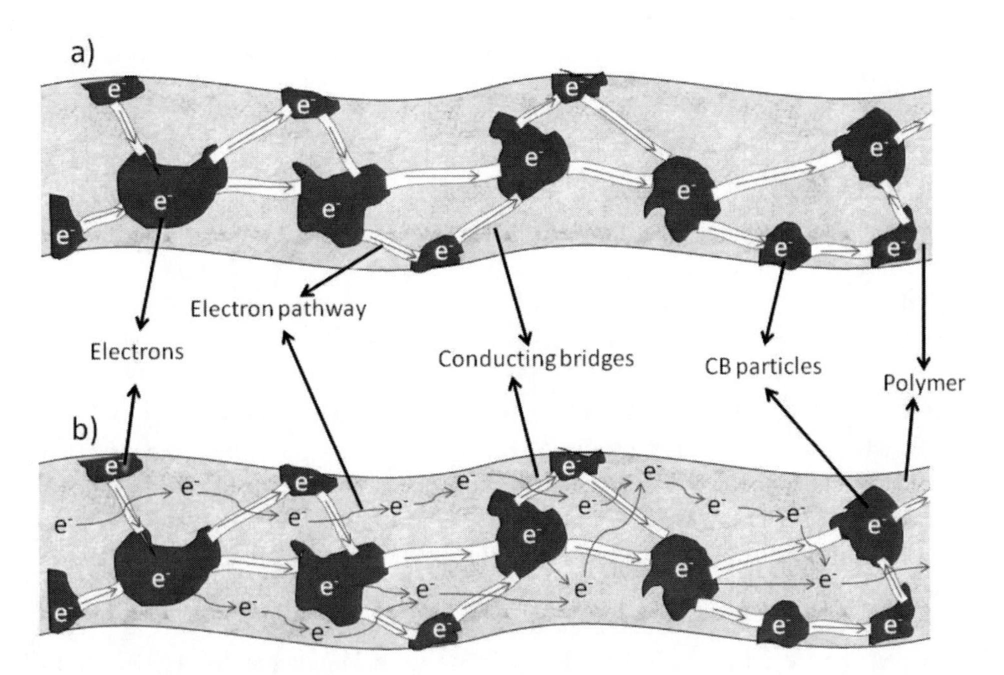

Figure 16. Schematic representation of possible electron transfer pathways in PANI-CB composite (a) intial current raise and (b) after Schottky barrier overcome.

There were many reports based on the influence of metallic features on the electrical conductivity of various polymeric materials. For example, the decrease in the electrical conductivity upon heating was observed in polyaniline nanofibers [68,69]. However, it should be noted that the electron hopping is not the only transport mechanism responsible for the metallic characteristics in the electrical conductivity of electroactive polymers. Report by

Prigodin and Epstein demonstarted that the electron tunneling through the intergrains through intermediate states of the polymer back chains connecting them are strongly responsible for electron transport mechanism [70]. This method was used to explain the electron transfer mechanism in PANI fibers [71] providing results compatible with the previous transport mechanism experiments [72]. In the case of electroactive conducting polymeric materials, metal phase takes initiation and the conducting bridge in between is provided by intermediate sites. In such electron transport mechanism, the effect of phonons can be very prominent. These phonons bring an inelastic component to the intergrain current and underlie the interplay between the elastic transport by the electron tunneling and the thermally assisted dissipative transport. Also, they may cause some other effects, as was shown while developing the theory of conduction through molecules [73–80].

However, in the present case, it is hypothesized that the presence of "metallic" carbon black (work function = 4.1 eV [81]) in the PANI-CB polymer matrix possibly generates nanoscaled Schottky barrier contacts with the "semiconducting" PANI (band gap Eg = 1.4 eV, LUMO = -3.8 eV [82]). The possibility of carbon black forming nanoscaled Schottky contacts with the PANI may play a key role in determining the conduction pathways. The schematic representation of this possible mechanism is shown in Figure 16. For the initial current rise, the carbon black phases bridged together serve to conduct the charges across the composite. Once the voltage is increased, the nanoscaled Schottky barriers are overcome and the conduction mechanism is primarily guided through polymer networks of PANI through a hopping mechanism as well as synergistically through the carbon black bridges [83]. However, further studies related to understanding of the interfacial mechanisms needs to be performed.

CONCLUSION

The idea of incorporating nanoscale fillers such as CNT, nanoscale carbon black in polymeric solution for electrospinning composite nanofibers has been extended to prepare composite solution of organic and inorganic materials for electrospinning [84]. Several researchers have tried the electrospinning of conductive polymers such as polyaniline, polypyrrole etc., to make highly efficient semiconductor and electronic devices such as schottky nanocontacts diodes, electrochemical sensors, flexible electronics and photovoltaic devices [85-87]. Further, nanofibers also will eventually find application in processing nanocomposites. This may be because employing nanofibers offer better mechanical properties and hence enhances the structural properties of nanocomposites [88]. Other applications which have been proposed includes energy storage devices, dye-sensitized solar cells, electromagnetic interference shielding, composite delamination resistance, ultra-light weight spacecraft armour applications, higher efficient and functional catalysts, and liquid crystal devices [89-92]. For example, electrospinning of PVdF, PVdF-PVC composite in the form of nanofibers have been used as an electrolyte for polymer lithium ion secondary batteries resulting in high power density [93, 94]. Electrospinning of PVDF–HFP and PVDF–HFP/PS blend, liquid crystal embedded PVDF–HFP as nanofibers have been processed for use as electrolytes in dye-sensitized solar cells resulted in high performance [95, 96]. In supercapacitors, such electrospun nanofibers can increase the surface of the electrodes and hence the capacitance values. For instance, PBI based activated carbon nanofibers were electrospun and used as electrodes for supercapacitors resulting in improved specific capacitance values [96, 97]. However it should be noted that these applications are at lab scale level which can show potential results for their industry and product level applications.

REFERENCES

[1] Do Nascimento, G. M. In *Nanofibers*; Kumar, A.; Ed.; Spectroscopy of polyaniline nanofibers; Intech: Rijeka, HR, 2010; Vol.2, 349-366.

[2] Stejskal, J.; Gilbert, R. G. *Pure Appl. Chem.* 2002, *74*, 857-867.

[3] Fong, H.; Chun, I.; Reneker, D. H. *Polymer* 1999, *40*, 4585-4592.

[4] Meixiang, W.; Jiping, Y. *J. Appl. Polym. Sci.* 1995, *55*, 399-405.

[5] Chiang, J. C.; MacDiarmid, A. G. *Synth. Met.* 1986, *1*, 193-205.

[6] Feast, W. J.; Tsibouklis, J.; Pouwer, K. L.; Groenendaal, L.; Meijer, E.W. *Polymer* 1996, *37*, 5017-5047.

[7] Hui, D.; Alexandrescu, R.; Chipara, M.; Morjan, I.; Aldica, G.; Chipara, M. D.; Lau, K. T. *J Optoelectron Adv. M* 2004, *6*, 817-824.

[8] Shabnam, V.; Jiaxing, H.; Richard, B. K.; Bruce, H. W. *Nano. Lett.* 2004, *4*, 491-496.

[9] Specification of polyaniline salt. http://www.polyaniline.net/polyaniline _salt.html

[10] Properties and description of polyaniline (emeraldine salt). http://www.sigmaaldrich.com/catalog/product/aldrich/428329?lang=enandregion=IN

[11] Stephen, S. H.; Richard, V. G. In *Polymer Data Handbook*; James, E. M.; Ed.; Polyaniline; Oxford University Press, Inc: New York, New York, 1999; 271-275.

[12] Neeraj, J.; Dinesh, P.; Saxena, N. S.; Saraswat, Y. K. *Indian J. Pure AP* 2008, *46*, 385-389.

[13] Sambhu, B.; Dipak, K. *Synth. Met.* 2009, *159*, 1141-1146.

[14] Ching, P. C.; Cheng, D. L.; Sung, W. H.; Dean, Y. C.; Sung, N. L. *Synth. Met.* 2004, *142*, 275-281.

[15] Vinodini, S.; Dinesh, P.; Kananbala, S.; Narendra, S. S.; Thansewar, P. S. *Cent. Eur. J Chem.* 2009, *7*, 769-773.

[16] Jaydeep, K.; Ioan, N.; Efstathios, I. M. *Wear* 2002, *252*, 361-369.

[17] Kakarla, R. R.; Byung, C. S.; Kwang, S. R.; Jaegeun, N.; Youngil, L. *Synth. Met.* 2009, *159*, 1934-1939.

[18] Yang, C. C.; Wu, K. H.; Gu, W. T.; Peng, Y. H.; Liu, Z. H. *Polyaniline/carbon black composite and preparation method thereof*, United States patent US2009/0314999 A1, 1, 2009.

[19] Wu, K. H.; Ting, T. H.; Wang, G. P.; Ho, W. D.; Shih, C. C. *Polym. Degrad. Stabil.* 2008, *93*, 483-488.

[20] Jumi, Y.; Ji, S. I.; Hyung-Il, K.; Young, S. L. *Appl. Surf. Sci.* 2012, *258*, 3462-3468.

[21] Yang, C. C.; Wu, K. H.; Gu, W. T.; Peng, Y. H. Organic siloxane composite material containing polyaniline/carbon black and preparation method thereof, United States patent US7,956,106 B2, 2011.

[22] Kun, D.; Xiang, B. X.; Zhong, M. L. *Polymer* 2007, *48*, 849-859.

[23] Murray, C. B.; Kagan, C. R.; Bawendi, M. G. *Annu. Rev. Mater. Sci.* 2000, *30*, 545-610.

[24] Krans, J. M.; van Rutenbeek, J. M.; Fisun, V. V.; Yanson, I. K.; de Jongh, L. J. *Nature* 1995, *375*, 767-769.

[25] Likharev, K. K. K. *IBM J. Res. Dev.* 1988, *32*, 144-158.

[26] Markovich, G.; Collier, C. P.; Henrichs, S. E.; Remacle, F.; Levine, R. D.; Heath, J. R. *Acc. Chem. Res* 1999, *32*, 415-423.

[27] Younan, X.; Peidong, Y.; Yugang, S.; Yiying. W.; Brian, M.; Byron, G.; Yadong, Y.; Franklin, K.; Haoquan, Y. *Adv. Mater.* 2003, *15*, 358-389.

[28] Utama, M. I. B.; Jun, Z.; Rui, C.; Xinlong, X.; Li, D.; Handong, S.; Xiong, Q. *Nanoscale* 2012, *4*, 1422-1435.

[29] Long, Y.; Chen, Z.; Wang, N.; Li, J.; Wan, M. *Physica B* 2004, *344*, 82-87.

[30] Heeger, A. J. *J. Phys. Chem. B.* 2001, *105*, 8475-8483.

[31] Huang, Z. M; Zhang, Y. Z; Kotaki, M.; Ramakrishna, S. *Comp. Sci. Technol.* 2003, 63, 2223-2253.

[32] Pham, Q. P.; Sharma, U.; Mikos, A. G. *Tissue Eng*. 2006, 12, 1197-1211.

[33] Hua, B.; Lu, Z.; Canhui, L.; Chun, L.; Gaoquan, S. *Polymer* 2009, *50*, 3292-3301.

[34] Li, D.; Xia, Y. *Adv. Mater.* 2004, *16*, 1151-1170.

[35] Sill, T. J.; von Recom, H. A. *Biomaterials* 2008, 29, 1989-2006.

[36] Collins, G.; Federici, J.; Imura, Y.; Catalani, L. H. *J. Appl. Phys.* 2012, 111, 044701-044719.

[37] Garg, K.; Bowlin, G. L. *Biomicrofluidics* 2011, 5, 013403-013422.

[38] Ramakrishna, S.; Fujihara, K.; Teo, W. E.; Lim, T. C.; Ma, Z. *An Introduction to Electrospinning and Nanofibers*; World Scientific: Singapore, SG, 2005; 1-382.

[39] Genovese, J. A.; Spadaccio, C.; Rainer, A.; Covino, E. *Stud Mechanobiol. Tissue Eng Biomater.* 2011, 6, 215-242.

[40] Kahol, P. K.; Pinto, N. *Synth. Met.* 2004, *140*, 269-272.

[41] Hui, M. H.; Li, Z. Y.; Wang, C. *Solid State Phenom*. 2007, *121*, 579-582.

[42] Rajesh, K. P.; Sundaray, B.; Jagadeeshbabu, V.; Santhamoorthy, G.; Natarajan, T. S.; Anantha, K. *IJENA* 2007, *1*, 173-191.

[43] Sujith, K.; Asha, A.M.; Anjali, P.; Sivakumar, N.; Subramanian, K. R. V.; Nair, S. V.; Balakrishnan, A. *Mater. Lett.* 2012, *67*, 376-378.

[44] Mikrajuddin, A.; Khairurrijal, K. *Indonesian J. Phys.* 2009, 20, 37-40.

[45] Eufinger, K.; Poelman D.; Poelman, H.; De Gryse, R.; Marin, G.B. *Thin Solid Films: Process and Applications* (2008) 189-227.

[46] Deitzel, J. M.; Kleinmeyer, J.; Harris, D; Tan, N. C. B. *Polymer* 2001, 42, 261-272.

[47] Demir, M. M.; Yilgor, I.; Yilgor, E.; Erman, B. *Polymer* 2002, 43, 3303-3309.

[48] Zhong, X. H.; Kim, K. S.; Fang, D. F.; Ran, S. F.; Hsiao, B. S.; Chu, B. *Polymer* 2002, 43, 4403-4412.

[49] Krishnappa, R. V. M.; Desai, K.; Sung, C. M. *J. Mater Sci.* 2003, 38, 2357-2365.

[50] Rutledge, G. C.; Li, Y.; Fridrikh, S.; Warner, S. B.; Kalayci, V. E.; Patra, P. *National Textile Centre, 2000 Annual Report (M98-D01),* National Textile Centre,1-10.

[51] Yuan, X.; Zhang, Y.; Dong, C.; Shang, *J. Polym. Int.* 2004, 53, 1704-1710.

[52] Zhao, S. L.; Wu, X. H.; Wang, L. G.; Huang, Y. *J. Appl. Polym. Sci.* 2004, 91, 242-246.

[53] Reneker, D. H.; Yarin, A. L.; Fong, H.; Koombhongse, S. *J. Appl. Phys.* 2000, 87, 4531-4547.

[54] Lee, J. S.; Choi, K. H.; Ghim, H. D.; Kim, S. S.; Chun, D. H.; Kim, H. Y.; Lyoo, W. S. *J. Appl. Polym. Sci.* 2004, 93, 1638-1646.

[55] Casper, C. L.; Stephens, J. S.; Tassi, N. G.; Chase, D. B.; Rabolt, J. F. *Macromolecules* 2004, 37, 573-578.

[56] Srinivasarao, M.; Collings, D.; Philips, A.; Patel. *Polymer Film Science* 2001, 292, 79-83.

[57] Lida, D.R.; Frederise, H.P.R. *Handbook of chemistry and physics*; CRC press: Boca Raton, FL, 1993; Vol.89, 1-2712.

[58] Bhadra, J.; Sarkar, D. *Bull Mater. Sci.* 2010, 33, 519-523.

[59] Chen, J.; Guo, Y.; Jiang, B.; Zhang, S. J Disp. Sci. Technol. 2008, 29, 97-100.

[60] Yang, Y.; Wan, M. *J. Mater Chem.* 2002, 12, 897-901.

[61] Hussain, A. M. P.; Kumar, A. *Bull Mater Sci.* 2003, 26, 329-334.

[62] Allen, J. B.; Fu-Ren, F. F.; Kwak, J.; Lev, O. *Anal. Chem.* 1989, 61, 132-138.

[63] Martin, A. E.; Martin, S.; Whitworth, A. L.; Macpherson, J. V.; Unwin, P. R. *Physiol. Meas.* 2006, 27, R63-R108.

[64] Allen, J. B.; Fu-Ren, F. F.; David, T. P.; Patrick, R. U.; David, O. W.; Feimeng, Z. *Science* 1991, 254, 68-74.

[65] Fong, H.; Reneker, D. H. *J. Polym. Sci. B, Polym. Phys.* 1999, 37, 3488-3493.

[66] Schneider, S. J.; *Engineered Materials Handbook: Ceramics and Glasses*; ASM International: Metals Park, OH, 1991; Vol.4, 1-1217.

[67] Zimbovskaya, N. A. In *Nanofibers;* Kumar, A.; Ed.; *On the electron transport in conducting polymer nanofibers*; Intech: Rijeka, HR, 2010; Vol.2, 329-348.

[68] Subramanian, C. K.; Kaiser, A. B.; Gibberd, R. W.; Liu, C-J.; Westling, B. *Solid State Comm.* 1996, 97, 235-238.

[69] Fuhrer, M. S.; Cohen, M. L.; Zettl, A.; Crespi, V. *Solid State Comm.* 1999, 109, 105-109.

[70] Prigodin, V. N.; Epstein, A. J. *Synth. Met.* 2002, 125, 43-53.

[71] Zimbovskaya, N. A.; Johnson, A. T.; Pinto, N. J. *Phys. Rev. B* 2005, 72, 024213-024216.

[72] Zhou, Y.; Freitag, M.; Hone, J.; Stali, C.; Johnson Jr., A. T.; Pinto, N. J.; MacDiarmid, A. G. *Appl. Phys. Lett.* 2003, 83, 3800-3802.

[73] Galperin, M.; Ratner, A.; Nitzan, A. *J. Chem. Phys.* 2004, 121, 11965-11979.

[74] Egger, R.; Gogolin, A. O. *Phys. Rev. B* 2008, 77, 113405-113408.

[75] Gutierrez, R.; Mandal, S.; Cuniberti, G. *Phys. Rev. B* 2005, 71, 235116-235124.

[76] Ryndyk, D. A.; D'Amico, P.; Cuniberti, G.; Richter, K. *Phys. Rev. B* 2008, 78, 085409-085414.

[77] Endres, R. G.; Cox, D. L.; Singh, R. R. P. *Rev. Mod. Phys.* 2004, 76, 195-214.

[78] Bolton-Neaton, C. J.; Lambert, C.J.; Falko, V. I.; Prigodin, V. M.; Epstein, A. J. *Phys. Rev. B* 1999, 60, 10569-10572.

[79] Ryndyk, D. A.; Cuniberti, G. *Phys. Rev. B* 2007, 76, 155430-155434.

[80] Lortscher, E.; Weber, H. B.; Riel, H. *Phys. Rev. Lett.* 2007, 98, 176807-176810.

[81] Muller, K.; Burkov, Y.; Mandal, D.; Henkel, K.; Paloumpa, I.; Goryachko, A.; Schmeiber, D. In *Physical and Chemical Aspects of Organic Electronics: From Fundamentals to Functioning Devices*; Christof, W.; Ed.; Wiley-VCH, Inc.: Morlenbach, HE, 2009; pp. 454-456.

[82] Sergawie, A.; Yohannes, T.; Gunes, S.; Neugebauer, H.; Sariciftci, N. S. *J. Braz. Chem. Soc* 2007, 18, 1189-1193.
[83] Rajendran, S.; Prabhu, M. R.; Rani, M. U. *Int. J. Electrochem. Sci.* 2008, 3, 282-290.
[84] Angammana, C. J.; Jayaram, S. H. *Prosc. ESA Annual Meeting on Electrostatistics* 2010, L3.
[85] Laskarakis, A.; Logothetidis, S.; Kassavetis, S.; Papaioannou, E. *Thin Solid Films* 2008, 516, 1443-1448.
[86] Xiong, J.; Huo, P.; Ko, F. K. *J Mater Res* 2009, 24, 2755-2761.
[87] Chand, S. *J. Mater Sci.* 2000, 35, 1303-1313.
[88] Desai, K.; Sung, C. *MRS Proceedings* 2002, 736, 121-126.
[89] Baji, A.; Mai, Y. W.; Wong, S. C.; Abtahi, M.; Chen, P. *Compos Sci. Technol.* 2010, 70, 703-718.
[90] Fang, J.; Niu, H. T.; Lin, T.; Wang, X. G. *Chin Sci. Bull* 2008, 53, 2265-2286.
[91] Miao, J.; Miyauchi, M.; Simmons, T. J.; Dordick, J. S.; Linhardt, R. J. *J. Nanosci. Nanotechno.* 2010, 10, 5507-5519.
[92] Zhong, Z.; Cao, Q.; Jing, B.; Wang, X.; Li, X.; Deng, H. *Mat. Sci. Eng. B.* 2012, 177, 86-91.
[93] Praveen, J. S.; Dhakshnamoorthy, M.; Kader, M. A. *International Conference on Advances in Polymer Technology* 2010, 163-166.
[94] Park, S. H.; Won, D. H.; Choi, H. J.; Hwang, W. P.; Jang, S.; Kim, J. H.; Jeong, S. H.; Kim, J. U.; Lee, J. K.; Kim, M. R. *Sol. Energ. Mat. Sol. C* 2011, 95, 296-300.
[95] Ahn, S. K.; Ban, T.; Sakthivel, P.; Lee, J. W.; Gal, Y. S.; Lee, J. K.; Kim, M. R.; Jin, S. H. *ACS Appl. Mater. Interfaces* 2012, 4, 2096–2100.
[96] Kim, C.; Park, S. H.; Lee, W. J.; Yang, K. S. *Electrochim Acta* 2004, 50, 877-881.
[97] Kim, C. *J. Power Sources* 2005, 142, 382-388.

In: Trends in Polyaniline Research
Editors: T. Ohsaka, Al. Chowdhury, Md. A. Rahman et al.
ISBN: 978-1-62808-424-5

Chapter 9

Designing of Conducting Ferromagnetic Polyaniline Composites for EMI Shielding

Kuldeep Singh[1,3], Anil Ohlan[2] and S. K. Dhawan[1,*]

[1]Polymeric and Soft Material Section, CSIR – National Physical Laboratory, New Delhi, India
[2]Department of Physics, Maharshi Dayanand University Rohtak, India
[3]School of Chemical Engineering and Bioengineering, University of Ulsan, Ulsan, Republic of Korea

Abstract

In recent times, the developments made in the field of conducting polymers have driven the progress of nano electronics and contributed to make electronic systems smaller, faster and more efficient. Among the conducting polymers, polyaniline (PAni) has been of great interest to many researchers because of its reasonably good conductivity, processibility, environmental stability and ease of forming composites. In the present chapter, we have tried to give in depth detail about interesting physics and chemistry of some special properties of PAni ferromagnetic composites for their applications as an electromagnetic interference (EMI) shielding material. This chapter also evaluates the recent progress and underlines the complex interplay of its intrinsic properties with EMI shielding. The bulk properties are dependent on the shape and size of the materials and can be tuned by controlling the reaction parameters, such as polymerization conditions and controlled addition of filler or ferrite nanoparticles. EMI attenuation is the function of reflection, absorption and multiple reflections which requires the existence of mobile charge carriers (electrons or holes), electric and/or magnetic dipoles, usually provided by materials having high dielectric constants (ε) or magnetic permeability (μ) and large surface or interface area respectively. Therefore, it is productive to make PAni composites with magnetic (Fe_2O_3/Fe_3O_4) and dielectric fillers titanium dioxide (TiO_2) nanomaterials to achieve better absorbing properties which

* Corresponding author: S. K. Dhawan. Polymerc & Soft Materials Section, CSIR-National Physical Laboratory, New Delhi – 110012, India; E-mail: skdhawan@mail.nplindia.ernet.in.

makes them futuristic shielding material. The present chapter has tried to cover the facets related to synthesis, conduction mechanism, magnetic and dielectric properties of conducting ferromagnetic PAni nanocomposites and associated phenomenon to EMI attenuation.

1. INTRODUCTION

The advances made in science and technology has added ease to our lives and electronic equipment based on the technology driven by nano-sciences has become more dense, fast and efficient at one end. On the other end, the electronic systems have become smaller with increased density of electrical components within an instrument and operating at higher frequency. Due to the upsurge use of high operating frequency and band width in electronic systems, especially in X-band and broad band frequencies, there are concerns and more chances of deterioration of the radio wave environment known as electromagnetic interference (EMI). EMI may cause malfunction to medical apparatus, industry robots or even cause harm to human body and become one of civic pain and continues to be a serious concern in society [1-3]. Therefore, considerable attention has been devoted to the development of effective EMI shielding and absorption materials that are cost efficient, lightweight, and effective over a broad frequency range. The primary mechanism of EMI shielding is usually reflection. For reflection of the radiation by the shield, the shield must have mobile charge carriers (electrons or holes) which interact with the electromagnetic fields of the radiation. As a result, the shield tends to be electrically conducting and therefore, traditionally metals are used as shielding materials. A secondary mechanism of EMI shielding is absorption. For significant absorption of the radiation, the shield should have electric and/ or magnetic dipoles which interact with the electromagnetic fields in the radiation.

Thus, having both conducting and magnetic components in a single system could be used as an EMI shielding material. Generally, magnetic or metal particles are used for the microwave absorption materials [4-5]. However, high specific gravity and difficult formulation have limited their practical applications.

Thus, there remains a need for an efficient microwave-absorbing material that is relatively lightweight, structurally sound and flexible, and efficient in absorption in a wide band range. The combination of the magnetic nanoparticles with conducting polymer leads to formation of ferromagnetic conducting polymer composites possessing unique combination of both electrical and magnetic properties.

This property of the nanocomposites can be used an electromagnetic shielding material since the electromagnetic wave consist of an electric (E) and magnetic field (H) perpendicular to each other. The ratio over E to H factor (impedance) has been subjugated in the shielding purpose. The conducting ferromagnetic type of materials can effectively shield electromagnetic waves generated from an electric as well as magnetic source. Polymer-matrix composites containing electrically conductive fillers are widely used for the EMI shielding of electronics [6-7]. Magnetic particles encapsulated within carbon-nanotube (CNT) composites and coated with carbon have been studied [8-11].

Yang et al. successfully fabricated two types of carbon nanofiber and CNT-polystyrene shielding composite [12]. Unfortunately, the complex preparation processes of the CNT based nano-composites are unfavourable for practical applications as such absorbing candidates.

On the other hand, conjugated polymers like polyaniline (PAni), polypyrrole and their composites are the potential candidates for microwave shielding because of the their light weight, corrosion resistance and flexibility beside this chemistry of conducting polymers offers a great variety of methods of synthesis and absorb radar waves to match new environmental constraints (mechanical properties for example).

The driving idea is based on the growing process at molecular scale of the conducting entity leading to a uniform macroscopic network in the material capable of switchable microwave absorption, which is called as intelligence stealth materials.

2. CONDUCTING POLYMERS

Organic polymer possesses the properties of metal (electrical, electronic, magnetic, and optical properties) along with the properties of conventional polymer (mechanical properties and processibility) are termed as intrinsically conducting polymer (ICP) or synthetic metals [13]. The unique property which differentiates them from conventional polymer is highly delocalized π-electron system with alternate single and double bonds in the polymer backbone. The π-conjugation of the polymer chain generates high energy occupied molecular orbitals and low energy unoccupied orbitals leading to a system that can be readily oxidized or reduced [14]. This leads to tunable level of conductivity from insulator to metallic by varying the doping level with a suitable dopant [13]. The significance of ICPs can be understood, as the discovery has awarded noble prizes in the year 2000 to Heeger, MacDiarmid and Shirakawa. ICPs have shown great potential for commercial applications such as rechargeable batteries [15-17], light emitting diodes (LEDs) [18-20], field-effect transistors (FETs) [21], solar cells [22-24], EMI shielding [25-29], electrostatic charge dissipation [30, 31], electrochromic devices [32-36], super capacitors [37-39] artificial muscles [40-42], corrosion control [43-45], and sensors [46-48].

Among the ICPs like polyacetylene, polypyrrole (PPY), polythiophene, PAni, poly(p-phenylenevinylene), and poly(p-phenylene), as well as their derivatives (Figure 1), PAni is well-known for its ease of synthesis, environmental stability, and unique acid/base doping/dedoping and oxidation/reduction chemistry and thought to be most promising conducting π-conjugated polymers compared to other ICPs [49].

It exists in three different discrete redox forms, which include the fully reduced leucoemeraldine, the fully oxidized pernigraniline and the semioxidized emeraldine base form as shown in figure 2. Emeraldine base ($\sigma \sim 10^{-10}$ S/cm) is the most interesting, since it can transport electrons when doped with aqueous acids (pH < 1) and yield conductive emeraldine salt form of PAni ($\sigma \sim 10^{1}$ S/cm). This occurs through proton induced spin unpairing. It is a reversible process, with deprotonating renovating the conductive emeraldine salt of PAni into an insulating emeraldine base [50]. Thus, the resulting conductivity can vary based on both the extent of oxidation and the extent of protonation, which makes PAni unique among ICPs.

In most of the cases, polymers are insulators in their neutral state and they become conducting only after introduction of electron acceptors/donors by a process known as 'doping'. The conductivity of a polymer can be tuned by chemical manipulation such as the nature of the dopant, the degree of doping and blending with other polymers.

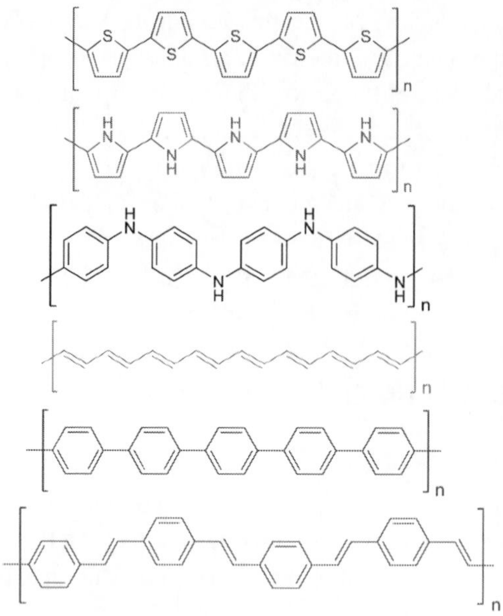

Figure 1. Chemical structure of some undoped conjugated polymers.

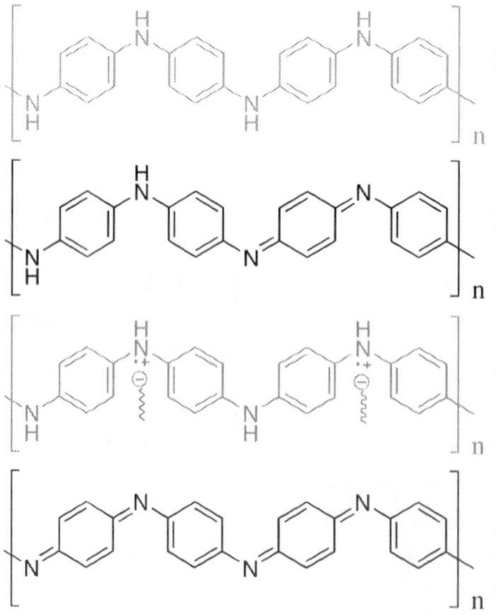

Figure 2. Different forms of PAni; leucoemeraldine (completely reduced polymer); emeraldine base (half-oxidized half-reduced polymer); pernigraniline (fully oxidized polymer).

Figure 1 shows the chemical structure of conjugated polymers in their neutral insulating state. Undoped conjugated polymers are semiconductors with band gaps ranging from 1 to 4eV, therefore their room temperature conductivities are very low, typically of the order of 10^{-8} S/cm or lower. However, doping can leads to an increase in conductivity of polymer by many orders of magnitude [51]. During the doping process, an undoped polymer having low conductivity, typically in the range of 10^{-10} to 10^{-5} S/cm, is converted to doped polymer, which is in a 'metallic' conducting regime (1 to 10^{4} S/cm). The highest value reported till date has been obtained for iodine-doped polyacetylene (PA) (>10^5 S/cm) whereas the predicted theoretical value of conductivity is about 2 × 10^7S/cm for doped PA, which is more than that of copper. Conductivity of other conjugated polymers reaches up to 10^3 S/cm as shown in Figure 3. Much of the combined research efforts of industrial, academic and government researchers have been directed toward developing materials that are stable (mechanically and electronically) for their use in nano-science and technology which are easily processible and can be produced by simple process at low cost. To make conjugated polymer electrically conducting, it is necessary to introduce mobile carriers into the conjugated system by doping. The electrical conductivity has resulted from the existence of charge carriers and the ability of charge carriers to move along the π-bonded "highway''. Doped conjugated polymers are good conductors for two reasons:

- Doping introduces charge carriers into the electronic structure.
- The attraction of an electron in one repeat unit to the nuclei in the neighboring units leads to carrier delocalization along the polymer chain and to charge carrier mobility, which is extended into three dimensions through interchange electron transfer.

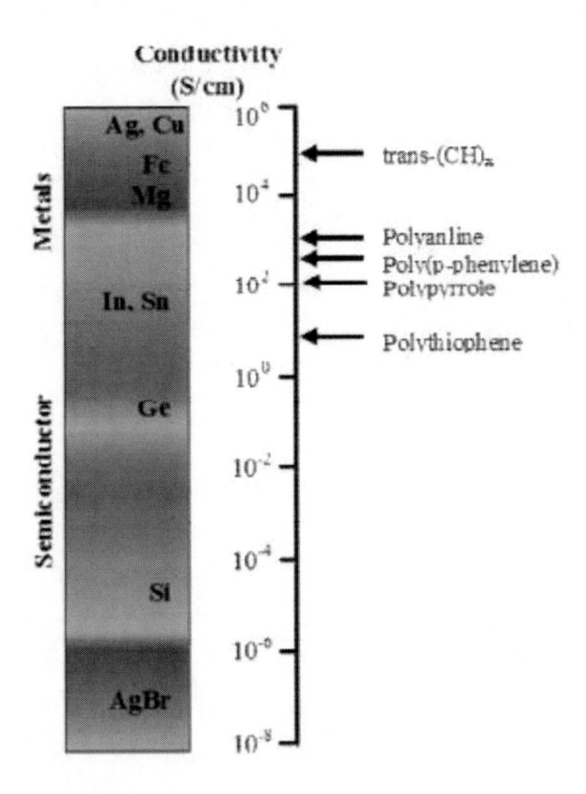

Figure 3. Conductivity of some metals and doped conjugated polymers.

To explain the electronic conduction phenomena in organic conducting polymers, new concepts including solitons, polarons and bipolarons [52-56] have been proposed by solid-state physicists. First, polymer having degenerate ground states, solitons is the important and dominant charge storage species. In this category, polyacetylene $(CH)_x$ is the only known polymer with a degenerate ground state due to its access to two possible configurations being energetically equivalent.

Secondly, Polymer with non-degenerate ground states such as poly-p-phenylene (PPP), polypyrrole (ppy), and PAni. These polymers do not support soliton-like defects because the ground state energy of the quinoid form is substantially higher than the aromatic benzenoid structure (Figure 4).

As a result, the charge defects in these polymers are different [57]. The removal of one electron from the π-conjugated system results in the formation of a radical cation which is partially delocalized over a segment of the polymer and known as polaron in solid-state physics. Since it is a radical cation, a polaron has spin 1/2. The radical and cation are coupled to each other via local resonance of the charge and the radical.

The presence of a polaron induces the creation of a domain of quinone-type bond sequence within the chain exhibiting an aromatic bond sequence. The lattice distortion produced is of higher energy than the remaining portion of the chain.

The creation and separation of these defects cost energy, which limits the number of quinoid-like rings that can link these two species, i.e., radical and cation, together. It is believed that the distortion extends over four rings. Upon further oxidation, the subsequent loss of another electron can result in two possibilities: the electron can come from either a different segment of the polymer chain thus creating another independent polaron, or from a polaron level (removal of an unpaired electron) to create a dication separating the domain of quinone bonds from the sequence of aromatic-type bonds in the polymer chain, referred to as a bipolaron. This is of lower energy than the creation of two distinct polarons; therefore, at higher doping levels it becomes possible for two polarons to combine and form a bipolaron, thereby replacing polarons with bipolarons [58]. Bipolarons can also be extended over four rings. The two free radicals combine leaving behind two cations separated by a quinoidal section of the polymer chain. At low doping levels, the defects are polarons, which tend to combine at higher doping levels to form bipolarons.

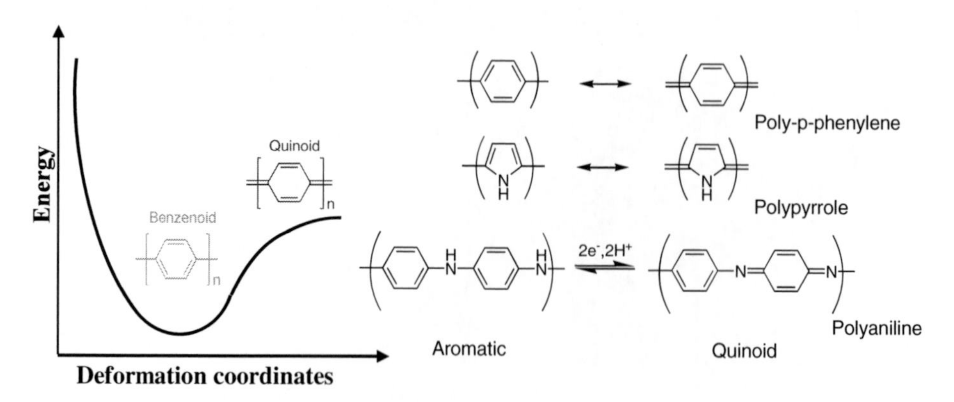

Figure 4. Resonance forms in conjugated polymers. The quinoid form has the higher energy than the benzenoid form.

The two bands inside the band gap are empty in the case of bipolarons while the lower polaron band is half-filled. Doping results in the appearance of mid-gap transitions. Three in-gap transitions are characteristic of polaron states while bipolaron states are noted for the two transitions.

In the conventional conductive polymers like polypyrrole, polythiophene, polyacetylene and poly p-phenylene etc., oxidative doping results in the removal of electrons from the bonding π-system while in PAni, the initial removal of electrons is from the non-bonding nitrogen lone pairs. Its formation requires reduction and deprotonation so that it actually differs in chemical composition from the benzenoid form (Figure 4). This peculiarity of the PAni structure makes doping by an acid-base reaction (Figure 5) possible.

In addition, the constituent parts of both the polaron and the bipolaron are very tightly bound owing to valence restrictions. The radical and cation of the polaron are confined to a single aniline residue. The bipolaron is confined to, and identical with a (doubly protonated) quinone-diimine unit. This narrow confinement may destabilize bipolarons with respect to polarons owing to the coulomb repulsion between the cations.

As shown in figure 5 doping of PAni can be achieved via two routes. Doping by oxidation of the leucoemeraldine form, results in the formation of radical cations which may then convert to bipolarons. Alternatively, protonation of the emeraldine base form leads to the initial formation of bipolarons that may rearrange with neutral amine units to form radical cations.

Conductivity (σ) is usually measured by a four-probe method and is an important property of conducting polymers. It is usually expressed as $\sigma = ne\mu$, where e is charge of electron, n and μ are density and mobility of charge carriers, respectively.

The carrier concentration, its mobility and the type of carrier of semiconductors are usually determined by the Hall effect measurement.

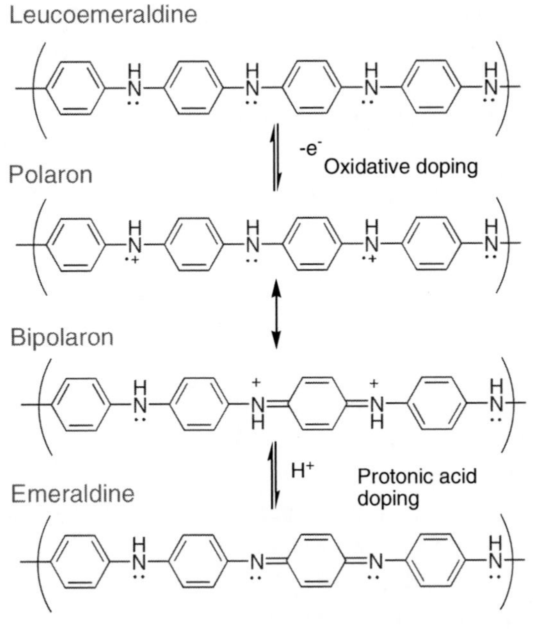

Leucoemeraldine

-e⁻ Oxidative doping

Polaron

Bipolaron

H⁺ Protonic acid doping

Emeraldine

Figure 5. Oxidative and protonic acid doping in the PAni system.

For the negatively charged carriers (n -type semiconductors) the Hall voltage is positive and for the positively charged carriers (p-type semiconductors) the Hall voltage is negative.

The experimental results show that the majority charge carriers in PAni are holes indicating PAni to be a p type semiconductor [59].

However the exact nature of the charge carriers in conductive polymers is uncertain, the physics of charge transport is controversial at best. Various models have been proposed for the electrical conduction mechanism in PAni. From the temperature dependent conductivity, the models proposed are quasi one-dimensional (1D) variable range hopping (VRH), 3D VRH, [60-68], a combination of 1D and 3D VRH [69], charging energy limited tunneling (CELT) for granular metals [70] and a heterogeneous model where temperature dependence of conductivity shows from nonmetallic to metallic [70]. The potential applications of PAni have been limited due to its infusibility, very low solubility in most of the available solvents, hygroscopic and relatively low conductivity [71].

To utilized full potential of PAni and achieve higher conductivity, different concentration of the dopant and the method of doping have been used to control its structure. Main parameters which affect the conductivity [72] as follows:

I Chain structure includes π-conjugated structure and length, crystalline and substituted grounds and bounded fashion of the polymeric chain. Regarding polymeric chain structure, the maximum value of the conductivity in iodine-doped PA was on the order of 10^3 S/cm while in case of doped PAni and PPy were below 200 S/cm. In general, the electrical conductivity at room temperature is proportional to the crystalline degree because of closer intermolecular distance in crystalline phase. Therefore, the conducting polymers with a branched chain have a low conductivity at room temperature is expected due to less crystallinity.

II The degree of doping level and structure of dopant are important parameter to convert an insulating π-conjugated polymer to conducting one. Molecular structure of dopants not only affects electrical properties, but also solubility in organic solvent or water. For example PAni doped by camphorsulfonic acid (CSA) has high conductivity (200 S/cm), but also can soluble in m-cresol while the PAni doped with HCl is insoluble in most of the organic solvent

III Polymerization conditions including concentration of monomer, dopant and oxidant, the molar ratio of dopant and oxidant to monomer and polymerization temperature and time are other important parameters that affects the conductivity, because these are contributed to chain conformation, morphology and crystallinity of the final product [73, 74].

To improve processibility and solubility, usually introduction of substituent in the polymeric chain or copolymerization are versatile techniques, alkyl substituents on the polymer backbone leads improvement in solubility in doped and undoped forms in common organic solvents and even in water [75-77] but on the cost of its electrical conductivity. A polymeric dopant such as sulfonated polystyrene (PSS) is used to obtain a better processable and dimensionally stable free-standing film [78].

Several functional sulfonic acids have been found to render PAni soluble in organic solvents in its doped state. Cao et al. reported that the PAni doped with dodecylbenzene sulfonic acid (DBSA) or camphor sulfonic acid (CSA) could be soluble in various organic solvents such a s m-cresol and xylene [79]. Dhawan et al. has reported the effect of DBSA concentration on its electrical and microwave absorption properties [74]

3. Electromagnetic Shielding and Microwave Absorption

Conceptually, a shield is a barrier to the transmission of the electromagnetic fields. It limits the amount of electromagnetic interference (EMI) radiation from the external environment that can penetrate the electronic circuit and, conversely, it influences how much EMI energy generated by the circuit can escape into the external environment. The effectiveness of a shield may be deduced by the ratio of the magnitude of the electric (magnetic) field that is incident on the barrier to the magnitude of the electric (magnetic) field that is transmitted through the barrier.

3.1. Shielding Effectiveness

Shielding can be specified in the terms of reduction in magnetic (and electric) field or plane-wave strength. The effectiveness of a shield and its resulting EMI attenuation are based on the frequency, the distance of the shield from the source, thickness of the shield, and shield material. Shielding effectiveness (SE) is normally expressed in decibels (dB) as a function of the logarithm of the ratio of the incident and transmitted electric (E), magnetic (H), or plane-wave field intensities (F): SE (dB) = 20 log (E_o/E_l), SE (dB) = 20 log (H_o/H_l), or SE (dB) = 20 log (F_o/F_l), respectively. With any kind of electromagnetic interference, there are three mechanisms contributing to the effectiveness of a shield. Part of the incident radiation is reflected from the front surface of the shield, part is absorbed within the shield material, and part is reflected from the shield rear surface to the front where it can aid or hinder the effectiveness of the shield depending on its phase relationship with the incident wave, as shown in figure 6. Therefore, the total shielding effectiveness (SE) of a shielding material equals the sum of the absorption factor (SE_A), the reflection factor (SE_R), and the correction factor to account for multiple reflections (SE_M) in thin shields [79- 82]:

$$SE = SE_A + SE_R + SE_M \tag{1}$$

All the terms in equation (1) are expressed in dB. The multiple reflection factor SE_M can be neglected if the absorption loss SE_A is greater than 10 dB. In practical calculation, SE_M can also be neglected for electric fields and plane waves.

3.2. Absorption Loss

Absorption loss, A, is a function of the physical characteristics of the shield and are independent of the type of source field. Therefore, the absorption term SE_A is the same for all three waves.

When an electromagnetic wave passes through a medium, its amplitude decreases exponentially, as shown in Figure 6. This decay or absorption loss occurs because currents induced in the medium produce ohmic losses and heating of the material, and E_l and H_l can be expressed as $E_1 = E_o e^{-t/\delta}$ and $H_1 = H_o e^{-t/\delta}$ [83]. The distance required by the wave to be attenuated to 1/e or 37% is defined as the skin depth.

Therefore, the absorption term SE_A in decibel is given by the expression:

$$SE_A = 20(t/\delta)\log e = 8.69(t/\delta) = 131 t\sqrt{f\mu\sigma} \quad (2)$$

where t is the thickness of the shield in mm; f is frequency in MHz; μ is relative permeability (1 to copper); σ is conductivity relative to copper. The skin depth δ can be expressed as:

$$\delta = \frac{1}{\sqrt{\pi f \mu \sigma}} \quad (3)$$

The absorption loss of one skin depth in a shield is approximately 9 dB. Skin effect is especially important at low frequencies, where the fields experienced are more likely to be predominantly magnetic with lower wave impedance than 377 Ω.

From the absorption loss point of view, a good material for a shield will have high conductivity and high permeability, and sufficient thickness to achieve the required number of skin depths at the lowest frequency of concern.

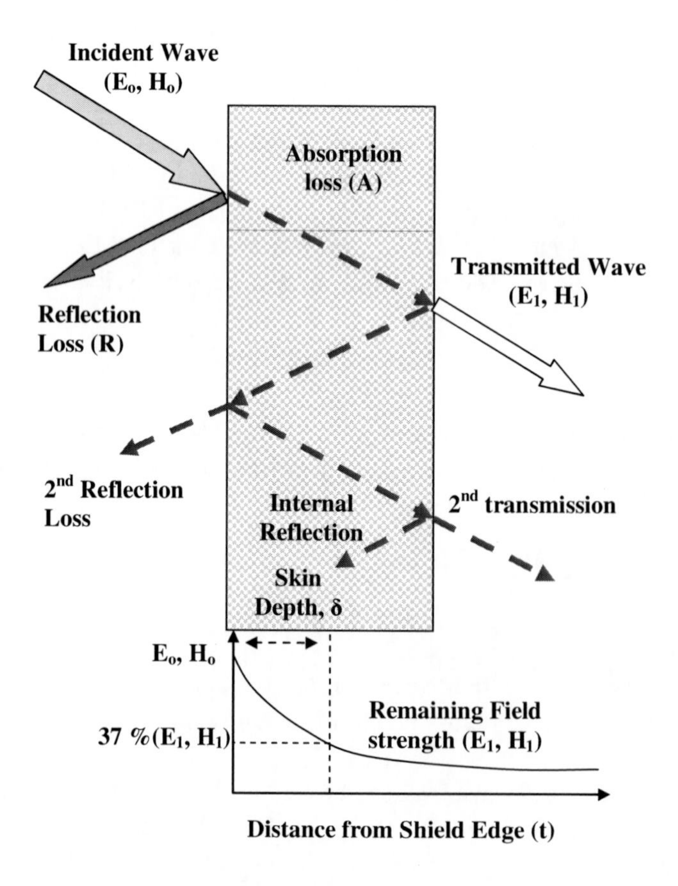

Figure 6. Graphical representation of EMI shielding.

3.3. Reflection Loss

The reflection loss is related to the relative mismatch between the incident wave and the surface impedance of the shield. The computation of reflection losses can be greatly simplified by considering shielding effectiveness for incident electric fields as a separate problem from that of electric, magnetic, or plane waves. The equations for the three principle fields are given by the expressions [84]

$$R_E = K_1 10\log\left(\frac{\sigma}{f^3 r^2 \mu}\right) \tag{4}$$

$$R_H = K_2 10\log\left(\frac{fr^2\sigma}{\mu}\right) \tag{5}$$

$$R_P = K_3 10\log\left(\frac{f\mu}{\sigma}\right) \tag{6}$$

where R_E, R_H, and R_P are the reflection losses for the electric, magnetic, and plane wave fields, respectively, expressed in dB; σ is the relative conductivity referred to copper; f is the frequency in Hz; μ is the relative permeability referred to free space; r is the distance from the source to the shielding in meter.

4. Synthesis of Conducting PAni and Its Ferromagnetic Composite

Although PAni is being known for over 150 years but actual research took off in the 1980s. The main focus is to understand chemical structure and electronic conduction mechanisms, designing polymerization techniques, and developing chemical or physical modification methods for attempting to make PAni processable.

With the rapid emergence of nanoscience and nanotechnology in recent years, synthesizing nanostructures of PAni like nanotubes, nanospheres, nanofibers, hollow and solid structures of core shell type have attracted growing attention with a hope that nanostructured PAni will offer better performance or new properties compared with its conventional bulk counter-part. A number of methods for producing PAni nanostructured have been reported via chemical or electrochemical oxidation of a monomer where the polymerization reaction is stoichiometric in electrons. However, number of methods such as photochemical polymerization [85], pyrolysis [86, 87], metal-catalyzed polymerization [88-90], solid-state polymerization [91], plasma polymerization [92], ring-forming condensation [93], step-growth polymerization [94, 95], and soluble precursor polymer preparation [96, 97], have been reported in literature for synthesis of conjugated polymers. In this section, discussion is mainly focused on the preparation of conducting ferromagnetic nanocomposite of aniline by emulsion polymerization. The deliberate modifications in chemical and super molecular structure of polymer matrix by incorporating nano ferromagnetic particles can lead to the formation of conducting ferromagnetic materials which can be suitably designed for electromagnetic shielding applications. For the absorption of electromagnetic radiations, ferrites are incorporated in the polymer matrix as they possess high magnetization values [98-100]. Many attempts to produce the colloidal PAni composites containing the ferrite have been made, using different dopants [101-107]. However, the resultant polymer composite loses its conductivity and has low magnetization value. A key factor in developing a nanocomposite is homogeneous dispersion of nanoparticles in the polymer matrix. It is therefore essential to develop a method for effectively dispersing nanoparticles in the solution and to prevent their agglomeration through Vander Waal's forces.

4.1. Synthesis of Conducting Ferromagnetic Nanocomposites by Emulsion Polymerization

Chemical oxidation of aniline in aqueous acidic media using ammonium persulfate oxidant is the most widely employed method [108, 109]. However, for the synthesis of ferrite nanocomposites, the chemical oxidation of aniline in the presence of inorganic acid has some limitations like, dissociation of ferrite moieties at lower pH value and settling of ferrite particles due to higher density. Therefore, emulsion polymerization is an appropriate polymerization reaction which takes place in a large number of loci dispersed in a continuous external phase. Emulsion polymerization methods have several distinct advantages over the conventional in situ polymerization.

The physical state of the emulsion system makes it easier to control the process. In a typical synthesis process, functional protonic acid such as dodecyl benzene sulphonic acid (DBSA) is used which being a bulky molecule, can act both as a surfactant and as dopant. Surfactants are amphiphilic compounds containing polar (hydrophilic) head and non-polar (hydrophobic) tails [110]. The polymerization of aniline monomer in the presence DBSA (dodecyl benzene sulfonic acid) leads to the formation of emeraldine salt form of PAni (Figure 7). When the ferrite particles are homogenized with DBSA in aqueous solution, micelles are formed as surfactant form aggregates in which the hydrophobic tail are oriented towards the interior of the micelles leaving the hydrophilic group in contact with the aqueous medium [111].

In emulsion polymerization of intrinsic conducting polymers, the monomer and the ferrite particles assemble themselves in to micelles. The location of the monomer in the micelles is important as it can dictate the reaction mechanism and the properties of the final product [112].

The semi-polar aniline monomer locates at the palisade layer (the region between the hydrophilic groups and the first few carbon atoms of the hydrophobic groups) of the micelle with the polar group at the micellar surface and the nonpolar hydrocarbon groups in the micellar. Anilinium cations sit between the individual DBSA molecules near the shell of the micelle complexed with sulfonate ion. When polymerization proceeds, anilinium cations are polymerized within the micelle together with DBSA and ferrite particles resulting in the formation of PAni–ferrite nanocomposites. Schematic representation for the formation of PAni-ferrite composite is shown in figure 8.

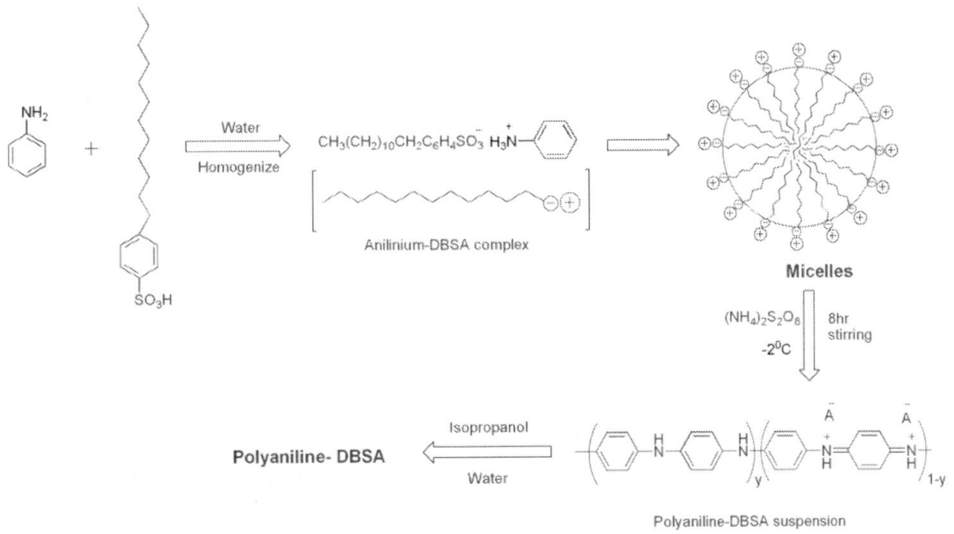

Figure 7. Schematic representation of the polymerization of aniline by formation of micelles.

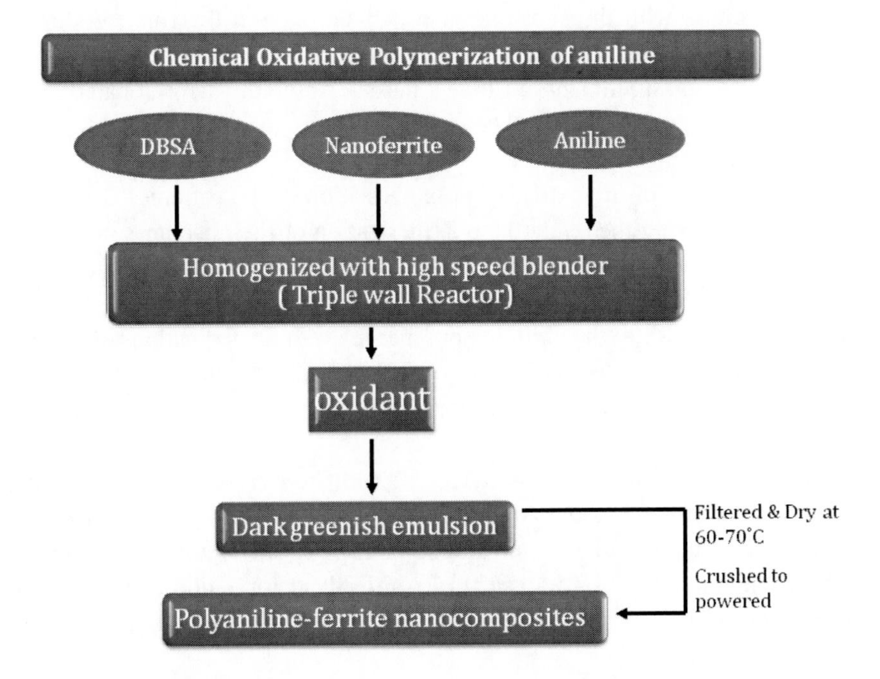

Figure 8. Pictorial representation for the formation of PAni–ferrite nanocomposite by chemical oxidative polymerization.

Polymerization methods generally play a very vital role in polymer morphology, internal structure, defects and degree of doping. Micro emulsion polymerization has very precise control over the properties of the polymer as it has high degree of polymerization then the normal suspension and precipitation method. In the PAni-DBSA formation in aqueous medium, water is the continuous phase and DBSA is a surfactant that acts as discontinuous phase.

Monomer aniline is emulsified to form the micro micelles of oil in water type. Emulsion polymerization has high degree of polymerization than those prepared by suspension and precipitation method. A typical micelle in aqueous solution forms a roughly spherical or globular aggregate with the hydrophilic "head" regions in contact with surrounding solvent, sequestering the hydrophobic tail regions in the micelle center. The shape of a micelle is a function of the molecular geometry of its surfactant molecules and solution conditions such as surfactant concentration, temperature, pH and ionic strength. Generally, in micellar solution there are the chances of formation of macroscopic particles that can be prevented by adding the steric stabilizers like poly (vinyl alcohol), poly (N-vinylpyrrolidone) and cellulose ethers, but in this present system the bulky surfactant dodecyl benzene sulphonic acid itself acts to prevent the formation of the macroscopic precipitation. When monomer aniline is added to the DBSA micelle, it occupies the place in between the micelle and the hydrophilic sulphonate unit, subsequently on addition of oxidant like APS, the polymerization takes place at the interface boundary. The oxidative polymerization is a two electron chain reaction and therefore persulphate requires one mole of monomer. However, a smaller quantity of oxidant is used to avoid oxidative degradation of the polymer formed. The principal function of the oxidatant is to withdraw a proton from the aniline molecule, without forming a strong coordination bond either with the substrate/intermediate or with the final product. Aniline oxidized to radical-cations can be oxidized to dications in a consecutive reaction. The dication molecule, because of its significant positive charge, is rapidly deprotonated, yielding a nitrenium cation. Dimerization and further oxidation leads to the formation of trimer and finally PAni with linear structure in successive steps [113-115].

A possible mechanism of the oxidative polymerization of the cationic form of aniline is presented in figure 9. During the oxidative polymerization of the micellar solution of aniline carried out by using ammonium peroxydisulphate, the color of the solution started changing from white to light green and finally to dark black green after the complete polymerization in 8 hours. Addition of the APS to the aniline monomer leads to the formation of cation radicals which combine with another monomer moiety to form a dimer, which on further oxidation and combination with another cation radical forms a termer and ultimately to a long chain of polymer.

The emulsion polymerization of aniline to PAni in the presence of Fe_2O_3 may bring certain changes in the properties of PAni because conduction mechanism in PAni involves protonation as well as ingress of counter anions in the polymer matrix to maintain charge neutrality. Protonation and electron transfer in PAni leads to formation of radical cations by an internal redox reaction, which causes the reorganization of electronic structure to give two semiquinone radical cations.

In the doping process, ingress of anions occurs to maintain charge neutrality in the resultant doped PAni matrix.

Insitu emulsion polymerization of aniline in the presence of γ-Fe_2O_3 constituents leads to the formation of ferromagnetic conducting polymer. In order to avoid phase segregation, the γ-Fe_2O_3 nanoparticles were functionalized with the surfactant DBSA that ensure its compatibility with the polymer. Cyclic Voltammetry is a good tool to know the effect of ferrite particles on the electrochemical properties, for this electrochemical polymerization of aniline with DBSA in aqueous medium was carried out using cyclic potential sweep method by switching the potential from -0.20 V to 0.95 V vs. SCE at a scan rate of 20 mV/sec.

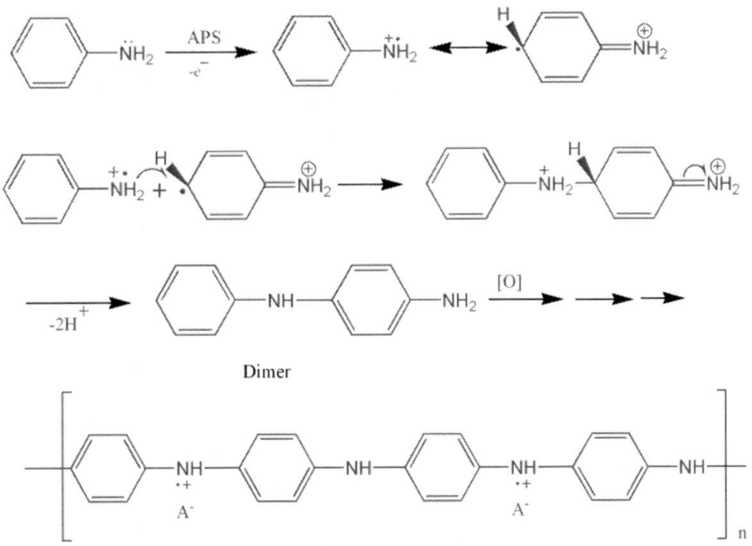

Figure 9. Polymerization mechanism of aniline using ammonium peroxydisulphate as oxidant.

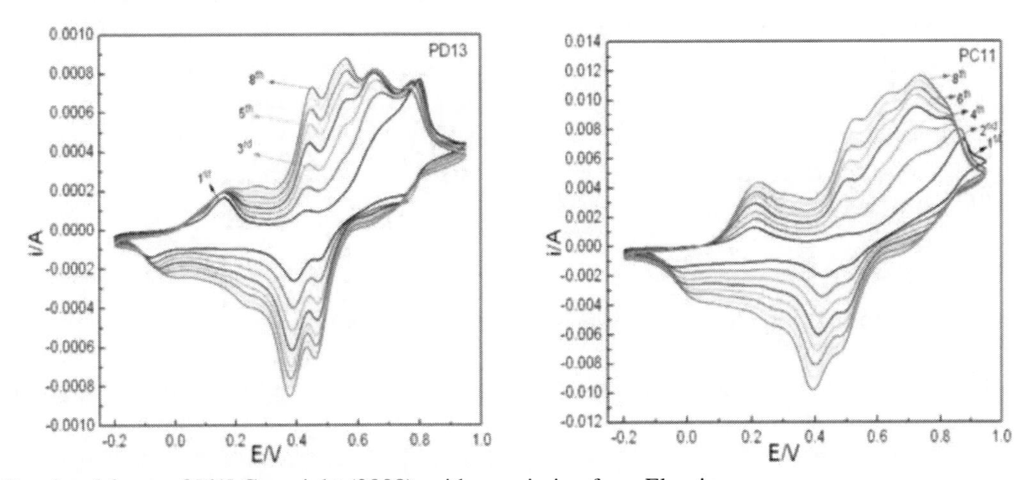

Reprinted from ref [61] Copyright (2008), with permission from Elsevier.

Figure 10. Electrochemical growth behavior of aniline in DBSA medium (PD13) and aniline in DBSA medium containing ferrite particles (PC11) on cycling the potential between -0.2 V to 0.95 V, taking eight successive scans, on platinum electrode vs. SCE at a scan rate of 20 mV/sec.

The rise in current value at 0.78 V in the first cycle corresponds to the oxidation of aniline leading to generation of anilinium radical cations (Figure 10).

In the subsequent cycles, new oxidation peaks appear which indicate that these radical cations undergo further coupling to form benzenoid structure and combination of benzenoid and quinoid structure. The peak current increases continuously with successive potential scans to build up electroactive PAni on the electrode surface.

However when polymerization of aniline was carried out in the presence of γ-Fe_2O_3 particles entrapped in the surfactant medium, electrochemical growth behaviour shows shifting of peak potential values, which indicates the incorporation of γ-Fe_2O_3 in the polymer backbone.

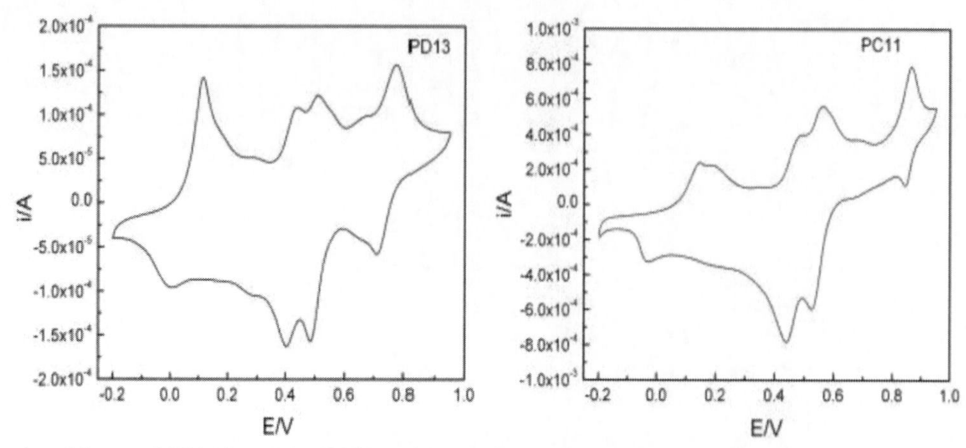

Figure 11. Cyclic voltammogram of PAni film in DBSA medium (PD13) and PAni-ferrite composite in DBSA medium (PC11) on Pt electrode vs. SCE at a scan rate of 20mV/sec.

Cyclic voltammogram of PAni film obtained by potential sweeping technique in blank DBSA medium shows characteristic peaks at 0.11 V, 0.44V, 0.54 V and 0.77 V (Figure 11) because of the radical cations at first peak potential values which subsequently oxidized to dications [116]. However, the cyclic voltammogram of PAni embedded with γ-Fe_2O_3 particle in DBSA medium shows characteristic peaks at 0.14 V, 0.48V, 0.56 V and 0.86V (Figure 11). The shifting of characteristic redox peaks can be assigned to the incorporation of γ-Fe_2O_3 particles in the PAni matrix.

4.2. Incorporation of Dielectric Filler in PAni Composite

Polymeric materials with high dielectric constants are highly desirable for use in electromagnetic shielding application. To raise the dielectric constant of polymers, high-dielectric constant ceramic powders such as $BaTiO_3$ and $PbTiO_3$ were added to the polymers to form nanocomposites [117-122]. The electrical properties of PAni can be modified by the addition of inorganic fillers [123-127]. The insertion of nanoscale fillers may improve the electrical and dielectric properties of host PAni materials. A high-dielectric constant up to 3700 for PAni–TiO_2 nanocomposite materials was reported by Dey et al. [126]. However, high-dielectric constants are usually accompanied with high-dielectric losses [127]. Xu et al. [128] noticed that the electrical conductivity of a PAni–TiO_2 nanocomposite with a low TiO_2 content is much higher than of PAni. Su et al. [129] reported that the PAni–TiO_2 nanocomposite has suitable conductivity (1–10 S/cm) and increases after thermal treatment at 80°C for 1 h. Dhawan et al. has recently shown that A ternary of PAni with TiO_2 and barium ferrite synthesized *via in situ* emulsion polymerization lead to form an array of nanoparticles encapsulated PAni. This ternary PAni has shown excellent microwave absorption of 58 dB (>99.999% attenuation) in 12.4-18GHz range. This high value had aroused due to synergistic effect of magnetic losses in barium ferrite and dielectric losses in TiO_2 and PAni [130].

A schematic presentation for creating multicomponent PAni- ferromagnetic composite with nanosize TiO_2 (~70-90nm) and γ-Fe_2O_3 (~10-15nm) particles via insitu emulsion polymerization using dodecyl benzene sulfonic acid (DBSA) as dopant has been shown in figure 12. The nano sized γ-Fe_2O_3 along with TiO_2 nanoparticles was homogenized in 0.3 M aqueous solution of dodecyl benzene sulfonic acid (DBSA) to form a whitish brown emulsion. To this an appropriate amount of aniline (0.1 M) was added and again homogenized for 2-3hrs to form micelles of aniline with γ-Fe_2O_3 and TiO_2. The micelles so formed were polymerized below 0 °C through chemical oxidization polymerization by $(NH_4)_2S_2O_8$ (0.1 M). The product so obtained was demulsified by treating with equal amount of isopropyl alcohol. The precipitates were filtered out and washed with alcohol and dried at 60-65 °C. Different formulations of polymer composite having different weight ratio of monomer to ferrite/TiO_2, An: γ-Fe_2O_3: TiO_2; 1:1:1 (PTF11); 1:1:2(PTF12), were synthesized in DBSA medium to check the effect of ferrite constituents on the properties. Beside this, PAni-Fe_2O_3 (PF12) with monomer to Fe_2O_3 weight ratio of 1:2 were also synthesized and the results were compared with pure PAni doped with DBSA.

5. CHARACTERIZATION

5.1. X-Ray Diffraction Studies

X-ray diffraction patterns of γ- Fe_2O_3, PAni doped with DBSA and it composites with different compositions of γ- Fe_2O_3 is shown in Figure 13. The main peaks for γ- Fe_2O_3 are observed at 2θ = 30.281(d = 2.949Å), 35.699 (d=2.513 Å), 43.435 (d=2.081 Å), 53.805 (d=1.702 Å), 57.437 (d=1.603 Å), 63.0460 (d= 1.473 Å) corresponding to the (2 0 6), (1 1 9), (0 0 12), (2 2 12), (1 1 15), (4 4 1) reflections. The peaks present in γ-Fe_2O_3 were also observed in all the compositions of PAni composite with γ- Fe_2O_3 which indicates the presence of ferrite particles in the polymer matrix and the intensity of peaks increase with the increases in the ratio of iron oxide.

While the presence of PAni and its semi crystalline nature is confirmed by the broad peaks at 2θ = 19.795 (d=4.481 Å) and 25.154 (d=3.537 Å) [131, 132], it is also observed that the intensity of these peaks increases with the decrease in iron oxide composite.

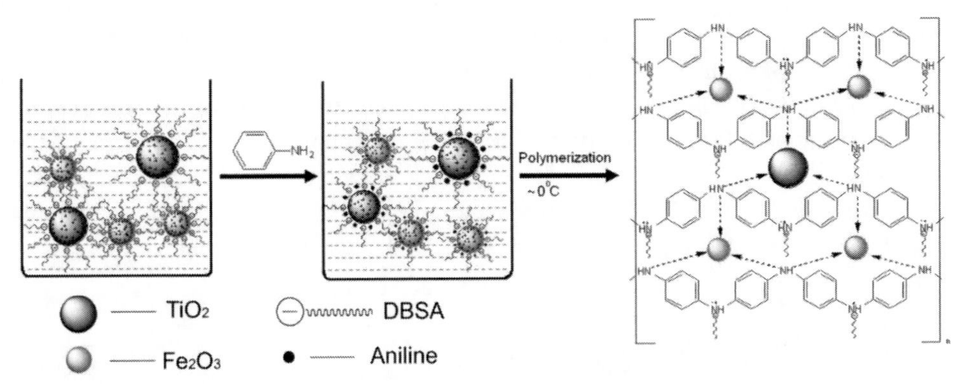

Figure 12. Schematic representation of synthesis of TiO_2 and γ- Fe_2O_3 ferromagnetic conducting nanocomposite.

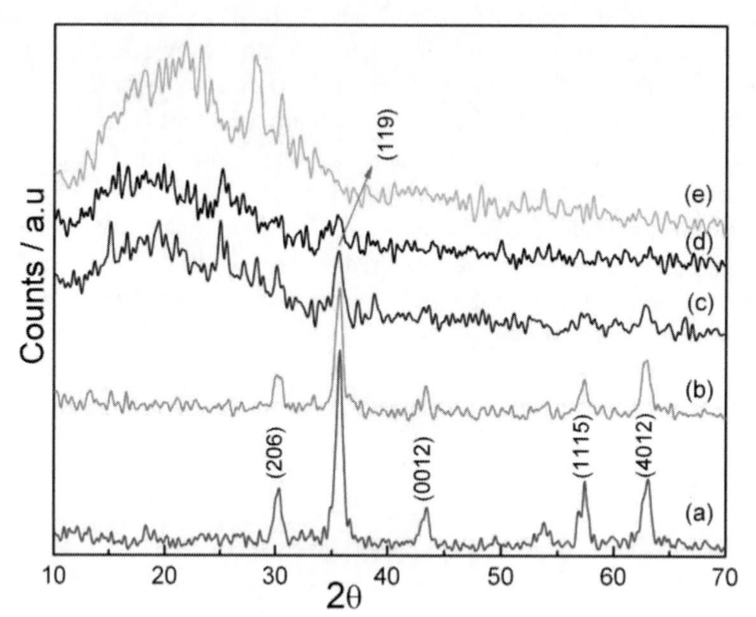

Figure 13. X-ray diffraction plots of γ-Fe_2O_3 (a), PC12 (b), PC11(c), PC21 (d) and PD13 (e) with arbitrary no of counts vs. 2θ.

The line broadening of the peaks in the entire patterns of PAni composite indicates the small dimensions of the iron oxide particles. The average size of γ- Fe_2O_3 particles was found to be 8.99 nm for pure γ-Fe_2O_3 and 9.87 nm for PAni composite with iron oxide having aniline: γ-Fe_2O_3:: 1:2 (PC12).

Figure 14 shows the X-ray diffraction patterns of TiO_2, γ- Fe_2O_3 and composites of PAni with TiO_2 and γ-Fe_2O_3 (PTF). The main peaks for TiO_2 were observed at 2θ value 25.283 (d=3.520 Å), 37.784 (d=2.379 Å), 38.530 (d=2.335 Å), 48.032 (d=1.893 Å), 53.874 (d=1.700 Å), 55.025(d=1.667 Å), and 62.660 (d=1.481 Å) corresponding to (1 0 1), (0 0 4), (1 1 2), (2 0 0), (1 0 5), (2 1 1), and (2 0 4) reflections (curve a). The peaks of γ-Fe_2O_3 were found in all the compositions of PTF composites which designate the presence of ferrite particles in the polymer matrix while the increase in intensity of peaks demonstrates the increase in the ratio of iron oxide. Distinguished sharp peaks were observed for the TiO_2 nanoparticles as compared to the iron oxide because of its larger crystallite size. The crystallite size of γ-Fe_2O_3 particles was estimated to be 10.3 nm for pure γ-Fe_2O_3 while it is 13.7 nm in PTF12. The crystallite size of TiO_2 is 36.6 nm for pure TiO_2 sample and 29.6 nm in PTF12 nanocomposite. The presence of peaks of TiO_2 and γ-Fe_2O_3 shows the formation of composite having separate phase of both the compounds properly dispersed in the polymer matrix.

5.2. TEM and HRTEM Analysis

Figure 15 shows the TEM images of γ- Fe_2O_3, TiO_2, and PAni composite with TiO_2 and Fe_2O_3 (PTF 12) From the graphs well-dispersed spherical particles of γ- Fe_2O_3 and TiO_2 particles are observed.

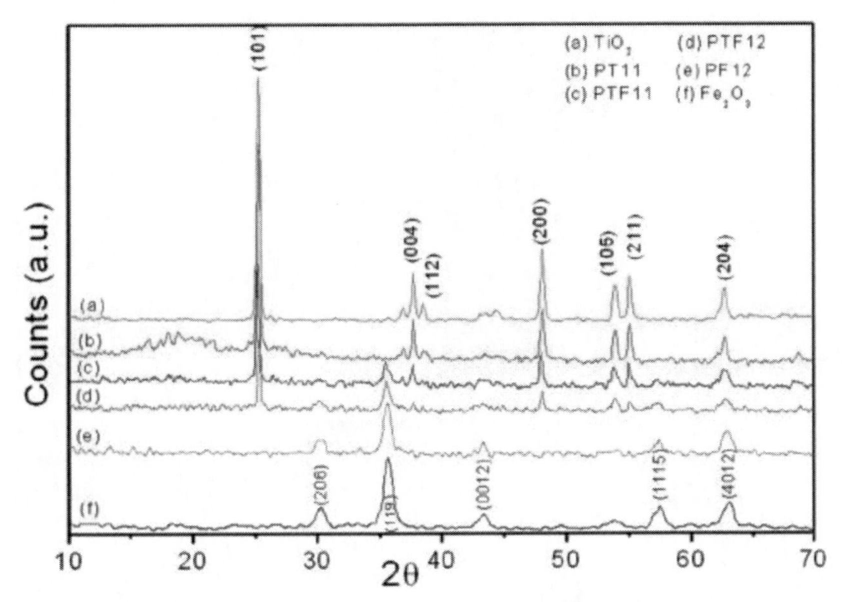

Reprinted from ref [62] Copyright (2010), with permission from Elsevier.

Figure 14. XRD plots of (a) TiO_2, (b) PT11, (c) PTF11, (d) PTF12, (e) PF12, and (f) γ-Fe_2O_3.

The particle size of γ- Fe_2O_3 is estimated to be 10-15 nm while TiO_2 has slightly larger particles of 70-90 nm (Figure 15 a andb). When these nanoparticles are incorporated in the polymer matrix, they show an agglomerated morphology (Figure 15c). The dispersion of γ-Fe_2O_3 and TiO_2 particles in the polymer matrix is confirmed by the HRTEM image (Figure 15d). The lattice fringes of γ- Fe_2O_3 with lattice spacing 0.25 nm corresponding to (119) plane and 0.35 nm corresponding to (101) plane for TiO_2 are matched with the XRD pattern of the polymer composite.

5.3. Thermo Gravimetric Analysis

The thermograms of PAni doped with different concentration of DBSA are shown in figure 16 shows. Thermogram of polymer shows three major weight losses; initial weight loss at 100 °C is attributed due to the loss of water contents. From 230-380 °C is supposed to be due to the decomposition of DBSA from the polymer chain because, in case of Emeraldine base (undoped form of PAni) no major step is observed in this temperature range. The degradation of the polymer chains starts at 400 °C and continues up to 700 °C. The extent of protonation can be estimated from the total amount of weight loss from the different samples. From the thermograms, it has been observed that the total weight loss increases as the concentration of the dopant DBSA increase. Thermo gravimetric analysis of the PAni composites shows the effect of the γ- Fe_2O_3 content on the thermal stability of the composite (Figure 17). PAni doped with DBSA (PD13) is thermally stable up to 230 ^{0}C. However, when a conducting polymer has been synthesized by incorporating ferric oxide moieties in the reaction system along with the surfactant, it was observed that the thermal stability of the polymer increased to 260 ^{o}C. This showed that *in situ* polymerization of aniline in the presence of ferrite particles leads to a better, thermally stable conducting polymer.

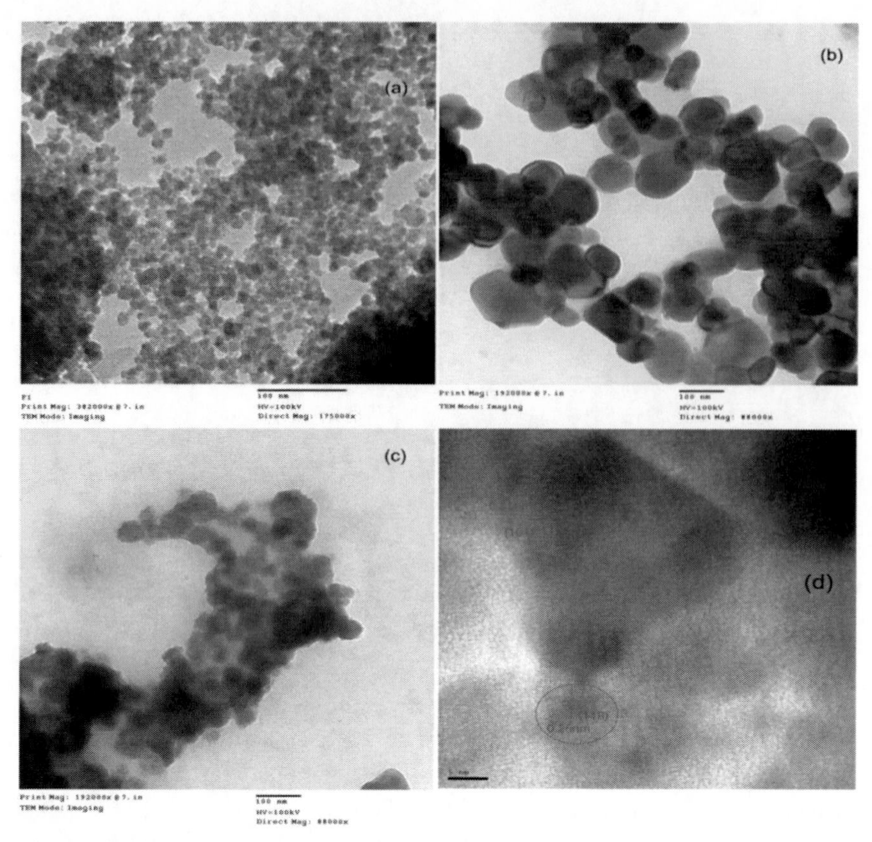

Reprinted from ref [62] Copyright (2010), with permission from Elsevier.

Figure 15. TEM images of (a) γ- Fe_2O_3, (b) TiO_2, (c) PTF12, and (d) HRTEM results of PTF12, showing the well dispersed nanoparticles of γ- $Fe_2O_3$and TiO_2 in the polymer matrix.

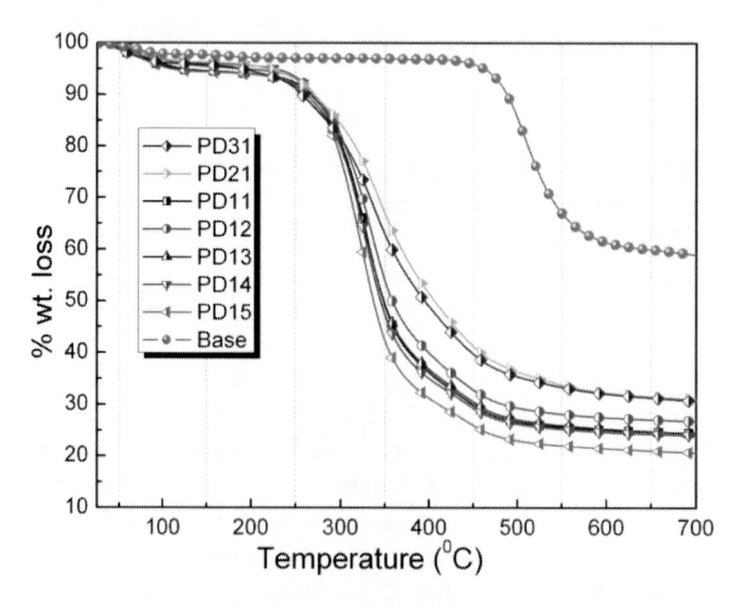

Figure 16. Thermograms of PAni doped with [aniline]/[DBSA]molar ratio of 3:1(PD31), 2:1(PD21), 1:1(PD11), 1:2(PD12), 1:3(PD13) along with its undoped form (emeraldine base).

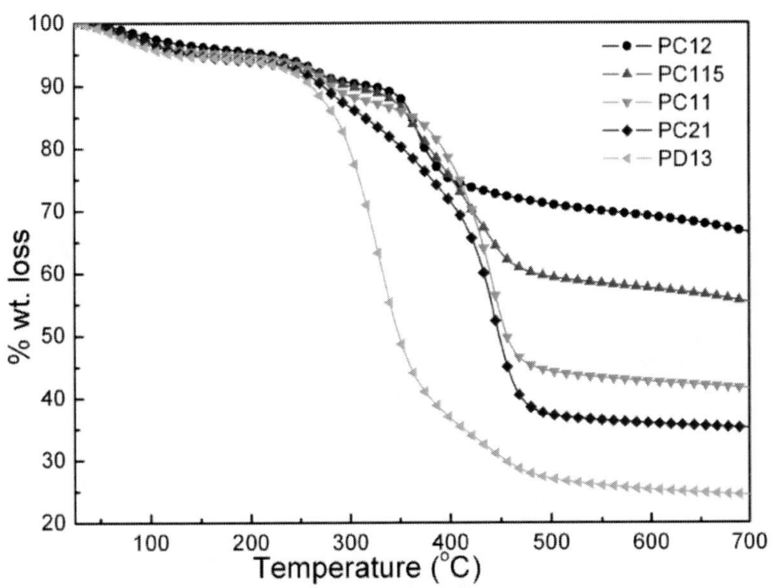

Figure 17. Thermal gravimetric analysis of PAni doped with DBSA and PAni composite with γ-Fe_2O_3 with increasing content of γ-Fe_2O_3; (●) PC12, (▲)PC115, (▼) PC11, (♦)PC21 and (◀)PD13.

The approximate amount of iron oxide for different polymer ferrite composites was calculated by subtracting the residual weight of the blank polymer from residual weight of composite at 700 ^{0}C. It was observed that the for different compositions, PC21; PC11; PC115 and PC12 the weight percent of γ-Fe_2O_3 was estimated to be 10.7%, 16.1%, 29.9% and 42.1%.

5.4. UV/Visible Spectroscopy

The UV absorption spectra of PAni and its composite with γ- Fe_2O_3 are shown in figure 18. Emeraldine base form of PAni in N-methyl pyrrolidione (NMP) shows two characteristic bands at 326 nm and 630 nm while the conductive form of PAni doped with DBSA has shown the red shift to 353 and 739 nm which were assigned to the π-π^* transition of the benzenoid ring and polaronic transition respectively. In case of PAni composite, two changes were observed: first a blue shift was observed for the band from 739 to 726 nm, which was ascribed to polaronic transition.

The reason behind this shifting may be the possible interaction of the γ-Fe_2O_3 with PAni ring leading to the formation of ferromagnetic composite.

The pictorial representation of interaction between Fe_2O_3 and nitrogen of the ammine group in PAni matrix is shown in Figure 18B.

Secondly, when the γ-Fe_2O_3 content increases in different samples, absorption spectra shows a bathochromic shift for the band 353 to 349 nm. The optical band energy of the polymer was obtained using the relation [133] $\alpha hv = (hv - E_g)^{1/2}$, where α is the absorption coefficient, hv is the photon energy and E_g is the optical band gap.

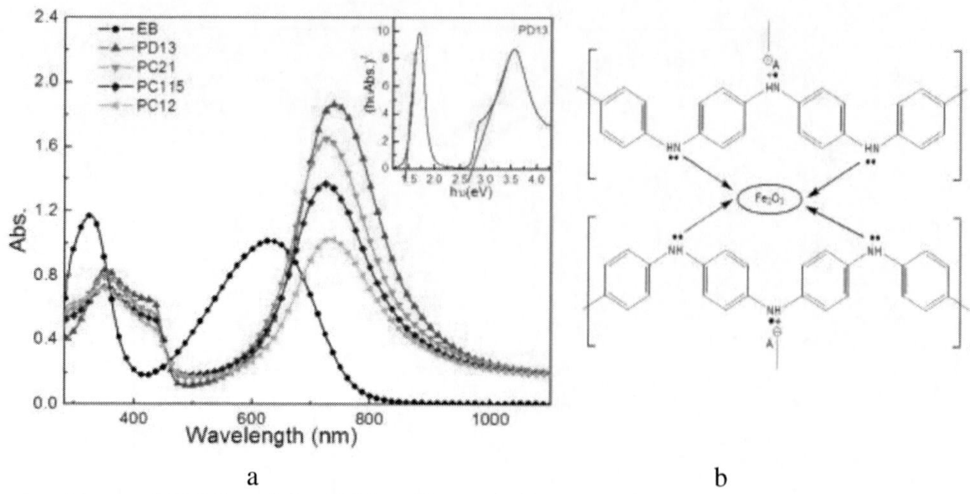

a b

Reprinted from ref [61] Copyright (2008), with permission from Elsevier.

Figure 18. (A) UV/Visible spectra of (●) EB, (▲) PD13, (▼) PC21, (♦) PC115 and (◀) PC12. Inset shows the calculation of band gap plots in $(\alpha h\nu)^{1/2}$ vs. photon energy (hν).(B) Scheme.2: The pictorial representation of interaction between Fe_2O_3 and nitrogen of the ammine group in PAni matrix.

The band gaps calculated were found to vary from 1.45-1.49 eV and 2.87-2.75 eV for the polaronic transitions and π-π^* transition of the benzenoid ring respectively. Figure 19 demonstrates the transition occurring in PAni and its composites with TiO_2 incorporated PAni-Fe_2O_3 (PTF) composites. For the composition PTF112 the main transitions are observed at 353, 444, and 668 nm. The shift in the polaronic band toward the lower wavelength is attributed to the presence γ-Fe_2O_3 and TiO_2 nanoparticles that interact with the –NH group of PAni ring, which contributes to the decrease in conductivity of the PTF12 composite as compared to the PAni doped with DBSA. From the UV-Vis spectra, the degree of doping is estimated from the ratio of absorption at polaronic transition to π-π^* transition (A_{735}/A_{350}). In case of PF12, the polaronic band is observed at 735nm but the value of A_{735}/A_{350} decreases, which results in the decrease of conductivity. The band gap is 1.6 eV for polaronic transition and 2.7 eV for π-π^* transition of the benzenoid ring, respectively.

5.5. FTIR Spectroscopy

Figure 20 shows the FTIR spectra of PAni doped with DBSA and composite PTF112. The main characteristics bands for the PAni doped with DBSA are found at 1515 and 1460 cm^{-1} which are assigned for the C=C bond stretching of quinoid and benzenoid ring respectively. Bands at 1257 and 1164 cm^{-1} are due to C=N stretching and in-plane bending of the C-H bond while peak at 1026 cm^{-1} is due to –SO_3H group [134, 135].

In case of PTF composite the main peaks have shifted to 1020, 1277, and 1233 cm^{-1} while the peak at 698cm^{-1} can be ascribed to Ti-O-Ti stretching and at 558 cm^{-1} is the characteristic band of Fe-O band stretching is observed. This clearly indicates the presence of γ-Fe_2O_3 and TiO_2 in the polymer matrix and the shift in the main peaks from PAni-DBSA mainly arises due to the interaction of the Fe_2O_3 and TiO_2 with –NH group of the aniline ring.

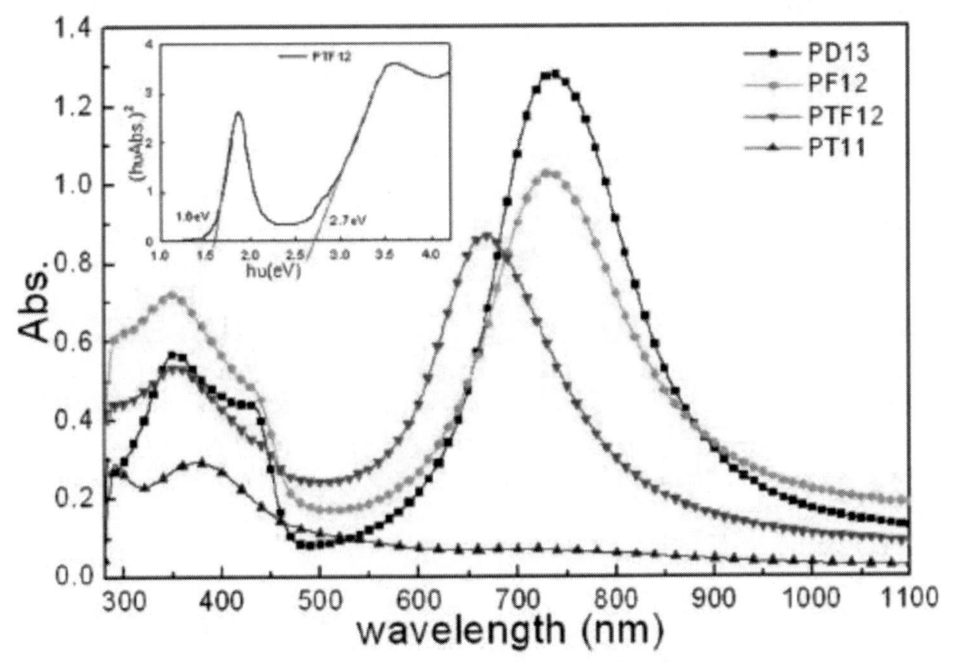

Reprinted from ref [62] Copyright (2010), with permission from Elsevier.

Figure 19. UV-Visible plots of PD13 (■), PF12 (●), PTF12 (▼), and PT11 (▲) vs. wavelength while the inset shows the (Abs. hν)^2 vs. hν plot of PTF12 for the calculation of band gap.

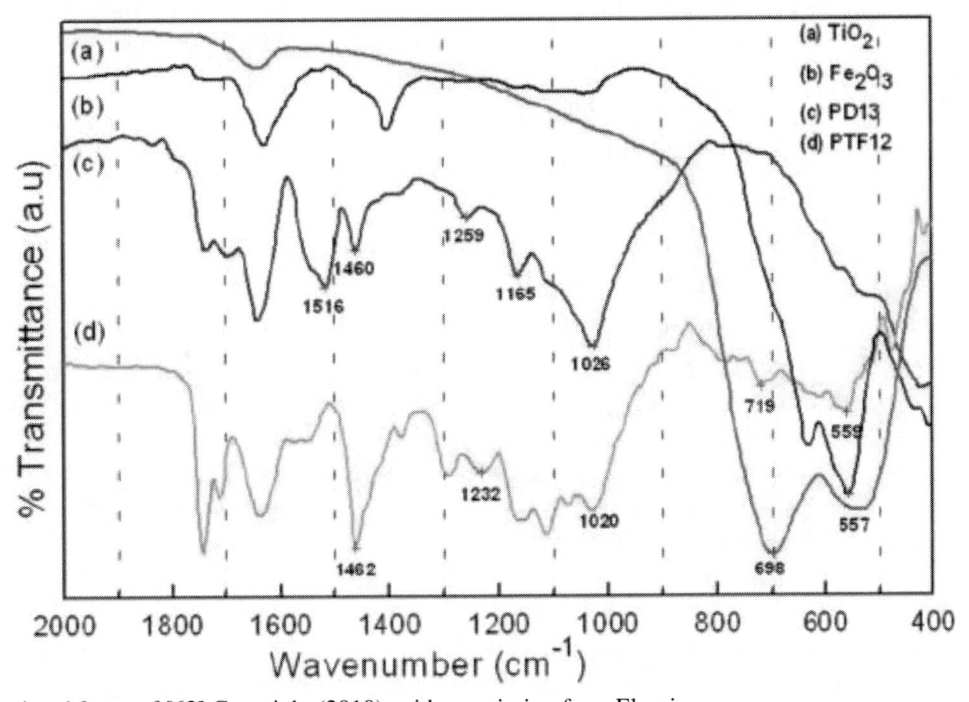

Reprinted from ref [62] Copyright (2010), with permission from Elsevier.

Figure 20. FTIR plot of (a) TiO_2, (b) Fe_2O_3, (c) PD13, and (d) PTF12 taken in KBr.

5.6. Conductivity Measurements

The temperature dependent d.c. conductivity (σ_{dc}) of the PT11 and PTF12 composites having different weight ratio of ferric oxide contents was measured in temperature ranging from 30-300 K. The variation of logσ_{dc} as function of $T^{-1/4}$ was plotted in Figure 21, whereas inset shows the logσ_{dc} vs. 1000/T plot. It shows that the conductivity tends to saturate at lower temperature.

To check the effect of nanoparticles on conductivity, the room temperature conductivity measurements of PAni and its composite were performed. It is observed that with the addition of nanoparticles of Fe_2O_3 (σ~10^{-9} S/cm) and TiO_2 (σ~10^{-11} S/cm) in the polymer matrix the conductivity of the PAni doped with DBSA decreased from 2.2 S/cm to 0.46 S/cm for the composite.

The expected decrease in conductivity was due to the incorporation of insulating nanoparticles in the polymer matrix which hinder the conduction path. Several models were established in order to explain the conductivity variations of conducting polymers.

Band Conduction: According to band conduction model conductivity is given by the

$$\sigma = \sigma_O \exp\left[-E_A / K_B T\right] \tag{7}$$

where σ_0 is a constant and E_A is the activation energy and is found to be temperature dependent for many materials. In normal semiconductors, at high temperature, the number of charge carriers in the conduction band increases exponentially with temperature and these thermally excited charge carriers are predominantly responsible for the conduction mechanism.

Hopping Conduction: In case of hopping conduction, given by Mott and Davis, the conductivity is express by the relation,

$$\sigma \sim \exp\left(-2\alpha R - \frac{\Delta W}{K_B T}\right) \tag{8}$$

where α is the coefficient of exponential decay of localized states involved in the hopping process and ΔW is the amount of energy needed to transfer a charge carrier from one site to another at the hopping distance R. Hopping is considered to take place between nearest neighboring sites. Mott suggested that the conduction at low temperature is due to the hopping of localized charge carriers but the charge carriers may not hop to the nearest neighbor sites. The charge carrier prefers to hop in a variable range in order to minimize the energy required for hopping.

Tunneling Conduction: When the size of the highly conducting regions or islands is sufficiently small i.e., less than ~ 20 nm, then the energy required to remove an electron from an electrically neutral is land is significant. If the voltage between two adjacent islands is small as compared with K_BT/e, the charge carriers can be generated only by thermal activation, making the conductivity temperature dependent on the charging energy.

The charge carriers then percolate along the path with least resistance. In such a situation, conductivity varies with temperature as,

$$\sigma(T) = \sigma_O \exp\left[-\left(\frac{T_O}{T}\right)^{1/2}\right] \tag{9}$$

where T_0 and σ_0 are the material constants. In the high field regime, the field induced tunneling becomes important and the conduction becomes non-ohmic. At a sufficiently high field, charge carrier generation by the field replaces thermal activation energy as the dominant mechanism and the resistivity starts behaving exponentially.

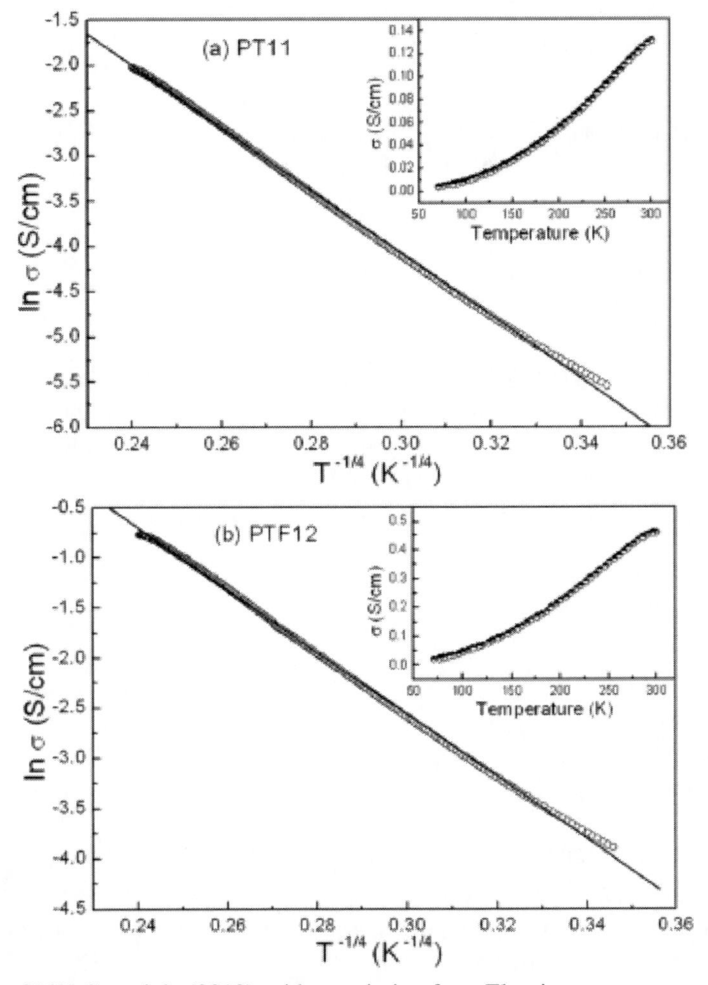

Reprinted from ref [62] Copyright (2010), with permission from Elsevier.

Figure 21. Temperature dependence of logσ as function of $T^{-1/4}$ for the samples PT11 (a) and PTF12 (b) in the temperature range 300-70 K whereas the inset shows the conductivity (σ) variation vs. temperature.

But it is observed that the conductivity studies are best explained by VRH (Variable range hopping) model which follows Mott's equation [136-138]

$$\sigma(T) = \sigma_O \exp\left[-\left(\frac{T_O}{T}\right)^{1/\gamma}\right] \tag{10}$$

where σ_o and T_o are constants and exponent γ is the dimensionality factor having values 2, 3, 4 for 1 dimension, 2 dimensions and 3-dimension conduction mechanism respectively. In order to calculate the exponent γ, $\log\sigma_{dc}$ vs. $T^{-1/4}$ is plotted which yields a straight line for the temperature range of 70-300K. To satisfy the Mott's equation, activation energy [139, 140],

$$E_A = \left\{-\partial \ln \sigma \big/ \partial(1/K_B T)\right\} \tag{11}$$

of the samples was calculated from the slope of $\log\sigma_{dc}$ vs. 1000/T plot. Equation (11) can be correlated with the with Mott's equation by the following expression,

$$E_A = mK_B T_o \left(\frac{T_o}{T}\right)^{\gamma-1} \tag{12}$$

It is evident from the expression (12) that a plot of log E_A vs. log T should give a straight line of slope (γ-1). The straight line corresponding to γ =1/4 indicates that the variable range hopping mechanism of the type $T^{-1/4}$ explains the conduction mechanism in the polymer composite PT11 and PTF12. For 3-D conduction mechanism, the values of Mott characteristic temperature (T_o) and σ_o (conductivity at T= ∞) are given by

$$T_O = 16\alpha^3 / [k_B N(E_F)] \tag{13}$$

$$\sigma_O = e^2 R^2 \nu_{ph} N(E_F) \tag{14}$$

$$R = [9/8\{\pi\alpha K_B T N(E_F)\}]^{1/4} \tag{15}$$

is the average hopping distance, α^{-1} is the localization length, $N(E_F)$ is the density of states at the Fermi level, and ν_{ph} is the phonon frequency (~10^{13} Hz). The average hopping energy W can be estimated by knowing the average hopping distance R and the density of states at the Fermi level $N(E_F)$ using the following relation

$$W = 3 \big/ 4\pi R3N(E_F) \tag{16}$$

The values of various Mott's parameters T_0, $N(E_F)$, R, and W for the composites PT11 and PTF12 are given in the table 1 which are calculated by using the above equations (6.6-6.12). The results are consistent with the Mott's requirement that $\alpha R>>1$ and $W>>K_BT$ for conductivity for hopping to distant sites [141, 142]. The conductivity data fits for the 3D-VRH model with $\gamma = 4$ having the linearity factor of 0.9996 for PT11 and 0.9997 for PTF12 composite from 300-70K. Thus it was concluded that 3D–VRH model is suitable for explaining the conduction mechanism wherein the charge transport occurs by phonon aided hopping or by thermally stimulated jumps between the localized sites.

Conductivity in the polymer composite is due to semi-quinone radical cations formed by H-bonding between neighboring polymers. The new states are generated between the valence and conduction bands by doping and are responsible for conduction and lead to the variation in activation energy. The overlapping of π-delocalized wave orbital of aniline ring with the d-orbital of metal ion in polymer composite forms the charge transfer complex site which acts as localized states from where the hopping of charge carrier takes place. Below 70 K it is observed that conductivity data deviates from the linear behavior because in low temperature region, charge conduction is mainly dominated by the thermally stimulated tunneling through the localized sites, as reported earlier for the other conjugated polymers [143-145].

5.7. Magnetic Studies

The magnetic properties of the PAni-γ- Fe_2O_3 composite and γ-Fe_2O_3 were explained with the help of M-H curve (Figure 22). The saturation magnetization (M_S) value of the γ-Fe_2O_3 was found to 69.77emu/g at an external field of 10kOe having a small value of coercivity and negligible retentivity with no hysteresis loop, indicating the super paramagnetic nature. When these nano ferrite particles were incorporated in the PAni matrix in weight ratio of 1:1(PC11), the magnetization saturation (M_S) value was found to be 4.13emu/g. However, on changing the weight composition of An/γ-Fe_2O_3 to 1:2, the Ms value drastically increased from 4.13emu/g to 48.9emu/g, keeping the external applied field at 10kOe.

When TiO_2 nano particles were incorporated along with Fe_2O_3 in the PAni matrix in weight ratio of 1:1 (PTF11), the saturation magnetization (M_S) value was found to be 11.7emu/g.

However, on changing the weight composition of An/γ-Fe_2O_3 to 1:2, (PTF112) the Ms value was increased from 11.7emu/g to 26.9 emu/g, keeping the external applied field at 6kOe as shown in figure 23. M_S value increased due to high polydispersivity of the γ-Fe_2O_3 in PAni matrix.

Table 1. Linearity factor and various Mott's parameters for the samples PT11 and PTF12

Sample Name	Linearity $T^{-1/4}$	T_0 (K)	N (E_F) ($cm^{-3}eV^{-1}$)	R (Å)	W (meV)	(αR)
PT11	0.9997	1.43×10^6	1.76×10^{18}	164	18.6	4.62
PTF12	0.9996	9.3×10^5	2.7×10^{18}	174	16.6	4.15

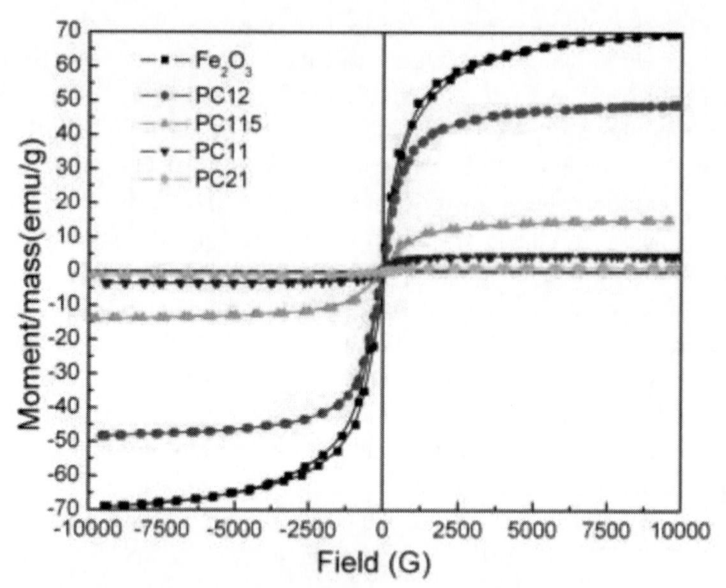

Figure 22. Magnetization curves of (■) γ-Fe_2O_3, (●) PC12, (▲) PC115, (▼) PC11 and (♦) PC21 showing decrease in saturation magnetization with the decrease in γ-Fe_2O_3 content.

The surface area, number of dangling bonded atoms and unsaturated coordination on the surface of polymer matrix were all enhanced. These variations lead to the interface polarization and multiple scattering, which is useful for the absorption of large number of microwaves [146]. The shielding properties of the composite depend upon the permeability of the material which in turn depend on the saturation magnetization of the material given by Wallace,

$$\mu = 1 + \frac{(4\pi M_S)^2}{(4\pi M_S)Ha - (f/2.8)^2 + j\alpha(4\pi M_S)(f/2.8)} \tag{24}$$

where Ms is the saturation magnetization, Ha is Anisotropy field, f is the frequency and α is the damping factor [147]. It has been observed that with the increase in concentration of Fe_2O_3 in the polymer matrix, saturation magnetization (M_S) increases from 11.7 to 26.9 emu/g which consequently leads to increase in the shielding effectiveness (SE_A) value from 35 dB to 45dB.

5.8. Electromagnetic Shielding Studies

The EMI shielding effectiveness (SE) of a material is defined as the ratio of transmitted power to incident power and is given by

$$SE(dB) = -10\log\left(\frac{P_T}{P_O}\right) \tag{17}$$

where P_T and P_o are the transmitted and incident electromagnetic powers respectively. For a shielding material, total $SE = SE_R + SE_A + SE_M$, where, SE_R is due to reflection, SE_A is due to absorption and SE_M is due to multiple reflections.

In two port network, S-parameter S_{11} (S_{22}), S_{21} (S_{12}) represents the reflection and the transmission coefficients given by,

$$T = \left|\frac{E_T}{E_I}\right|^2 = |S_{21}|^2 = |S_{12}|^2 , \tag{18}$$

$$R = \left|\frac{E_R}{E_I}\right|^2 = |S_{11}|^2 = |S_{22}|^2 , \tag{19}$$

and

$$\text{Absorption coefficient (A)} = 1 - R - T \tag{20}$$

Here, it should be noted that absorption coefficient is given with respect to the power of the incident EM wave. If the effect of multiple reflection between both interfaces of the material is negligible, the relative intensity of the effectively incident EM wave inside the material after reflection is based on the quantity as 1-R. Therefore, the effective absorbance (A_{eff}) can be described as $A_{eff} = (1 - R - T)/(1 - R)$ with respect to the power of the effectively incident EM wave inside the shielding material.

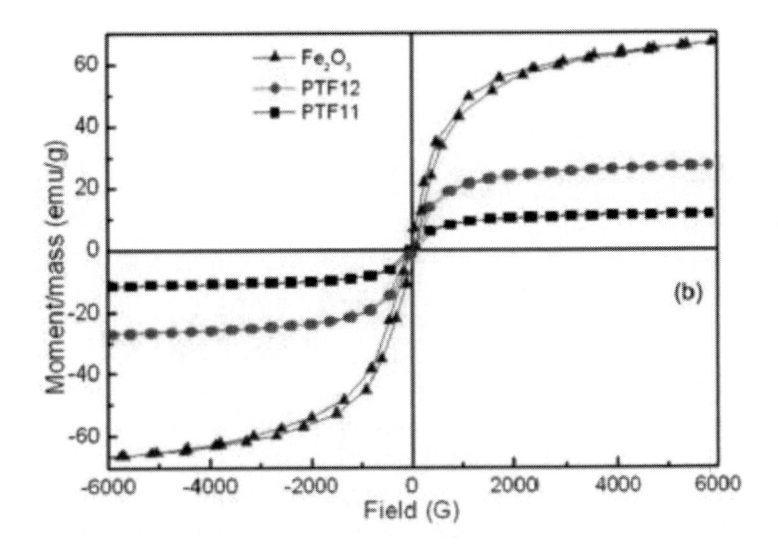

Figure 23. Variation of in magnetic moment per unit mass with applied field for Fe_2O_3 (▲), PTF11(●), and PTF12(■).

It is convenient to express the reflectance and effective absorbance in the form of -10log (1-R) and -10log (1-A_{eff}) in decibel (dB) respectively, which give SE_R and SE_A as SE_R = -10 log (1-R) and

$$SE_A = -10\log(1 - A_{eff}) = -10\log\frac{T}{1-R} \tag{21}$$

For the material the skin depth (δ) is the distance up to which the intensity of the electromagnetic wave decreases to 1/e of its original strength. The skin depth is related to the attenuation constant (β) of the wave propagation vector $\delta = 1/\beta = \sqrt{2/\omega\mu\sigma_{AC}}$ with the approximations that σ >> ωε. As $\delta \propto \omega^{-1/2}$, therefore, at low frequencies for the electrically thin samples (d << δ), the shielding effectiveness of the sample is described as

$$\text{SE (dB)} = 20\log\left(1 + \frac{1}{2}Z_0 d\sigma\right) \tag{22}$$

where σ is the a.c. conductivity, Z_o is free space impedance and d is the sample thickness.

Whereas for the higher frequencies, sample thickness (electrically thick samples) is sufficiently greater than skin depth and EMI shielding effectiveness for the plane electromagnetic wave [148] is given as

SE (dB) = SE_R (dB) + SE_A (dB),

$$SE_R(dB) \approx 10\log\left(\frac{\sigma_{AC}}{16\omega\varepsilon_o\mu_r}\right), \tag{23}$$

and

$$SE_A(dB) = 20.\frac{d}{\delta}.\log e \tag{24}$$

where $\sigma_{a.c.}$ depends upon the dielectric properties [149] ($\sigma_{a.c} = \omega\varepsilon_0\varepsilon''$) of the material, ω is the angular frequency (ω=2πf), ε_o is the free space permittivity and μ_r is the relative magnetic permeability of the sample. In equation (23), the first term is related to the reflection of the EM wave and contributes as the shielding effectiveness due to reflection.

The second term expresses the loss due to the absorption of the wave when it passes through the shielding material. In microwave range, the contribution of the second part becomes more as compared to the reflection term. From figure 24 the maximum SE of -11.2 dB was recorded for the sample PC21 while SE of -2.5 dB was observed for γ- Fe_2O_3 proving that the polymer composite is a better EMI shielding material. It has been observed that conducting ferromagnetic composite of PAni with Fe_2O_3 and TiO_2 has shielding effectiveness (SE) mainly due to absorption (Figure 25).

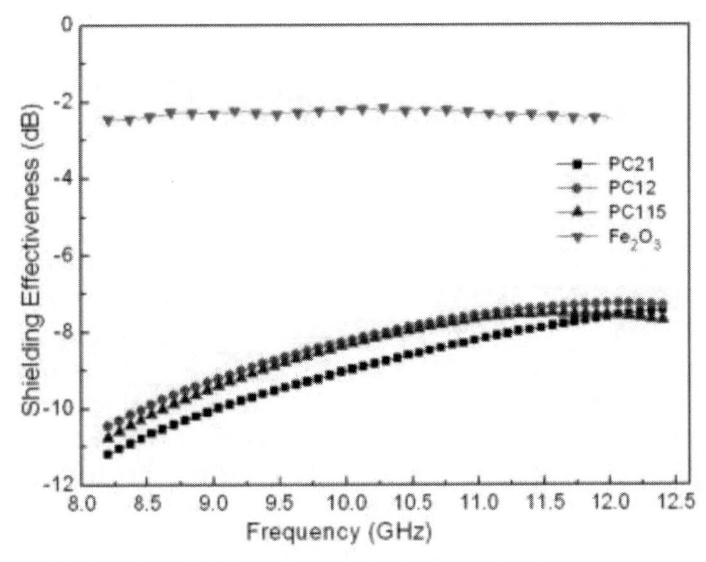

Reprinted from ref [61] Copyright (2008), with permission from Elsevier.

Figure 24. EMI shielding effectiveness (SE) of γ-Fe_2O_3 (▼), PC12 (●), PC115 (▲), PC21 (■) measured in X-band (8.2-12.4GHz).

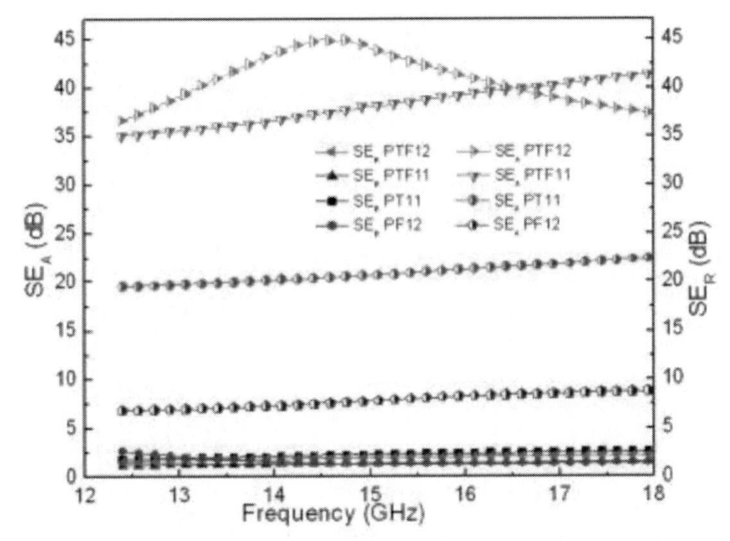

Reprinted from ref [62] Copyright (2010), with permission from Elsevier.

Figure 25. Variation of SE_A and SE_R for the samples PTF12, PTF11, PT11, and PF12 with frequency in 12.4-18 GHz range.

From the experimental measurement, the shielding effectiveness due to absorption (SE_A) was found to be 8.8, 22, 35, and 45 dB for PF12, PT11, PTF11, and PTF12 samples respectively, while the shielding effectiveness due to reflection (SE_R) was nominal and contributed very little. The higher value of SE_A of PTF composites was mainly due to combined effect of TiO_2 and Fe_2O_3.

5.8. Dielectric Properties

The electromagnetic absorption behavior of the material depends on complex permittivity ($\varepsilon_r = \varepsilon' - j\varepsilon''$) and permeability ($\mu_r = \mu' - j\mu''$). The real ($\varepsilon'$) and imaginary ($\varepsilon''$) parts of complex permittivity vs. frequency are shown in figure 26. The real part (ε') is mainly associated with the amount of polarization occurring in the material and the imaginary part (ε'') is related with the dissipation of energy.

In all the samples, ε' is found to decrease with frequency. In PAni, strong polarization occurs due to the presence of polaron/bipolaron and other bound charges, which leads to high value of ε' and ε''. With the increase in frequency, the dipoles present in the system cannot reorient themselves along the applied electric field and as a result dielectric constant decreases. The main characteristic feature of TiO_2 is that it has high dielectric constant with dominant dipolar polarization and the associated relaxation phenomenon constitutes the loss mechanism [150]. With the addition of γ-Fe_2O_3 and TiO_2 in PAni matrix, significant increase in real and imaginary part of complex permittivity was observed.

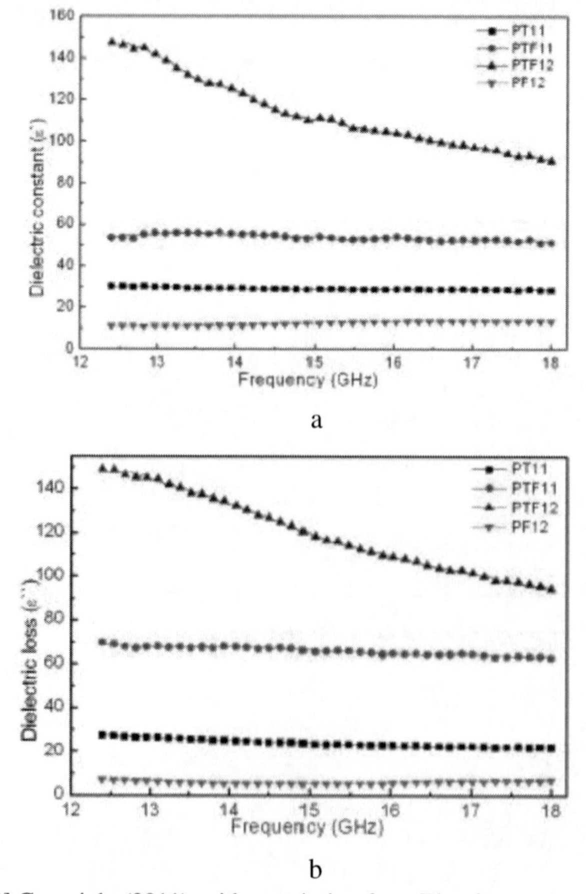

Reprinted from ref [62] Copyright (2011), with permission from Elsevier.

Figure 26. Dielectric constant (a) and dielectric loss (b) behavior with the variation of frequency for PT11 (■), PTF11 (•), PTF12 (▲), and PF12 (▼).

The higher values of dielectric constant and loss is owing to more interfacial polarization due to the presence of insulating γ-Fe_2O_3 particles and high dielectric TiO_2 particles, consequently leading to more shielding effectiveness due to absorption.

Figure 26a shows the variation of real part of magnetic permeability (μ') with frequency while change in the induced magnetization with the applied field is shown in Figure 26b. The magnetic permeability of all the samples decreases with the increase in frequency. The superior permeability was observed for higher percentage of iron oxide in the polymer matrix. The PTF12 nanocomposite has relative permeability value of 5.4, which decreases to 1.1 for PTF11as the surface area, number of dangling bonded atoms and unsaturated coordination on the surface of polymer matrix were all enhanced. These variations lead to the interface polarization and multiple scattering, which is useful for the absorption of large number of microwaves.

CONCLUSION

PAni is one of the most versatile conjugated polymer systems. Owing to the ease of synthesis of its composite with nanoparticles and ferrites can lead to new engineered materials. PAni will continue to lead the way to new unique materials, to highly stable, to highly conductive plastics, and to new microwave absorbing material. In spite of these interesting developments, a lot remains to be done with regard to both fundamental understanding and the much needed improvement of the method of the designing of electromagnetic shielding materials to operate at higher frequencies. The enhancement in the microwave shielding and absorption properties of the PAni nanocomposite has been achieved by the incorporation of dielectric filler (TiO_2) along with the magnetic Fe_2O_3 in the PAni matrix. TEM and HRTEM images demonstrate that in the PAni-ferrite-TiO_2 (PFT) nanocomposites forms the array of nanoparticles connected via conducting PAni system. These conducting paths of PAni between the magnetic and dielectric nanoparticles increase the absorption of the electromagnetic wave to a large extent. The contribution to the absorption value comes mainly due the magnetic losses (μ'') in ferrite and dielectric losses (ε'') in TiO_2 and PAni. The dependence of SE_A on magnetic permeability and conductivity demonstrates that better absorption value can be obtained for materials with higher conductivity and magnetization. Therefore, from the present studies, it can be concluded that the incorporation of magnetic and dielectric fillers in the polymer matrix lead to better absorbing material which make them futuristic radar absorbing material.

In spite of these interesting developments, a lot remains to be done with regard to both fundamental understanding and the much needed improvement of the method of the designing of electromagnetic shielding materials to operate at higher frequencies for their application.

REFERENCES

[1] M. Hu, J. Gao, Y. Dong, K. Li, G. Shan, S. Yang, R. K.-Y. Li, *Langmuir* 28 (2012) 7101.

[2] J. Hamilton, *Canadian Medical Association Journal* 154(1996) 373.

[3] W. R. Hendee, J. C. Boteler, *Health Physics* 66 (1994) 127.
[4] B. Lu, X. L. Dong, H. Huang, X. F. Zhang, X. G. Zhu, J. P. Lei, J. P. Sun, *Journal of Magnetism and Magnetic Materials* 320 (2008)1106.
[5] X. Li, B. Zhang, C. Ju, X. Han, Y. Du, P. Xu, *The Journal of Physical Chemistry* C 115 (2011) 12350.
[6] Z. An, S. Pan, J. Zhang, *The Journal of Physical Chemistry* C 113 (2009) 2715.
[7] G. Mu, N. Chen, X. Pan, K. Yang, M. Gu, *Applied Physics Letters* 91 (2007) 043110.
[8] H.-B.Yao, G. Huang, C.-H. Cui, X.-H. Wang, S.-H. Yu, *Adv. Mater.* 23 (2011) 3643.
[9] R. C. Che, L. M. Peng, X. F. Duan, Q. Chen, X. L. Liang, *Advanced Materials* 16 (2004) 401.
[10] Y. Yang, M. C. Gupta, K. L. Dudley, R. W. Lawrence, *Nano Letters*, 5 (2005) 2131.
[11] A. Ohlan, K. Singh, A. Chandra, S. K. Dhawan, *Applied Physics Letters* 93 (2008) 053114.
[12] S. Varshney, K. Singh, A. Ohlan, V. K. Jain, V. P. Dutta, S. K. Dhawan, *Journal of Alloys and Compounds* 2012, 538 (0), 107-114.
[13] Alan G. MacDiarmid, *Angew. Chem. Int. Ed.* 40 (2001) 258.
[14] A. F. Diaz, J. F. Rubinson, Mark, Jr. H. B. *Adv. Polym. Sci.*, 84 (1988)113
[15] Y. Cao, P. Smith, A. J. Heeger; *Synth. Met.* 48 (1992) 91.
[16] M. Geng, Z. Cai, Z. Tang; *J. Mater. Sci.* 39 (2004) 4001.
[17] F. Cheng, W. Tang, C. Li, J. Chen, H. Liu, P. Shen, S. Dou, *Chem. Eur. J.* 12 (2006) 3088.
[18] M. D. Levi, Y. Gofer, D. Aurbach; *Polym. Adv. Technol.* 13 (2002) 697.
[19] A. Kraft, A. C. Grimsdale, A. B. Holmes; *Angew. Chem. Int. Ed.* 37 (1998) 403.
[20] U. Mitschke, P. J. Bauerle; *Mater. Chem.* 10 (2000) 1471.
[21] A. Greiner, C. Weder; Light-emitting diodes, In: Kroschwitz, J. I. Ed. Encyclopedia of Polym. *Sci. Technol.* 3rd Ed. Wiley-Interscience, New York, 3 (2003) 87.
[22] G. Horowitz; *Adv. Mater.* 10 (1998) 365.
[23] W. U. Huynh, J. J. Dittmer, A. P. Alivisatos; *Science* 295 (2002) 2425.
[24] B. Sun, E. Marx, N. C. Greenham, *Nano Lett.* 3 (2003) 961.
[25] G. Yu, J. Gao, J. C. Hummelen, F. Wudl, A. J. Heeger; *Science* 270 (1995) 1789.
[26] E. Hakansson, A. Ammet, A. Kaynak; *Synth. Met.* 156 (2006) 925.
[27] A. Ohlan, K. Singh, A. Chandra, S. K. Dhawan, *ACS Applied Materials and Interfaces* 2 (2010) 927.
[28] N. Gandhi, K. Singh, A. Ohlan, D. P. Singh, S. K. Dhawan, *Composites Science and Technology* 71(2011) 1754.
[29] E. Hakansson, A. Amiet, S. Nahavandi, A. Kaynak, *Euro. Polym. J.* 43 (2007) 213.
[30] K. K. Satheesh Kumar, S. Geeta, D. C. Trivedi, *Curr. Appl. Phys.* 5 (2005) 608.
[31] M. A. Soto-Oviedo, O. A. Araujo, R. Faez, M. C. Rezende, M. A. D. Paoli, *Synth. Met.* 156 (2006) 1249.
[32] A. Ohtani, M. Abe, M. Ezoe, T. Doi, T. Miyata, A. Miyake, *Synth. Met.* 57 (1993) 3696.
[33] A. A. Argun, A. Cirpan, J. R. Reynolds, *Adv. Mater.* 15 (2003) 1338.
[34] A. Cirpan, A. A. Argun, C. R. G. Grenier, B. D. Reeves, J. R. Reynolds, *J. Mater. Chem.* 13 (2003) 2422.
[35] J. L. Boehme, D. S. K. Mudigonda, J. P. Ferraris, *Chem. Mater.* 13 (2001) 4469.

[36] P. Chandrasekhar, B. J. Zay, G. C. Birur, S. Rawal, E. A. Pierson, L. Kauder, T. Swanson, *Adv. Funct. Mater.* 12 (2002) 2137.
[37] M. A. De Paoli, W. A. Gazotti, J. Braz, *Chem. Soc.* 13 (2002) 410.
[38] Y. Y. Wang, H. Q. Li, Y. Y. Xia, *Adv. Mater.* 18 (2006) 2619
[39] A. Karina Cuentas-Gallegos, Monica Lira-Cantu', Nieves Casañ-Pastor, and Pedro Gómez-Romero, *Adv. Funct. Mater.* 15 (2005) 1125
[40] K. S. Ryu, S. K. Jeong, J. Joo, K. M. Kim, *J. Phys. Chem.* B. 11 (2007) 731.
[41] T. F. Otero, I. Boyano, M. T. Cortes, G. Vazquez, *Electrochimica. Acta,* 49 (2004) 3719.
[42] T. Mirfakhrai, John D. W. Madden, R. H. Baughman, *Mater. Today* 10 (2007) 30
[43] S. Skaarup, L. Bay, K. West, *Synth. Met.* 157 (2007) 323
[44] C. K. Tan, D. J. Blackwood, *Corros. Sci.* 45 (2003) 545.
[45] J. E. P. Da Silva, S. I. C. De Torresi, R. M. Torresi, *Corros. Sci.* 47 (2005) 811.
[46] H. Bhandari, S. Sathiyanaranayan, V. Choudhary, S. K. Dhawan, *J. Appl. Polym. Sci.* 111 (2008) 2328
[47] T. M. Swager, *Chem. Res. Toxicol.* 15 (2002) 125.
[48] X. B. Yan, Z. J. Han, Y. Yang, B. K. Tay, Sens. *Actuators.* B 123 (2007) 117.
[49] D. Li, J.Huang, R. B. Kaner, *Accounts of Chemical Research* 42 (2008) 135-145.
[50] E. T. Kang, K. G. Neoh, K. L. Tan, *Progress in Polymer Science*, 23 (1998) 277.
[51] T. A. Skotheim, R. L. Elsenbaumer, J. R. Renolds; *Handbook of Conducting Polymers,* 2nd Ed., Marcel Dekker, New York 1998.
[52] W. P. Su, J. R. Schrieffer, A. J. Heeger, *Phys. Rev. Lett.* 42 (1979) 1698.
[53] W. P. Su, J. R. Schrieffer, A. J. Heeger, *Condens. Matter Mater. Phys.* 22 (1980) 2099.
[54] W. P. Su, J. R. Schrieffer, *Proc. National Academy Sci. US* 77 (1980) 5626.
[55] A. R. Bishop, D. K. Campbell, K. Fesser, *Mol. Cryts. Liq. Cryst.* 77 (1981), 253.
[56] J. L. Bredas, R. R. Chance, R. Silbey, *Mol. Cryst. Liq. Cryst.* 77 (1981) 319.
[57] J. L. Bredas, R. R. Chance, R. Silbey, *Phys. Rev.* B 26 (1982) 5843.
[58] P. Sheng, B. Abeles, Y. Arie, *Phys. Rev. Lett.*, 31 (1973) 44
[59] S. Bhadra, S. Chattopadhyay, N. K Singha, D. Khastgir, *Journal of Applied Polymer Science* 108 (2008) 57.
[60] N. F. Mott, E. A. Davis, *Electronic Processing Non-Crystalline Materials,* 2nd Ed., Clarendon press, Oxford (1979).
[61] K. Singh, A. Ohlan, R. K. Kotnala, A. K. Bakhshi, S. K. Dhawan, *Materials Chemistry and Physics* 112(2008) 651
[62] K. Singh, A. Ohlan, R. K. Kotnala, A. K. Bakhshi, S. K. Dhawan, *Materials Chemistry and Physics*, 119(2010) 201
[63] Z. H. Wang, E. M. Scherr, A. G. MacDiarmid, A. J. Epstein, *Phys. Rev.* B 45 (1992) 4190
[64] W. Lee, G. Du, S. M. Long, A. J. Epstein, S. Shimizu, T. Saitoh, et al. *Synth. Met.* 84 (1997) 807.
[65] B. I. Shklovskii, A. L. Efros. Berlin: Springer Verlag 1985.
[66] M. Campos, Jr. B. Bello. *J. Phys. D: Appl. Phys.* 30 (1997) 1531.
[67] V. M. Mzendaa, S. A. Goodman, F. D. Auret, L. C. Prinsloo, *Synth. Met.* 127(2002)279.
[68] Q. Li, L. Cruz, P. Phillips, *Phys. Rev.* B 47 (1993) 1840–5.
[69] J. Li, K. Fang, H. Qiu, S. Li, W. Mao. *Synth. Met.* 142 (2004)107.

[70] P. Sheng, B. Abeles, *Phys. Rev. Lett.* 28 (1972) 34.
[71] T. Nakajima, T. Kawagoe. *Synth. Met.* 28 (1989) 629.
[72] M. Wan, *Conducting Polymers with Micro or Nanometer Structure*, Tsinghua University Press, Beijing and Springer-Verlag GmbH Berlin Heidelberg 2008
[73] K. Singh, A. Ohlan, P. Saini, S. K. Dhawan, *Polymers for Advanced Technologies* 19 (3), (2008) 229.
[74] A. Ohlan, K. Singh, S. K. Dhawan, *Journal of Applied Polymer Science* 115 (2010) 498.
[75] M. T. Nguyen, P. Kasai, J. L. Miller, A. F. Diaz. *Macro-molecules* 27 (1994) 3625–31
[76] A. Ohlan, K. Singh, A. Chandra, S. K. Dhawan, *Journal of Applied Polymer Science* 108 (2008), 2218.
[77] J. W. Chevalier, J. Y. Bergeron, L. H. Dao, *Macromolecules* 25 (1992) 3325.
[78] J. Jang, J. Ha, K. Kim, *Thin Solid Films* 516 (2008)3152.
[79] S. A. Schelkunoff, *Electromagnetic Waves*, Van Nostrand, NJ, (1943).
[80] R. B. Schulz, V. C. Plantz, D. R. Brush, Shielding theory and practice, IEEE Trans. *Electromagn. Compat.* EMC-30 (1988) 187.
[81] C. R. Paul, *Electromagnetics for Engineers*, Wiley, Hoboken, NJ, 2004.
[82] H. W. Ott, *Noise Reduction Techniques in Electronic Systems*, 2nd ed. New York: John Wiley and Sons.1988.
[83] A. M. Nicolson, G. F. Ross, *IEEE Trans. Instrum. Meas.* 19 (1970) 377
[84] W. B. Weir, *Proc. IEEE* 62 (1974) 33
[85] H. Segawa, T. Shimadzu, M. Honda, *J. Chem. Soc. Chem. Commun.* (1989)132
[86] M. M. Coleman, R. J. Petanck, *J. Polym. Sci.* 16 (1986) 821.
[87] C. K. Chen, R. Liepins, *Electrical Properties of Polymers*, Hanser Publisher Munich, (1987) 274.
[88] G. Natta, G. Mazzanti, P. Corradini, *Atti Accad. Naz Linceicl. Sci. Fis. Mat. Nat. Rend.* 2 (1958) 25.
[89] A. M. Saxman, R. Liepins, M. Aldissi, *Prog. Polym. Sci.* 11 (1985) 57.
[90] P. Kovacic, A. Kyriakis, *Tetrahedron Lett.* (1962) 467.
[91] G. M. Carter, M. K. Thakur, Y. J. Chen, J. V. Hryniewicz, *Appl. Phys. Lett.* 47 1985) 457.
[92] B. Thomas, M. G. K. Pillai, S. Jayalakshmi, *J. Phys. D: Appl. Phys.* 21 (1988) 503.
[93] A. W. Snow, *Nature* 292 (1981) 40.
[94] T. Yamamoto, Y. Hayashi, A. Yamamoto, *Bull. Chem. Soc. Japan* 51 (1978) 2091.
[95] S. Rahman, M. Mahapatra, M. M. Maiti, S. Maiti, *J. Polym. Mater.* 6 (1989) 135.
[96] K. Soga, M. Nakamura, Y. Kobayashi, S. Ikeda, *Synth. Met.* 6 (1983) 275.
[97] F. E. Karasz, J. D. Capistran, D. R. Gagnon, R. W. *Lenz Mol. Cryst. Liq. Cryst.* 118 (1985) 327.
[98] J. Qiu, H. Shen, M. Gu, *Powder Technology* 154 (2005) 116.
[99] J. L. Kirschvink, *Bioelectromagnetics* 17 (1996) 187.
[100] N. E. Kazantseva, J. Vilcakova V. Kresalek, *J. Magn. Magn. Mater.* 269 (2004)
[101] G. Li, S. Yan, E. Zhou, Y. Chen, Colloids and Surfaces A: Physicochem. Eng. *Aspects* 276 (2006) 40.
[102] M. Wan, J. Li, *J. Polym. Sci. A: Polym. Chem.* 36 (1998) 2799.
[103] Qiu, Q. Wang, N. Min, *J. Appl. Poly. Sci.* 102 (2006) 2107.

[104] S. E. Jacobo, J. C. Aphesteguy, A. R. Lopez, N. N. Schegoleva, G. V Kurlyandskaya, *Eur. Poly. J.* 43 (2007) 1333.

[105] W. Xue, K. Fang, H. Qiu, J. Li, W. Mao, *Synth. Met.* 156 (2006) 506.

[106] L. Li, J. Jiang, F. Xu, *Eur. Poly. J.* 42 (2006) 2221.

[107] M. Wan, W. Li, *J. Polym. Sci. A: Polym. Chem.* 35 (1997) 2129.

[108] A. G. MacDiarmid, A. J. Heeger, *Synth. Met.* 1(1987/88) 101

[109] T. A. Skotheim, *Handbook of Conducting Polymers*, Vol. 1 and 2, Marcel Dekker, New York,1986.

[110] H. L. Huang, W. M. G. Lee, *Chemoshere*, 44 (2001) 963.

[111] D. J. Shaw, *Introduction to Colloid and Surface Chemistry*, Oxford Butterworth-Heinemann Ltd.1991.

[112] M. G. Han, S. K. Cho, S. G. Oh, S. S. Im, *Synth. Met.* 126 (2002) 53

[113] Seddique, M. *Ahmed: Polymer Degradation and Stability* 85 (2004) 605-614.

[114] Dongxue Han, Ying Chu, Likun Yang, Yang Liu, Zhongxian Lv, Colloids and Surfaces A: Physicochem. *Eng. Aspects* 259 (2005) 179-187.

[115] Chul Hyun Lim, Young Je Yoo, *Process Biochemistry* 36 (2000) 233-241

[116] D. E. Stilwell, S. M. Park, *J. Electrochemical Society* 135 (1988) 2254

[117] H. Banno, K. Ogura, *Ferroelectrics* 95 (1989) 111.

[118] X. X. Huang, Z. F. Chen, W. Q. Zou, Y. S. Liu, J. D. Li, *Ferroelectrics* 101(1990) 111.

[119] J. B. Ngoma, J. Y. Cavaille, J. Paletto, J. Perez, F. Macchi, *Ferroelectrics* 109 (1990) 205.

[120] K. D. Dilip, *Ferroelectrics* 118 (1991) 165.

[121] B. Wei, Y. Daben, *Ferroelectrics* 157 (1994) 427.

[122] H. L. W. Chan, M. C. Cheung, C. L. Choy, *Ferroelectrics* 224 (1999) 113.

[123] W. J. Bae, K. H. Kim, W. H. Jo, *Macromolecules* 37 (2004) 9850.

[124] W. M. A. T. Bandara, D. M. M. Krishantha, J. S. H. Q. Perera, R. M. G. Rajapakse, D. T. B. Tennakoon, *J. Compos. Mater.* 39 (2005) 759.

[125] P. Aranda, M. Darder, R. Fernandez-Saavedra, M. Lopez-Blanco, E. Ruiz-Hitzky, *Thin Solid Films* 495 (2006) 104.

[126] A. Dey, S. De, A. De, S. K. De, *Nanotechnology* 15 (2004) 1277.

[127] C. Huang, Q. M. Zhang, *Adv. Mater.* 17 (2005) 1153.

[128] J. C. Xu, W. M. Liu, H. L. Li, *Mater. Sci. Eng.* C 25 (2005) 444.

[129] S. J. Su, N. Kuramoto, *Synth. Met.*114 (2000) 147.

[130] A. Ohlan, K. Singh, A. Chandra, V.N. Singh, S. K. Dhawan, *Journal of Applied Physics* 106 (2009) 044305.

[131] M. G. Han, S. K. Cho, S. G. Oh, S. S. Im, *Synth. Met.* 126 (2002) 53.

[132] T. D. Castillo-Castro, M. M. Castillo-Ortega, I. Villarreal, F. Brown, H. Grijalva, M. Perez-Tello, S. M. Nuno-Donlucas, J. E. Puig, *Composites: Part A* 38 (2007) 639.

[133] F. Yakuphanoglu, E. Basaran, B. F. Suenkal, E. Sezer, *J. Phys. Chem.* B 110 (2006) 16908.

[134] S. Kim, J. M. Ko, I. J. Chung, *Polym. Adv. Technol.* 7 (1996) 599.

[135] J. C. Aphesteguy, S. E. Jacob, *Physica* B 354 (2004) 224.

[136] N. F. Mott, E. A. Davis, *Electronic Processes in Non-Crystalline Materials,* 1st Edition, Clarendon Press, Oxford, 1971.

[137] B. Sanjai, A. Raghunath, T. S. Natrajan, G. S. Rangarajan, P. V. P. Thomas, S. Venkatachalam, *Phys. Rev.* B 55 (1997) 10734.

[138] R. Singh, A. K. Narula, R. P. Tandon, A. Mansingh, S. Chandra, *J. Appl. Phys.* 79 (1996) 1476.
[139] R. Singh, V. Arora, R. P. Tondon, S. Chandra, N. Kumar, A. Mansingh, *Polymer*, 38 (1997) 4897.
[140] R. Singh, V. Arora, R. P. Tondon, A. Mansingh, S. Chandra, *Synth. Met.* 104 (1999) 137-144.
[141] N. J. Pinto, P. K. Kahol, B. J. McCormick, N. S. Dalal, H. Han, *Phys. Rev.* B 49 (1994) 13983.
[142] M. Reghu, Y. Cao, D. Moses, A. J. Heeger, *Phys. Rev.* B 47 (1993) 1758.
[143] R. S. Kohlman, A. J. Epstein, *Handbook of Conducting Polymers*, edited by T. A. Skotheim, R. L. Elsenbaumer, J. R. Reynolds, 2nd ed. Marcel Dekker, New York, 1998, p. 85.
[144] R. Menon, S. V. Subramanyam, *Solid State Commun.* 72 (1989) 325.
[145] P. Sheng, J. K. Lafter, *Phys. Rev.* B 27 (1983) 2583.
[146] X. F. Zhang, X. L. Dong, H. Huang, Y. Y. Liu, W. N. Wang, X. G. Zhu, B. Lv, J. P. Lei, C. G. Lee, *Appl. Phys. Lett.* 89 (2006) 053115.
[147] J. L. Wallace, *IEEE Trans. Magnetics* 29 (1993) 4209
[148] N. F. Colaneri, L. W. Shacklette, *IEEE Trans. Instru. Meas.* 41 (1992) 291.
[149] R. Singh, J. Kumar, R. K. Singh, R. C. Rastogi, V. Kumar, *New J. Phys.* 9 (2007) 40.
[150] L. L. Diandra, D. R. Reuben, *Chem. Mater.* 8 (1996) 1770.

In: Trends in Polyaniline Research
Editors: T. Ohsaka, Al. Chowdhury, Md. A. Rahman et al.
ISBN: 978-1-62808-424-5

Chapter 10

BIPHASIC ELECTROPOLYMERIZATION OF POLYANILINE-MULTIWALLED CARBON NANOTUBE NANOCOMPOSITE

Monika Srivastava[1]*, Ashish Kumar[1], Bhavana Gupta[1], Sanjay K. Srivastava[2] and Rajiv Prakash[1]

[1]School of Materials Science and Technology, Indian Institute of Technology, Banaras Hindu University, Varanasi

[2]Department of Physics (MMV) Banaras Hindu University, Varanasi

ABSTRACT

Despite the tremendous research on conjugated polymers to date, polyaniline, PAni and its derivatives are the most interesting conjugated polymers due to their unique electrochemical properties, easy polymerization, low cost and wide range of application. In the recent years polymer composites formation with various nanofillers are explored to enhance the various properties of the polymers. Multi-walled carbon nanotube, MWNT is one of the well-known nanofillers for its excellent electrical, mechanical and thermal properties. Besides all properties, the processability, stability and solubility hinder its commercial utilization and fabrication of devices. To overcome these limitations, coating of PAni over MWNTs is one of the possible ways to obtain the high performance of the resulted nanocomposite towards device applications. Although various strategies have been developed to produce such nanocomposites (i.e. polymer supported by conducting templates) but still the homogeneity/uniformity of nanocomposite and stability are the major challenges. Though the electrochemical polymerization technique suffers from various limitations, it is still a good approach for the uniform and thin film formation. However, formation of composites is rather difficult as the fillers may not be electroactive and also may not form homogeneous suspension (or colloidal state) in the electrolytes in which monomers are dissolved. Therefore, the distribution of fillers in the polymer matrixes is difficult to control using conventional electrochemical methods. We proposed a biphasic electropolymerization method as an efficient approach for polymerization of monomers with nanofillers to get uniform and homogeneous polymer

* E mail: monikabhu.srivastava@gmail.com.

composites. This technique not only provides a better selection of solvents for monomers and fillers but also better control of diffusion and kinetics.

In this chapter, electrochemical techniques used for synthesis with the latest development in the electrochemical synthesis of PAni nanocomposite is described with our novel biphasic electropolymerization technique for multi-walled carbon nanotube-polyaniline (MWNT-PAni) nanocomposite. MWNT-PAni nanocomposite formation, based on the biphasic electropolymerization technique, provides a general method for the formation of uniform nanocomposites for other polymers and fillers.

Keywords: Biphasic electropolymerization, Electrochemistry, Multi-walled carbon nanotube, Polyaniline, Nanocomposite

1. INTRODUCTION

Amongst various conducting polymers viz. Polypyrrole (Ppy), Polythiophene (PTh), Polycarbazole (PCz), Polyindole (PIn) Polyaniline (PAni) got the most attention in the scientific group. PAni has been used in those areas of research where high conductivity of conducting polymers is required. However, poor stability towards environmental exposure and processability of PAni opens the arena for nanocomposites, where the fillers help in increasing its conductive stability and processability too [1]. In addition, recently, research in morphology control synthesis of conducting polymer nanocomposites are triggered due to various specific applications [2-5]. Various types of morphologies of PAni nanocomposites have been achieved such as nanotube, nanorod, spherical, nanofibrous, etc. as shown in Scheme 1.

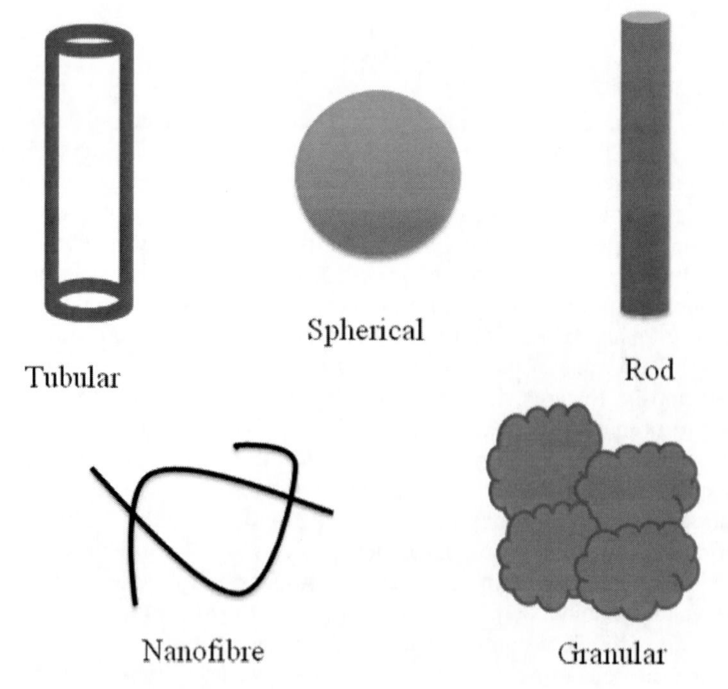

Scheme 1. Some common shapes of PAni.

Morphology achieved by the nanocomposite mainly depends on the morphology of the fillers. Various types of nanofiller have been explored for the formation of PAni nanocomposite viz. MWNT, metal oxides, nano-clay, recently graphene, etc.

Oxidative chemical polymerization is common for the formation of conducting polymers as well as nanocomposites. Probably hydrophobic nucleates help in anchoring the polymer chains to the solid support (anode). The adsorption of nucleates at solid surfaces immersed in the reaction mixture leads to polymer nanofilms or coatings on anode substrates. The competition between nucleate adsorption and nucleate self-assembly may lead to more complex morphologies combining one-dimensional to three-dimensional features, resulting in nanobrushes, spheres, sheets, etc. The control of nucleates self-assembly and of polymer growth depends on the pH of the medium as also observed for the typical case of PAni [6]. In the chemical synthesis process, there is least control over the self-assembly of nucleates, which results into amorphous bulk conducting polymer without any specific morphology. Moreover, with the chemical polymerization, uniform composite or thin film of nanocomposite is also a difficult task. Opposite to that in electrochemical synthesis uniformly distributed (mainly suspension or colloidal form) nanofillers in the electrolyte participate in nanocomposite formation almost as a single particle. Such process of synthesis results in a nanocomposite with a uniform dispersity of filler in the polymer matrix [7]. Thus it is worth studying electrochemical synthesis and the effect of fillers loading on the polymer and its physicochemical and electrochemical properties for future applications. The present chapter covers the details of the electrochemical techniques used for synthesis with the latest development in the electrochemical synthesis of PAni nanocomposites using different types of fillers to achieve controlled morphology and our proposed gradation in the procedure of better nanocomposite formation.

1.1. Electrochemical Synthesis

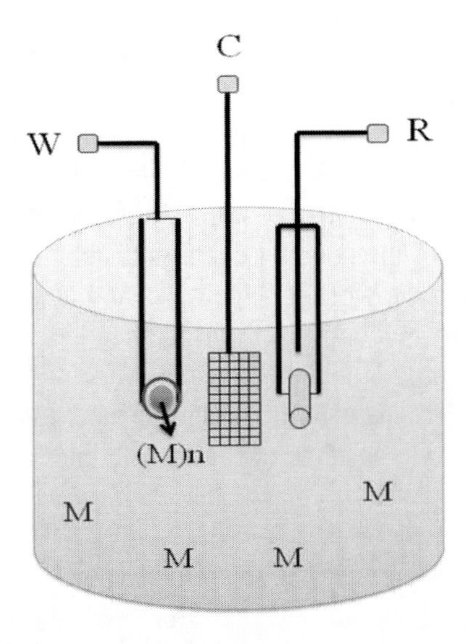

Scheme 2. Three electrode system for electrochemical synthesis of conducting polymers. W-Working electrode, C-Counter electrode, R- Reference electrode and M-Monomer.

A typical three-electrode (working, reference and counter) system is used for most of the electrochemical polymerization of monomers as shown in Scheme 2. In a more simplified system, polymerization with two electrodes (no reference electrode) using a current source can also be done. However, the absence of a reference electrode limits the correct information of the potential at the working electrode. In an electrochemical system, the transport of the monomer towards the electrode occurs by three processes: via convection, migration and diffusion. In an unstirred medium the convection is negligible and the presence of electrolytes in the solution supports the current through the cell by migration. During electrochemical oxidation the diffusion is mainly affecting the monomer concentration over the electrode [8].

1.2. Electrochemical Polymerization Condition

The simplest condition for polymerization is the application of some positive potential over the working electrode (anode). Apart from potential, several other factors, that affect various properties of the resulting polymers and also the rate of polymerization, are described in following subsections:

1.2.1. Electrode Material

Electrode substrates play an important role of the polymer synthesis process [9-11]. Electrode material property is related to two processes, one with oxidation of the monomers and second with the adsorption of monomer cations/oligomers followed by polymerization. Sometimes a polymer does not deposit on the electrode surface probably because the surface energy of the electrode surface is unable to control the hydrophilic/hydrophobic nature of the resultant polymer. It has been observed that oxide-containing electrodes create very strong adsorption of the polymer because of better interactions and also surface roughness [12]. For making a nanostructured polymer, the electrode can work as a hard template. The size of the electrode is also a very important parameter to control the quality and quantity of the polymers. The large size of the electrode affects the electropolymerization related to conductivity (decrease) and depletion effect. Using small working electrodes can minimize both. By using smaller (tens to hundred μm diameter) electrodes, electropolymerization can be carried out in low-conductivity media also. The rate of transport to and from the electroactive center (electrode) is relatively high. In some cases, the transport from the electrode affects the deposition of the polymer even with smaller size. The auxiliary electrode size should be much larger than the working electrode, and it should be chemically inert and facilitate the cathodic reactions. Selection of the reference electrode is carried out according to the medium of polymerization i.e. aqueous or non-aqueous. Polymerization potential may differ with the variation of reference electrodes due to the change in the standard electrode potential. However, two electrode systems (without reference) are sometimes used for the polymerization.

1.2.2. Solvent

Solvents not only act as medium but also play a crucial role in deciding the conformation of the polymers. The polymer can change its conformation in an aqueous medium (coiled) to protect its hydrophobic group and uncoiled in non-polar solvents to expose the hydrophobic groups. The choice of solvent is critical in terms of (i) Purity of solvents and its inertness to not participate in any unwanted side reactions; (ii) solubility of monomer, electrolyte and fillers in an appropriate concentrations; (iii) solubility of resulted polymer; (iv) interaction with the electrode, substrate, monomer, counter ion and filler [13] and (v) nucleophilicity as more nucleophilic solvent is likely to react with the free-radical intermediates [14]. In addition, morphology of the polymer is also dependent on solvents especially with ionic liquids [15]. In case of two solvents, polymerization is affected by the diffusion coefficient of the monomer.

1.2.3. The Counter Ion/Cation Effect: Choice of Electrolyte

Various workers have studied the effect of the counter ion on the electropolymerization process [16, 17]. The electrolyte may influence the conductivity of the solution, the polymer properties and, hence, the rate of polymerization. The anions incorporated in the polymer matrix affect the morphology or surface property and conductivity of the polymer, which also affects (indirectly) the rate of polymerization [18]. Therefore the anion should be readily incorporated into the polymer matrixes for efficient and large amounts of polymer synthesis with better conductivity. By using electrolytes it is possible to have small volume and higher charge density. Furthermore, electrolytes should be chemically and electrochemically stable, otherwise the decomposition may interfere with the polymerization. Sometimes, it has also been observed,that the electrolyte can work as a catalyst in the polymerization process. The oxidation potential of the polymerization reaction also changes with the change in electrolyte viz. para-toluenesulfonic acid (pTS), tetrabutylammonium perchlorate (TBAP), tetrabutylammonium tetrafluoroborate ($TBABF_4$), and tetrabutylammonium hexafluorophosphate ($TBAPF_6$) etc. [19].

1.2.4. Monomer

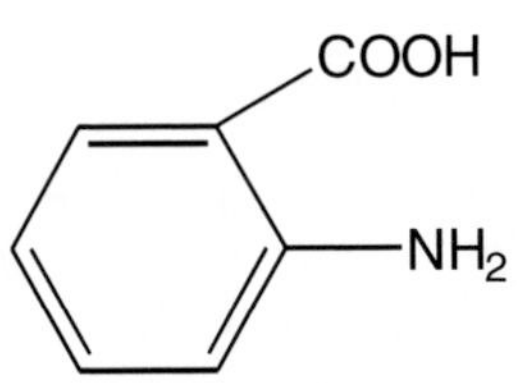

Anthranilic acid 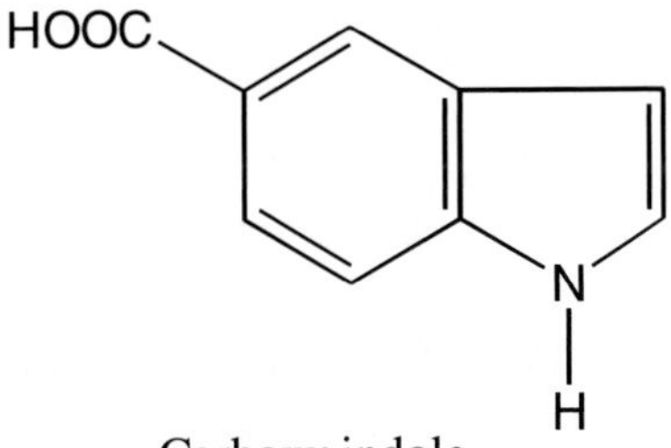 Carboxy indole

Figure 1. Chemical structure of anthranilic acid and carboxy indole.

Monomer concentration affects the quality, chain length and morphology of the polymer. The monomer-counter ions ratio is also important for polymers doping and stability. Monomers functionalization by the covalent functionalization of some group is a common way to control structural, electronic and electrochemical properties of the polymer [20]. However, the functional groups may affect the polymerization because of the electronic and steric effect. It is observed that when the monomer is covalently linked with some electron-withdrawing group, the oxidation potential of the monomer increases. An increase in oxidation potential causes a decrease in the polymerization rate or sometimes an inhibition of polymerization, which is opposite to that of an electron-donating group. In case of heterocyclic monomer, viz. having nitrogen or sulphur as heteroatoms, the conductivity of the polymer reduces when it is substituted at the nitrogen or sulphur site. Halogen containing monomer (3-bromo-thiophene) allows the polymerization but the resultant polymer is much less conducting in nature. Bulky groups, due to a steric effect, may create blocking in the polymerization process but at the same time affect the solubility of the monomer as well as polymers. Ionizable functional groups (viz. carboxy and sulfonic group) alter the solubility of the monomers and finally polymers [21]. In addition the ionizable groups, such as $-SO_3H$ or $-COOH$, show "self doping" of the polymer. Anthranilic acid and carboxy indole (structure shown in Figure 1) are good examples in this regard, in which the ionizable –COOH group dope the polymer during oxidation and polymerization.

1.3. Template

The template, in electrochemical synthesis, is not a required parameter as we have discussed previously. However, in the presence of template, electrochemical synthesis produces the conducting polymer, which retains the properties of bulk conducting polymers and have the characteristics of nanomaterials for the various applications [22]. Apart from this, templates are also used to get specific morphology of the polymer like wire, rod, spheres, etc. Several methods have been developed to synthesize the conducting polymer in the presence of a template, including chemical oxidative polymerization, electrochemical synthesis, etc. Mainly three types of templates are known for the synthesis of conducting polymers with controlled morphology i.e., hard templates [22], soft templates [23] and reactive templates. For electrochemical synthesis, only soft and hard templates are known. Reactive templates are only related to oxidative chemical synthesis; where the reactive template is itself acts as an oxidant for the monomers viz. gold chloride. The forthcoming section will describe the synthesis of PAni by the electrochemical method, using various types of soft and hard templates.

1.3.1. Electrochemical Synthesis of PAni in the Presence of Soft-Template

Soft template based electrochemical synthesis of conducting polymers got the interest of the scientific community, not only because of its environmentally friendly nature, but also due to its simple and effective way of controlling surface morphology of synthesized polymers via self-assembled micelles. Surfactants, colloidal particles, structure-directing molecules, oligomers, soap bubbles, polyelectrolytes, and bulky dopants have been employed as a source of soft templates to synthesize conducting polymer nanostructures. The surfactant based soft template and its role in polymer synthesis has been studied intensively. A mechanism of self-organization of conducting polymers through the surfactant is shown in Scheme 3. Various methods of surfactant self-assembly have been reported (theoretically and experimentally) for the synthesis of micro to nanoscale structures of conducting polymers with controlled morphology and dimensions [24-26].

Aggregate surfactant morphology in a solution may be of various types, such as spherical, cylindrical, or a flat bi-layer with minimal surface energy depending on these parameters [27]. The organization of surfactant aggregates are decided by the volume and length of the surfactant tail reflecting the hydrophobic core of the aggregate and the effective area occupied by each surfactant head group at the surface of the aggregate [27]. Other molecules also follow almost the same type of mechanism. Molecular template-guided synthesis is an efficient approach in the area of soft template synthesis for preparation of nanomaterials with well-controlled morphologies and size. An example is PAni, doped with sodium alginate nanofibers, with 40–100 nm diameters is prepared by a molecular template-guided process in a dilute solution of sodium alginate [28]. This morphology is changed in spherical nanoparticles simply by varying the reaction condition in the absence of sodium alginate when $FeCl_3$ is used as the oxidant.

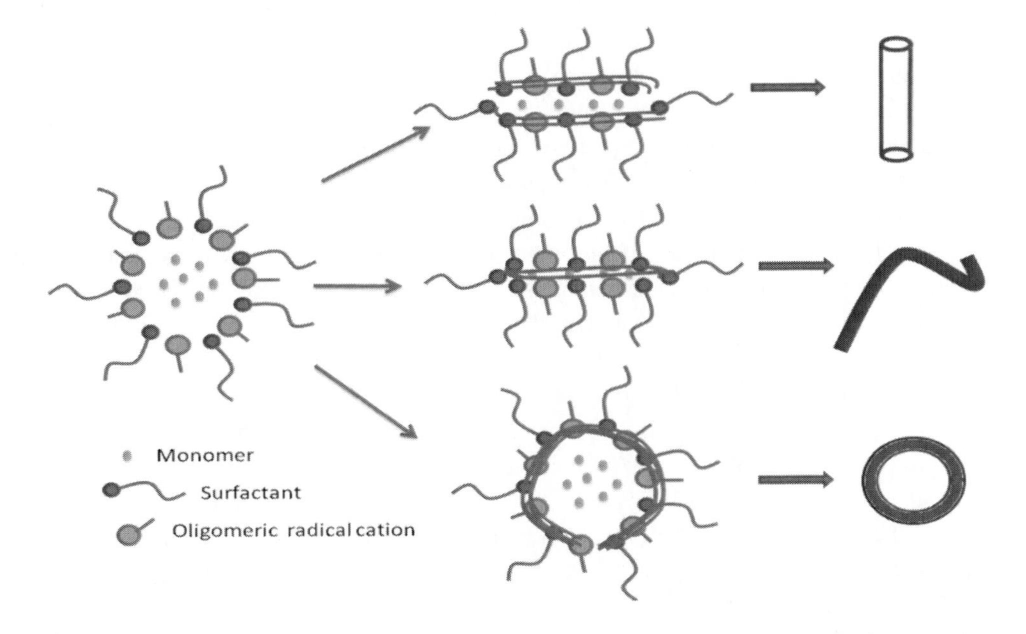

Scheme 3. Mechanism of conducting polymer formation through surfactant based template.

Table 1. Soft template electrochemical synthesis of PAni

Template	Polymerization Technique	Shape	Chemical name of template	Reference
Polyelectrolyte	Potentiodynamic (-0.1 to 0.8V) (versus Ag/Ag^{+})	Egg-shell	Poly(styrene-sulfonated)	[29]
Polyelectrolyte	Potentiostatic (0.75V)	Nano grain	poly(amidosulfonic acid)s	[30]
Surfactant	Potentiodynamic −0.2 to 1.2 V (versus SCE)	Nano grain	CSA, CTAB TritonX 100	[31]
Dopant	Potentiodynamic −0.2 to 0.9 V (versus SCE)	twisted nanofibers	d-CSA and l-CSA	[32]
Dopant	Potentiostatic (0.6V) (versus SCE)	Helical fiber	CSA	[33]

Organic acids containing chiral groups are interesting soft-templates for the synthesis of conducting polymer nanofibers. Aniline oligomer is used as a guiding template to produce PAni nanofibers with high optical activity in the presence of enantiomeric camphor sulfonic acid (CSA) [33]. Recently, Wei et al. [32] prepared PAni nanostructures with two different morphologies, sub-micrometer tubes and helical nanofibers. They used an electrochemical polymerization technique in the presence of chiral D- or L-CSA as a dopant. In this method, monomer-filled micelles acted as soft-templates in the formation of sub-micrometer tubes, while nanofibers were produced by preventing the overgrowth of PAni nanofibers by a high concentration of CSA. Table 1 describes the formation of PAni nano structures in the presence of different types of soft templates.

1.3.2. Electrochemical Synthesis of PAni in the Presence of Hard-Template

Hard-templates generally require a template membrane or template dispersion to guide the growth of the nanostructures within the pores or channels of the membrane. This hard-template provides controlled nanostructures in terms of morphology and diameter. C. R. Martin is a pioneer of this template-synthesis method and explored the synthesis of various structure-like nanofibers or nanotubes and nanowires of conducting polymers [34]. The main advantage of the hard-template method is significant control in the shape and size of the conducting polymer. The diameter of the nanostructure is controlled by the dimension and porous nature of the hard template, whereas the length and thickness of the nanostructure is usually adjusted by changing the polymerization time and length of the filler. The hard-template method is the most suitable approach for the preparation of well-controlled and defined nanostructures of conducting polymers [35]. Hard templates are removed or separated from the polymer after formation and the pure form of polymers are further used for various applications. But the templates may remain inside the polymer and interaction with the polymer may enhance the properties or create some additional properties. This type of polymer synthesis comes under the class of polymer composites, when templates are nothing they are fillers viz. metal/metal particles, carbon materials, clay, or inorganic materials, etc. Various fillers from micro to nano sizes are explored as hard template in electrochemical synthesis to form PAni-composites such as metal oxide, graphene and carbon nanotubes, etc. Among the conducting polymers PAni is one of the high redox active conducting polymers, therefore, its composite makes it more redox and environmentally stable. In addition to this, some additional properties may be introduced by using suitable fillers viz. magnetic properties by magnetic metal oxides, etc. The above objective can be fulfilled by incorporating fillers in the PAni matrix, or coating PAni over the fillers or PAni penetrated into the filler materials. Homogeneous and interactive composites show synergistic effects exhibiting better conductivity, thermal and mechanical properties, long redox cycle-life and improved specific capacitance. In the following sub-section some important nano-fillers are described for the above purpose.

1.3.2.1. Transition Metal and Metal Oxide

Usually, inorganic-organic nanocomposite hybrid materials are prepared by electrochemical polymerization through the redox intercalation of conducting polymer monomers in a redox active lamellar host material. The most comprehensive studies of the electrochemical synthesis of PAni on Ti surfaces have been undertaken for the synthesis of macroporous interconnected open networks of polymer fibers. Further, these structures were explored as a template for the deposition of a semiconducting transition metal oxide such as vanadium oxide. Karatchevtseva et al. reported a novel two-step one-pot all-electrochemical method for the preparation of interpenetrating conducting-polymer semiconducting oxide (V_2O_5) nanocomposites [36]. Incorporation of metal ions during the synthesis and further conversion in elemental metal on reduction or metal oxides by oxidation are the common routes for the formation of such composites. However, electrochemical methods are less known for metal and metal oxide composites in comparison to chemical methods using metal salts as oxidants or co-precipitation of polymers as well as metal hydroxides.

1.3.2.2. Graphene

Now a days graphene is one of the most attractive carbon materials for the scientific community because of its unique properties, viz. high electrical and thermal conductivities, great mechanical strength, large surface area and low manufacturing cost [37-40]. Graphene and carbon nanotubes have nearly similar structure, having sp^2 hybridized carbon in the form of two-dimensional sheets and hollow tubular structures, respectively. These structures introduce the electron-accepting tendency of graphene and carbon nanotubes. Thus, it is possible that donor type aromatic molecules can be π-stacked on the graphene surface. Aniline with planar aromatic structures can easily interact onto the hydrophobic surface of graphene sheets, and at elevated temperatures a charge-transfer complexation takes place [41]. In addition, the negative charge of the aniline helps in the separation of the graphene in the solution, which also avoids aggregation of graphene. Formation of the nanocomposite also takes place based on the same principle. Better interaction of monomers with graphene and intercalation of monomers between the graphene sheets result in uniform and interactive composites. As the polymerization proceeds, separation of the graphene layers increases and finally, separated sheets dispersion in polymer matrix may occur as shown in Scheme 4.

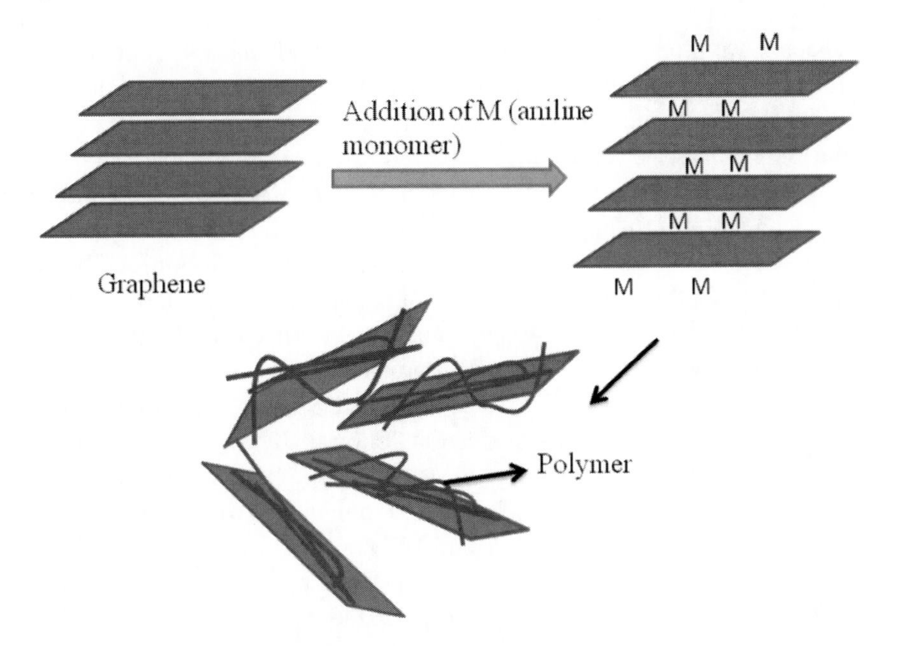

Scheme 4. Synthesis of graphene-PAni nanocomposite.

Recently, various polymerization methods and selection of precursors have been tried to make graphene-PAni nanocomposite toincrease its interaction and create the possibility of charge transfer complexation between the components. Feng et al. have reported PAni-graphene nanocomposite films using graphite oxide (GO) and aniline as the starting materials [41]. Nanocomposite is formed by oxidation of aniline monomer and reductions of graphene oxide in the potential range -1.3 to 1.0 V. Khan et al. synthesized a PAni-graphene-lemeller nanocomposite in aqueous medium by potentiodynamic polymerization between -2 to 0 V for several cycles [42]. Efthekari et al. made a nanocomposite by modifying working electrodes with graphene oxide-aniline paste cycling the potential in the range of -2.0 to 1.0V [43].

1.3.2.3. Carbon Nanotube (CNT)

Due to various novel properties, easy synthesis with better properties than other carbon materials, CNTs become promising fillers for the development of future nanotechnology, resulting into advanced materials and devices of great practical interests. However, poor processability of CNTs and also to get better surface properties, CNTs are generally modified via a proper functionalization. Nanocomposite formation is also a type of functional modification of CNT surface, where the polymer is coated over the CNTs [44]. For the preparation of PAni-CNT composites, a number of methods have been tried so far, such as non-covalent solution coating processes, wrapping of CNTs with conjugate polymers or *in-situ* polymerization of aniline in the presence of CNTs etc. [44, 45]. However, electrochemical synthesis is less explored and tedious also. Abalyaeva et al. developed PAni-CNT composite by an electrochemical method using 1-65 wt. % of filler [46]. Recently PAni-CNT nanocomposite formation was reported where a covalent interaction between the polymer chains and CNTs are achieved to overcome the problem of detachment for long-term application [47]. However, the synthetic process is two steps and results into small polymer chains. Control morphology, uniform distribution of CNTs and quality of polymer are difficult to achieve using conventional electrochemical methods.

In this chapter, we propose a synthetic approach named biphasic electropolymerization for the development of PAni-MWNT nanocomposite film. This is an approach in which the experimental setup is almost similar to conventional electro-chemical polymerization method. However, the single-phase electrolyte is replaced by two immiscible solvent systems (biphase). Here monomer and filler are distributed in both the phases based on their solubility and interaction with solvents (mainly in the case of filler) and electrochemical polymerization is carried out in one of the phases. The advantage of this method is to kinetically control the polymerization and retain the filler in the vicinity of the anode therefore having better control on the nanocomposite formation [48, 49]. During the experiment a 100 μl of pre-distilled aniline (Merck, India) was dissolved in 2 ml dichloromethane in a vial. In another vial 0.5 mg multi-walled carbon nanotubes, MWNT (Sigma-Aldrich, USA; outer diameter 10-15 nm) was dispersed in 5ml of 0.2 N HCl under sonication followed by vigorous stirring. Both the dispersed MWNT and the monomer mixed with each other and transferred into the electrochemical cell. This mixture was left for 5 minutes to obtain a complete phase separation and formation of bi-phase. Dispersed MWNTs were mainly present in the aqueous phase acquired upper layer; however, aniline monomer was present in the non-aqueous phase acquired down layer (*cf.* Scheme 5 a). Polymerization was carried out using both potentiostatic and potentiodynamic techniques in one of the phases (preferred in aqueous phase). Potentiostatic polymerization was carried out at 1.1V vs. Ag/Ag^+. Selection of potential was done using the potentiodynamic technique performing in a wide range of potential. A suitable range of potential for polymerization was monitored when there was a successive increase in current due to efficient coupling of radical cations, as shown in Figure 2. The highest potential of the range is usually selected for potentiostatic polymerization. Time of polymerization was fixed for 30 minutes when a green film formed over the electrode surface. The film formed at the working electrode was washed (first with dichloromethane to remove the monomer and then distilled water to remove some bi-products), dried and used as such for further investigations. Working electrodes, used for the synthesis, was Au plate of 1x1 cm^2 with Pt gauge counter and Ag/Ag^+ reference electrode.

Since MWNT dispersion in acid was made by extensive sonication and stirring, followed by decantation, it is expected that there will not be any agglomeration of MWNT and it will form a stable dispersion of MWNTs during the oxidation polymerization of aniline over the working electrode. Schematic representation of biphasic electropolymerization and mechanism of polymerization is shown in Scheme 5a,b. Aniline, present in lower phase, has the tendency to move in the upper aqueous layer because of its partition co-efficient. Aniline forms anilinium ion in acid medium, which helps its solubility in aqueous medium. As the anodic potential applied, anilinium ion oxidized and moved toward the working electrode along with MWNT (due to interaction with the surface of MWNTs) where the polymerization took place in accordance to the mechanism shown in scheme 5b.

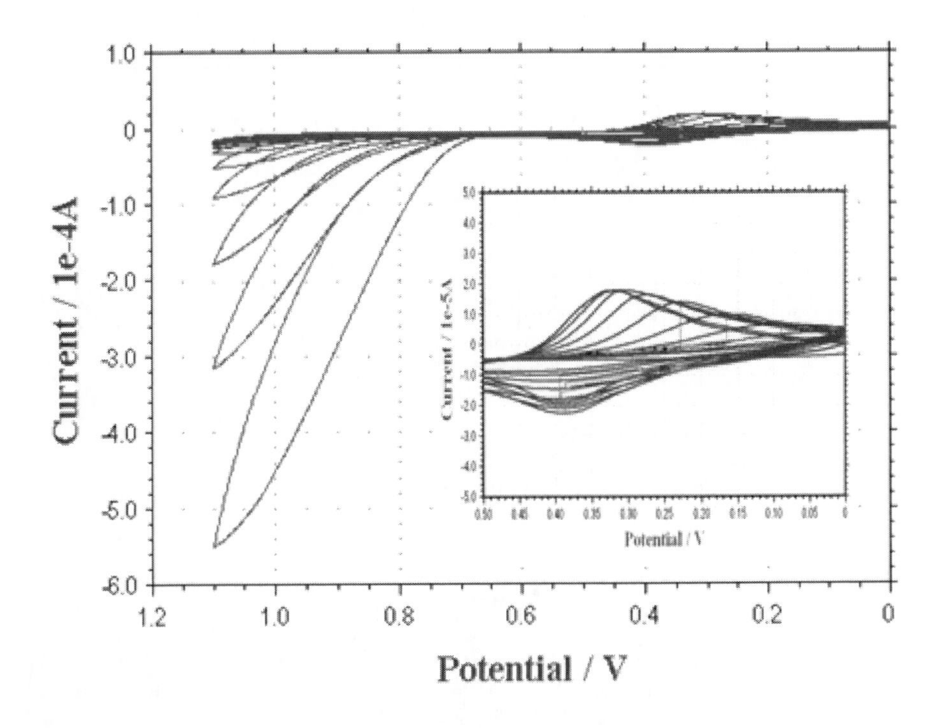

Figure 2. Electrochemical synthesis of MWNT–PAni nanocomposite under potentiodynamic condition.

One more interesting aspect of this synthesis is the movement of the monomer from the down non-aqueous phase to the upper aqueous layer, which provided control polymerization and interaction of monomers with MWNTs to form uniform arrangement of polymer chains over MWNTs surface over the working electrode. Film formed on the gold electrode is dissolved in THF for absorbance measurement. Fourier transform infrared (FT-IR) and X-ray diffraction (XRD) characterizations have been used to prove the interaction between the components and homogeneity of the filler in the PAni matrix.

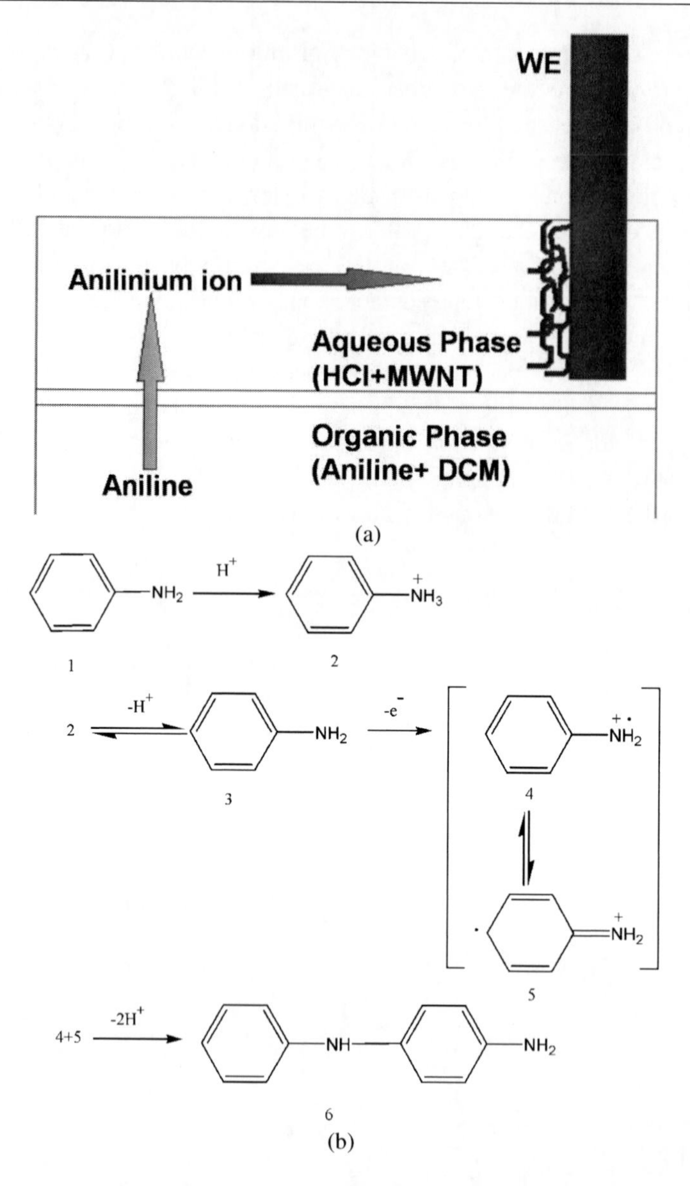

Scheme 5. (a) Schematic representation of biphasic electropolymerization at working electrode (WE) and (b) Mechanism of PAni formation at interface and electrode surface [1 to 2; at inter-phase diffusion, 2 to onwards; polymerization at electrode surface].

1.4. UV-vis, FT-IR and XRD Characterizations

The UV-vis spectra of pure PAni and MWNT-PAni composite are shown in Figure 3. Both the spectra showed characteristics of absorbance peaks of PAni. They are due to π-π^* transition (strong band at 284 nm), polaron lattice or radical cations (weak band at 406 nm), exciton absorption (strong band at 530 nm) and shifting of electrons from benzenoid ring to quinonoid ring or localized polarons (weak band at around 730 nm) [50]. However, in the case of MWNT-PAni composite, the blue shift of exciton absorption and well resolved peak at 730 nm is an indicative evidence of interaction of PAni with MWNT. MWNTs act as an electron acceptor and facilitate the electron transfer from the benzenoid ring to the quinonoid ring [50, 51] as also discussed under heading 1.3.2.2.

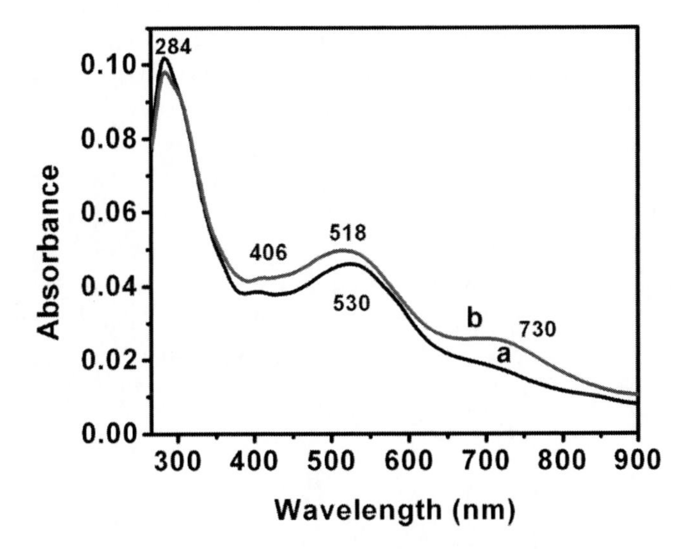

Figure 3. UV-Vis. of (a) PAni and (b) MWNT-PAni composite.

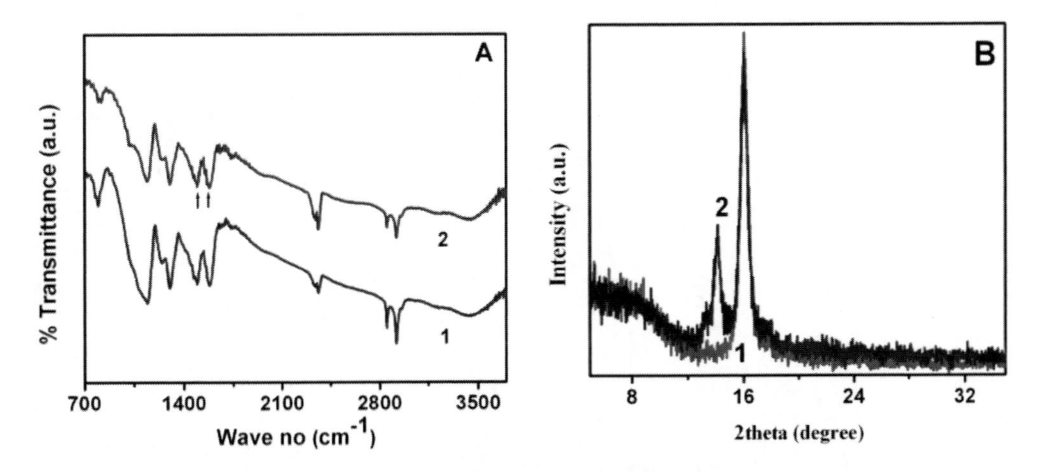

Figure 4. A) FT-IR and B) XRD of 1) PAni and 2) PAni-MWNT nanocomposite.

In FT-IR(Figure 4A), the stretching vibration of quinonoid and benzenoid rings fall at about 1580 cm^{-1} and 1500 cm^{-1}. The ratio of these two peaks attributes to the oxidation state of the polymer chain. PAni in nanocomposite shows a higher intensity of the quinonoid C=C bond in comparison to benzenoid, similar to most of the cases reported earlier [52].

XRD showed the presence of both components as well as an idea about the surface of MWNTs. If the surface of MWNTs densely covered by polymer then intensity of the peaks belonging to MWNT gets reduced and vice versa.

In our synthesized nanocomposite there was much less reduction in the MWNT (002) peaks corroborated by a very thin coating of PAni over the surface as shown in Figure 4B, contrary to the other nanocomposites synthesized by different polymerization methods [53].

1.5. Electrochemical Characterization

To evaluate the effect of MWNT on PAni, an electrochemical characterization of nanocomposite was done by carrying out cyclic voltammetric (CV) measurements in 1M HCl at room temperature. The difference in electrochemical behavior is illustrated between pure PAni and MWNT- PAni composite, which is shown in Figure 5A. Oxidation of PAni takes place in four steps (similarly reductions also) indicating various oxidation states of PAni (leucoemeraldine, emeraldine, nigraniline and pernigraniline) [10]. These states are not clear in the case of pure PAni as it showed two broad peaks, one for oxidation at 0.5V and another for reduction at 0.15V vs. Ag/Ag+ [54, 55]. However, the MWNT-PAni composite film demonstrated well-defined peaks indicating various states of PAni. This difference is due to the π-π interaction between the MWNT π-bonded surface and the highly conjugated π-system of PAni [56] that facilitated the electron transfer and showed better electroactivity of PAni.

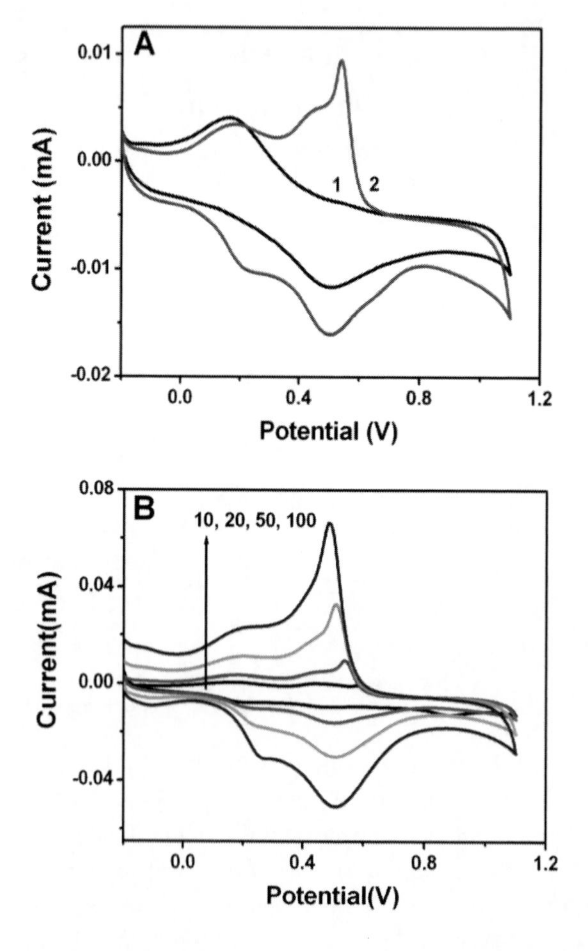

Figure 5. CV of (A) (1) PAni and (2) MWNT-PAni nanocomposite, (B) MWNT-PAni at different scan rates.

To check the stability of composite film, CV is further performed at different scan rates, which are shown in Figure 5B. The increase in redox current was observed with an increase in scan rates and they were found to be directly proportional to the square root of the scan rate, indicating that the reaction kinetics is controlled by a diffusion step [57]. Furthermore, MWNT-PAni exhibited the high-normalized current response with respect to the mass in comparison to pure PAni because of the charge transfer phenomenon between the PAni and MWNT. Electrochemical studies clearly indicated the coating of PAni over MWNTs and also the strong interaction of the polymer with CNT surfaces. This interaction further enhanced the properties, which were also observed during other characterizations.

1.6. Morphological Characterization

Scanning electron microscopy (SEM) is used to investigate the morphology of the MWNT-PAni composite. The micrograph of pure PAni and MWNT-PAni composite is shown in Figure 6. The thin film of pure PAni was composed of compact globular morphology (Figure 6A). However, MWNT-PAni composite film consisted of interwoven fibrous morphology with a diameter of 20-25 nm (pure MWNT diameter was 10-15 nm) (Figure 6E). This again supported the coating over MWNTs and formation of uniform nanocomposite.

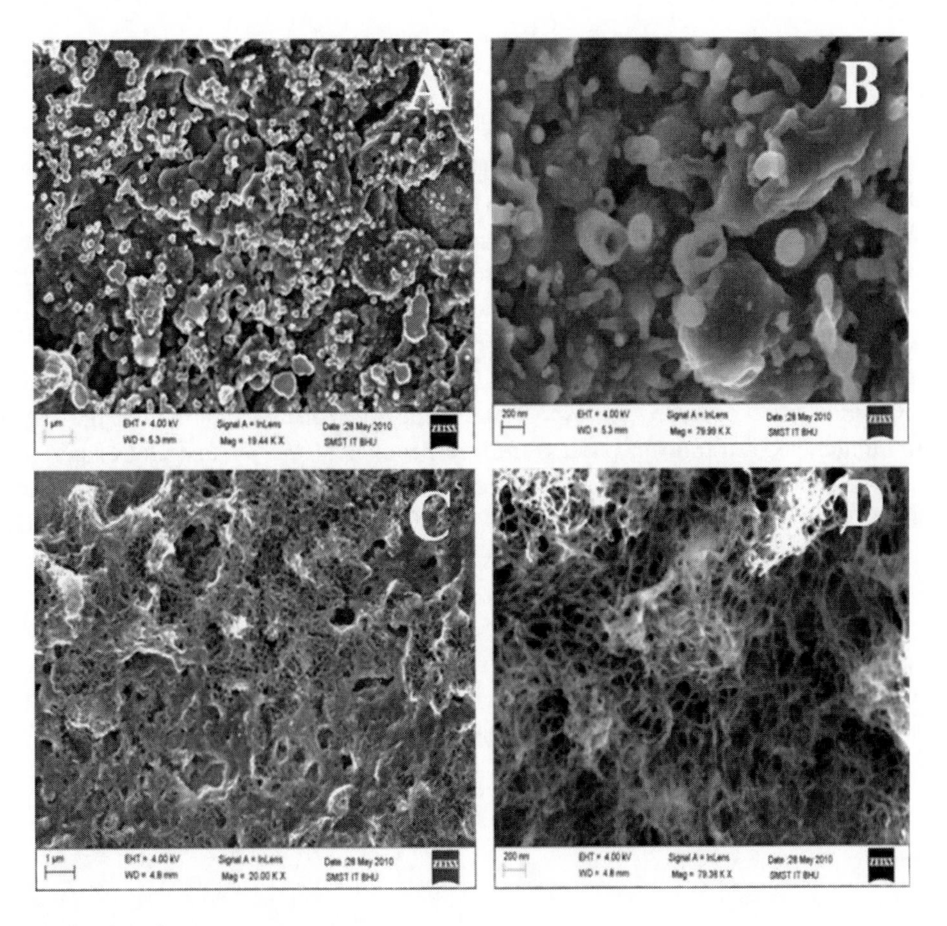

Figure 6. Continued on next page.

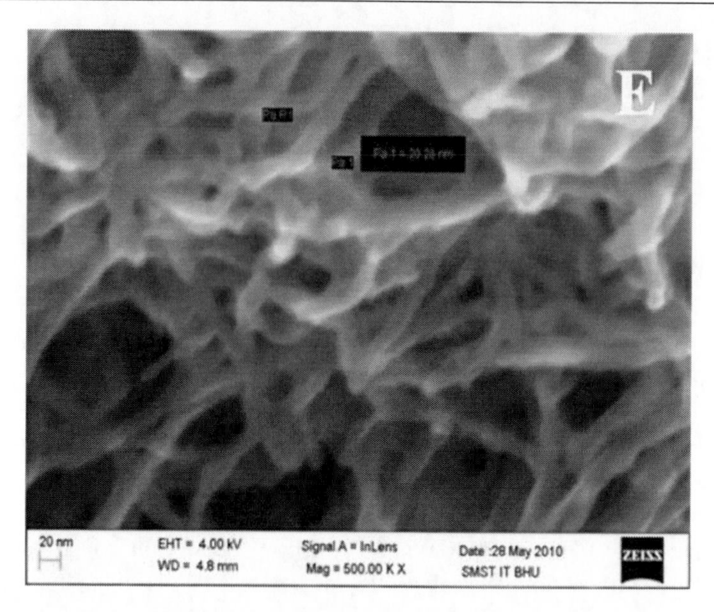

Figure 6. SEM of (A), (B) PAni and (C), (D) MWNT-PAni nanocomposite at two different magnifications (E) Further magnified image of MWNT-PAni nanocomposite to resolve the thickness of PAni modified MWNTs.

1.7. Impedance Spectroscopy

Supercapacitor application of PAni is widely studied in the presence of carbon materials due to their extraordinary stability and charge storage. To study the specific capacitance, an impedance measurement was carried out at open circuit potential on platinum disc electrodes having PAni and MWNT-PAni nanocomposite film for 300 sec (at 1.1V). The impedance analysis (Nyquist plot) was studied for the conductivity, structure and charge transport at the film-electrolyte interface.

In this plot, a single semicircle at the high frequency region and a straight line at the low frequency region were observed corresponding to the electron transfer limited process and diffusion limited electron transfer process [58], respectively. The Nyquist plot of pure PAni and MWNT-PAni composite is shown in Figure 7A. In the case of pure PAni, charge transfer resistance was higher in comparison to the MWNT-PAni composite and PAni layers coated over MWNTs offered high capacitance due to porosities and large contact with the electrolyte. High conductivity of MWNT-PAni composite was evident by a small semi-circle that showed less charge transfer resistance of the film.

Specific capacitances in the case of PAni and MWNT-PAni nanocomposites were calculated as 10 and 80 F/g, respectively. Specific capacitance was stable for more than 500 cycles of charging-discharging as shown in Figure 7B. Such property makes nanocomposite suitable for supercapacitor and other electronic applications.

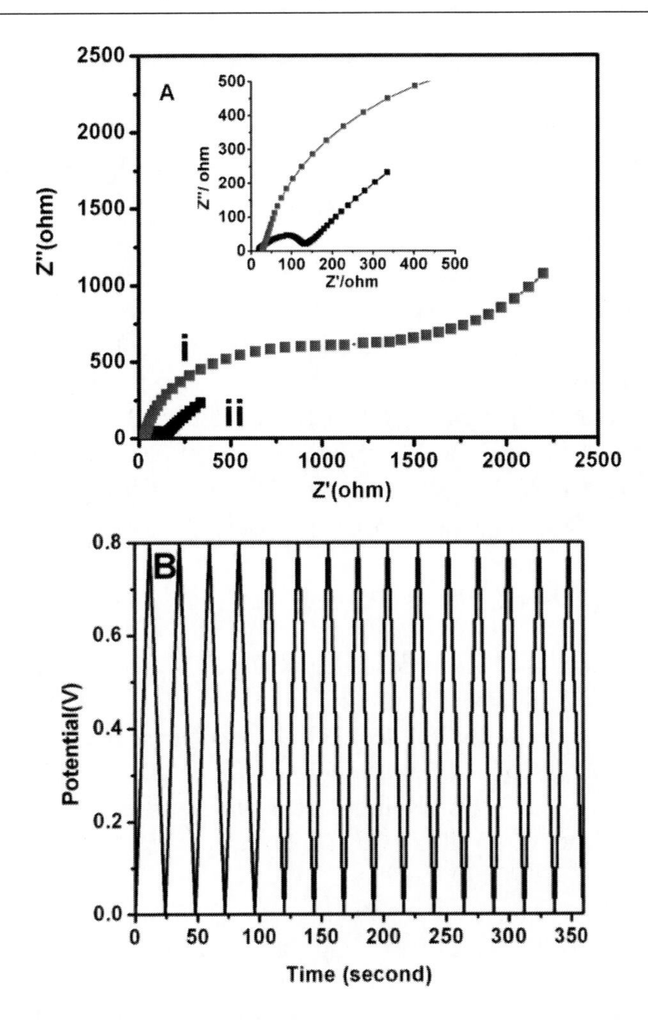

Figure 7. (A) Nyquist plot for i) PAni and ii) MWNT-PAni nanocomposite and (B) charging-discharging plot of MWNT-PAni nanocomposite.

CONCLUSION

Conducting polymer nanocomposites are interesting materials with improved and unanticipated properties that will continue to keep the pace in the field of science and technology. PAni has been of great interest to many researchers because of its excellent conductivity, easy synthesis, low cost and redox properties in comparison to other conducting polymers. But processibility in common solvents, environmental and thermal stabilities are the major issues that limit its applications. Modifications in synthetic process, functionalization and by the use of conducting fillers, these problems have been addressed in the recent years. The proposed biphasic electropolymerization process is likely to receive more attention as less attention is presently focused on electrochemical synthesis of nanocomposites of conducting polymers. This will lead to emergence of a new technique to get better composites with improved properties that will be superior to some of the presently known methods for nanocomposites. This chapter is an attempt to summarize electrochemical synthesis methods of conducting polymers putting emphasis on polyamine and also exploring a new biphase electrochemical method for the formation of polymer nanocomposites.

REFERENCES:

[1] Markovic, M.G.; Matisons, J.G.; Cervini, R.; Simon, G.P.; Fredericks, P.M. *Chemistry of Materials* 2006, *18*, 6258-6265.

[2] Huang, J.; Virji,S.; Weiller, B.H.; Kaner, R.B. *Journal of American Chemical Society* 2003,*125* , 314-315.

[3] Gupta, B.; Prakash, R. *Synthetic Metals* 2010, *160*, 523–528.

[4] Joshi, L.; Gupta, B.; Prakash, R. *Thin Solid Films* 2010, *519*, 218–222.

[5] Li, W.; Zhu, M.; Zhang, Q.; Chen, D. *Applied Physics Letter* 2006, *89*, 103-110.

[6] Stejskal, J.; Sapurina, I.; Trchova, M. *Progress in Polymer Science* 2010, *35*, 1420–1481.

[7] Srivastava, M.; Prakash, R. *Journal of Nanoscience and Nanotechnology,* 2012, *12 (1)*, 489-493.

[8] Bard, A.J.; Fauknar, L.R. *Electrochemical methods, Fundamental and Application," 2nd Ed., Wiley and Sons, New York* 2001.

[9] Rodriguez, I.I.; Marcos, M.L.; Velasco, J.G. *Electrochimica Acta* 1987, *32*, 1181-1185.

[10] Prakash, R. *Journal of Applied Polymer Science,* 2002, 83, 378-385.

[11] Gregory, R.V.; Kimbrell, W. C.; Kuhn, H. H. *Synthetic Metal* 1989, *28,* C823-C835.

[12] Tallman, D.E.; Vang, C.; Wallace, G.G.; Bierwagen, G.P. *Journal of Electrochemical Society* 2002, *149*, C173-C179.

[13] Visy, Cs.; Lukkari, J.; Pajunen, T.; Kankare, J. *Synthetic Metal* 1989, *33*, 289-299.

[14] Ko, J. M.; Rhee, H.W.; Park, S.M.; Kim, C.Y. *Journal of Electrochemical Society* 1990, *137,905*-909.

[15] Pringle, J.M.; Efthimiadis, J; Howlett, P.C.; Efthimiadis, J.; MacFarlane, D.R.; Chaplin, A.B.; Hall, S.B.; Officer, D.L.; Wallace, G.G.; Forsyth, M. *Polymer* 2004, *45*, 1447-1453.

[16] Zinger, B. *Journal of Electroanalytical Chemistry* 1998, *244*, 115-121.

[17] Sun, B.; Jones, J.J.; Burford, P.R.; Skyllas-Kazaces, M. *Journal of Material Science,* 1989, *24*, 4024-4029.

[18] Shen, Y.; Qiu, J.; Qian, R. *Macromolecular Chemistry and Physics* 1987, *188*, 2041-2045.

[19] Ge, H.; Ashraf, S.A.; Gilmore, K.; Too, C.O.; Wallace, G.G. *Journal of Electroanalytical Chemistry* 1992, *340*, 41-52.

[20] Masuda, H.; Tanaka, S.; Kaeriyama, K.J. *Journal of Polymer Science Part A Polymer Chemistry* 1990, *28*, 1831-1840.

[21] Neoh, K.G.; Kang, E.T.; Tan, T.C. *Journal of Applied Polymer Science* 1989, *38*, 2009-2017.

[22] Balamurugan, A.; Lin, C.Y.; Nien, P.C.; Ho K.C. *Electroanalysis* 2012, *24*, 325-331.

[23] Xia, L.; Wei, Z.; Wan, M. *Journal of Colloid and Interface Science* 2010, *341*, 1-11.

[24] Israelachvili, J.; Mitchell, D.J.; Ninham, B.W. *Journal of Chemical Society Faraday Transactions 2* 1976, *72*, 1525-1568.

[25] Manne, S.; Gaub, H.E. *Science* 1995, 270, 1480-1482.
[26] Grant, L.M.; Ederth, T.; Tiberg, F. *Langmuir* 2002, *16*, 2285-2291.
[27] Huang, W.S.; Humphrey, B.D.; MacDiarmid, A.G. *Journal of Chemical Society Faraday Transactions 1* 1986, *82*, 2385-2400.
[28] Yu, Y.; Zhihuai, S.; Chen, S.; Bian, C.; Chen, W.; Xue, G. *Langmuir* 2006, *22*, 3899-3905.
[29] Briseno, A. L.; Han, S.; Rauda, I. E.; Zhou, F. *Langmuir* 2004, *20*, 219-226.
[30] Bhandari, H.; Bansal, V.; Choudhary, V.; Dhawan, S.K. *Polymer International* 2009, *58*, 489–502.
[31] Raj, J.A.; Mathiyarasu, J.; Vedhi, C.; Manisankar, P. *Materials Letters* 2010, *64* , 895–897.
[32] Weng, S.; Lin, Z.; Chen, L.; Zhou, J. *Electrochimica Acta* 2010, *55*, 2727–2733.
[33] Lei, W.; Wang H.L. *Advance Functional Material* 2005, 15, 1793-1798.
[34] Parthasarathy, R.; Martin, C.R. *Nature* 1994, *369*, 298-301.
[35] Wu, C.G.; Bein, T. *Science* 1994, *264*, 1757-1759.
[36] Karatchevtseva, I.; Zhang, Z.; Hanna, J.; Luca, V. *Chemistry of Materials* 2006, *18*, 4908-4916.
[37] Novoselov, K.S.; Geim, A.K.; Morozov, S.V.; Jiang, D.; Zhang,Y.; Dubonos, S.V.; Grigorieva, I.V. ; Firsov, A.A. *Science* 2004, *306* , 666-669.
[38] Geim, A.K. *Science* 2009, *324*, 1530-1534.
[39] Pandey, H; Parashar, V; Parashar, R; Prakash, R; Ramteke, P W; Pandey, A. C.; *Nanoscale* 2011, 3, 4104-4108.
[40] Parashar, V.; Kumar, K.; Prakash, R.; Pandey, S.K.; Pandey, A. C.; *Journal of Materials Chemistry*, 2011, 21, 6506-6509.
[41] Feng, X.M.; Li, R.M.; Ma, Y.W. ; Chen, R.F.; Shi, N.E.; Fan, Q.L.; Huang, W. *Advance Functional Material* 2011, *21*, 2989–2996.
[42] Khan, J.M.; Kurchania, R.; Sethi, V.K. *Thin Solid Films* 2010, *519*, 1059–1065.
[43] Eftekhari, A.; Yazdani, B. *Journal of Polymer Science Part A Polymer Chemistry* 2010, *48,* 2204–2213.
[44] Srivastava, R.K.; Srivastava, A.; Prakash, R.; Singh, V. N.; Mehta, B. R. *Journal of Nanoscience and Nanotechnology*, 2009, 9, 1-7.
[45] Chiang, L.Y.; Anandakathir, R.; Hauck,T.S.; Lee, L.; Canteenwala, T.; Padmawar, P.A.; Pritzker, K.; . Brunoc, F.F; Samuelson L. A. *Nanoscale* 2010, *2*, 535–541.
[46] Abalyaeva, V.V.; Vershinin, N.N.; Yu. M.; Shul'ga, O.N. Efimov, *Elektrokhimiya* 2009, *45,* 1367–1376.
[47] Gupta, B.; Prakash R. *Materials Science and Engineering C* 2009, *29*, 1746–1751.
[48] Mei, L.; Zhixiang, W.; Lei J. *Journal of Materials Chemistry* 2008, *18*, 2276–2280. Nedungadi, P.A.K.; Baweja, S.; Zutshi,K. *Bulletin of Material Science* 1988, *10(4)* ,361-366.
[49] Alexander, A.N.; Victor F.I.; Anatoly V.V. *Electrochimica Acta* 2001, *46* ,4051-4057. U.P.B. *Scientific Bulletin Series A- Applied Mathematics and Physics* 2009, *71 (4)*, 21-30.DUN SUNITA BAWEJA and K ZUTSHI K NEDUNGADI,
[50] Yoon, S.B.; Yoon, E.H.; Kim, K.B. *Journal of Power Sources* 2011, *196*, 10791–10797.
[51] Nguyen, V.H.; Shim, J.J. *Synthetic Metals* 2011, *161*, 2078– 2082.

[52] Singh, V.; Mohan, S.; Singh, G.; Pandey, P.C.; Prakash, R. *Sensors and Actuators B: Chemical* 2008, 132, 99- 106.

[53] Hussain, A.M.P.; Kumar, A. *Bulletin of Material Science* 2003, *26*, 329-.334.

[54] Zhang, J; Kong, L.B.; Wang, B; Luo, Y.C.; Kang, L. *Synthetic Metals* 2009, *159*, 260-266.

[55] Zhang, T; Fu, L.; Gao, J.; Yang, L.; Wu, Y.; Wu, H. *Pure and Applied Chemistry* 2006, *78*, 1889-1896.

[56] Kumar, A.; Prakash, R. *Chemical Physics Letter* 2011, *511*, 77-81.

In: Trends in Polyaniline Research ISBN: 978-1-62808-424-5
Editors: T. Ohsaka, Al. Chowdhury, Md. A. Rahman et al.

Chapter 11

POLYANILINE: A PROMISING PRECURSOR MATERIAL FOR MEMBRANE TECHNOLOGY APPLICATIONS

Evangelos P. Favvas and Sergios K. Papageorgiou
Institute of Physical Chemistry, NCSR "Demokritos"
Terma Patriarchou Grigoriou & Neapoleos
Aghia Paraskevi Attikis, Greece

ABSTRACT

Polymeric membranes have been used in various industrial applications such as reverse osmosis (RO), gas separation, microfiltration, microfiltration, desalination, water purification, heavy metal removal, hemodialysis etc. For gas separation, the permeability and selectivity coefficients are the main properties which determine their efficiency and suitability for a specific process, while porosity, temperature resistance, swelling phenomena and electrical conductivity play an important role and depending on the intended application must be optimal. Polymers such as cellulose acetate, polyimides, polysulfone, polycarbonates and poly(phenylene oxide) have been the most extensively studied and applied materials in the membrane gas separation field. However, there is always room for innovation, with the evaluation of materials with unique properties that may be well studied and applied in other technological fields but remain relatively new with respect to their application in membrane technology. One of these materials has been polyaniline (PAni), one of the most studied conducting polymers of the past 50 years, that has recently emerged in the literature for the preparation of gas separation membranes.

In this chapter, both single phase and mixed matrix polyaniline based membranes are presented with reference to their properties and potential application based on the recent literature. A special focus is given to the recent advances in the area of gas separation, reverse osmosis and pervaporation processes.

1. INTRODUCTION

Polyaniline (PAni) is a long studied polymer but always up to date. In fact, the compound itself was discovered some 150 years ago, as early as 1862, when H. Letheby of the College of London Hospital, by anodic oxidation of aniline in sulphuric acid, obtained a partly conductive material which was probably PAni. Since then, PAni, has been extensively studied and found applications as part of the Intrinsically Conducting Polymers (*ICP*) family. The first Intrinsically Conducting Polymer (*ICP*), polyacetylene $(CH)_\chi$, was discovered by Shirakawa et al., 1977 who found that its conductivity can increase by several orders of magnitude after chemical doping, reaching values comparable to those of metals. For this work, Alan J. Heeger, Alan G. MacDiarmid and Hideki Shirakawa were awarded jointly the Nobel prize in chemistry "for the discovery and development of conductive polymers" in 2000 [1,2]. Among the ICPs available, PAni has been found to be the most promising because of its ease of synthesis, its monomers low cost, its tunable properties, and its better stability compared to other ICPs [3,4]. Moreover, the electrical properties of PAni can be reversibly controlled by charge–transfer doping and protonation.

Because of its rich chemistry [5], PAni, has been in the past 50 years one of the most studied conductive polymers. Its referred electrical conductivity [6–12] spreads in the range of 10^{-8} to 400 Scm^{-1} (*the conductivity of a typical metal is* $>10^4$ Scm^{-1}) and is likely to increase as better processing methods are developed reducing structural defects. Its conductivity can be tuned to cater for specific end uses for a variety of applications. PAni is reasonably stable under ambient conditions and with the proper selection of dopants, it retains its conductivity properties over long periods of time (*i.e., five years and longer*). Moreover, it can easily be converted from its conductive (*emeraldine salt*) to its insulating form (*emeraldine base*) depending on the pH. Under acidic conditions the polymer becomes conductive but if exposed to higher pH values the polymer switches to the insulating form and this facile switching can be cycled many times [13]. Figure 1 shows the chemical structure of PAni.

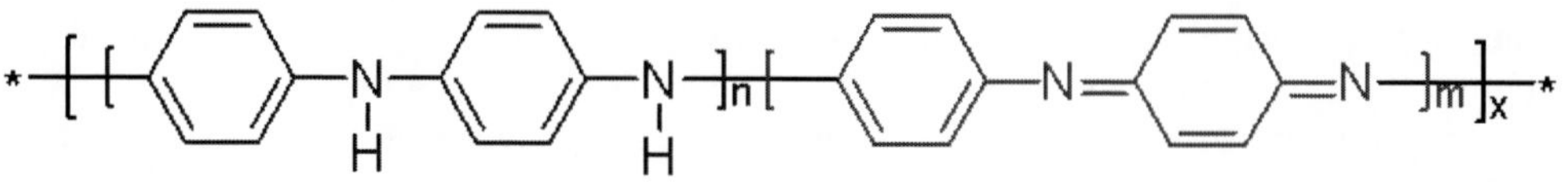

Figure 1. PAni chemical structure.

PAni is one of the most extensively used conductive polymers in a range of applications such as [14]: Packaging Industry (*antistatic films, general antistatic products*), fenestration (*electrochromic "smart" windows, electrochromic automobile rear vision systems*), textile industry (*conductive fabrics, "smart" cloths*), automotive industry (*antistatic charge dissipation, paint primers, electrochromic rear vision systems*), construction (*antistatic floors, antistatic work surfaces*), mining (*conductive pipes for explosives, antistatic packaging*). Especially owing to its conductivity properties PAni has been extensively used in the field of electronics finding applications in lightemitting diodes (*LED*) [15], electroluminescence diodes [16], metallic corrosion resistance, organic rechargeable batteries, biological and environmental sensors, composite structures, textile structures for specialized applications or static dissipation, membrane gas–phase separation, actuators, Electromagnetic interference

(*EMI*) shielding [17], organic semiconductor devices for circuit applications, blends with insulative host polymers to impart a slight electrical conductivity, bioelectronic medical devices, and a variety of other applications where tunable conductivity in an organic polymer is desirable [13].

Apart from the applications related directly to its conductivity properties, PAni has also played a significant role in membrane technology and applications. This chapter is limited to the PAni based membranes field and will mainly focus on their gas/vapor separation applications. Unfortunately the polyanilines conductivity properties have not yet been exploited in this field despite the abundant research, but it is not unlikely that the combination of its gas separation properties with its tunable conductivity could reveal new paths in gas separations research and applications.

2.1. Synthesis

PAni can be synthesized mainly by two methods: (a) *Chemically*, by direct oxidation of the monomer (*aniline*), using chemical oxidants (*labelled PAniC*) and (b) *Electrochemically* by anodic oxidation on an inert electrode (*Labeled PAniE*) [18].

- *Chemical Synthesis*

The chemical synthesis of PAni (*PAniC*) can be achieved by oxidative polymerization reactions carried out in an acidic solution (*e.g., 1 M HCl*). Usually the final product is a precipitate in aqueous solution containing typical reagents: acids like hydrochloric, sulphuric, nitric or perchloric, ammonium peroxydisulphate (*persulphate*), and aniline [18]. According to this method, PAni is synthesized from its monomer, aniline, which is converted into the conjugated polymer via a condensation process. During the polymerization of aniline the solution becomes coloured, step by step, resulting black at the final step of the condensation process. The violence as well as the progress of the colour transition depends on the solubility factor of the oligomers, the concentration of the oxidant and the nature of the medium. This behaviour is common for any typical chemical polymerization route. A series of different chemical routes has been reported in the literature for the preparation of PAni.

- *Electrochemical Synthesis (Polymerization)*

The Electrochemical polymerization (*ECP*) is a method used extensively for PAni (*PAniE*) synthesis. This method offers better control of the reaction system and often yields products much purer than those obtained via chemical synthesis because of the absence of oxidants, surfactants etc [4]. In ECP, both initiation and termination steps can be fine controlled. The electrochemical polymerization of aniline can be attained by three methods [4]:

1. under a constant current (*galvanostatic*)
2. under a constant potential (*potentiostatic*)
3. under a scanning/cycling or sweeping potential.

In a typical route, the process takes place in an electrolyte bath containing the electrolyte solution and the monomer. Two assembled electrodes are dipped into the reaction solution and PAniE is formed on the surface of the platinum plate electrode. There are differences between the abovementioned methods, for example in the adhesion strength of the final product on the electrode as well as its form. The constant potential method, gives PAni in the powder form adhering weakly on the electrode [19], the constant current technique yields a polyanilin film weakly adhered on the surface of the electrode while in the third case the PAni forms as a strongly adherent film on the electrode surface [20, 21].

Apart from these two basic methods, several other synthesis routes are mentioned in the literature [4] for PAni synthesis, some listed below:

1. *Heterophase polymerization*: a method producing high quality, pure and well defined polymers with accurately controlled properties from small to large scales [22–24]. It should be noted that the heterophase polymerization technique includes a series of different polymerization methods such as precipitation, suspension, microsuspension, emulsion, miniemulsion, microemulsion, dispersion, reverse micelle and inverse polymerization [4].
2. *Solution Polymerization of aniline*: a typical route for PAni synthesis involving polymerization of aniline in chloroform as well as the electropolymerization in acrylonitrile [4, 25, 26].
3. *Interfacial polymerization of aniline*: the polymerization takes place on the interface of two immiscible solvents. For the case of PAni two immiscible solvents usually used are water and chloroform. The final product is isolated by centrifugation [4, 27].
4. *Seeding polymerization of aniline*: a typical template approach where foreign materials are used as seeds and the polymerization reaction is carried out in their presence. PAni nanofibers have been reported as the product of the seeding polymerization of aniline [28].
5. *Metathesis polymerization of aniline*: where the aniline monomer is not required [4]. Instead, p–dichlorobenzene is heated at 220 °C for 12 hours in an organic medium (*benzene*) the resulting in the formation of PAni according to a metathesis reaction [29].
6. *Self-assembling polymerization*: PAni copolymer film can be produced from aniline and m–aminobenzene sulfonic acid using APS as an oxidant on indium tin oxide (*ITO*) as a substrate [30]. PAni films can be directly grown on the polymeric film substrate by polymerization of the aniline monomer in the vapour phase [4].
7. *Sonochemical synthesis of PAni*: taking place under ultrasonic irradiation, with an acidic ammonium persulfate (*APS*) solution added dropwise into an acidic aniline solution. This technique has also been used for the preparation of PAni nanofibers [31].
8. *Template synthesis of PAni*: one of the most popular simple synthesis techniques for nanostructure formation. The synthesis takes place into the pores of a template. After polymerization, the template is dissolved and the produced material is obtained. This method is used in both chemical and electrochemical synthesis routes and is referred in the literature as a successful route to produce conducting polymer nanotubes [32, 33].

9. *Plasma polymerization of aniline*: the monomer, or a dispersion of monomer–droplets, is injected directly in the stream of plasma induced by a DC–glow discharge reactor. In a second step, the reactor is appropriately reconfigured to give a deposition of polymer particles at mild conditions [4]. With this method different PAni nanocomposites can be prepared [34].
10. *Photo–induced polymerization of aniline*: involves the photo–excitation of aniline monomer to obtain PAni. PAni has been synthesized by using a Nd:YAG laser to irradiate an Au electrode in a solution, obtaining aniline under an applied external bias [35]. Only photons in the UV or visible region are used to induce polymerization of aniline in an aqueous solution of transition metal salts [4, 36].
11. *Enzymatic synthesis of PAni*: has attracted the attention of the scientific community because it is carried out under milder conditions. Horseradish peroxidase (*HRP*) and soybean peroxidase (*SBP*) are oxidoreductase enzymes capable of oxidizing aromatic amines in the presence of hydrogen peroxide [37, 38]. These enzymes can be derived from non–contaminant renewable sources and have high reaction selectivity to aromatic compounds converting oxidation by–products to water [4, 39].

2.2. Properties

The Intrinsically Conducting Polymers (*ICPs*) such as PAni are inherently conducting by nature due to the presence of a π electron system in their structure. ICPs also exhibit low energy optical transition, low ionization potential and high electron affinity [4, 40]. The intrinsic conductivity of the ICPs can be increased through oxidation–reduction reactions as well as by doping with a suitable dopant. This way ICPs are able to reach conductivity values comparable to metals [4, 41]. The conductivity of five conductive polymers in relation to liquid mercury and copper is presented in Figure 2(*a*) whereas the conductivity of seven doped ICPs is presented in Figure 2(*b*).

Few studies provide evidence for pure PAni [42–45], and only recently Valentona and Stejskal, 2010 published a paper reporting some of the physical and mechanical properties of PAni salt and base compared to polystyrene [46].

Table 1. Young's modulus of the polystyrene and PAni [52]

Sample	Shape	Deformation	E′, GPa
Polystyrene	Rectagular	Tension	2.1 ± 0.2
Polystyrene	Rectagular	Bending	2.3 ± 0.1
Polystyrene	Pellet	Bending	1.8 ± 0.1
Compressed polystyrene powder	Pellet	Bending	0.80 ± 0.02
PAni salt	Pellet	Bending	0.9 ± 0.2
Polyanyline base	Pellet	Bending	1.3 ± 0.2

For PAni, electrical conductivity is considered to be its most significant property and the main reason for its popularity. However information concerning other characteristics such as mechanical properties is relatively scarce in the literature with the majority of those few studies concerning blends containing PAni or other composite materials [47-52].

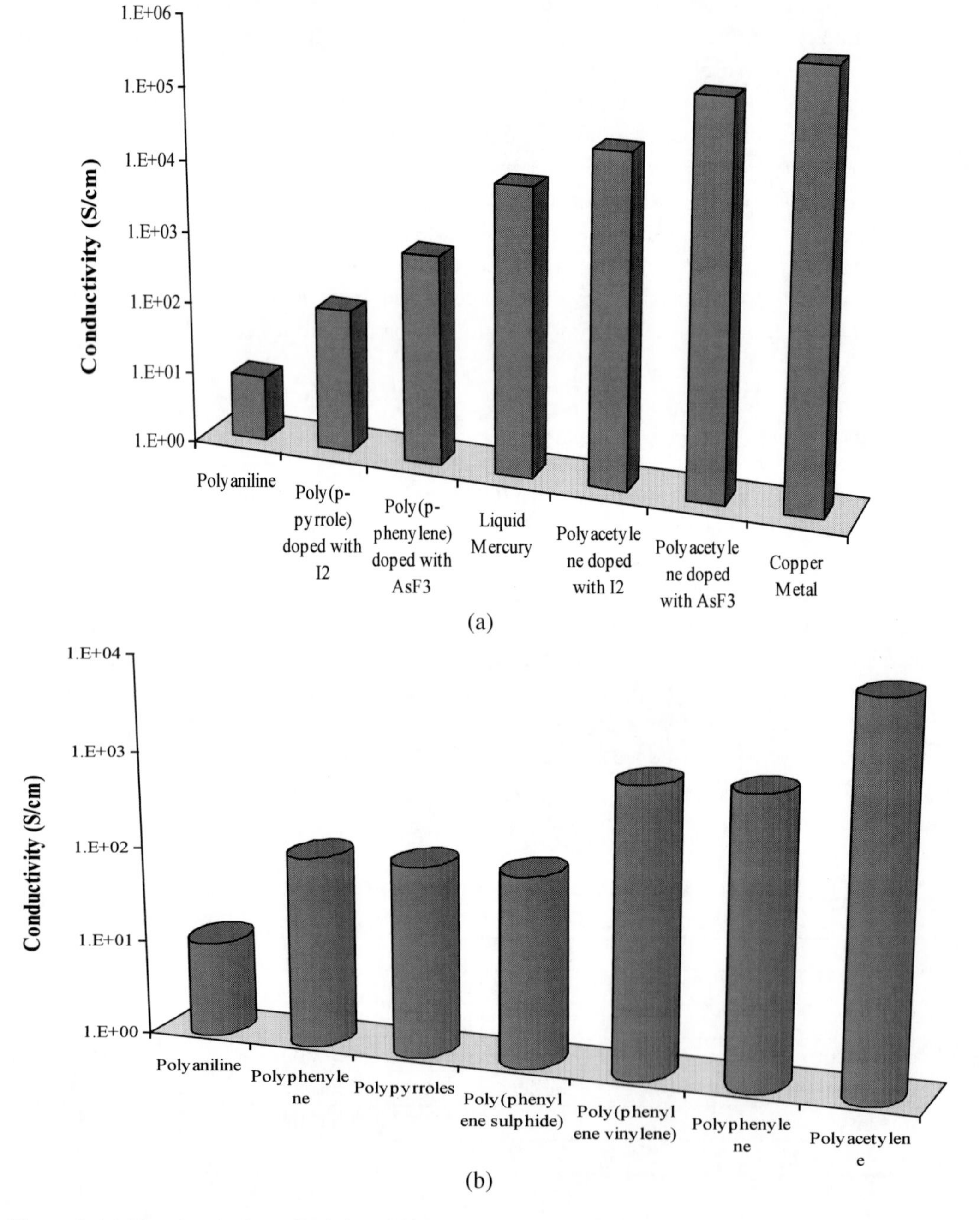

Figure 2. (a) The electrical conductivity of ICPs relative to metal copper and liquid mercury and (b) the conductivity of a number of doped ICPs.

4. PAni for Membrane Technology Applications

The golden age of membrane technology (1960–1980) began with the invention of the first asymmetric integrally skinned cellulose acetate RO membrane by Loeb and Sourirajan [53]. This development stimulated both commercial and academic interest, first in desalination by reverse osmosis, and later in other membrane applications and processes. During this period, significant progress was made in virtually all membrane technology areas: applications, research tools, membrane preparation processes, chemical and physical structures, configurations and packaging. Asymmetric membranes can be produced via the phase inversion technique, according to the spinodal decomposition [54] process a method that can be adapted to produce membranes in the form of thin films, tubes or fibers. Both organic and inorganic membranes have gained great attention for their use in the chemical and petrochemical technology and are nowadays used in a wide range of applications. However, advantages of inorganic membranes such as their high thermal and chemical stability as well as their resistance to corrosive liquids and gases can be negated by their high manufacturing and operation cost, hence research has recently been focusing on the development of improved polymeric membranes.

Polymeric membranes are already being used in various industrial applications such as reverse osmosis (*RO*) [55], gas separation [56], microfiltration [57], desalination [58], water purification [59], heavy metal removal [60], hemodialysis [61] etc. Especially for gas separation, the permeability and selectivity coefficients are the main properties which determine the applicability of the materials. Porosity, temperature resistance characteristics, swelling phenomena and electrical conductivity are properties which, in many cases, must be improved in order to produce membranes that can be good candidates for separation applications. Membrane separation processes can be divided into several categories according to the membranes pore sizes, i.e., conventional filtration (10–100 μm), microfiltration (0.1–10 μm), ultrafiltration (50–1000 Å), and the processes of reverse osmosis, gas separation, and pervaporation (<50 Å) [62–64]. Polymers such as cellulose acetate [65], polyimides [66], polysulfone [67], polycarbonates [68] and poly(phenylene oxide) [69] are the most referred materials in the membrane gas separation field. On the other hand new materials have recently been reported in the literature in the area of membrane technology, and PAni can be considered as one of the most promising.

Specifically, a large number of studies in the last fifteen years have evaluated PAni for its potential use in vapour, liquid and gas separation applications. In a very interesting work Ball et al., 1999 studied the temperature dependency of permeability and permselectivity (*in a range of 20°C to 100°C*) of water, acetic acid and 50 % acetic acid / 50% water solutions in undoped and doped PAni films [70]. One year later the same group published another work on the pervaporation process of carboxylic acids and water in undoped and doped PAni membranes [71]. They concluded that the fully HCl–doped PAni exhibits extremely high Separation factors, α, of >1300 for 50 wt % acetic acid / 50 wt % water contrary to the poor selectivity factor of the corresponding undoped films. Lee et al., 1999 studied the pervaporation of water/isopropanol mixtures in PAni doped with poly(acrylic acid) (*PAAc*) membranes [72]. They found that dopant content above 30 wt % PAAc in the blended membrane, resulted in water concentration of ca. 100 wt % in the permeate, with permeate flux about 300 g/m^2/h, measured at 80 ^{o}C. The permeate flux decreased with increasing isopropanol (*IPA*) feed concentration while water concentration in the permeate was maintained over 95 wt %. In 2000 Ball et al. published a new paper reporting permeability (*P*) values, diffusion coefficients (*D*) and time lags (Θ) for water, alcohols and carboxylic acids passed through undoped and HCl–doped PAni membranes [73]. The main conclusions of this study were that permeabilities through undoped PAni are stable and linear with time, but tend to be low. HCl–doped PAni membranes exhibit much shorter time lags, and much larger diffusion coefficients and permeabilities for several of the feed solvents used. However, larger molecules, such as acetic acid, propionic acid and isopropanol, do not yield measurable permeabilities through doped PAni. These differences in permeabilities between small and large molecules indicate that doped PAni membranes could be useful in separating solvent pairs such as water and acetic acid. On the other hand, undoped PAni is the most selective membrane, since for water/ethanol feeds the composition of the permeate through undoped PAni was 99.9% water as determined by ^{1}H–NMR. This further supports the idea that undoped PAni is essentially impermeable to ethanol. In a slightly different study, Wen and Kocherginsky, 2000 [74] conducted direct measurements of H^+ transport through a PAni membrane separating two aqueous solutions demonstrating that Fick's law could not describe the H^+ transport through the PAni membrane. It was found that the transport of H^+ from acidic to neutral solution through a PAni membrane (*~30 μm thickness*) takes a significant time to reach its maximum steady state value as time lag was 400 min at pH=2 in the donor solution. It was also concluded that the HC transport through the PAni membrane under a concentration gradient of HCL could not be explained by simple direct transport through the membrane pores. The large value of the time lag was attributed to slow structural changes in the film during the HC induced doping and absorption of acid in the polymer during the doping process. In the case of gases, Illing et al., 2001 studied the ideal gas separation of H_2/CO_2, O_2/N_2, H_2/N_2, CO_2/O_2, CO_2/N_2 and N_2/O_2 gas mixtures through dense PAni membranes [75]. The highest selectivities $a_{(A/B)}$ obtained were: 7.6 for the H_2/CO_2 gas pair for a de–doped membrane and 6 for O_2/N_2, 10 for H_2/CO_2, 200 for H_2/N_2 in the case of re–doped membranes. PAni membranes can outperform conventional membrane materials for gas separation purposes by over 600% in the case of hydrogen containing gas pairs. Equivalent separation factors are however observed for the CO_2/O_2, CO_2 /N_2 and O_2/N_2 gas pairs. Naidu et al. 2005 examined the efficiency of pervaporation separation of water and isopropanol mixtures through poly(vinyl alcohol)/PAni blend membranes [76]. PAni nanoparticles, with sizes ranging from 30 to 100 nm, blended into a PVA matrix to obtain a suspension of colloidal PAni particles in PVA. This reaction mixture was then poured onto a clean glass plate to cast the membranes. The dried PVA–PAni nanocomposite membranes were cross–linked with glutaraldehyde (*GA*) by dipping them in 200 ml aqueous acetone mixture containing 1ml of GA and 1ml of HCl for 12 h. Three membranes with different PAni powder compositions were studied, with the 40:60 PAni/PVA membrane exhibiting a selectivity value of 564, very high when compared to a value of 77 observed for pure PVA membranes even though flux was considerably lower. Furthermore Illing et al., 2005 produced and studied a defect–free polyaniuline thin film of a thickness ranging from 0.8 and 10.5 μm spread on a polyvinylidene difluoride (*PVDF*) support [77]. The CO_2 permeability was higher in all PAni ranges of thickness for both O_2 and N_2. Despite having a thickness of less than 1 μm the dense supported PAni maintained its previously reported intrinsic selectivity for different gases. A similar work was published one year later by Gupta and co–

workers reporting a novel method for reproducible production of defect free self–supported PAni films with thicknesses between 2 and 6 μm, and PAni nano–membranes with a selective layer thickness of 300 nm supported on a porous polyvinylidene difluoride (*PVDF*) substrate [78]. For dense PAni membranes, ideal separation factors for all tested gas pairs were independent of the membrane thickness. The ideal selectivity factors, for self–supported undoped PAni films, for H_2/N_2, H_2/O_2, H_2/CO_2, CO_2/O_2, CO_2/N_2 and O_2/N_2 reached 348, 69.5, 8.6, 8.1, 40.4 and 7.1 respectively. The reported values were considerably higher than those obtained by other researchers while the ideal separation factors for PAni nano–membranes were similar to those obtained for self–supported PAni films. In 2008, Fan et al. prepared composite PAni/polysulfone membranes for aqueous separation processes [79] and reported that the PAni/PS nanocomposite membrane exhibited good hydrophilicity, which resulted in great enhancement in permeability. Specifically, during the filtration of a bovine serum albumin (*BSA*) solution, the nanocomposite membranes antifouling performance was much higher than the PS membranes because of the hydrophilicity and steric hindrance effects of its nanofiber layer. Concerning the polyanilines antifouling performance, one year later Loh et al., 2009 evaluated the effectiveness of crosslinked PAni membranes for organic solvents removal [80]. The studied membranes were prepared by phase inversion and then crosslinked using two different chemical crosslinkers, *a,a′-dichloro-p-xylene* and *glutaraldehyde* and exhibited good stability in various organic solvents including acetone, methanol, ethyl acetate, tetrahydrofuran and N,Ndimethyl formamide. They also gave stable permeate fluxes and good separation performances in the nanofiltration range during experiments carried out in acetone and dimethyl formamide showing high stability even at elevated operating temperatures up to 70°C and possibly higher. Following this work the same group examined in 2010 skinned asymmetric PAni nanofiltration membranes in spiral –wound modules for organic solvent nanofiltration [81]. The resulting membranes and modules, were also stable in various organic solvents including acetone, tetrahydrofuran and N,N–dimethyl formamide and gave stable permeate fluxes and good separation performances in the nanofiltration range, in acetone, tetrahydrofuran and dimethyl formamide. Molecular weight cut–off (*MWCO*) of the PAni membranes in different solvents was found to lie between 150 and 300 g mol−1 at 30 ◦C.

Gas separation performance of PAni membranes has also been studied in the hollow fiber configuration. Recently, Hasbullaha et al., 2011 investigated the effects of air–gap distance on nascent fiber morphology produced via the dry–jet wet spinning process and studied their gas permeation and mechanical properties. They found that the spin–line stresses resulted in the molecular orientation of the polymer which improved the mechanical properties of the resulting membranes, while due to a synergistic effect the gas separation performance of the PAni hollow fiber membranes improved as well. The PAni based hollow fibers membranes showed a selectivity of 10.2 for O_2/N_2, 105.6 for H_2/N_2 and 7.9 for H_2/CO_2 with H_2 and O_2 permeance of about 5.0 and 0.49×10^{-6} cm^3 (STP)/cm^2 s cm Hg, respectively [82]. In a follow up in 2012 by the same group the gas transport properties of asymmetric hollow fibre membranes based on ring–substituted PAni was studied [83]. This time, a more soluble PAni derivative, i.e. poly(o–anisidine), (*POAn*), was synthesized by using an ortho substituted aniline monomer. The POAn membranes showed selectivities of O_2/N_2, CO_2/N_2, H_2/N_2 and H_2/CO_2 of about 13, 89, 370 and 4.2, respectively with permeation rates of H_2, CO_2 and O2 about 10.4, 2.5 and 0.37×10^{-6} cm^3 (STP)/cm^2 s cmHg, respectively. Presence of a methoxy group in POAn showed substantial enhancement in gas permeation properties compared to previously developed PAni membranes. The H_2/N_2 and CO_2/N_2 selectivities were improved by 3 and 7 times, respectively as compared to PAni membranes.

4.1. Hollow Fiber Membranes, a Preferential Membrane Configuration

An important advantage of membranes in the hollow fiber form is that they enable the production of compact modules with very high membrane surface areas. However, this advantage is offset by the lower fluxes of hollow fiber membranes compared to flat–sheet membranes made from the same materials [57]. Nonetheless, the development of hollow fiber membranes by Mahon and the Dow Chemical group in 1966 [84] and their later commercialization by Dow, Monsanto, Du Pont, and other large manufacturers represents one of the major events in membrane technology. A good review of the early development of hollow fiber membranes is given by Baum et al., 1976 [55]. Reviews of more recent developments are given by Moch, 1995 [85] and McKelvey et al., 1997 [86].

Recent advances in membrane materials research have resulted in various polymers with potential applicability in gas separation applications [87, 88]. The performance of membrane units in the separation of gaseous mixtures is highly dependent on the intrinsic physicochemical characteristics of the utilized polymeric materials. The most important properties taken into account when selecting a polymeric membrane material are: (i) its gas permeability and selectivity coefficients, (ii) its mechanical properties, (iii) its glass transition or melting temperature, (iv) its critical pressure of plasticization, (v) the material availability and processability and (vi) its cost. In general, polymeric materials used industrially for the preparation of gas separation membranes, do not absolutely and simultaneously meet all of these criteria. For example, highly permeable polymers, normally exhibit moderate to low selectivity values [89]. Asymmetric hollow fibers are attractive for membrane–based gas separations, since they provide high active surface area–to–volume ratios, low resistance to gas flow, and the ability to withstand high trans–membrane pressure drops. Each of these factors could contribute to greater gas productivity [90, 91] making this type of membranes ideal candidates for use in popular gas separation applications such as natural gas purification or nitrogen production from air via several purification steps.

Hollow fibers are mainly prepared by dry/wet phase inversion spinning [92, 93] (Figure 3). Dopes consisting of polymer blends and a solvent (*or mixture of solvents*) are mixed in a thermostated vessel under stirring in order to achieve a homogeneous solution and are then passed through an appropriate filter to remove impurities existing in the raw polymers. After filtering, the dopes are allowed to degas inside a second vessel. After degassing, the polymer solution and bore fluid –typically a degassed solvent– are simultaneously pumped through a tube –in–orifice spinneret and the resulting extruded fibers pass trough an air gap varying from 0 to ~90 cm before entering a coagulation bath filled with a non–solvent.

All parts of the apparatus such as the mixing and degassing vessels, the spinneret and the coagulation baths are usually thermostated at the appropriate temperature depending on the polymer, in order to facilitate the polymer solution flow and optimise the coagulation process. The final morphology of such an asymmetric polymeric hollow fiber (*Polyimide BTDA–TDI/MDI (P84)*) is shown on Figure 4.

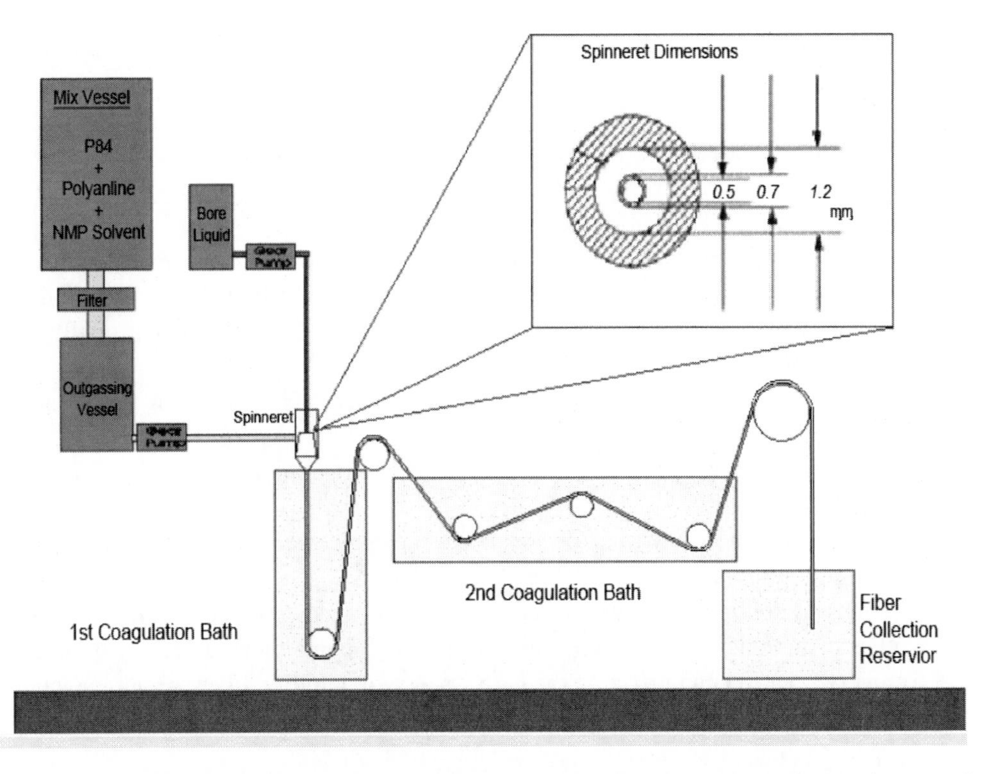

Figure 3. Hollow fiber spinning set up: (1) spinning dope tank; (2) bore liquid vessel; (3) spinneret; (4) air gap; (5) coagulation bath; (6) fiber guiding wheel; (7) pulling wheel; (8) spinning line; (9) fiber collecting reservoir.

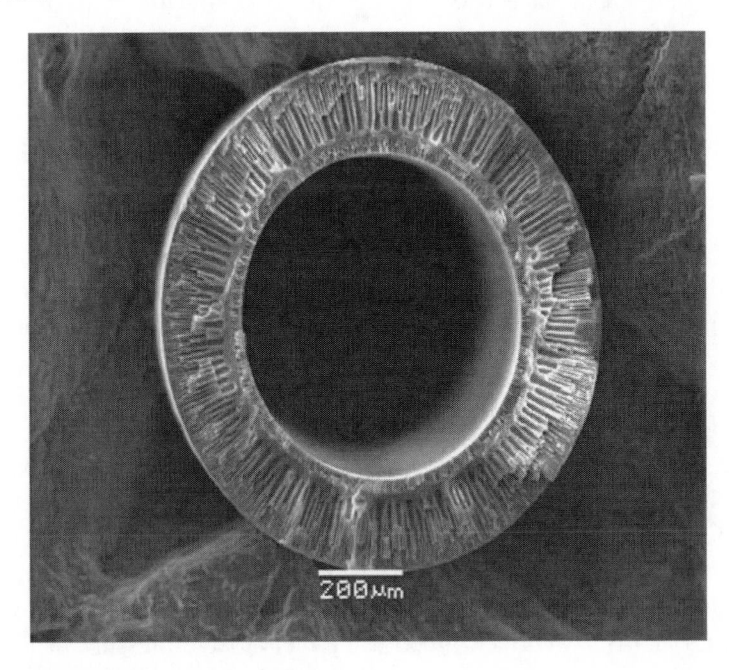

Figure 4. SEM picture of PI 100%, cross section.

Advantages and Disadvantages of Hollow Fiber Membranes

Hollow fibers are the most popular membranes used in the industry, due to the following beneficial features:

a) large surface per unit volume. Hollow fibers exhibit large membrane surface per module volume. Hence, the size of a hollow fiber module is smaller than other type of membranes, but can offer increased performance,
b) modest energy requirement: In hollow fiber filtration, no phase change and consequently no latent heat is involved. This gives hollow fiber membranes the potential to replace some unit operations which consume heat, such as distillation or evaporation columns,
c) flexibility: hollow fiber is a flexible membrane as it can carry out filtration either "inside–out" or "outside–in" and
d) low operation cost compared to other types of unit operations.

However, hollow fibers have also disadvantages posing constraints in their applications:

a) Membrane fouling: Hollow fibers are more prone to membrane fouling compared to other membranes due to their configuration and contaminated feed will have very quickly a detrimental effect on the performance of a hollow fiber membrane module.
b) Expensive to produce: Hollow Fibers, because of their production method, are more expensive than other membranes available on the market.
c) Lack of research: as hollow fiber membrane production is a relatively new technology,
d) Physical and Chemical constraints: Polymeric hollow fibers cannot be used with corrosive substances and in high temperature conditions.

4.2. Polyimide/PAni Membranes: Evaluation Study in Gas Separation Performance

The use of PAni in combination with polyimides for the preparation of membranes for gas separation has been reported previously [94] and the results indicate that such blends exhibit greater gas selectivity and increased permeability for all gases compared to the pure polyimide.

For the purpose of their evaluation as to their gas separation properties, highly permeable blend polyimide/PAni membranes were prepared, and their morphology and thermal properties were investigated by means of SEM and TGA. FTIR spectroscopy was used for the study of their chemical structure and permeability and selectivity measurements in different gases (He, H_2, CH_4, CO_2, O_2 and N_2) were performed in order to evaluate the performance of the membrane in gas separation applications [95].

Single gas permeation measurements: Permeation measurements of various gases (He, H_2, CH_4, CO_2, O_2 and N_2) were performed using the variable pressure method in a high–pressure (70 bar) stainless steel permeation rig. Fig 5 illustrates the design of the custom–built apparatus used for the permeation experiments [95], consisting of 3 different main parts: a) The high pressure side, b) the low pressure side and c) the permeation cell.

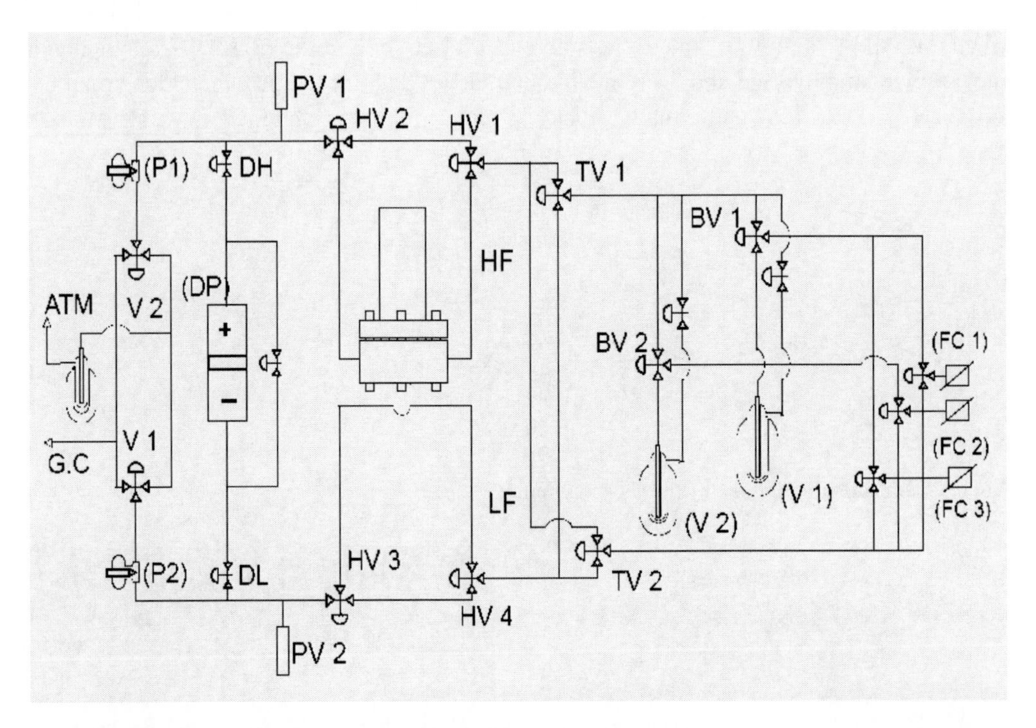

Figure 5. From high vacuum up to 70 bar permeability/selectivity apparatus.

The studied membranes, in both flat and hollow fiber form factor, were thoroughly outgassed for at least 24 hours at 10^{-6} mbar at 298 K. Sealing and loading of the membranes into the permeation cell was carried out in a glove box under inert (He) atmosphere. Gas was introduced into the high–pressure section of the rig, while the low–pressure side remained isolated under vacuum. For the permeance experiments the pressure increase in the low–pressure side of the rig was continuously monitored by means of an accurate differential pressure transducer.

Permeance, *Pe*, (GPU*) was determined using the following equation:

$$Pe = 6 \cdot 10^4 \cdot \frac{V_{low} \cdot (\delta P_{low} / \delta t)}{\Delta P \cdot U \cdot T_{\exp}} \quad (GPU)$$

where, V_{low} is the collection volume (cm^3), $\delta P_{low}/\delta t$ the pressure increase rate (*mbar/min*) in the collection volume, U the total active area of membrane sample (cm^2), ΔP (*mbar*) the pressure head and T_{exp} (*K*) the experimental temperature. During the experiments special care was taken in order to maintain the pressure boundary conditions constant.

$$^{*}1Barrer = 1 \cdot 10^{-10} \frac{cm^3(STP) \cdot cm}{cm^2 \cdot s \cdot chHg} \text{ and } 1GPU = 1 \cdot 10^{-6} \frac{cm^3(STP)}{cm^2 \cdot s \cdot chHg}$$

4.2.1. Materials, Synthesis of PAni and Preparation of Flat and Hollow Fiber Membranes

All reagents were of analytical grade (*Sigma Aldrich*) and were used without further purification unless otherwise stated. Commercial P84 co–polyimide (3,3´4,4´–benzophenone tetracarboxylic dianhydride and 80 % methylphenylene–diamine + 20 % methylene diamine), [P84 (BTDA\–TDI/MDI] a thermally stable co–polyimide was obtained from Lenzing AG (Figure 6).

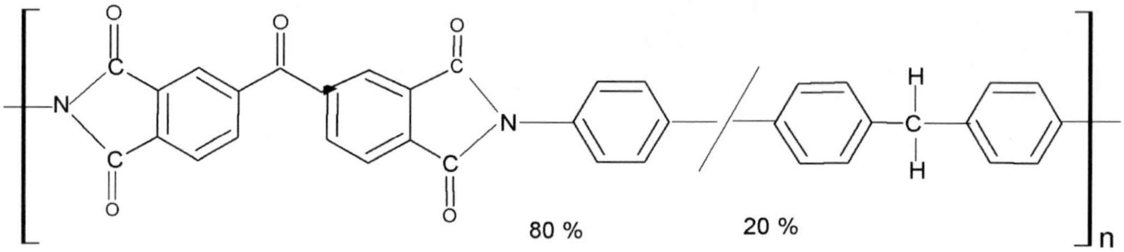

Figure 6. Chemical structure of PI (P84) co-polyimide.

(a) (b) (c) (d)

Figure 7. PAni powder synthesized by aniline, $NH_4S_2O_8$ 1,5M HCl aqueous solution: (a) photo of the PAni powder and (b), (c) and (d) the SEM pictures of the same material.

PAni synthesis route: The aniline (*monomer*) used for the preparation of PAni (*Merck, pro–analysis*) was purified by distillation before use. Thirty (30) g of $NH_4S_2O_8$ were dissolved in 326 ml of 1,5M HCl aqueous solution [Solution I] which was added slowly while stirring into a 313 ml distilled water containing 26 ml of distilled aniline and a 13 ml of concentrated HCl solution (35%), previously cooled to 0° using a temperature regulated bath. After all oxidant was added over a period of 3.5 hours, the reaction mixture was left stirring at 0° for thirty additional minutes.

The precipitated PAni was filtered and washed with warm distilled water (3.4 l) until the washing liquids were completely colorless (PH ≈ 5.5 – 6.0), and in order to remove oligomers and other organic by–products it was subjected to further washing, first with methanol (1.3 l) and finally with ethyl–ether (1.0 l) until the washing solution was colorless. The precipitated product was dried in dynamic vacuum for 48 hours at room temperature until constant mass was reached [96] and finally was ground into a fine powder (Figure 7).

In the FTIR spectra (Figure 8), the band observed at 1721 cm^{-1} can be attributed to C=O symmetrical stretching of imide groups, whereas the bands at 1362 cm^{-1} and 721 cm^{-1} are associated with the C–H vibrational modes, present in the polyimide. The band occurring at 1511 cm^{-1} can be attributed to the vibration of aromatic groups C–C, present in both polyimide and PAni.

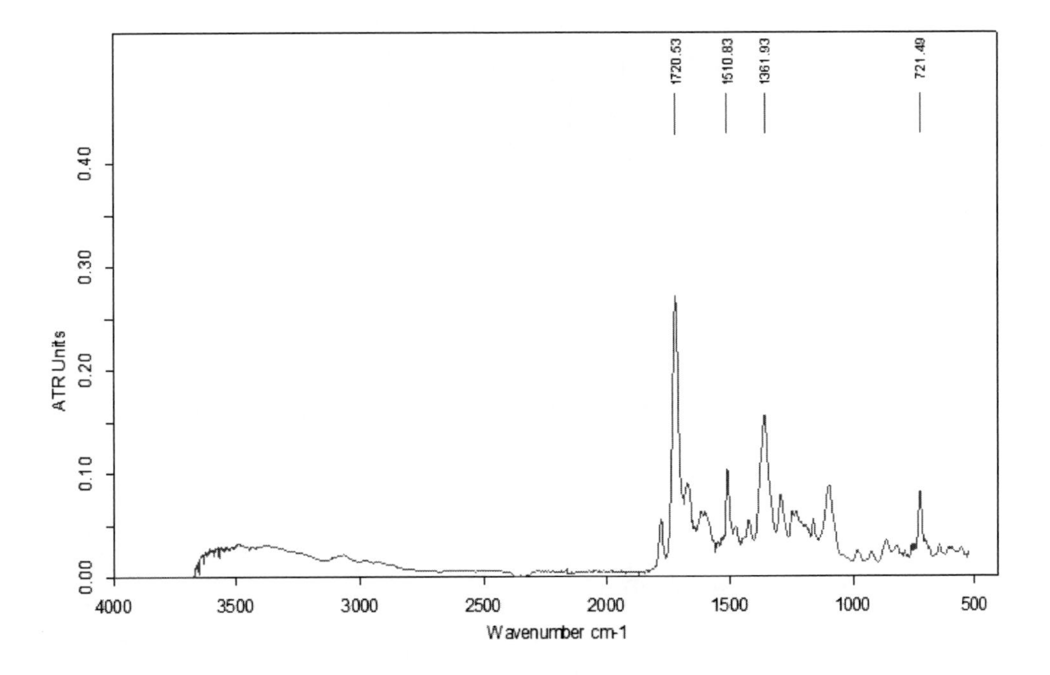

Figure 8. FTIR spectrum of PI/PAni hollow fiber membrane.

Flat Membranes: Using the produced PAni powder and commercial co–polyimide P84, three different membranes were prepared in the form of films. Prior to its use, PAni was treated with an ammonium hydroxide (*NH_4OH*) solution (32 %). In all cases n–methyl–2 pyrolidone (*NMP*) was used as organic solvent common for both polymers.

(a) (b)

(c)

Figure 9. Pictures of (a) PAni, (b) PI and (c) PAni/PI films.

Three types of films were obtained: pure PAni film (1.5 % w.w. PAni/NMP), pure polyimide film (1.5 % w.w. PI/NMP) and blended PAni/PI film (20 % w.w. in NMP with a ratio of PAni/PI 1/3). Figure 9 shows the pictures of the three flat films.

Table 2. Permeability/permselectivity values for the prepared PAni, PI and PAni/PI flat membranes

Membrane	Gas Permeability (*Barrer*) at 298 K					
	He	H_2	CH_4	CO_2	O_2	N_2
PAni	11.5	10.6	0.61	1.59	0.17	0.03
PI	25	17.5	0.2	7.15	1.87	0.28
PAni/PI	65.2	129	76	50.5	42	46.9
Membrane	Gas Permselectivities at 298 K					
	H_2/CH_4	CH_4/CO_2	He/N_2	H_2/N_2	H_2/CO_2	CO_2/N_2
PAni	17.38	0.38	383.33	353.33	6.67	53.00
PI	87.50	0.03	89.29	62.50	2.45	25.54
PAni/PI	1.70	1.50	1.39	2.75	2.55	1.08

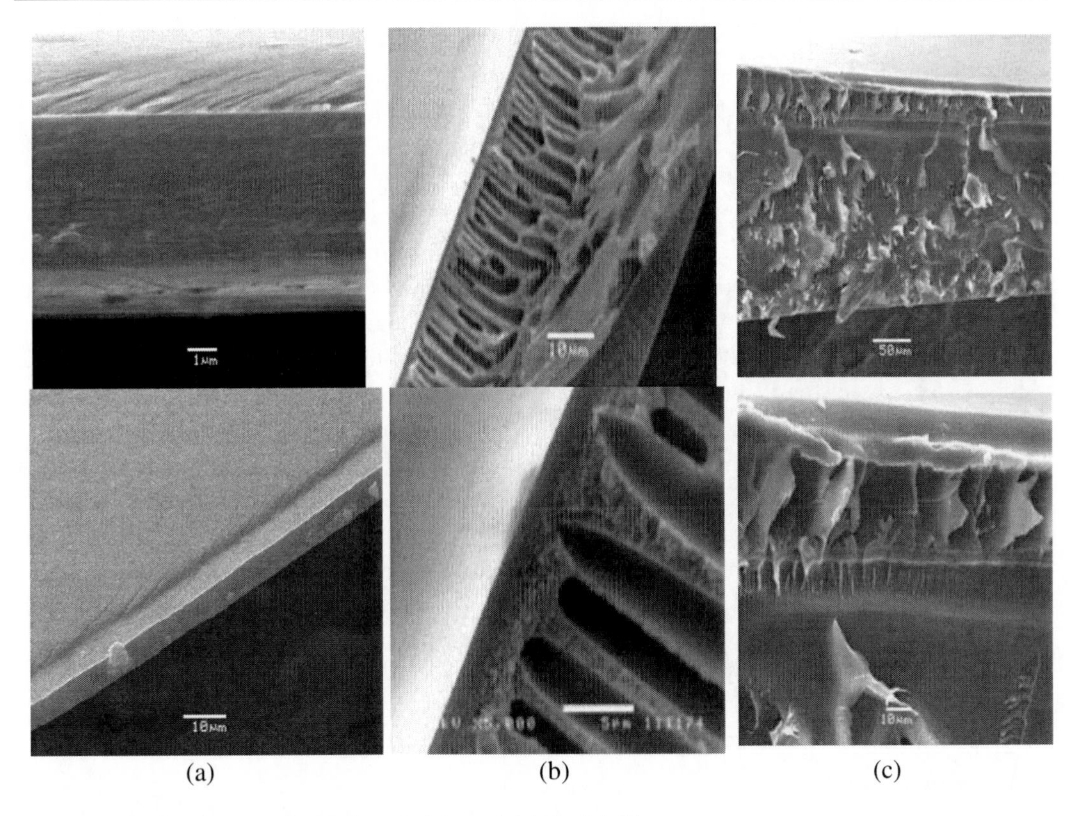

Figure 10. SEM pictures of (a) PAni, (b) PI and (c) PAni/PI films.

Although macroscopically the three films seem similar, the differences in the microscopic level are significant. As depicted in the SEM photographs (Figure 10) the thickness of the films is increasing in the order of $PANI < PI \cong PANI / PI$. In the case of PAni films the structure is homogeneous and more "dense" with a thickness of about 6.1 μm. Pure polyimide films, as shown in Figure 10(*b*), exhibit a sponge like porous structure of 47.5 μm thickness. Finally, the thickness of the PI/PAni film was about 250 μm. The film is composed of two discernable phases, a bottom one consisting of flake–like formations and a top one with oriented large voids. This structure has also been reported in the literature for composite PAni/PVDF flat membranes [77].

The gas permeability values of He, H_2, CH_4, O_2 and N_2 for PAni, PI and PI/PAni films as well as the ideal selectivities for H_2/CH_4, CH_4/CO_2, He/N_2, H_2/N_2, H_2/CO_2 and N_2/CO_2 gas couples are shown in table 2. PAni/PI film was the most permeable with a permeability of 65.2 Barrer for He at 298K while in the cases of PI and PAni films the He permeabilities were 11.5 and 25 respectively. These differences can be attributed to the different packing of the PAni/PAni, polyimide/polyimide and PAni/polyimide polymer chains. The higher ideal selectivity values for PAni membranes were observed for the He/N_2 (383) and H_2/N_2 (353) mixtures, while PI membranes performed better in the separation of H_2/CH_4 (87.5) and He/N2 (89.29) and PAni/PI membranes for H_2/N_2 (2.75) and H_2/CO_2 (2.55). The PAni/PI membranes exhibited the lowest selectivities but at the same time permeability values were two orders of magnitude higher rendering the PAni/PI membranes good candidates for H_2/N_2 and H_2/CO_2 separations.

Hollow Fiber Membranes: Two types of polymeric hollow fibers, namely the PI (*P84*) and composite PI/PAni hollow fibers, were prepared by dry/wet phase inversion process in a spinning set–up as described previously (Figure 3). Specifically, polymer solutions in NMP were prepared after the polymers were dried overnight at 120° under vacuum. For the PI hollow fiber preparation the spinning dope consisted only of PI in NMP, while for the PI/PAni fiber preparation a polyimide/PAni blend in NMP was prepared at a ratio 12.5 / 4.2 / 83.3 % w/w co–polyimide BTDA–TDI/MDI (*P84*) PI/PAni/n–methyl–2–pyrolidone (*NMP*). The spinning dope was mixed overnight at 50° in a stainless 3L vessel in order to obtain a homogeneous solution which was subsequently passed through a metal 25 μm sieve diameter filter to remove impurities. After filtering, the dopes were allowed to degass inside a second stainless steel vessel for 12 hours. The bore liquid was a degassed 1:3 solution of NMP and deionized water. The polymer solution and bore fluid were simultaneously pumped through a tube–in–orifice spinneret using gear pumps. The i.d. (*inner diameter*) of the spinneret was 700 μm and the o.d. (*outer diameter*) 1200 μm. Both vessels as well as the spinneret were thermostated at 50°. The extruded fibers entered the coagulation bath, which was filled with tap water at room temperature. The produced fibers were oriented by means of two guiding wheels and pulled by a third wheel into a collecting reservoir. In order to remove residual NMP, the produced fibers were left in tap water overnight and then solvent exchanged in plastic containers with ethanol for 6 hr.

4.2.2. Morphology Study, Thermal and Permeability Properties of PI and PI/PAni Hollow Fiber Membranes

The morphology of the PI (*P84, polyimide*) and PI/PAni hollow fibers was studied using a JEOL JSM–5600 Scanning Electron Microscope (*SEM*), after gold sputtering. The morphological characteristics of the asymmetric polymeric PI hollow fiber and PI/PAni blend hollow fiber membranes are shown in Figure 11.

The macro–voids in the hollow fibers, seen in Figure 11(*a–c*) and 11(*a′–c′*), were a result of the phase inversion between the polymer dope and coagulation liquid during the wet spinning process. The dense layer appearing towards the internal surface of the hollow fiber membranes was created due to the different solvent NMP/non–solvent (H_2O) ratios between the bore liquid and the coagulation solution. The main separating layer is the skin layer at the outer surface of the membrane for both PI and PI/PAni membranes. The effect of the addition of PAni on the skin layer of the PI/PAni membrane can be seen in Figure 11c′ where it appears "weathered" accounting for the dramatic increase in all permeation values for the gases under study (Table 3). On the other hand the effects of the air gap on the structural and permeation characteristics have also been studied, in the case of BTDA–TDI/MDI (P84) co–polyimide hollow fiber membranes [97]. Specifically, three polymeric hollow fiber membranes (AG0-AG6-AG10) are presented (Figure 12) prepared at the same experimental conditions (polymer concentration, bore fluid composition) but different air gaps (Table 3).

All hollow fibers membranes exhibit a cylindrical symmetry with dense separating layers without defects or cracks. Apart from the inner and outer separating layers a third continuous "dense" layer can be observed between the inner and outer wall. The formation and characteristics of these separating layers depend not only on bore liquid composition, but also on the air gap. The existence of elongated voids (finger–like pores) is the common trend for all studied samples.

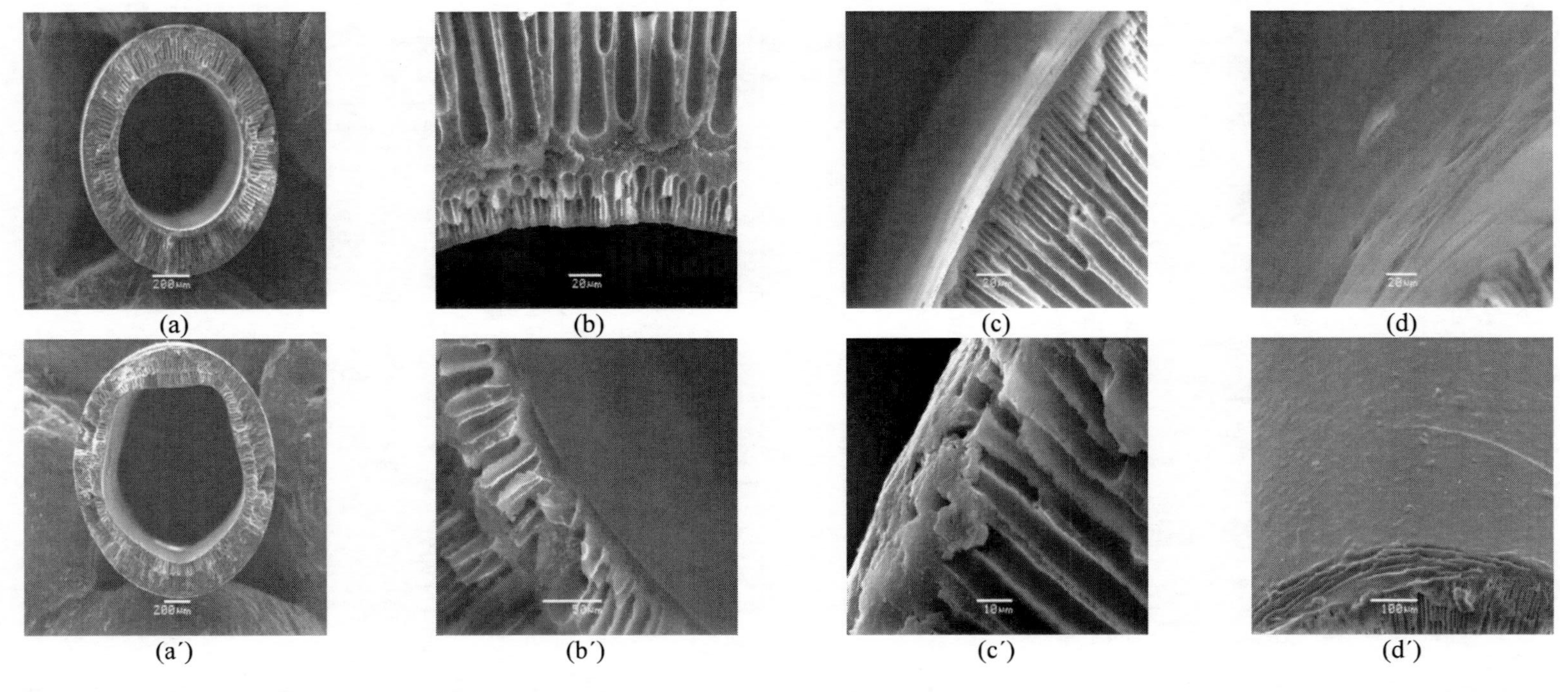

Figure 11. SEM pictures (a) Cross section, (b) section at bore side (c) outer side – skin layer and substrate and (d) outer side of PI hollow fiber membrane and (a′) Cross section, (b′) section at bore side (c′) section at outer side and (d′) outer side of P8I/PAni hollow fiber membrane.

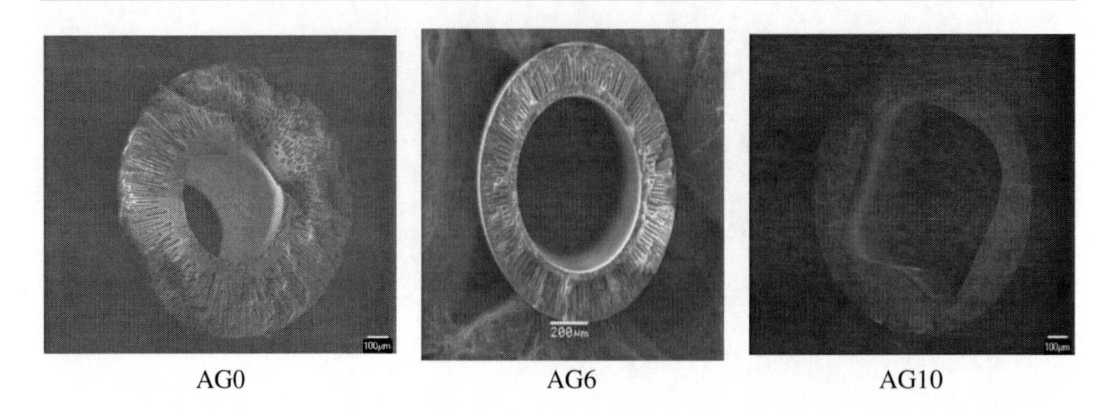

AG0 AG6 AG10

It is evident that air gap mainly affects the wall thickness characteristics of the prepared hollow fiber membranes. Specifically, the AG0 membrane prepared at zero air gap exhibits the highest wall thickness reaching 300μm compared to the AG6 (≈200μm) while at 10cm air gap AG10 exhibits irregular wall thickness.

In general, gravity and elongation stress, as well the surface tension, are the main factors affecting fiber dimensions. As other researchers have observed, increasing the air gap results in smaller wall thicknesses in hollow fiber membranes [98–100].

Table 3. Preparation conditions of six (S1–S6) BTDA-TDI/MDI (P84) co-polyimide hollow fiber membranes

	S6	**S1**	**S4**
Bore fluid composition (NMP/H_2O)	70/30	70/30	70/30
Air Gap (cm)	0	6	10
He Permeance (GPU) at 373 K	33.4	48.5	85.8

For membranes produced via the wet spinning technique air gap mainly affects wall thickness but affects the thickness of the separating layers as well. As shown in Table 3, increasing the air gap has a positive effect in gas permeance properties. Specifically, when comparing He permeance of the AG0, AG6 and AG10 membranes, all prepared using a 70/30 of bore liquid composition (NMP/H_2O), it is evident that larger air gap results in more permeable membranes. Of course wall thickness has an effect on permeance values but permeance differences reveal changes on the thickness of the separating layers as well.

Thermogravimetric analysis (*TGA*) for PI/PAni and PI hollow fibers, was performed in order to obtain information on the thermal stability of both fiber types. As seen in Figure 13, at 440°C both fiber types are subject to only 5 % weight loss and up to about 620°C they exhibit the same thermal behavior (*34 %weight loss*). However, at 750°C, differences in thermal stability become obvious as PI/PAni looses 45 % of its weight whereas PI 42 %, and become more pronounced at 1090°C, where weight loss for PI/PAni and PI is 55 and 47 % respectively. Although this relatively small difference reflects the lower thermal stability of the PI/PAni blend, the results indicate that both fiber types could be used in future trials for high temperature separation processes.

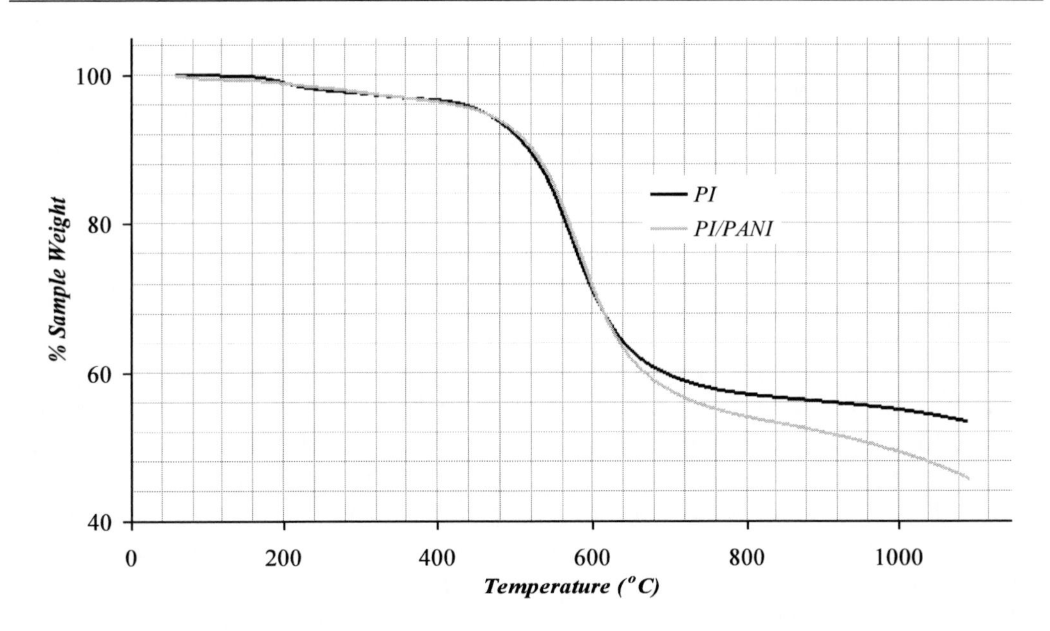

Figure 13. TGA curves of PI (P84) hollow fiber and PI/PAni blends.

Gas permeability measurements performed on the single and blend hollow fiber membranes are summarized in table 4. The polyimide/PAni (*PI/PAni*) blend hollow fiber membrane shows high permeance properties for all gases with values 2 to 3 orders of magnitude higher than its pure polyimide counterpart while the calculated permselectivity values for binary gaseous mixtures of high industrial interest such as N_2/CO_2 and CH_4/CO_2 is significantly improved offering the possibility for the membrane to be used for applications such as natural gas purification.

Table 4. Permeance/permselectivity values for the prepared PI and blend PI/PAni hollow membranes

Membrane	Gas Permeance (*GPU*) at 298 K					
	He	H_2	CH_4	CO_2	O_2	N_2
PI	1.17	1.03	0.07	0.18	0.08	0.05
PI/PAni	74.44	76.06	29.28	19.64	26.44	28.61
Membrane	Gas Permselectivities at 298 K					
	H_2/CH_4	CH_4/CO_2	He/N_2	H_2/N_2	H_2/CO_2	N_2/CO_2
PI	16.71	0.38	23.4	20.6	5.72	0.28
PI/PAni	2.6	1.34	2.62	2.66	3.87	1.46

On the other hand the permselectivity ratio for H_2/CH_4 and H_2/CO_2 decreased for the PI/PAni composite membrane [101]. As expected, an increase in the permeabilities of two gases leads to a decrease in selectivity between those gases (table 4). Although polymeric PI/PAni membranes have been previously studied [102,103] in the form of thin films, this was the first time that hollow fiber membranes have been prepared from this material.

Although, in this type of membrane configuration, typical skin layer thickness is about 1 μm [104], the calculated permeance values are given in GPU and not in Barrer as the exact selective skin layer size is unknown for the membranes studied in this work. The hollow fiber membranes presented, exhibited higher permeance values than their flat membrane counterparts as well as fibers previously reported in the literature [105]. Based on their performance they could be promising candidates for industrial use in gas separation applications

Conclusion

Even though PAni has been long known and extensively studied there are still applications that would benefit from further research on this versatile polymer. Apart from the several technological fields in which it has already been established because of its conductivity properties, there are also areas where PAni has shown great potential as a polymer. Especially in the field of membrane technology, PAni has proved its value for a multitude of applications including gas separations. Undoubtedly, it would be important for further research to keep exploring its potential in this field not only to improve its application as a polymeric precursor for the preparation of membranes but also to take advantage of its conductive properties that could lead to novel conductive membranes and new, more efficient separation processes.

Acknowledgments

Evangelos P. Favvas would like to thank the research project “NANOSKAI”. This research, The “NANOSKAI” Project (Archimedes Framework) of the Kavala Institute of Technology, has been co-financed by the European Union (European Social Fund – ESF) and Greek national funds through the Operational Program "Education and Lifelong Learning" of the National Strategic Reference Framework (NSRF) - Research Funding Program. Investing in knowledge society through the European Social Fund.

References

[1] Shirakawa, L. H. E. J.; MacDiarmid, A. G.; Chiang, C. K.; Heeger, A. J. *J. Chem. Soc. Chem. Commun.* 1977, 16, 578–80.

[2] Chiang, C. K.; Fincher, C.R.; Park, Y. W.; Heeger, A. J.; Shirakawa, H.; Louis, E. J.; Gau, S. C.; MacDiarmid, A. G. *Physical Review Letters* 1977, 39, 1098–1101.

[3] Yilmaz, F. *Polyaniline: synthesis, characterization, solution properties and composites*; Ph.D. Dissertation, Middle East Technical University, Ankara/Turkey, 2007.

[4] Bhadra, S.; Khastgir, D.; Singha, N. K.; Lee, J. H. *Progr. Polym. Sci.* 2009, 34, 783–810.

[5] Okamoto, Y.; Brenner, W. *Organic Semiconductors*; Reinhold: 7th Chapt. Polymers, 1964, pp 125–158.

[6] Gregory, R. V. In *Handbook of Conducting Polymers*; Marcel Dekker, Skotheim, T. A.; Elsenbaumer, R. L.; Reynolds, J. R.: New York, 1998, p. 437.
[7] Hsu, C. H.; Epstein, A. J. *SPE ANTEC'96*, 1996, 54(2), 1353.
[8] Hardaker, S. S.; et al. *SPE ANTEC'96*, 1996, 54(2), 1358.
[9] Kohlman, R. S., and A. J. Epstein. *In Handbook of Conducting Polymers*; Skotheim, T. A.; Elsenbaumer, R. L.; Reynolds, J. R. (eds), Marcel Dekker: New York, 1998, p. 85.
[10] Wang, Z. H. ; Scherr, E. M.; MacDiarmid, A. G.; Epstein, A. J. *Phys. Rev. B* 1992, 45(8), 4190–4202.
[11] Hardaker, S. S.; Eaiprasertsak, K.; Yon, J.; Gregory, R. V.; Tessem, G. X. *Mat. Res. Soc. Symp. Proc.* 1998, 488, 365–370.
[12] Menon, R. et al. *In Handbook of Conducting Polymers*; Skotheim, T. A.; Elsenbaumer, R. L.; Reynolds, J. R. eds: New York, 1998, p. 27.
[13] Mark, J. E. *Polymer data handbook*; Oxford University Press, 1999.
[14] Skotheim, T. A.; Elsenbaumer, R. L.; Reynolds, J. R. *Handbook of conducting polymers*; Marcel Dekker, INC: New York, 2nd edition, 1998.
[15] Yang Y.; Heeger, A. *J. Appl. Phys. Lett.* 1994, 64, 1245–1247.
[16] Friend, R. H.; Gymer, R. W.; Holmes, A. B.; Burroughes, J. H.; Marks, R. N.; Taliani, C.; Bradley, D. D. C.; Santos, D. A. D.; Brédas, J. L.; Lögdlund M.; Salaneck, W. R. *Nature* 1999, 397, 121–128.
[17] Mäkelä, T.; Pienimaa, S.; Taka, T.; Jussila, S.; Isotalo, H. *Syntheitc Metals* 1997, 85, 1335–1336.
[18] Akheel, A. S.; Maravattickal, K. D. *Talanta* 1991, 38, 815–837.
[19] Diaz, F.; Logan, J. A. *J. Electroanal. Chem.* 1980, 111, 111–114.
[20] Genies, E. M.; Syed, A. A.; Tsintavis, C. *Mol. Cryst. Liq. Cryst.* 1985, 121, 181–186.
[21] Genies, E. M.; Tsintavis, C. J. *Electr. Chem.* 1985, 195, 109–128.
[22] Vidotto, G.; Crosato-Arnaldi, A.; Talamini, G. *Makromol. Chem. Phys.* 1969, 122, 91–104.
[23] Dowding, P. J.; Vincent, B. *Coll. Surf. A* 2000, 161, 259–269.
[24] Kim, N.; Sudol E. D.; Dimonie, V. L.; El-Aasser, M. S. *Macromolecules* 2006, 36, 5573–5579.
[25] Kuramoto, N.; Tomita, A. *Synth. Met.* 1997, 88, 147–151.
[26] Miras, M. C.; Barbero, C.; Haas, O. *Synth. Met.* 1991, 43, 3081–3084.
[27] Dallas, P.; Stamopoulos, D.; Boukos, N.; Tzitzios, V.; Niarchos, D.; Petridis, D. *Polymer* 2007, 48, 162–169.
[28] Chen, J.; Chao, D.; Lu, X.; Zhang, W. *Mater. Lett.* 2007, 61, 419–423.
[29] Guo, Q.; Yi, C.; Zhu, L.; Yang, Q.; Xie, Y. *Polymer* 2005, 46, 3185–3189.
[30] Yang, C. H.; Huang, L. R.; Chih, Y. K.; Lin, W. C.; Liu, F. J.; Wang, T. L. *Polymer* 2007, 48, 3237–3247.
[31] Jing, X.;Wang, Y. Y.; Wu, D.; Qiang, J. P. *Ultrason Sonochem* 2007, 14, 75–80.
[32] Sauer, G.; Brehm, G.; Schneider, S.; Nielsch, K.; Wehrspohn, R. B.; Choi, J.; Hofmeister, H.; Gösele, U. *J. Appl. Phys.* 2002, 91, 3243–3247.
[33] Choi, J.; Sauer, G.; Nielsch, K.; Wehrspohn, R. B.; Gosele, U. *Chem. Mater.* 2003, 15, 776–779.
[34] Nastase, C.; Nastase, F.; Dumitru, A.; Ionescu, M.; Stamatin, I. *Compos. Part. A. Appl. Surf.* 2005, 36, 481–485.
[35] Kobayashi, N.; Teshima, K.; Hirohashi, R. *J. Mater. Chem.* 1998, 8, 497–506.

[36] Khanna, P. K.; Singh, N.; Charan, S.; Viswanath, A. K. *Mater. Chem. Phys.* 2005, 92, 214–219.

[37] Kobayashi, S.; Uyama, H.; Kimura, S. *Chem. Rev.* 2001, 101, 3793–3818.

[38] Arias-Marin, E.; Romero, J.; Ledezma-Perez, A. S.; Kniajansky, S. *Polym. Bull.* 1996, 37, 581–587.

[39] Gross, R. A.; Kumar, A.; Kalra, B. *Chem. Rev.* 2001, 101, 2097–2124.

[40] Unsworth, J.; Lunn, B. A.; Innis, P. C.; Jin, Z.; Kaynak, A.; Booth, N. G. *J. Intel. Mat. Syst. Str.* 1992, 3, 380–395.

[41] Bhadra, S.; Singha, N. K.; Khastgir, D. *Synth. Met.* 2006, 156, 1148–54.

[42] Kitani, A.; Yoshioka, K.; Sasaki, K. *Synthetic Metals* 1993, 55, 3566–3570.

[43] Li, Z. F.; Kang, E. T.; Neoh, K. G.; Tan, K. L. *Synthetic Metals* 1997, 87, 45–52.

[44] Yadav, J. B.; Puri, R. K.; Puri, V. *Appl. Surf. Sci.* 2007, 253, 8474–8477.

[45] Wei, Y.; Jang, G. W.; Hsueh, K. F.; Scherr, E. M.; MacDiarmid, A. G.; Epstein, A. J. *Polymer* 1992, 33, 314–322.

[46] Valentováa, H.; Stejskal, J. *Synthetic Metals* 2010, 160, 832–834.

[47] Soundararajah, Q. Y.; Karunaratne, B. S. B.; Rajapakse, R. M. G. *Mater. Chem. Phys.* 2009, 113, 850–855.

[48] Mirmohseni, A.; Dorraji, M. S. S. *J. Polym. Res.* 2012, 19, 9852–9861.

[49] Jin, X.; Xiao, C.; Wang, W. *Synthetic Metals* 2010, 160, 368–372.

[50] Castillo-Castro, T. D.; Castillo-Ortega, M. M.; Herrera-Franco, P. J. *Composites: Part A* 2009, 40, 1573–1579.

[51] Tsotra, P.; Friedrich, K. *Compos. Sci. Technol.* 2004, 64, 85–91.

[52] Tsotra, P.; Friedrich, K. *Synth. Metals* 2004, 143, 237–42.

[53] Loeb, S.; Sourirajan, S. *Advan. Chem. Ser.* 1962, 38, 117–132.

[54] Favvas, E. P.; Mitropoulos, A.Ch. *J. Eng. Sci. Tech. Rev.* 2008, 1, 25–27.

[55] Baum, B.; Holley, W. J.; White, R. A. *Hollow Fibres in Reverse Osmosis, Dialysis, and Ultrafiltration, in Membrane Separation Processes*; P. Meares (ed.); Elsevier: Amsterdam, 1976, pp. 187–228.

[56] Strathmann, H. *Membranes and Membrane Separation Processes*; Wiley–VCH Verlag GmbH & Co. KGaA, Ullmann's Encyclopedia of Industrial Chemistry, 2005.

[57] Baker, R. W. *Membrane Technology and Applications*, John Wiley & Sons Ltd, 2nd ed.: England, 2004.

[58] Curcio, E.; Drioli, E. *Membranes for Desalination, in "Seawater Desalination, Conventional and Renewable Energy Processes"*; Springer–Verlag, Berlin Heidelberg, Cipollina, A.; Micale, G.; Rizzuti, L. Eds., 2009, pp. 41–77.

[59] Shannon, M. A.; Bohn, P. W.; Elimelech, M.; Georgiadis, J. G.; Mariñas, B. J.; Mayes, A. M. Nature 2008, 452, 301–310.

[60] Qdais H. A.; Moussa, H. Desalination 2004, 164, 105–110.

[61] Kolf, W. J. *New Ways of Treating Uremia*; J. and A. Churchill: London, 1947.

[62] Baker, R. B.; Cussler, E. L.; Eykamp, W.; Koros, W. J.; Riley, R. L.; Starthmann, H. *Membrane Separation Systems*; Noyes Data Corporation: Park Ridge, NJ, 1991.

[63] Huang, R. Y. M. *Pervaporation Membrane Separation Processes*; Elsevier: New York, 1991.

[64] Haggin, J. *Chem. Engin. News.* 1990, 1, 22–26.

[65] Gantzel, P.; Merten, U. *Ind. Eng. Chem. Process Des. Develop.* 1970, 9, 331–332.

[66] Favvas, E. P.; Romanos, G. E.; Papageorgiou, S. K.; Katsaros, F. K.; Mitropoulos A. Ch.; Kanellopoulos, N. K. *J. Membr. Sci.* 2011, 375, 113–123.
[67] Ismail, A. F.; Lai, P. Y. *Separ. Purif. Techn.* 2003, 33, 127–143.
[68] Şen, D.; Kalıpçılar, H.; Yilmaz, L. *J. Membr. Sci.* 2007, 303, 194–203.
[69] Yoshimune, M.; Haraya, K. *Separ. Purif. Techn.* 2010, 75, 193–197.
[70] Ball, I. J.; Huang, S. C.; Su, T. M.; Kaner, R. B. *Synthetic Metals* 1997, 84, 799–800.
[71] Huang, S. C.; Ball, I. J.; Kaner, R. B. *Macromolecules* 1998, 31, 5456–5464.
[72] Lee, Y. M.; Nam, S. Y.; Ha, S. Y. *J. Membr. Sci.* 1999, 159, 41–46.
[73] Ball, I. J.; Huang, S. C.; Wolf, R. A.; Shimano, J. Y.; Kaner, R. B. *J. Membr. Sci.* 2000, 174, 161–176.
[74] Wen, L.; Kocherginsky, N. M. *J. Membr. Sci.* 2000, 167, 135–146.
[75] Illing, G.; Hellgardt, K.; Wakeman, R. J.; Jungbauer, A. *J. Membr. Sci.* 2001, 184, 69–78.
[76] Vijaya, B.; Naidu, K.; Sairam, M.; Raju, K. V. S. N.; Aminabhavi, T. M. *J. Membr. Sci.* 2005, 260, 142–155.
[77] Illing, G.; Hellgardt, K.; Schonert, M.; Wakemana, R. J.; Jungbauer, A. *J. Membr. Sci.* 2005, 253, 199–208.
[78] Gupta, Y.; Hellgardt, K.; Wakeman, R. J. *J. Membr. Sci.* 2006, 282, 60–70.
[79] Fan, Z.; Wang, Z.; Duan, M.; Wang, J.; Wang, S. *J. Membr. Sci.* 2008, 310, 402–408.
[80] Loh, X. X.; Sairam, M.; Bismarck, A.; Steinke, J. H. G.; Livingston, A. G.; Li, K. *J. Membr. Sci.* 2009, 326, 635–642.
[81] Sairam, M.; Loh, X. X.; Bhole, Y.; Sereewatthanawut, I.; Li, K.; Bismarck, A.; Steinke, J. H. G.; Livingston, A. G. *J. Membr. Sci.* 2010, 349, 123–129.
[82] Hasbullah, H.; Kumbharkar, S.; Ismail, A. F.; Li, K. *J. Membr. Sci.* 2011, 366, 116–124.
[83] Hasbullah, H.; Kumbharkar, S.; Ismail, A. F.; Li, K. *J. Membr. Sci.* 2012, 38, 397–398.
[84] Mahon, H. I. *Permeability Separatory Apparatus, Permeability Separatory Membrane Element, Method of Making the Same and Process Utilizing the Same*; US Patent: 3,228,876, 1966.
[85] Moch, J. *Hollow Fiber Membranes, in Encyclopedia of Chemical Technology*; John Wiley–InterScience Publishing, 4th Edn: New York, 1995, Vol. 13, p. 312.
[86] McKelvey, S. A.; Clausi, D. T.; Koros, W. J. J. Membr. Sci. 1997, 124, 223.
[87] Maier G. Angew. Chem. Int. Ed. 1998, 17, 2960–2974.
[88] Wallace, D. *Crosslinked hollow fiber membranes for natural gas purification and their manufacture from novel polymers*; Ph.D. Dissertation, University of Texas. Austin, 2004.
[89] Kapantaidakis, G. C.; Koops, G. H. J. Membr. Sci. 2002, 204, 153–171.
[90] Chung, T. S.; Wang, R.; Ren, J. Z.; Cao, C.; Liu, Y.; Li, D. F. *Manufacture of polyimide hollow fibers*; WIPO Patents, WO/2003/040444.
[91] Niwa, M.; Kawakami, H.; Nagaoka, S.; Kanamori, T.; Shinbo, T. *J. Membr. Sci.* 2000, 171, 253–261.
[92] Favvas, E. P.; Kapantaidakis, G. C.; Nolan, J. W.; Mitropoulos, A. Ch.; Kanellopoulos, N. K. *J. Mater. Proc. Tech.* 2007, 186, 102–110.
[93] Favvas, E. P.; Kouvelos, E. P.; Romanos, G. E.; Pilatos, G. I.; Mitropoulos, A. Ch.; Kanellopoulos, N. K. *J. Por. Mater.* 2008, 15, 625–633.

[94] Su, T. M.; Ball, I. J.; Conklin, J. A.; Huang, S. C.; Larson, R. K.; Song, L.; Nguyen, S. L.; Lew, B. M.; Kaner, R. B. *Synthetic Metals* 1997, 84, 801–802.
[95] Chatzidaki, E. K.; Favvas, E. P.; Papageorgiou, S. K.; Kanellopoulos, N. K.; Theophilou, N. V. *Europ. Polym. J.* 2007, 43, 5010–5016.
[96] Angelopoulos, M.; Asturias, C. E.; Ermer, S. P.; Ray, E.; Scherr, E. M.; MacDiarmid, A. G.; Akhtar, M. A.; Kiss, Z.; Epstein, A. *J. Mol. Cryst. Liq. Cryst.* 1988, 160, 151–163.
[97] Favvas, E. P; Papageorgiou, S. K; Nolan, J. W; Stefanopoulos, K. L; Mitropoulos, A. C. under review.
[98] Khayet, *Chem. Engin. Sci.* 2003, 58, 3091–3104.
[99] Zhang, X; Wen, Y; Yang, Y; Liu, L. *J. Macromol. Sci. Part B: Physics* 2008, 47, 1039–1049.
[100] Qin, J. J ; Li, Y ; Lee, L. S ; Lee, H. *J. Membr. Sci.* 2003, 218, 173–183.
[101] Steriotis, T. A.; Katsaros, F. K.; Mitropoulos, A. C.; Stubos, A. K.; Galiatsatou, P.; Zouridakis, N.; Kanellopoulos, N. K. *Rev. Sci. Instrum.* 1996, 67, 2545–2458.
[102] Wang, H. L.; Mattes, B. R. *Synthetic Metals.* 1999, 102, 1333–1334.
[103] Ball, I. J.; Huang, S. C.; Wolf, R. A.; Shimano, J. Y.; Kaner, R. B. *J. Membr. Sci.* 2000, 174, 161−176.
[104] Yoshino, M.; Nakamura, S.; Kita, H.; Okamoto, K.; Tanihara, N.; Kusuki, Y. *J. Membr. Sci.* 2003, 212, 13–27.
[105] Mattes, B. R.; Anderson, M. R.; Conklin, J. A.; Reiss, H.; Kaner, R. B. *Synthetic Metals.* 1993, 57, 3655−3660.

In: Trends in Polyaniline Research
Editors: T. Ohsaka, Al. Chowdhury, Md. A. Rahman et al.
ISBN: 978-1-62808-424-5

Chapter 12

POLYANILINE-BASED SENSORS FOR GAS DETECTION

Shengxue Yang and Jian Gong
Key Laboratory of Polyoxometalates Science of Ministry of Education,
Department of Chemistry, Northeast Normal University,
Changchun, Jilin, P. R. China

ABSTRACT

Gas detection is important because it is necessary in many different fields, such as clinical assaying, industrial and vehicle emission control, agricultural storage and shipping, household and workplace security, and environmental monitoring. Recent studies showed that conducting polymers could substitute for inorganic materials used for gas detection because of diversity, ease of processing, mechanical flexibility, modifiable electrical conductivity and sensitivity at room temperature. Among the conducting polymers, polyaniline (PAni) has received wide-spread attention due to its high sensitivities, fast response times, reversible doping/dedoping chemistry, and room temperature operation. We describe the research made on PAni-based gas sensors that contain the PAni with different morphologies, PAni nanocomposite, and PAni-based junction. The review, then, does not aim to be a compendium of all the information available on the subject; rather, it selects some representative examples of various synthetic methods involved in PAni-based gas sensors and it highlights the reduction of the limit detection, the different mechanisms involved in gas sensors and enhancement of the selectivity to some uncommon gases. We start with how to design and synthesize different morphological PAni nanomaterials for gas sensors and then move on to the PAni-metal or other PAni nanocomposite sensors such as PAni/metal salt, PAni/metal, PAni/metal oxide, PAni/carbon black, PAni/graphene, PAni/polymer nanocomposite etc. for different gas detection. Finally, we discuss the design of the junction in PAni-based gas sensors and the effect on gas response.

1. INTRODUCTION

A gas sensor is a device that can convert the measured gas concentration to electric quantity output related to this concentration. Utilizing the physical property and chemical property of the measured gas, we can control and realize the automatic detection of different gases or different concentrations of gas. [1-5] In 1906, Cremer first discovered, as the precursor of the sensor, the hydrogen ion alternative response of the glass-film electrode. [6] With intensive research, many studies have focused on investigating and developing the gas sensors, which can be used to detect the presence of coal gas, liquefied petroleum gas, natural gas, and methane gas. 50 years ago, Bardeen found that the gas adsorption on the surface of the semiconductor could cause a change of the conduction. [7] From then on, many studies have been motivated with the hope that a commercialized semiconductor device can be applied in the detection of gas.

In recent years, the study and development of the gas sensor has constantly been on the increase, and the achievability of gas sensors prompts us to find new air-sensitive materials, discuss ways to increase the performance of the gas sensors, and more deeply study the present practicability of gas sensor materials. Gas detection is important because it is necessary in many fields, such as clinical assaying, industrial and vehicle emission control, agricultural storage and shipping, household and workplace security, and environmental monitoring.

The gas sensors, based on inorganic materials such as semiconductor oxide, have had considerable impact to date. [8-14] However, the low selectivity to specific target gases and the need to operate at high temperatures often increases power consumption and reduces gas sensor life, which limits the use of this type of detection. For these reasons, proposals have been made to overcome these issues. Recent studies show that conduction polymers can substitute for inorganic materials as a new gas sensor because of diversity, ease of processing, mechanical flexibility, modifiable electrical conductivity and sensitivity at room temperature. [15-19] Among conducting polymers, polyaniline (PAni) has received wide-spread attention due to its high sensitivities, fast response times, reversible doping/dedoping chemistry, and room temperature operation. [20, 21] PAni sensors have been investigated and used to detect a great deal of different chemical materials including ammonia (NH_3), hydrogen chloride (HCl), ethanol vapor, methanol vapor, 2-propanol vapor, 1-propanol vapor, hydrazine (N_2H_4), chloroform, toluene, triethylamine, water vapor, NO_2, CO etc. [22-30]

To reduce the limit of detection, enhance the sensitivity, widen the dynamic range, and increase the selectivity to different gases, several approaches have been explored, such as the synthesis of different morphological PAni nanomaterials using different methods, preparation of PAni nanocomposites, design of PAni-based Schottky junction or p-n junction. By electrodeposition, Dhawale et al. fabricated a CdS/PAni p-n heterojunction sensor. The sensor exhibited a high response towards liquefied petroleum gas (LPG) compared to N_2 and CO_2 at room temperature. [31] Using a suitable fiber as a template, Li et al. prepared a PAni/Ag nanocomposite through UV rays irradiation. [32] The introduction of metal into PAni film is effective in promoting the chemiresistor sensitivity to NH_3 at room temperature. Gong et al. reported an ultrasensitive NH_3 gas sensor by enchasing PAni nanograin on the TiO_2 fiber surface. [33] This sensor can detect 50 ppt of NH_3 gas in air. To deposit PAni nanofibers on either gold (Au) or platinum (Pt) electrodes, Fowler et al. designed a Schottky junction sensor for the detection of hydrogen (H_2), which showed that the greater sensitivity of Pt-PAni sensor could be used to detect H_2 at a concentration of 10 ppm. [34]

This chapter describes the research made on PAni-based gas sensors that contain the fabrication of various morphological PAni gas sensors, PAni/metal nanocomposite gas sensors, PAni/other nanocomposite gas sensors and PAni-based heterojunction type gas sensors.

The chapter, then, does not aim to be a compendium of all the information available on the subject; rather, it selects some representative examples of various synthetic methods involved in PAni-based gas sensors and highlights the reduction of the limit detection, the different mechanisms involved in gas sensors, and enhancement of the selectivity to some uncommon gases. We start with how to design and synthesize different morphological PAni nanomaterials for gas sensors and then move on to the PAni/metal or PAni/other materials nanocomposite sensors for different types of gas detection, such as PAni/metal salt, PAni/metal oxide, PAni/carbon black, PAni/graphene, PAni/polymer nanocomposite etc. Finally, we discuss the design of the heterojunction type PAni-based gas sensors and their gas response research.

2. PAni with Different Morphologies for Gas Sensors

Nanostructures of PAni such as nanofibers, nanorods, nanonoodles, nanospheres, nanowires, nanotubes, nanosheets, nanoleaves, and rose-like nanoplates, etc, have been prepared by using methods such as soft/hard templates, surfactants, interfacial polymerization, template-free, in-situ chemical oxidative polymerization, seeds, electrospinning, electrodeposition [35-50]. They have been attracting considerable attention as sensor materials because of controllable conductivity, good environmental stability, ease of preparation, high surface area, and fast diffusion of gas molecules into the structure.

2.1. PAni Sensors for NH_3 Detection

NH_3 is an important component of many industrial processes and power plants, thus, to detect NH_3 in air has recently achieved widespread importance for environmental monitoring and process control because of its high toxicity.

2.1.1. PAni Nanofiber Synthesized by Interfacial Polymerization

Huang et al. has prepared an ammonia sensor based on high-quality PAni nanofibers using aqueous/organic interfacial polymerization. [51] Figure 1A shows the real-time resistance change of the HCl-doped PAni nanofibers film exposed to 100 ppm of NH_3. The nanofibers film synthesized by this interfacial method responds much faster than the conventional PAni film even though the thickness of the film is twice that of the ordinary film. [51]

Virji et al. has constructed a gas sensor composed of PAni nanofiber with different diameters by interfacial synthesis. [52] The sensor response for NH_3 is related to the diameter of the PAni nanofibers. The smaller the diameter of the nanofiber is, the greater the response can be shown because of its higher surface area (Figure 1B). Using the interfacial polymerization method, Chen et al. has also reported a highly sensitive NH_3 gas sensor made of PAni nanofibers that is coated on the commercially ceramic substrates. [53] The typical response and recovery curves of the prepared PAni sensor when cycled by increasing NH_3 concentration (1-200 ppm) in ambient air at room temperature are shown in Figure 1C-a.

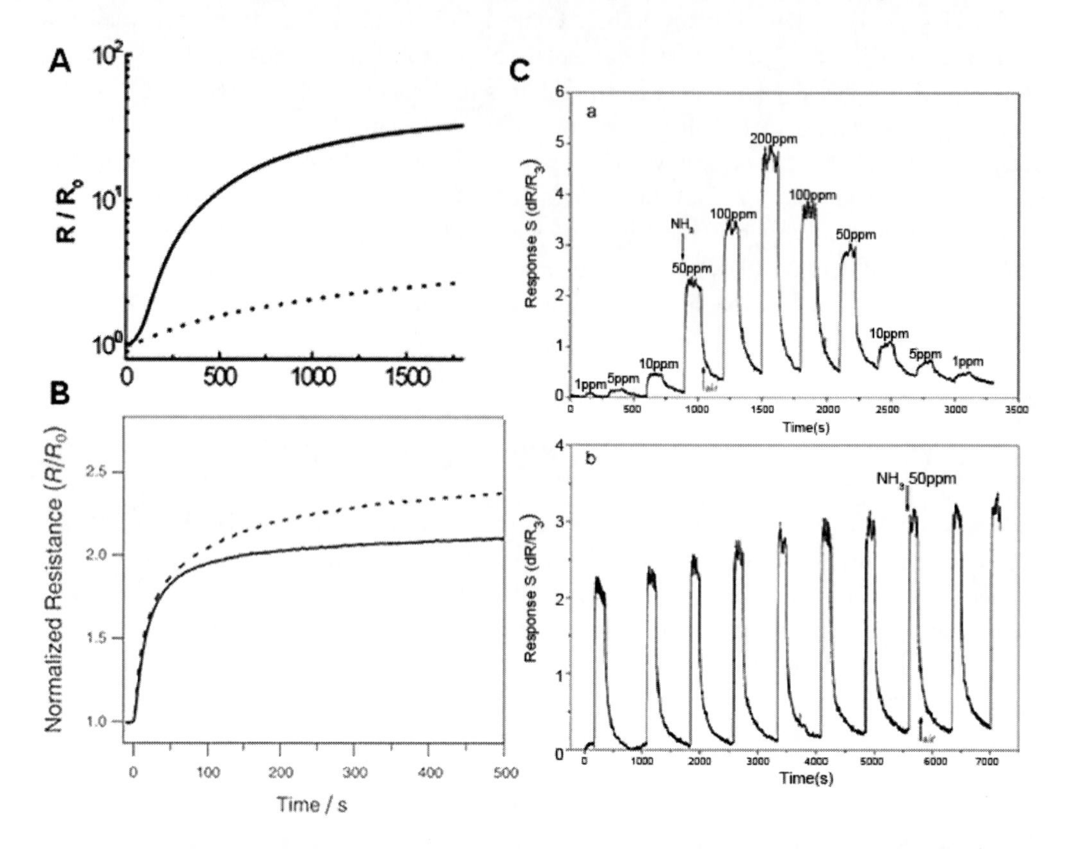

Figure 1. (A) Resistance changes of an HCl-doped PAni nanofibers film prepared by interface polymerization (solid line) and conventional film (dotted line) upon exposure to 100 ppm of NH_3 vapor in nitrogen. (Reproduced with permission of the American Chemical Society from ref. [51]) (B) Normalized resistance measured versus time upon exposure of HCl-doped PAni nanofibers with small (dotted line) and large (solid line) diameters to 50 ppm of NH_3. (Reproduced with permission of the American Chemical Society from ref. [52]) (C) The response curves of PAni sensors of (a) different concentrations (1-200 ppm) of NH_3 gas and (b) the reproducibility to 50 ppm of NH_3. (Reproduced with permission of the Elsevier B.V. from ref. [53]).

The sensor exhibits a good ability to detect NH_3 as low as 1 ppm and a long-term reproducibility tested at a fixed concentration of 50 ppm (Figure 1C-b).

2.1.2. PAni with Different Morphologies Obtained by the Template Method

Xue et al. has fabricated ultra-fine conducting PAni micro/nanotubes gas sensors using fiber templates prepared by electrospun nitrocellulose (NC). [54] Aniline polymerizes on the surface of NC template, then the NC template is dissolved, and the PAni micro/nanotubes are obtained. Figure 2A-a shows the resistance changes (R/R_0) of the PAni micro/nanotubes prepared with different ratios of aniline/ammonium persulfate (APS) as a function of time to 50 ppm of NH_3.

Obviously, the difference of gas sensitivity and response rate to NH_3 gas can be ascribed to the different morphology and conductivity of the PAni for different ratios of aniline/APS.

This PAni micro/nanotubes sensor can act as an "electronic nose" to detect NH_3 gas below 20 ppm, as shown in Figure 2A-b. Using copper wires and rings as a template, Gao et al. has successfully prepared PAni nanorods and hollow-microspheres sensors. [55]

Owing to the high surface areas, small diameters, and porous structure, the PAni hollow-microspheres sensor has better sensitivity than the nanorods sensor to NH_3 gas (Figure 2B-a). The resistance of the PAni hollow-microspheres exposed to 10 ppm of NH_3 is 1.25, which indicates that the sensor has superior performance as shown in Figure 2B-b. Using an easily removed ice-template method, PAni microflakes stacked with one-dimensional nanofibers have been synthesized. [56] The PAni microflakes appear to have super performance in both sensitivity and time response to NH_3 and reversible properties because of their small diameter, strongly adherent structure and porous nature (Figure 2C).

Ma et al. also synthesized PAni hemispheres through the ice-template method. [57] It is important that the doping acid and the addition of the second solvent (diethyl ether) have great influence on the morphology of PAni. Figure 3A-a shows the gas response of the PAni hemispheres to NH_3, and Figure 3A-b shows the response of PAni hemispheres to different concentrations of NH_3 from 5 to 100 ppm. The PAni hemispheres have better sensitivity than the PAni prepared in the ice bath upon exposure to 100 ppm of NH_3. Assisted by a secondary solvent, they prepared various morphologies of PAni through the ice-template method, such as microflakes, porous mocrowebs, hemispheres, and nests piled by nanoparticles. [58] The gas response to NH_3 examined at room temperature shows that the PAni microflakes have the best performance in both sensitivity and time response compared with PAni porous microwebs, hemispheres, or nests piled by nanoparticles as seen from Figure 3B. Recently, Li et al. presented a new method for synthesis of PAni nanotubes using Mn_2O_3 as an oxidant template prepared by the electrospinning technique. [59] After polymerization, PAni shells form on the surface of the Mn_2O_3 nanofiber template, and the Mn_2O_3 template is spontaneously removed. It was found that the PAni nanotubes sensor could detect as low as 25 ppb of NH_3 in the air at room temperature and exhibit good reversibility exposed to 50 ppb of NH_3 (Figure 3C).

2.1.3. PAni Crystal Obtained by Gas Phase Polymerization

PAni single-crystalline with dendritic nano/micro structures have been synthesized by a vapor-phase polymerization method. Major experiments carried out in this work are summarized in Table 1. [60] In this experiment, a petri dish with 1 g of aniline monomer in it was put into a glass container with a volume of about 5000 cm^3. When the glass container was full of aniline vapor a beaker, which was covered with an ordinary plastic membrane with many pores to contain the mixture solution of 2 mol/L of HCl and 0.2 g of APS, was put into the glass container. Then, the polymerization system was kept at 0-5°C for 5-15 days. Finally, a great amount of PAni crystals were found on the surface of the microporous membrane. The dark-green crystals of PAni were washed several times with distilled water and ethanol, respectively. The schematic drawing of the experimental setup is shown in Scheme 1. The single-crystalline PAni with dendritic nano/microstructures has higher gas sensitivity and more rapid response time to NH_3 than the PAni synthesized by a conventional method, as shown in Figure 4. Meanwhile, the reversible circulation response change of the PAni crystal has a reasonable reproducibility and has more potential applications in the area of sensor development.

2.1.4. Other Methods to Prepare PAni with Different Morphologies

Sutar et al. employed the amino groups (NH_2) of the amino-silane ($(CH_3O)_3$-Si-$(CH_2)_3NH(CH_2)_2$-NH_2) self-assembled monolayer as artificial seeds for self-organization of PAni on the silicon substrates surface. [61]

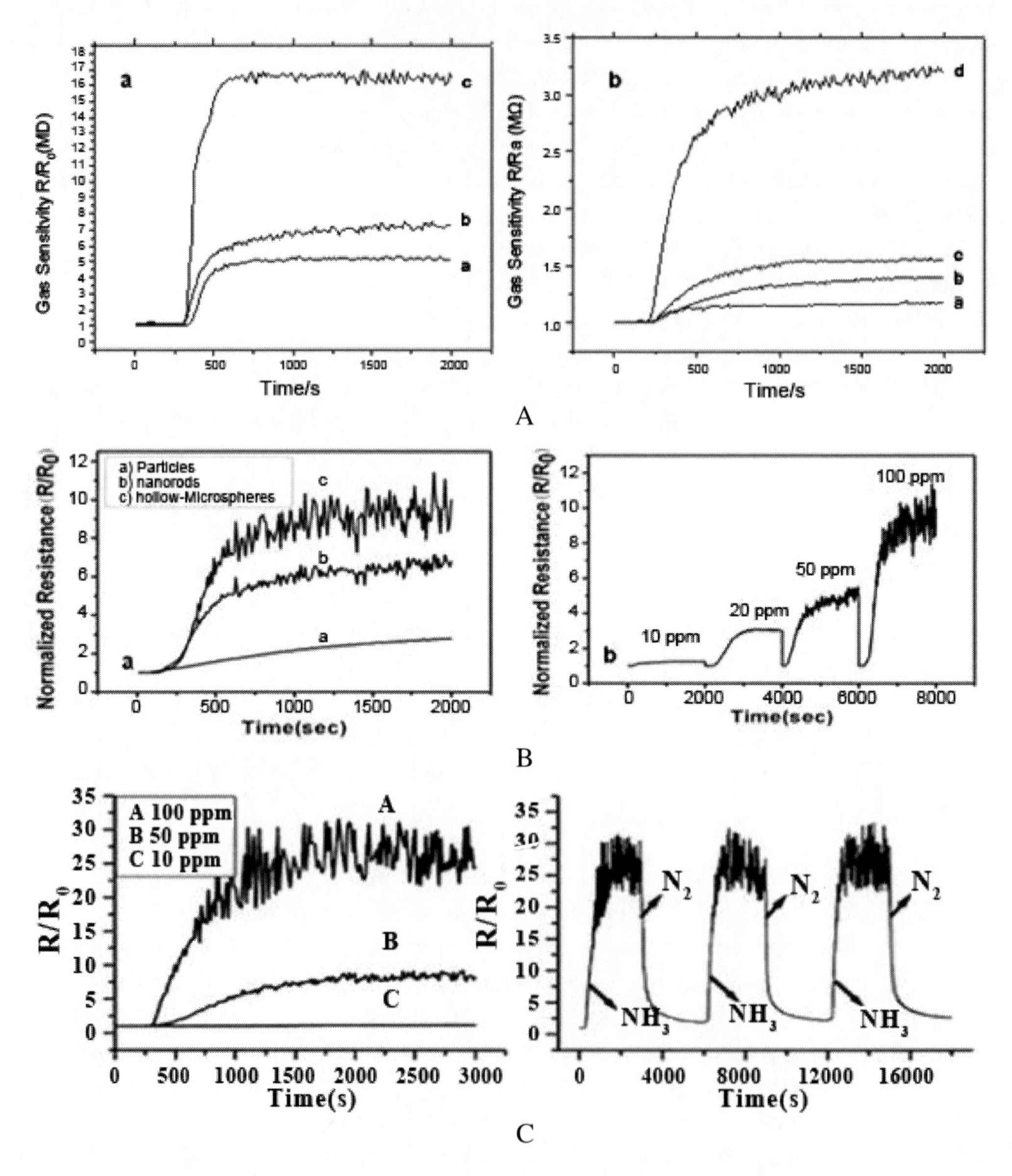

Figure 2. (A) (a) Gas sensitivity of PAni micro/nanotubes sensor prepared with different ratios of aniline/APS (a: 1:2, b: 1:1, c: 1:0.6) up to 50 ppm of NH_3 gas. (b) The PAni micor/nanotubes prepared with aniline/APS=1:2 exposed to different concentrations of NH_3 (a: 2 ppm, b: 5 ppm, c: 10 ppm, d: 20 ppm). [54] (B) Resistance change of (a) different morphological PAni sensors exposed to 100 ppm of NH_3 gas and (b) PAni hollow-microsphere sensors upon different concentrations of NH_3. [55] (C) (left) Resistance changes of PAni nanofiber microflakes upon exposure to different concentrations of NH_3 gas (A-100 ppm, B-50 ppm, C-10 ppm). (right) Reversible circulation response changes of PAni sensor exposed to 100 ppm of NH_3 gas. (Reproduced with permission of the American Chemical Society from ref. [56]).

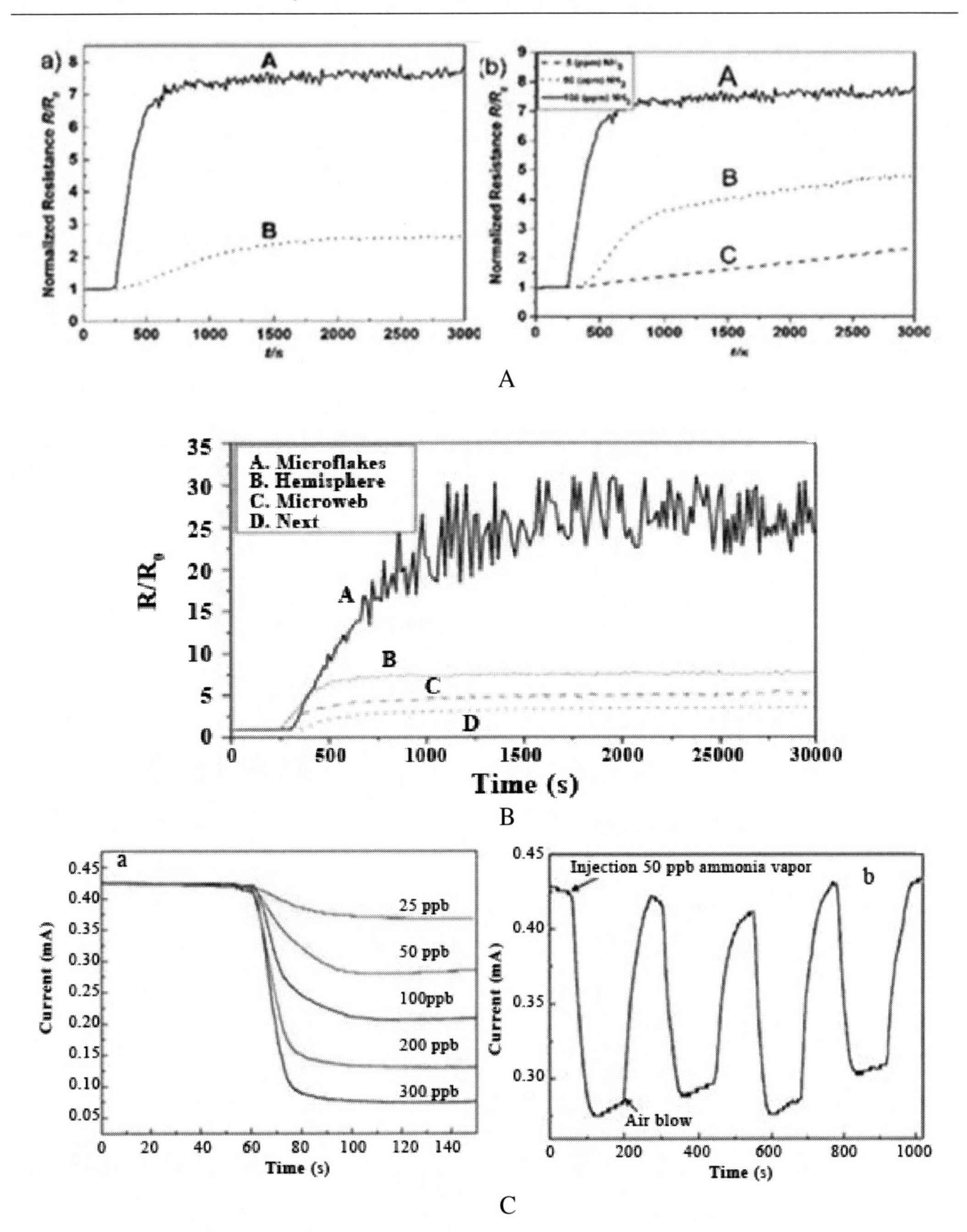

Figure 3. (A) (a) Resistance changes of PAni hemispheres (curve A) and PAni prepared in ice bath (curve B) upon exposure to 100 ppm of NH_3. (b) The hemisphere PAni exposed to different concentration of NH_3 gas (curve A-100 ppm, B-50 ppm, and C-5 ppm). (Reproduced with permission of Wiley-VCH from ref. [57]) (B) Resistance changes of PAni with various morphologies to 100 ppm of NH_3: curve A-microflakes, B-hemispheres, C-microwebs, and D-nests piled by nanoparticles. (Reproduced with permission of the American Chemical Society from ref. [58]) (C) (a) Current changes of the PAni nanotubes sensor upon exposure to NH_3 gas of different concentrations. (b) Reversible sensitivity of the PAni nanotubes to 50 ppb of NH_3 gas. (Reproduced with permission of the Elsevier B.V. from ref. [59]).

Table 1. Summary of the reaction conditions and product morphologies

Aniline (g)	HCl (mol/L)	APS (g)	Temperature (°C)	Morphologies
1	1	0.2	0-5	Well-dendritic morphology
1	2	0.2	0-5	Incomplete dendritic morphology
1	4	0.2	0-5	Dendrite-like morphology
1	2	0.2	25	No PAni

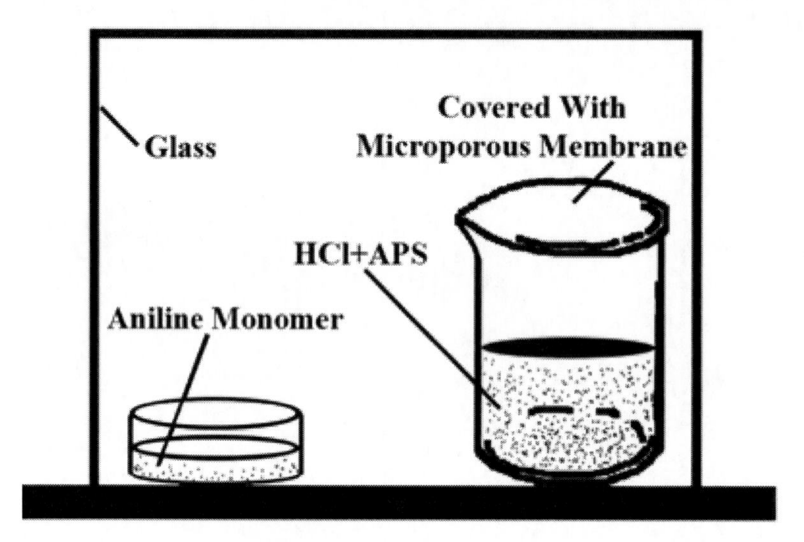

Scheme 1. Schematic drawing of the experimental installation for the synthesis of the PAni single-crystalline with dendritic nano/microstructures.

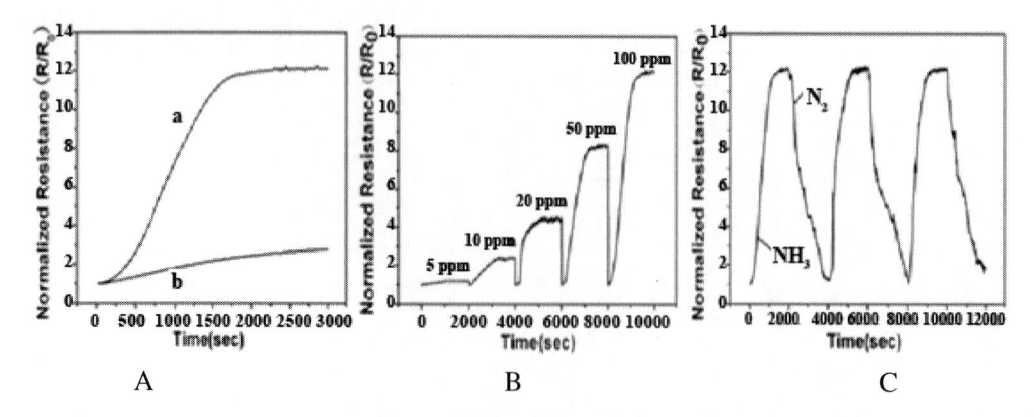

Figure 4. A) Resistance changes of PAni synthesized by (a) the vapor-phase polymerization method and (b) the conventional method upon exposure to 100 ppm of NH_3. B) Single-crystalline PAni with dendritic nano/microstructures exposed to different concentration of NH_3. C) The reversible circulation response change of the PAni single-crystalline with dendritic nano/microstructures up to 100 ppm of NH_3. [55].

Using the film as a sensitive layer, the chemiresistor sensor is formed. The sensor showed sensitivity, good response and recovery times to very low concentrations of 0.5 ppm NH_3. Crowley et al. fabricated a NH_3 sensor using an inkjet printing of dodecylbenzene sulfonate-doped PAni nanoparticles deposited onto a screen-printed carbon paste electrode. [62] The sensor was found to have stable sensitivity with an experimental detection limit of 20 μM (0.36 ppm) and a theoretical detection limit of 3.17 μM (0.054 ppm). Liu et al. developed a template-free in-situ approach to prepare PAni nanowires and nanofibers with horizontal orientation on solid substrates. [63] Gas sensing experiments confirmed that the horizontal orientation of the nanostructural PAni helped to improve the sensitivity and response time to NH_3 as shown in Figure 5. Sengupta et al. synthesized *para*-toluene sulfonic acid (PTSA) doped PAni through *in-situ* oxidative polymerization of aniline or aniline hydrochloride. [64] This could be directly used as a sensor by depositing the PAni film on maleic acid (MA) crosslinked poly vinyl alcohol (PVA) precoated glass slides. PAni-HCl-PTSA formed by using HCl as co-dopant has higher conductivity and better NH_3 gas sensing performance due to the higher intermolecular force and closer packing of molecular chains in PAni-HCl-PTSA than in PAni-PTSA. [64]

2.2. PAni Sensors for Detection of Other Gases

2.2.1. Alkaline and Reduced Gases

Redox active chemical vapors can be used to change the inherent oxidation state of PAni, thereby changing the degree of conjugation of the PAni backbone and the conductivity.

Gao et al. has used the inner eggshell membrane as well as electrospun PVA fiber mats as templates to prepare PAni nanotubes. [65] The gas response to trimethylamine ($(C_2H_5)_3N$), NH_3, and N_2H_4 shows that PAni nanotubes give significantly better performance in both gas sensitivity and time response because of their high surface areas, small diameter and porous nature (Figure 6A-a). The Pani nanotubes prepared using the PVA fiber mats template showed higher selectivity and quicker response to trimethylamine gas than the PAni prepared without a template (Figure 6A-b). As shown in Figure 6A-c, the reversible circulation response change of PAni nanotubes upon exposure to 100 ppm of $(C_2H_5)_3N$ has a reasonable reproducibility and can be applied in the sensor area. [65] They have also presented a novel approach using thin glass tubes as a single template to produce bulk quantities and pure $H_4SiW_{12}O_{40}$-doped PAni nanotubes in one step. [66] The gas responses of the PAni nanotubes to a series of chemical vapors such as NH_3, N_2H_4, and $(C_2H_5)_3N$ indicate that the PAni nanotubes have superior performance as a chemical sensor (Figure 6B-a). As shown in Figure 6B-b, it can be found that PAni films have a small increase in resistance when exposed to 10 ppm of NH_3.

The gas sensitivity can be obtained under this low concentration, which indicates that the PAni nanotubes can act as an "electronic nose" in chemical detection. Yang et al. examined the potential of PAni inverse opals as sensors synthesized via templating polystyrene (PS) colloidal crystals and then subjected them to four response tests: dry gas flow, ethanol vapor, hydrogen chloride, and ammonia. [67] The gas response to different conditions (such as NH_3 in Figure 7A) demonstrates that the three-dimensionally ordered macroporous PAni exhibits high sensitivity and fast response due to the porosity facilitating diffusion and large surface area interacting with substances.

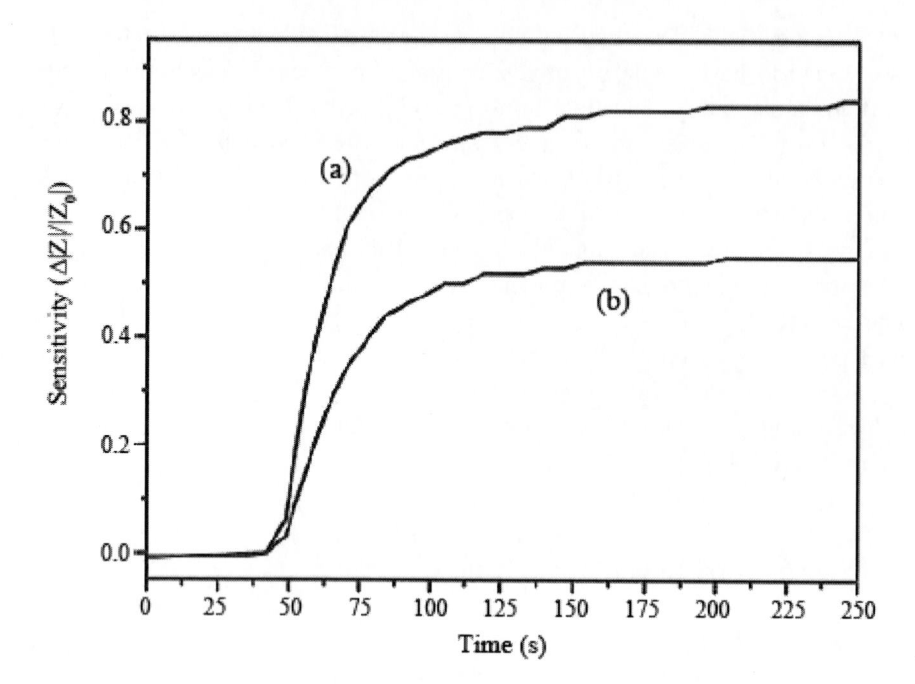

Figure 5. Real-time response of PAni sensor devices upon exposure to 1 ppm of NH_3 gas. Curves a and b are corresponding to the PAni films deposited on the 3-aminopropyl-triethoxysilane-treated electrodes with a current of 3 μ A and a time of 1800 s and 2700 s, respectively. (Reproduced with permission of the American Chemical Society from ref. [63]).

Huang et al. made resistive-type sensors from doped PAni nanofibers through interfacial polymerization outperforming conventional PAni on exposure to acid or base vapors, respectively.

When fully HCl-doped PAni is exposed to NH_3 gas, as expected, an interfacial polymerized PAni nanofiber film outperformed a conventional PAni film (Figure 7B). [51] Using an interfacial polymerization method, Virji et al. have developed nanofiber sensors, and five different response mechanisms are explored including acid doping, base dedoping, reduction, swelling, and polymer chain conformational changes. [68]

The PAni sensor has been compared to conventional PAni sensors when the sensors are exposed to NH_3 and N_2H_4 (Figure 7C) etc., which indicates that the high surface area, porosity and small diameters of the PAni nanofibers sensor enhance diffusion of molecules and dopants into the nanofibers.

2.2.2. Acidic Gases

Acidic and basic chemical vapors can change the conductivity of PAni through alternating the doping level of the emeraldine form of PAni. Huang et al. have synthesized PAni nanofibers by interfacial polymerization. [23] When the undoped PAni nanofibers are used as the sensitive layer up to HCl vapor, the time response and corresponding sensitivity of PAni films are strongly thickness dependent.

As the thickness is decreased from 1 um to 0.3 um, the response time increases significantly and the magnitude of the response at a fixed time increases by more than five orders of magnitude.

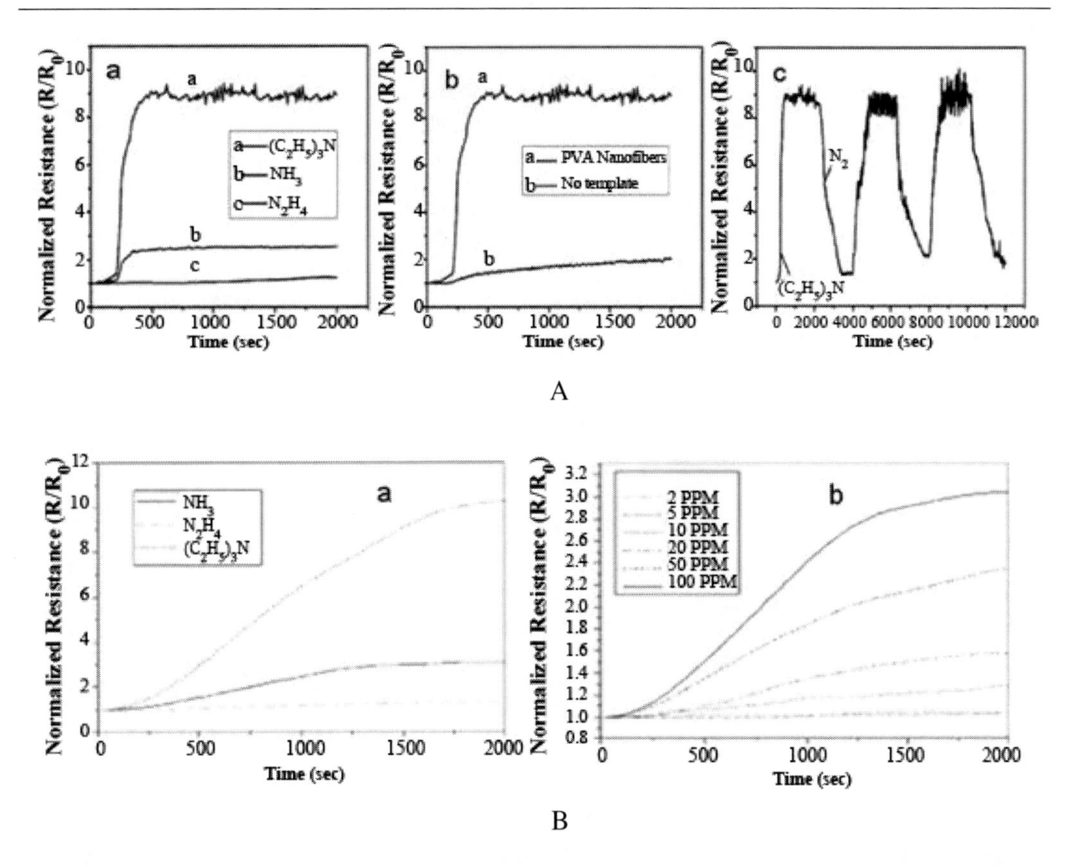

Figure 6. (A) (a) Resistance change of PAni nanotubes prepared using the electrospun PVA fiber mats template upon exposure to 100 ppm of different gases (NH_3, N_2H_4, and $(C_2H_5)_3N$). (b) Response of PAni prepared by using PVA fiber mats template and without a template exposed to 100 ppm of $(C_2H_5)_3N$. (c) The reversible circulation response change of PAni nanotubes upon exposure to 100 ppm of $(C_2H_5)_3N$. (Reproduced with permission of the American Chemical Society from ref. [65]) (B) (a) Resistance change of PAni nanotubes upon exposure to 100 ppm of different gases [NH_3 (solid line), N_2H_4 (dotted line), $(C_2H_5)_3N$ (dash line)]. (b) The PAni nanotubes exposed to different concentrations of NH_3. (Reproduced with permission of Wiley-VCH from ref. [66]).

When nanofibers are used as the selective layer, the performance is essentially unaffected by thickness, at least in the range of 0.2-2 μm (Figure 8A), which is due to the porous nature of nanofiber films, allowing vapor molecules to penetrate through the entire film and interact with all the fibers. [23] Virji et al. also measured the gas sensitivity of the PAni nanofibers obtained by interfacial reaction upon exposure to HCl vapor. [68] They found that the response time was measured to be -2 s for the nanofiber film and -30 s for the conventional PAni film (Figure 8B).

The HCl vapor sensitivity of PAni inverse opals was also measured by Yang et al. [67] The resistance of the PAni inverse opals placed in the opening of a bottle containing 37% hydrochloric acid solution showed a decrease of more than three orders of magnitude in 10 s.

2.2.3. Other Inorganic Gases

Surwade et al. obtained PAni nanofibers in one step using a simple and versatile high ionic strength aqueous system (HCl/NaCl) and permitting pure H_2O_2 as a mild oxidant.

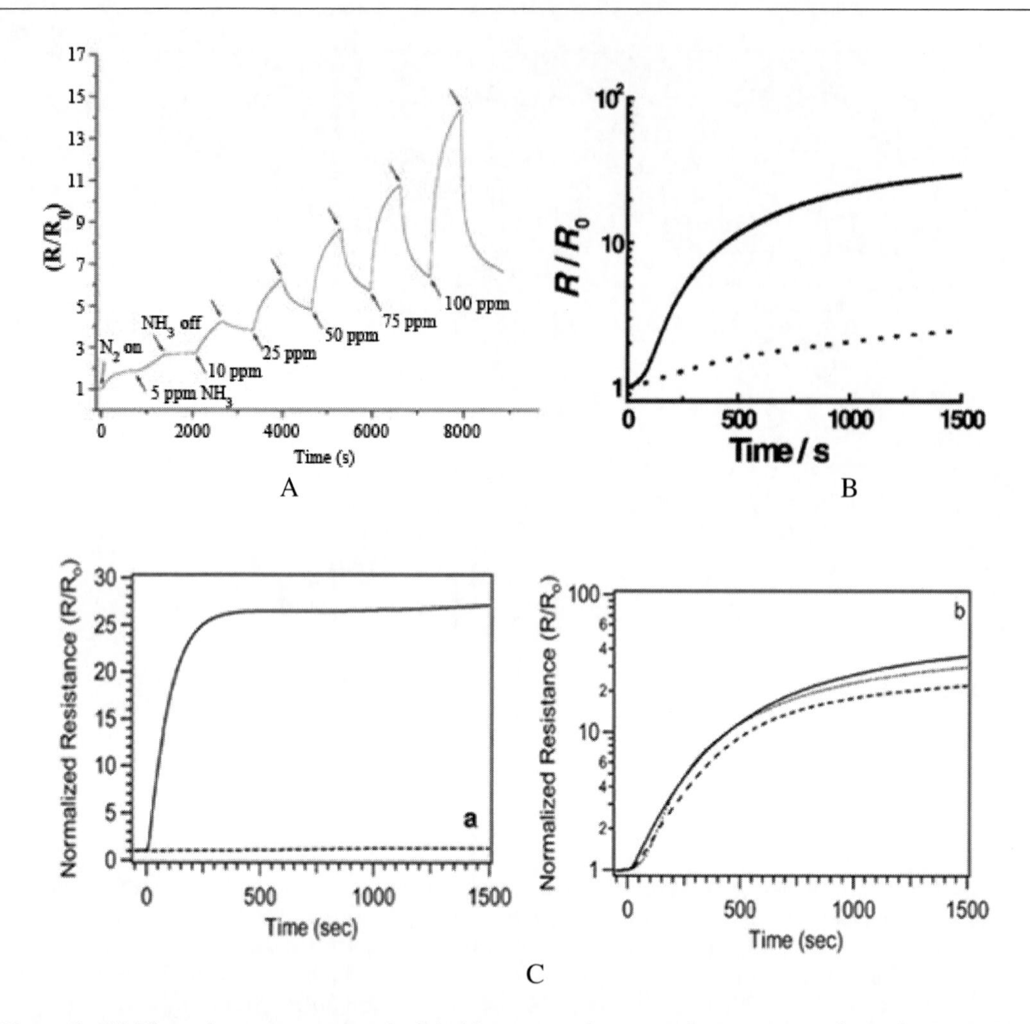

Figure 7. (A) The resistance response of a PAni inverse opal upon various concentrations (5, 10, 25, 50, 75, and 100 ppm) of NH_3. (Reproduced with permission of the Elsevier B.V. from ref. [67]) (B) Resistance changes of an HCl-doped PAni nanofiber thin film (solid line) and conventional PAni film (dotted line) upon exposure to 100 ppm of NH_3 in N_2. (Reproduced with permission of the American Chemical Society from ref. [51]) (C) (a) Response of 0.3 μm nanofiber (solid line) and conventional PAni (dash line) thin films to 3 ppm of N_2H_4, (b) Response of PAni films to 100 ppm of NH_3. Film thicknesses are as noted: 0.2 μm (solid line), 0.4 μm (dash line), 2.0 μm (dotted line). (Reproduced with permission of the American Chemical Society from ref. [68]).

Chemiresistors made of the obtained PAni thin films drop-cast on rugged and flexible plastic substrates could be capable of reversibly detecting NO_2 vapor (Figure 9A). [69] Li et al. presented a method for controlled deposition of PAni from colloidal suspensions formed by dispersing PAni/formic acid solution into acetonitrile on platinum and ITO substrate. [70] Figure 9B shows the relative changes of PAni film conductance in response to ppm levels of H_2O vapors in dry air at room temperature. Li et al. adopted a temperature controlling conductometric gas-sensing platform based on PAni. [71] The conducting PAni can serve as a donor or an acceptor when it is exposed to gaseous analytes. When PAni serves as a donor, the thermal excitation of electrons in PAni is beneficial to the charge transfer. On the contrary, at higher temperatures, the charge transfer becomes difficult (Figure 9C).

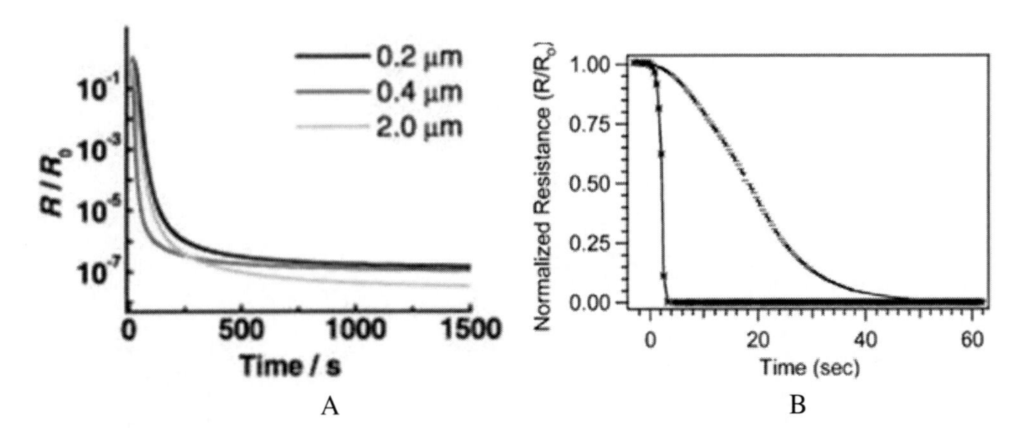

Figure 8. (A) Response of PAni nanofiber sensors of different thicknesses upon exposure to 100 ppm of HCl vapor. (Reproduced with permission of Wiley-VCH from ref. [23]) (B) Response of 0.3 μm nanofiber (solid line) and conventional PAni (dash line) thin films to 100 ppm of HCl. (Reproduced with permission of the American Chemical Society from ref. [68]).

2.2.4. Organic Gases

Some neutral and redox inactive volatile organic compounds, such as chloroform or toluene, have been reported to affect the conductance of PAni films through swelling effects by modifying the inter-chain hopping of charge carriers. Li et al. fabricated a sensor based on PAni nanofiber thin films by UV-irradiation of a precursor solution in a single-step and bottom-up process. The responses of bulk PAni and PAni nanofibers exposed to chloroform, and toluene were tested. The different mechanisms such as a weak proton donor for chloroform and a vapor that causes polymer swelling for toluene have been investigated and suggested. [27] Alcohol cannot alter the oxidation state of the doping level of PAni and hence has only a limited effect on the conductivity. One mechanism is through the formation of hydrogen bonds between the molecules and PAni; the manner of binding, PAni chain conformation, and conductance change are dependent on the doping level. Another is the swelling of PAni chains, which causes an increase in resistance. [72, 73] Ayad et al. fabricated a sensor of alcohol vapors based on an emeraldine base (EB) PAni thin films coating on the quartz crystal microbalance (QCM) electrode. The frequency shifts ($\triangle f$) of the sensor showed good reproducibility and reversibility upon exposure to alcohol vapors such as ethanol, methanol, 2-propanol and 1-propanol vapor. [30]

3. PANI/METAL NANOCOMPOSITES FOR GAS SENSORS

The potential applications of PAni have been obstructed because of the insolubility in the common solvent, low process ability and poor mechanical properties. [74] In order to meet the industrial demand, several methods have been developed to overcome such problems. Recently, from both an academic and industrial point of view, the multifunctionality of PAni/ metal composites with synergistic chemical and physical properties is particularly useful and has received considerable attention due to their enhanced gas sensing properties and electro catalytic activity, memory devices, and others, as compared to those of pure PAni. [75-79]

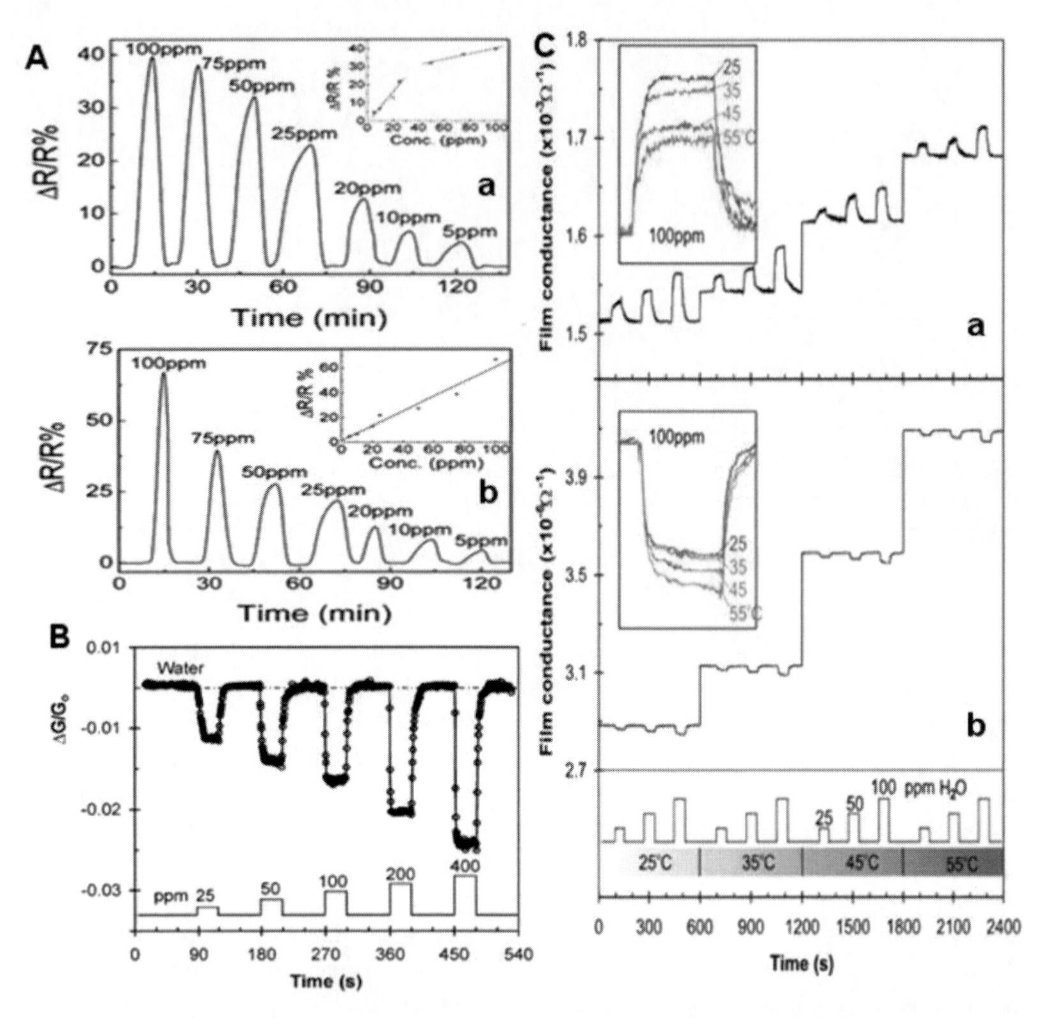

Figure 9. (A) Percent change in resistance vs. time of (a) aniline octamer and (b) aniline hexadecamer, when exposed to various concentrations of NO_2 gas (N_2 carrier gas). Insets: corresponding plots of percent change in resistance vs. vapor concn. (Reproduced with permission of the American Chemical Society from ref. [69]) (B) Relative changes in the conductance of the PAni film upon exposure to various concentrations of H_2O vapors in dry air. (Reproduced with permission of the American Chemical Society from ref. [70]) (C) Changes in conductance of PAni sensing films to various concentrations of H_2O in dry air at different temperatures, when PAni is acting (a) as an acceptor and (b) as a donor. Inset: temperature dependence of the sensor upon exposure to 100 ppm of H_2O vapor. (Reproduced with permission of the American Institute of Physics from ref. [71]).

Naturally, incorporation of the metal nanoparticles and the PAni molecule can generate a new family of materials, designated as PAni nanocomposite doped with metal, which can effectively improve the electrical, optical and dielectric properties. [80-82] These properties are very sensitive to small changes in the metal content and in the size and shape of the metal nanoparticles. [74] The metal nanoparticles can act as conductive junctions between the PAni chains that result in an increase of the electrical conductance of the PAni/metal composites. Moreover, the introduction of metal nanoparticles into organic films is effective in promoting the chemiresistor sensitivity to gas through the apparent creation of new chemisorption sites. [83]

The method of preparing the PAni/metal composites usually involves metal synthesis by chemical reduction of a metal cation in the presence of the aniline molecule, which can be carried out with water-soluble reducing agents as well as metal powder. [84-86] However, most of the distribution of the metal nanoparticles in the PAni matrix is not very uniform. In order to uniformly disperse metal nanoparticles in PAni composites, more new synthesis processes are required.

3.1. PAni/Metal Sensors for NH_3 Detection

Using NC fiber mats as a suitable template, Li et al. synthesized $H_2SiW_{12}O_{40}$-doped conducting PAni nanotubes encrusted with uniform dispersed Ag nanoparticles through UV rays irradiation for the first time. [32] The high surface areas, small diameter, and porous nature of the PAni/Ag composite nanotubes and the introduction of metal into PAni films are effective in promoting the chemiresistor sensitivity and fasting response and recovery rates to NH_3 at room temperature. Meanwhile, the stable reversible circulation response change of PAni/Ag composite nanotubes showed that they may have more potential applications in the area of sensor development (Figure 10A). [32] Gao et al. also successfully synthesized PAni/ Ag composite nanotubes by a self-assembly polymerization process using ammonium persulfate and silver nitrate as an oxidant without using any acid as dopant or hard template. The PAni/Ag composite nanotubes were immobilized on the surface of an indium tin oxide and applied to construct a sensor, which exhibited higher sensitivity upon exposure to NH_3 gas than pure PAni (Figure 10B). [87]

3.2. PAni/Metal Sensors for Detection of Other Gases

Chowdhury et al. successfully synthesized Ni-coated PAni nanowire using camphor sulfonic acid (CSA) as a dopant as well as a soft template. [28] Such material has potential for use in cigarette smoke detectors. As shown in Figure 11A, the PAni-CSA-Ni nanowire composite shows nearly four order decreases in AC impedance in the presence of cigarette smoke. Shirsat et al. reported a sensitive and fast responding room temperature chemiresistive hydrogen sulfide (H_2S) sensor based on PAni nanowires-gold nanoparticles (Au) hybrid network fabricated by facile templateless electrochemical polymerization using cyclic voltammetry technique. These chemiresistive sensors showed the lowest limit of detection 0.1 ppb, wide dynamic range 0.1–100 ppb, and very good selectivity and reproducibility when exposed to H_2S gas which is a toxic, corrosive, and inflammable gas produced in sewage, coal mines, oil, and natural gas industries (Figure 11B). The reaction of H_2S on the gold nanoparticles was obviously a critical step in the detection. In this reaction, the formation of AuS along with the enhancement of the doping level of PAni was assumed to change the resistance of the PAni-gold nanoparticles network. The transfer of electrons from the p-type PAni network as a donor to gold nanoparticles as an acceptor could decrease the resistance. [88] Sharma et al. reported a chloroform vapor sensor based on copper/PAni nanocomposite. It exhibited reversible and fairly good response to low concentrations of chloroform (<100 ppm) compared to pure PAni. The sensing mechanism mainly referred to adsorption-desorption of chloroform at metal cluster surfaces. [89]

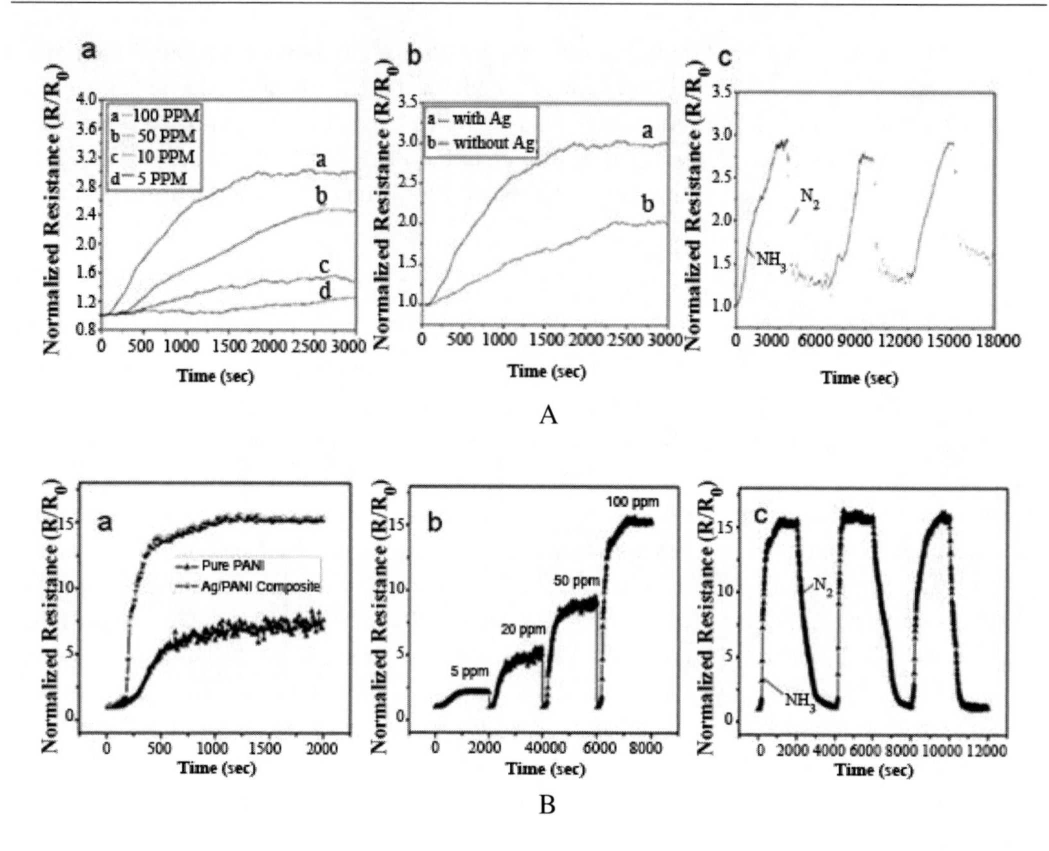

Figure 10. (A) (a) Resistance change of the PAni/Ag composite nanotubes upon exposure to different concentrations of NH_3 gas. (b) Response of PAni nanotubes with-a and without-b Ag exposed to 100 ppm of NH_3 gas. (c) The reversible circulation response change of PAni/Ag composite nanotubes up to 100 ppm of NH_3. (Reproduced with permission of the American Chemical Society from ref. [32]) (B) (a) Resistance changes of a-PAni/Ag composite nanotubes and b-pure PAni exposed to 100 ppm NH_3. (b) PAni/Ag composite nanotubes upon exposure to different concentrations of NH_3. (b) Reversible circulation response changes of PAni/Ag composite nanotubes upon exposure to NH_3 (100 ppm). (Reproduced with permission of the American Chemical Society from ref. [87]).

Choudhury et al. prepared PAni/Ag nanocomposites via *in-situ* oxidative polymerization of aniline monomer in the presence of different concentrations of Ag nanoparticles. It was found that the PAni/Ag nanocomposite possessed a superior ethanol sensing capacity compared to pure PAni and it showed that the significant enhancement of the ethanol vapor-sensing ability was in accordance with increasing Ag concentrations in the nanocomposite. [74]

4. Other PAni Nanocomposites for Gas Sensors

4.1. PAni/Metal Oxide Nanocomposite

Yao et al. has successfully prepared PAni/MnO_2 nanocomposites with special morphology by using special morphological urchin-like MnO_2 as a single template through solid-state methods. [90] The sensitivity of the PAni/MnO_2 nanocomposites to NH_3 has been increased by 4-5 times compared with other PAni material (Figure 12).

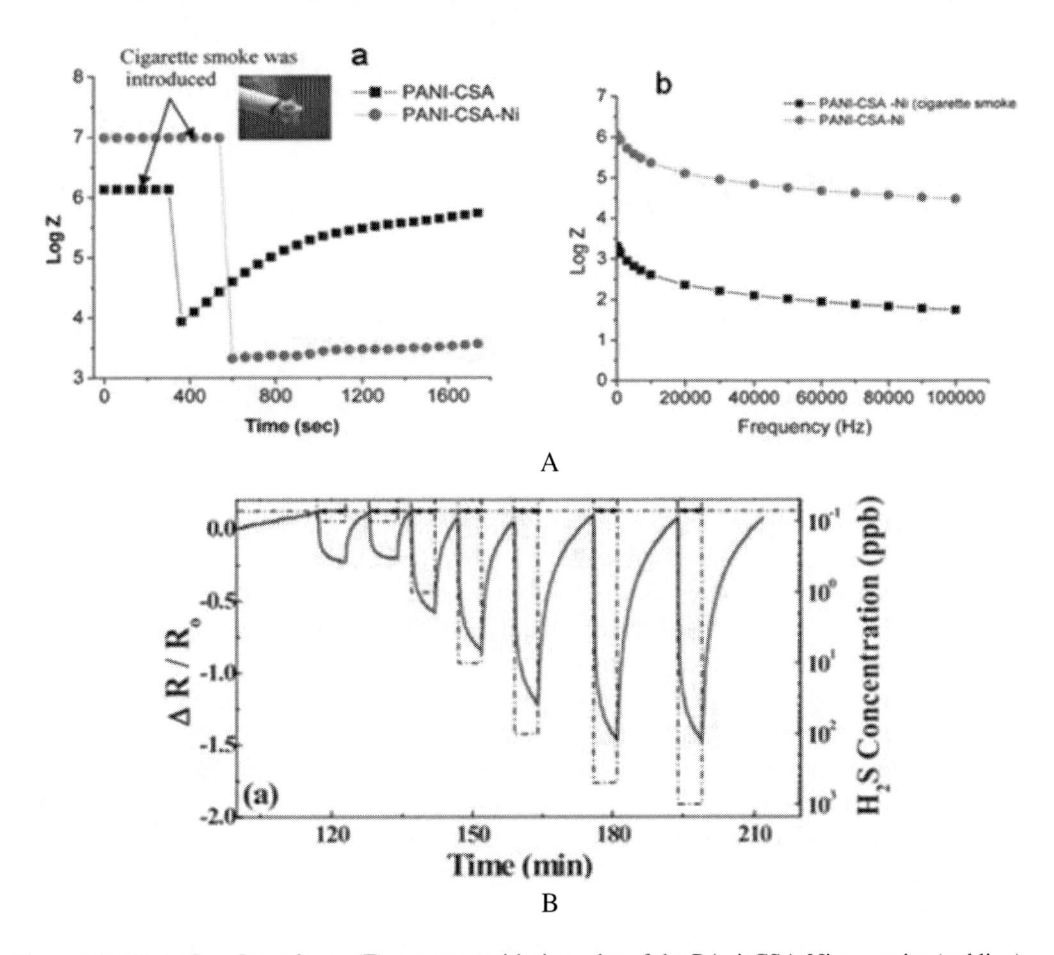

Figure 11. (A) (a) Log Impedance (Z) response with time plot of the PAni-CSA-Ni nanowire (red line) and Log Z versus time plot of PAni-CSA (black line) in the presence of cigarette smoke measured at 42 Hz and at 30°C. (b) Log Z versus frequency plot of PAni-CSA-Ni nanowire in the presence (black line) and absence (red line) of cigarette smoke in different frequencies from 42 Hz to 1MHz. (PAni: PAni). (Reproduced with permission of the Elsevier B.V. from ref. [28]) (B) Response and recovery transients (solid line) of Au-PAni nanowire network based chemiresistive sensors toward different concentrations of H_2S gas: 0.1 ppb, 1 ppb, 10 ppb, 100 ppb, 500 ppb, and 1 ppm (dashed line). (Reproduced with permission of the American Institute of Physics from ref. [88]).

Xie et al. made a methane (CH_4) gas sensor by depositing PAni/PdO thin films synthesized by chemical oxidative polymerization of aniline with PdO nanoparticles on a mass-type quartz crystal microbalance (QCM) device through a layer-by-layer self-assembly method. [91] This exhibited that the PAni/PdO based QCM gas sensor had a better response at room temperature than that at 50°C. Except for temperature, the humidity would also have a significant effect on the PAni/PdO nanocomposite thin film sensors. The relationship between the frequency of the PAni/PdO based QCM sensor and the relative humidity (RH) was also exhibited.

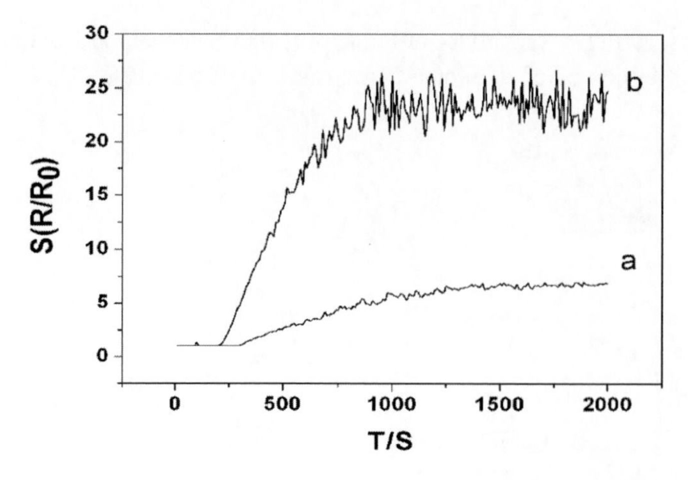

Figure 12. Response curve of gas sensor on (a) pure PAni, and (b) PAni/MnO_2 nanocomposites exposure to 100 ppm of NH_3.

Zheng et al. obtained a QCM chemical sensor by depositing a PAni/TiO_2 nanocomposite prepared by combining chemical polymerization and a sol-gel method on the electrode of quartz crystal. [92] The calibration curve towards trimethylamine exhibited long-term stability and selectivity. The thermal behavior of the sensing characteristics was investigated. A PAni/TiO_2 QCM sensor was compared with a PAni QCM sensor. The amplitudes of the sensor responses to the same concentration of the different gases were tested, including those of trimethylamine, triethylamine, ethanol, formaldehyde, and acetaldehyde gases in N_2, respectively, which showed that the response to trimethylamine was much higher than to other gases. Tai et al. obtained PAni/TiO_2 nanocomposite thin films processed on a silicon substrate with gold interdigital electrodes by an in-situ self-assembly approach. [93] The PAni/TiO_2 nanocomposite thin film gas sensor could be applied to detect NH_3, and the polymerization temperature had an effect on the gas response. The gas response of the PAni/ TiO_2 was superior to those prepared at other temperatures, and the thin film prepared at 10°C also exhibited long-term stability and good reproducibility. The responses of the thin film sensor prepared at 10°C to 23 ppm of CO and 500 ppm of H_2 were measured. It could be seen that the sensor showed very weak responses to CO as well as H_2 and exhibited high selectivity to NH_3. Scheme 2 illuminated the NH_3 gas sensing mechanism of PAni/TiO_2 nanocomposite thin films, the well energy band gap matching between the conduction band of TiO_2 and the lowest unoccupied molecular orbital (LUMO) level of PAni enhanced charge separation; such enhancement promoted the NH_3 gas-sensing ability of the PAni/TiO_2 nanocomposite. [93]

4.2. PAni/Metal Salt Nanocomposite

It is important to detect hydrogen sulfide (H_2S) because it is colorless, flammable, heavier than air, and dangerous at concentrations above 20 ppm. Virji et al. have shown that PAni nanofiber composites with transition-metal chlorides have a remarkable response to H_2S gas, which is four orders of magnitude greater than the corresponding unmodified PAni nanofibers. [94] Virji et al. have demonstrated that copper acetate and other salts could be incorporated into the PAni nanofibers matrix. [95]

As can be seen from Figure 13A, a copper chloride/PAni nanofiber composite film exposed to 10 ppm of H_2S exhibits a greater than 4 orders of magnitude decrease in resistance. A composite film formed from copper acetate and PAni nanofibers is more responsive than the copper chloride/PAni nanofiber composite film, and its response to H_2S is shown in Figure 13B. Arsine (AsH_3) is a very toxic gas used in the semiconductor industry with a permissible exposure level (PEL) of 50 ppb. This low detection requirement determines the need for a highly sensitive detector for AsH_3. Virji et al. also reported the use of PAni nanofiber/metal salt composites for the detection of AsH_3 gas using the resistance change of the composites including $CuCl_2$, $CuBr_2$, CuF_2, $Cu(O_2CCH_3)_2$, $Cu(NO_3)_2$, $EuCl_2$, $NiCl_2$, $FeCl_3$, and $CoCl_2$. The response is dependent on the metal content and the counterion of the metal salt, as well as temperature and humidity. Among the various composites, the $CuBr_2$/PAni nanofiber composite shows the best response with the greatest change in resistance upon exposure to AsH_3 gas (Figure 13C). They proposed the mechanism for the detection of AsH_3 by their PAni nanofiber/metal salt composites (Scheme 3). In the whole process, AsH_3 could reduce copper (II) to copper (I); the PAni served not only as a conducting backbone but also as a reducing agent to stabilize and make copper (I) available for the enhancement of this reaction. The acid byproduct of the redox reaction doped the PAni resulting in a decrease in resistance as seen upon exposure to AsH_3. [96]

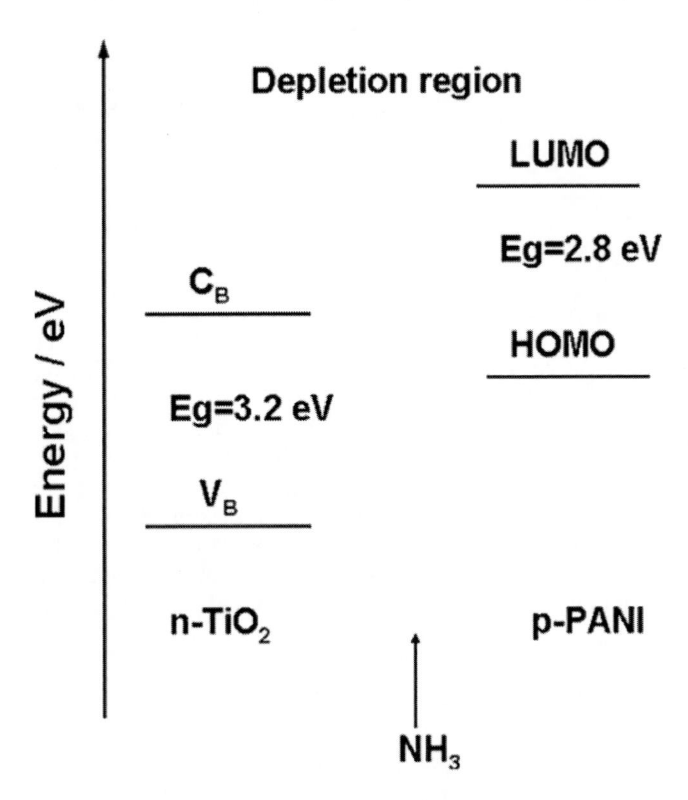

Scheme 2. Energy-band for PAni/TiO_2 nanocomposite (PAni: PAni) [93].

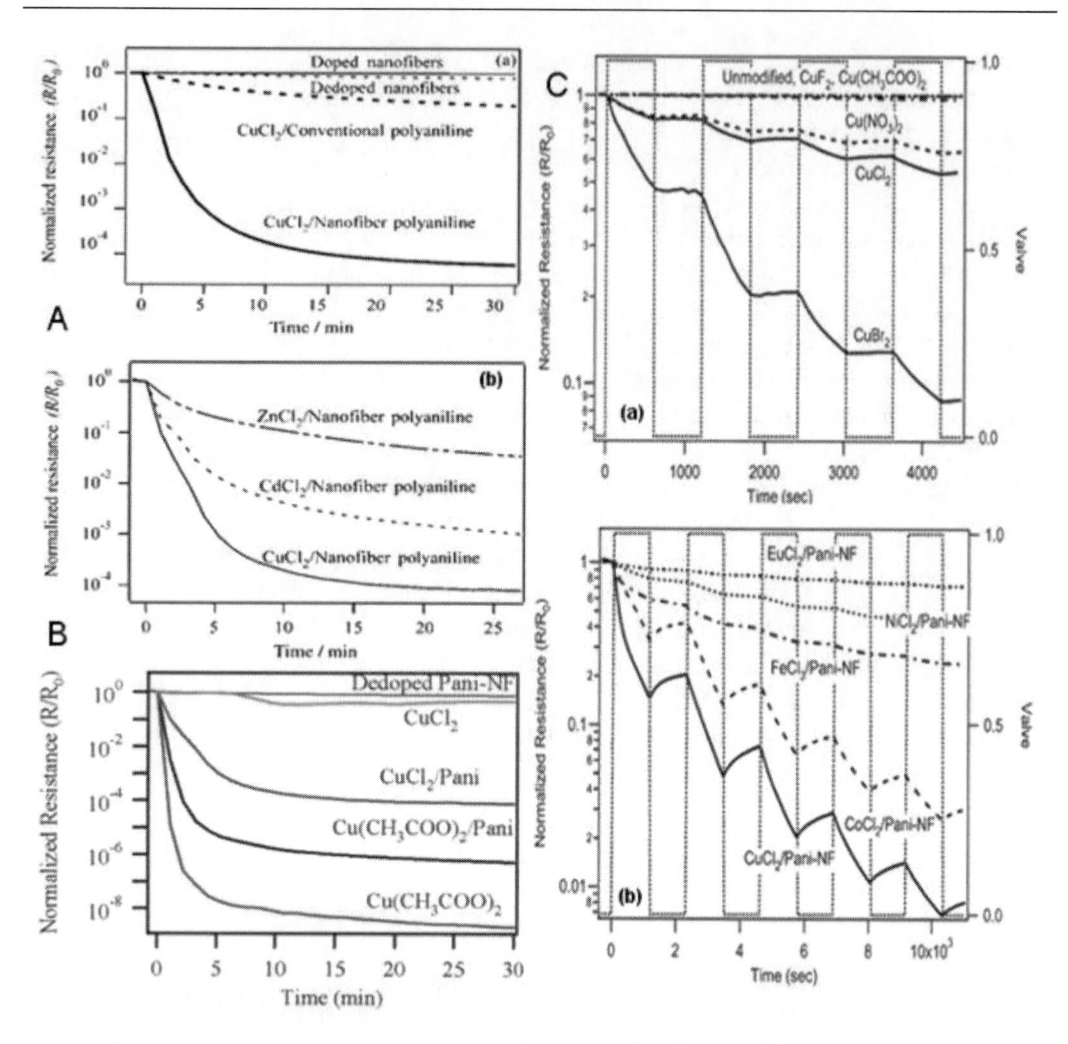

Figure 13. (A) (a) Resistance changes of the PAni films upon exposure to 10 ppm H_2S doped PAni nanofibers (solid line), dedoped PAni nanofibers (dotted line), conventional PAni containing copper chloride (thick dotted line), and PAni nanofibers containing copper chloride(thick solid line), and (b) Resistance changes of PAni nanofiber films containing $ZnCl_2$, $CdCl_2$, and $CuCl_2$ on exposure to 10 ppm H_2S under 45% relative humidity; all film thicknesses were 0.25 mm. (Reproduced with permission of Wiley-VCH from ref. [94]) (B) Electrical response of dedoped PAni nanofibers (green), copper chloride films (magenta), copper chloride/PAni nanofibers (dark blue), copper acetate/PAni nanofibers (black), and copper acetate films (red) exposed to 10 ppm of H_2S. (Reproduced with permission of the American Chemical Society from ref. [95]) (C) Unmodified and different metal-salt-modified PAni nanofiber films exposed to (a) 500 ppb AsH_3, and (b) 1 ppm AsH_3 at room temperature with 50% relative humidity. (Reproduced with permission of the American Chemical Society from ref. [96]).

4.3. PAni/Polymer Nanocomposite

Weng et al. designed a triethylamine (TEA) gas sensor based on one-dimensional PAni–polypyrrole coaxial nanofibers (PPCF) created through in situ polymerization of the polypyrrole layer on the surface of interface polymerized PAni nanofibers. [97]

$$CuX_2 \xrightarrow{PANi} CuX_2/PANi + CuX/PANi \quad (1)$$

$$AsH_3 + 3CuX_2/PANi \xrightarrow[cat.]{CuX/PANi} 3CuX/PANi + As^0 + 3HX \quad (2)$$

$$HX + PANi \longrightarrow \text{Doped PANi} \quad (3)$$

$$AsH_3 + 3CuX/PANi \longrightarrow 3Cu^0 + As^0 + 3HX/PANi \quad (4)$$

Scheme 3. Proposed mechanism of interaction of PAni nanofiber/metal salt composite upon exposure to AsH_3 gas. [96].

The comprehensive performance containing a rapid, sensitive and reversible conductance change upon exposure to TEA gas was better than the results obtained using PAni nanofibers and polypyrrole separately.

Liu et al. formed a nanowire chemical sensor utilizing individually oriented PAni/poly (ethylene oxide) nanowires deposited on lithographically defined gold microelectrodes through an electro-spinning process. [35] The sensor showed a rapid and reversible resistance change upon exposure to NH_3 gas at concentrations from 0.5 ppm to 50 ppm. The response times of polymeric nanowire sensors were different because of the various diameters upon the same concentration of NH_3 (Figure 14A).

Airoudj et al. designed a new multilayer integrated optical sensor based on PAni as a sensitive layer and polymethyl methacrylate as a passive layer, which was deposited between the waveguide core and the PAni sensitive layer to decrease optical losses. [98] This multilayer structural sensor showed fast response and recovery times, wide detection range (from 92 ppm to 4618 ppm), and good reversibility to NH_3 gas (Figure 14B).

4.4. PAni/Other Material Nanocomposites

Mashat et al. synthesized a graphene/PAni nanocomposite through a chemical synthetic route that graphene was prepared and ultrasonicated with a mixture of aniline monomer and ammonium persulfate to form PAni on its surface. [99]

They investigated the H_2 gas sensing performance of this nanocomposite, as shown in Figure 15A, and it was found that the graphene/PAni nanocomposite-based device sensitivity was much larger than the sensitivities of sensors based on only graphene sheets and PAni nanofibers. Yu et al. reported a QCM methane sensor based on a PAni and ionic liquid butylmethylimidazolium camphorsulfonate (BMICS) composite as a sensing material. [100]

The sensitivity for methane detection of the composite was relatively higher than those in pure BMICS and in PAni only (Figure 15B).

They also discussed the interactions within the PAni/BMICS composite and between the composite and methane. The hydrogen bonds formed from the anion of BMICS and the "nitrogen" sites of protic acid doped PAni aligned the camphorsulfonate anions in a comb-like manner along the PAni backbone and enhanced the long-range π-orbital conjugation of PAni.

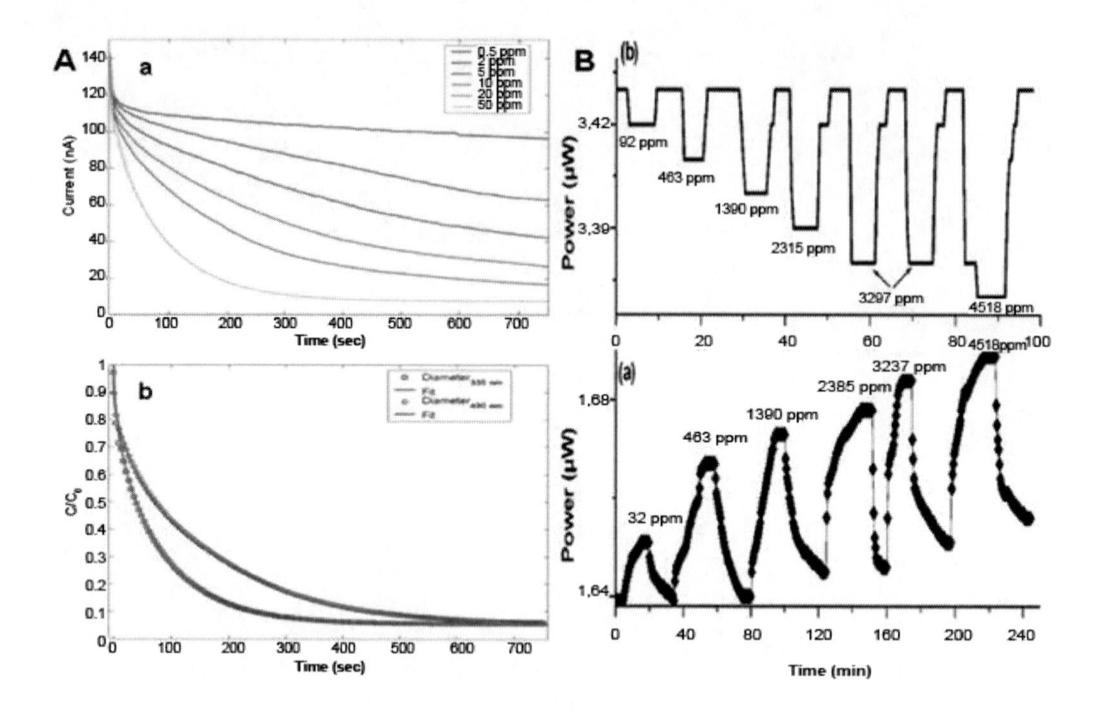

Figure 14. (A) (a) Measured time-dependent current through an individual nanowire sensor (the diameter was 335 nm) upon exposure to NH_3 gas at different concentrations. (b) Time-dependent conductance change for two nanowire sensors with wire diameters of 335 and 490 nm, measured upon exposure to 50 ppm NH_3. (Reproduced with permission of the American Chemical Society from ref. [35]) (B) TE_0 transmitted light power variations of a multilayer structural sensor exposed to different NH_3 concentrations (a) doped PAni; (b) dedoped PAni. PAni film thickness: 130 nm. Interaction length: 2 mm. (Reproduced with permission of the American Chemical Society from ref. [98]).

Methane molecules absorbed into the PAni/BMICS may sit in the "space" between the aligned anions and cations of BMICS. So the composite exhibited a significant increase in the sensitivity for methane detection. Virji et al. obtained an array of various amine-PAni nanofiber composite materials (including ethylenediamine, phenylenediamine and metanilic acid (3-aminobenzenesulfonic acid), and the amine salts phenylenediamine dihydrochloride and ethylenediamine dihydrochloride) made by drop-casting using disposable microliter pipettes for the detection of phosgene gas. [101] Amines could react with phosgene in a nucleophilic substitution reaction and in this reaction hydrochloric acid is formed, which can convert PAni from the emeraldine base oxidation state to the emeraldine salt oxidation state. This resulted in two orders of magnitude decrease in resistance. Sotzing et al. reported the construction of an "electronic nose" consisting of an array of PAni/carbon black composite to detect various organic vapors. The sensitivity of this composite was shown in Figure 15C. [102] Virji et al. reported a new sensor using conventional PAni thin films processed from hexafluoroisopropanol (HFIP) or hexafluoro-2-phenylisopropanol (HFPP). [103] The fluorinated alcohols could react with hydrazine to produce a strong hydrofluoric acid (HF) which protonated the emeraldine base form of PAni leading to large increases in conductivity (Figure 15D). In contrast, conventional PAni films processed from other solvents became more insulating upon exposure to hydrazine because the hydrazine served as a strong reducing agent and had the ability to convert the emeraldine PAni to the leucoemeraldine oxidation state.

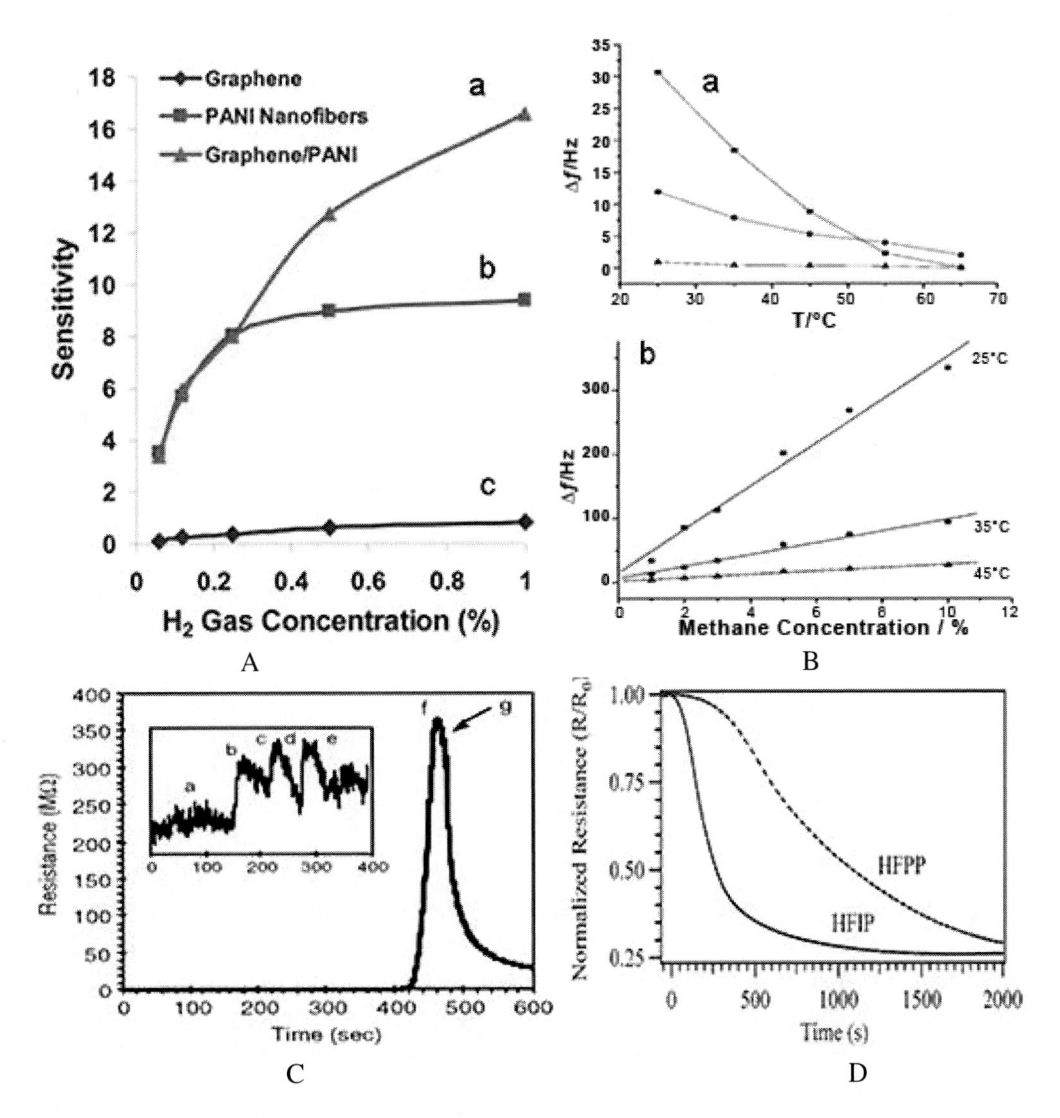

Figure 15. (A) The sensitivities of the H_2 gas sensors with sensitive layers of (a) graphene/PAni, (b) PAni nanofibers, and (c) grapheme sheets. (Reproduced with permission of the American Chemical Society from ref. [99]) (B) (a) Frequency changes at different temperatures on PAni (●), PAni/BMICS composite (■), and BMICS (▲), and (b) Calibration curves of methane absorption at 25, 35, and 45°C on a PAni/BMICS composite via QCM measurement. (Reproduced with permission of the American Chemical Society from ref. [100]) (C) Resistance response of a composite detector exposed to water (14.7 ppm) (a), acetone (145 ppm) (b), methanol (79 ppm) (c), ethyl acetate (59 ppm) (d), butanol (4 ppm) (e), and butylamine (58 ppm) (f), all at 0.05% of their saturated vapor pressure. (Reproduced with permission of the American Chemical Society from ref. [102]) (D) Response of PAni films processed using fluorinated alcohols diluted with N-methylpyrrolidinone (NMP) exposed to hydrazine: HFIP/NMP (solid line); and HFPP/NMP (dash line). The molar ratio is 1:4 in each case. (Reproduced with permission of the American Chemical Society from ref. [103]).

5. PANI-BASED SCHOTTKY JUNCTION OR P-N JUNCTION FOR GAS SENSORS

5.1. PAni-Metal Schottky Junction

In recent years, π-conjugated conducting polymer-based nanostructured materials have been utilized extensively in resistive sensors in order to shape the next generation of cheap and disposable electronic inventions.

The simplest and easiest configuration of an electronic sensor is a hybrid organic/ inorganic Schottky diode based on a resistive junction formed between a p-type polymer and a metal electrode or an n-type inorganic semiconductor. The polymers can be coated on the metal electrode or the n-type inorganic semiconductor via the electro-spinning method, electrochemical polymerization, mechanical stretching, and magnetic field-assisted assembly process. Pinto et al. reported an easy method to fabricate Schottky diodes by depositing the electrospun PAni nanofibers on the n-doped Si, and the schematic was shown in Scheme 4. The resulting Schottky diode was formed along the vertical edge of the substrate at the nanofiber-doped Si interface. [104]

Figure 16A shows the results of this sensor based on Schottky diode measured in a vacuum. After NH_3 exposure and after pumping out the NH_3, it can be seen that the change in the diode response is instantaneous upon exposure to NH_3 with nearly complete recovery of the current upon pumping because of the large surface to volume ratio of the PAni nanofiber. Zhang et al. demonstrated an Au/PAni/Au junction sensor through the magnetic-field-assisted assembly, as shown in Scheme 5. The Au/PAni/Au junction sensor exhibited good sensitivity and fast response time when exposed to HCl and butylamine vapors (Figure 16B). [105] By depositing PAni nanofibers on either gold or platinum electrodes, the resistive sensors for the detection of hydrogen (H_2) were formed. [35]

On gold electrodes, hydrogen interacts directly with PAni nanofibers to induce a small resistance decrease at 1% concentration of H_2. However, on platinum electrodes, H_2 interacts with platinum at the PAni-platinum interface to form platinum hydride, a Schottky barrier between platinum and PAni that causes a large increase in resistance (Figure 16C). Wang et al. fabricated individually addressable PAni nanoframework electrode junctions in a parallel-oriented array via an electrochemical method, and these junctions could be used for the chemical sensing of various vapors (Figure 17). [72]

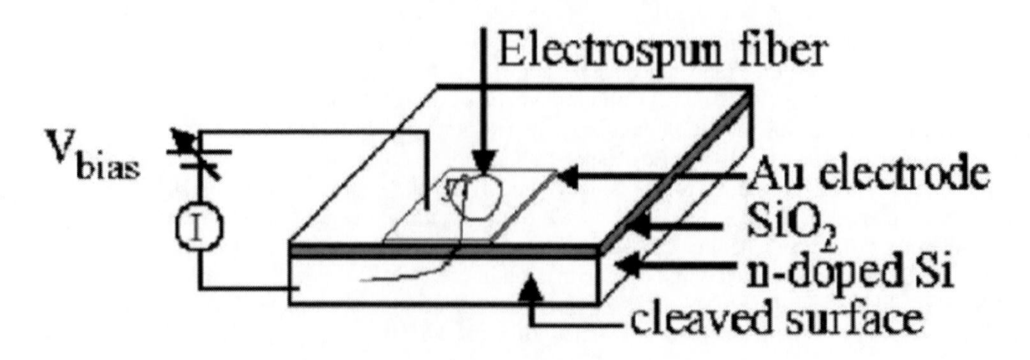

Scheme 4. The schematic of the device and the external electrical circuit. [104].

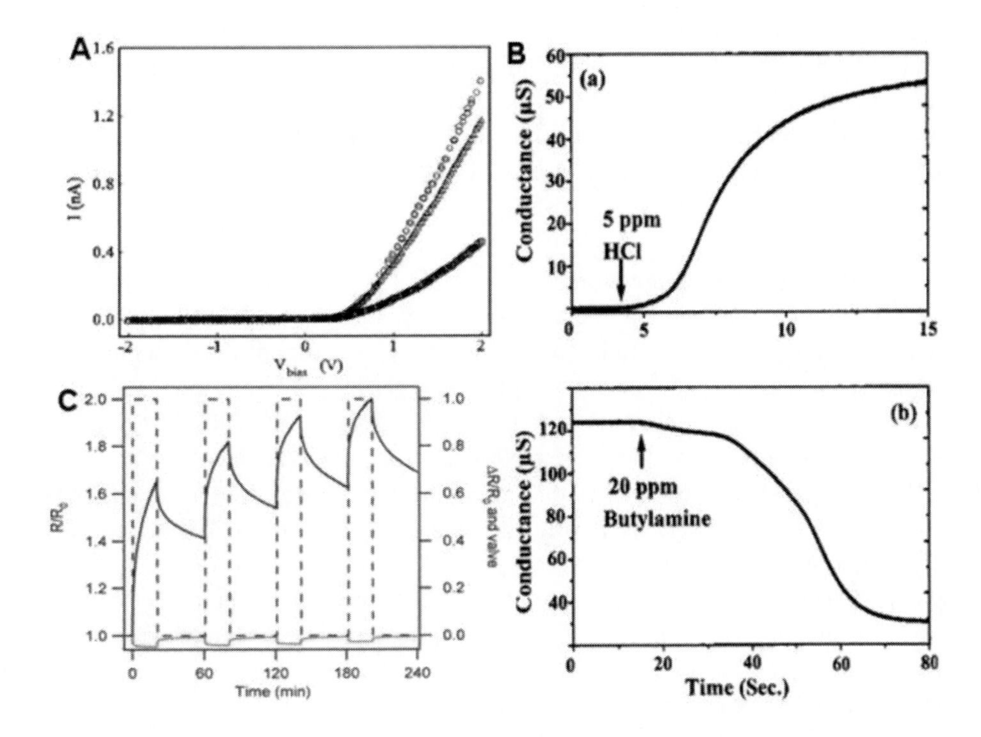

Figure 16. (A) I-V characteristic curves for a Schottky nanodiode measured in the following sequence: [○] in vacuum, [□] a few seconds after exposure to NH_3 vapor, and [△] after 5 h of pumping. (Reproduced with permission of the American Institute of Physics from ref. [104]) (B) Conductance response of Au/PAni/Au junction sensor to (a) HCl and (b) butylamine vapors. (Reproduced with permission of the American Chemical Society from ref. [105]) (C) Resistance change of the Pt-PAni sensor (black) and Au-based PAni sensor (red) when exposed to 1% H_2 in dry N_2. (Reproduced with permission of the American Chemical Society from ref. [34]).

5.2. Pani-Inorganic Semiconductor P-N Junction

It has been well known that when a p-type conductive polymer and a n-type inorganic semiconductor are in direct contact, a p-n junction can form at the interface of these two materials. In recent years, a heterojunction between the p-type polymer and n-type semiconductive inorganic materials has been developed for detecting small amounts of various toxic gases that exist in air, due to their low cost and simplicity of techniques used. The gas sensing mechanism of these semiconducting gas sensors is different from single oxide semiconductors. The heterojunction type sensors work on the principle of barrier mechanism and need no adsorption and desorption of oxygen for the detection of gas.

5.2.1. Heterojunction Sensor for Liquefied Petroleum Gas (LPG) Detection

Hazardous gases, specifically LPG, have been widely used for several industrial and domestic applications. But, at certain low concentrations of the toxic gases, the metal oxide-based sensors show poor performance including the sensitivity, long-term stability, and selectivity. As an option, the fabrication of heterojunctions between organic and inorganic materials can enhance the characteristics and mechanical strength of the sensor.

Using electrochemically deposited PAni on chemically deposited n-type inorganic semiconductors on a stainless steel substrate, different heterojunctions can be fabricated, such as PAni/TiO_2, PAni/CdTe, PAni/CdS p-n junctions. These heterojunctions show the maximum gas response of 63% upon exposure to 0.1 vol % LPG, 67.7% upon exposure to 0.14 vol % LPG, and 80% upon exposure to 1040 ppm LPG at room temperature, respectively. [31, 106, 107]

5.2.2. P-N Junction for NH_3 Detection

Gong et al. reported for the first time an ultrasensitive nanostructured sensor fabricated by encrusting nanograins of p-type conductive PAni on an electrospun n-type semiconductive TiO_2 microfiber surface that could detect 50 ppt of NH_3 gas in air (Figure 18a). The resistance of the p-n heterojunctions, combining with the bulk resistance of PAni nanograins, could function as electric current switches when NH_3 gas was absorbed by PAni nanoparticles (Figure 18b). Scheme 6a shows a simple case where an open voltage is applied to a semiconductive TiO_2 microfiber. Because of the high resistance of the TiO_2 microfiber, the current flowing through the microfiber is very low.

However, if a conductive PAni nanoparticle is attached to a TiO_2 microfiber, as shown in Scheme 6b, the situation is complicated due to the formation of a p-n junction between a p-type PAni and an n-type TiO_2 microfiber. On one hand, the current flows preferably through the conductive PAni particle due to the fact that PAni is much more conductive than the TiO_2 microfiber. On the other hand, the current flowing through the PAni nanoparticle may be not favorable because it must overcome the reverse-bias resistance of the p-n junction between the p-type PAni particle and n-type TiO_2 microfiber. The equivalent current circuit is shown on the bottom of Scheme 6b, where R_1 is the resistance of the forward bias of the p-n junction, and R_3 is the resistance of reverse bias of the p-n junction, and R_2 and R_4 are the bulk resistances of the PAni nanoparticle and TiO_2 microfiber, respectively.

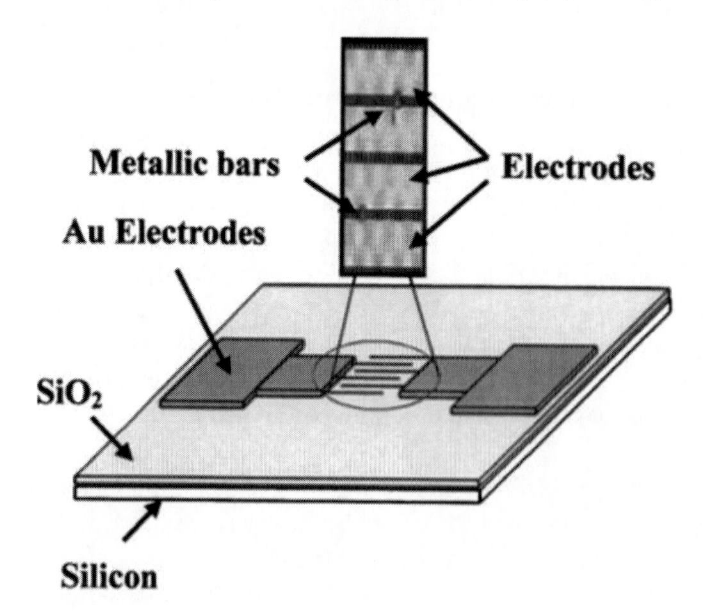

Scheme 5. Schematic diagram of the magnetic-field-assisted assembly of the Au/polymer/Au junctions. [105].

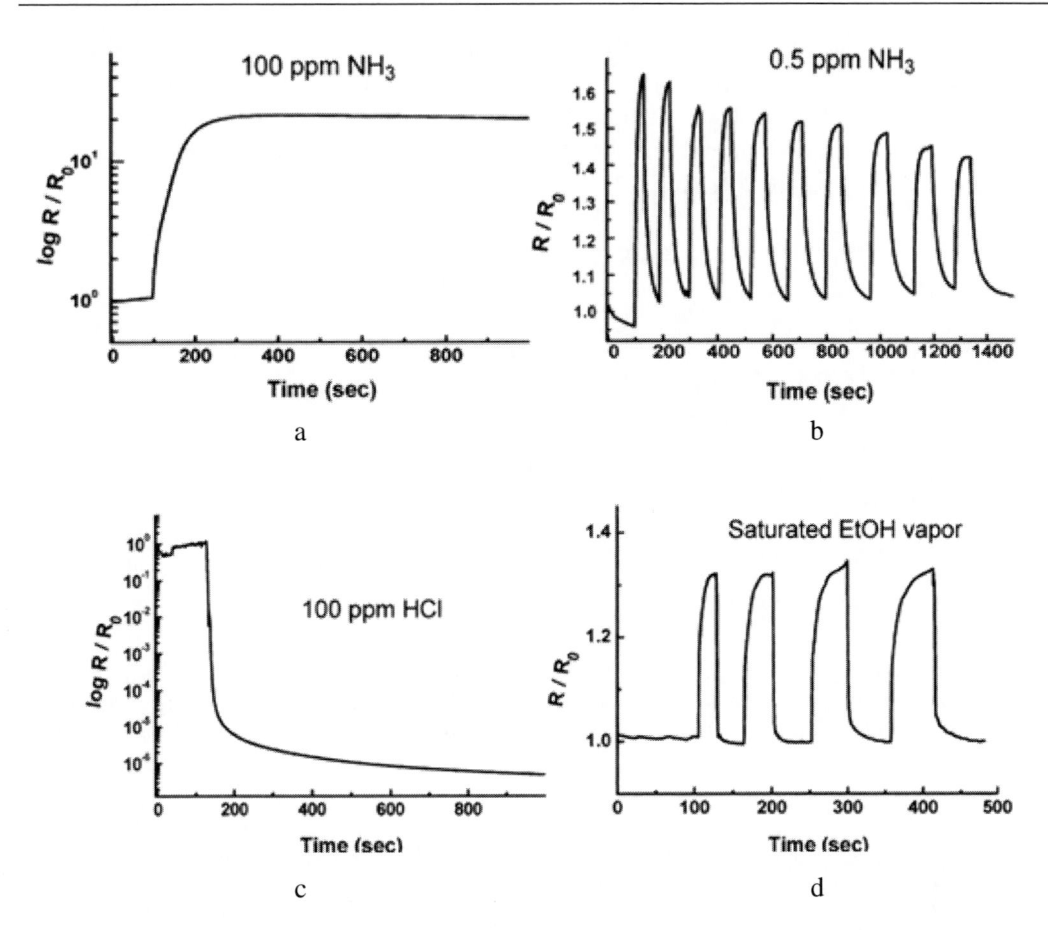

Figure 17. Real-time responses of a PAni nanoframework-electrode junctions (PNEJs) array to the presence of (a) 100 ppm NH_3 and (b) 100 ppm HCl under ambient conditions. Reversible and reproducible responses of a PNEJs array to (b) 0.5 ppm NH_3 and (d) saturated ethanol. (Reproduced with permission of the American Chemical Society from ref. [72]).

Obviously, because of R_4, the majority of the current would flow through the PAni nanoparticle if the electric field applied to the reverse bias of the p-n junction is higher than the breakdown voltage of the diode. At a constant open voltage, the electric field applied to the system must follow the following expression

$$V = I_1 (R_1 + R_2 + R_3) = I_2R_4 \quad (1)$$

Because H^+-doped PAni is a conductive polymer, both R_1 and R_2 are much smaller than R_3, eq. 1 can then be simplified as

$$V = I_1R_3 = I_2R_4 \quad (2)$$

If the open voltage is just above the breakdown voltage, the depletion layer in the p-n junction will be broken down, resulting in a sudden and dramatic decrease in R_3. As a result, I_1 must be much greater than I_2, and the conductivity of the entire system is high in this case, as shown in Scheme 6b (left). However, when the device is in contact with NH_3 gas, the PAni nanoparticles on the TiO_2 surface will be dedoped, so R_2 will increase dramatically. At a certain point, the resistance (R_2+ R_3) is high enough that we can almost completely turn off the route containing the PAni nanoparticle (Scheme 6b right). As a result, the current can only flow through a pure TiO_2 microfiber, resulting in a large drop in the entire current when NH_3 gas is absorbed on PAni nanograins. [33]

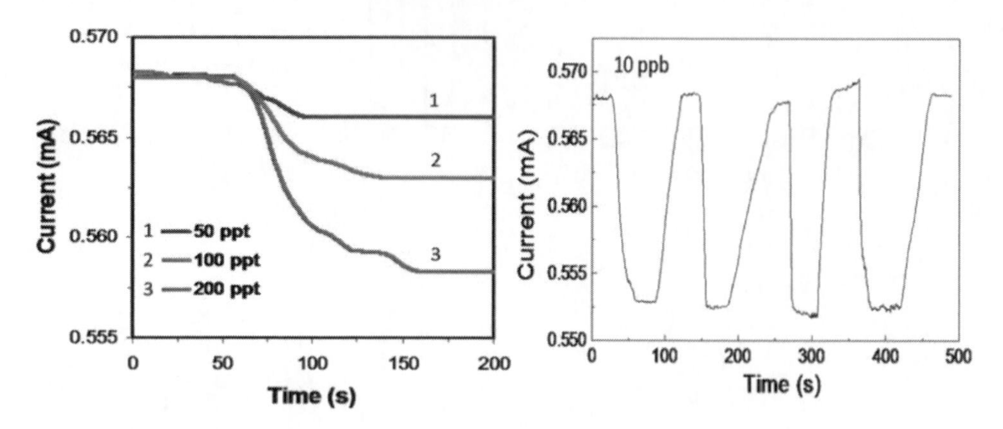

Figure 18. Current responses of a sensor made of TiO_2 microfibers enchased with PAni nanograins to different concentrations of NH_3 gas as a function of time (a) and reproducibility of the sensor exposed to 10 ppb NH_3 gas (b). (Reproduced with permission of the American Chemical Society from ref. [33]).

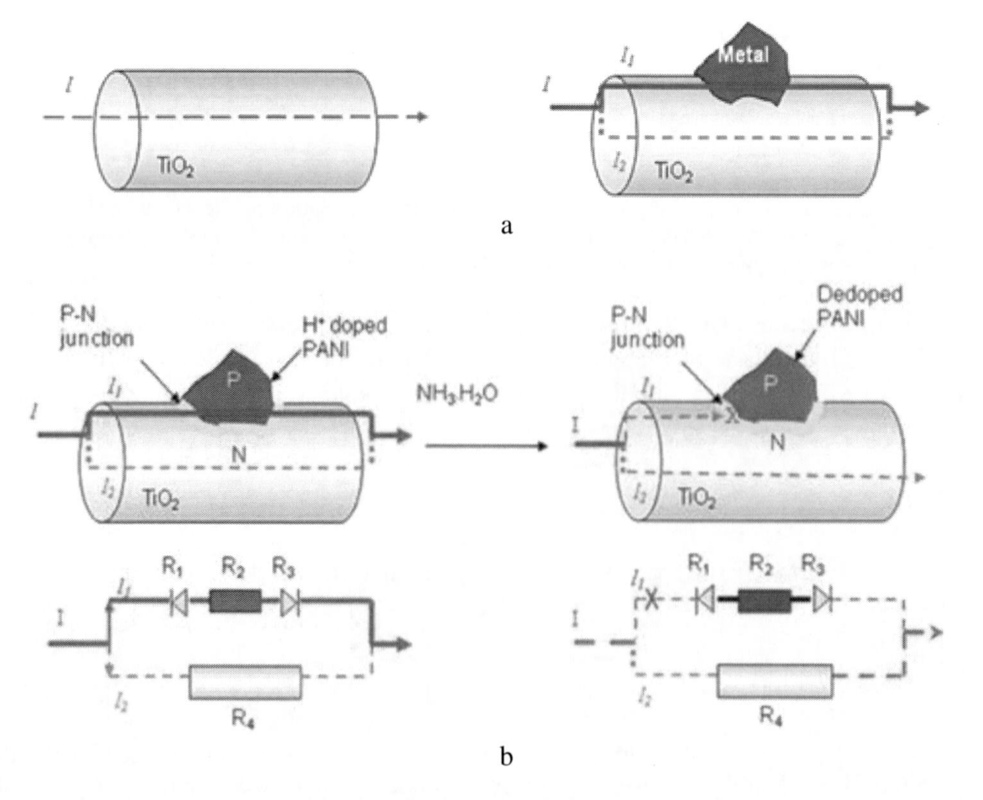

Scheme 6. Schematic of nanosized p–n heterojunction as a switch to control the electric current flow in TiO_2 microfibers. [33].

CONCLUSION

This chapter focuses on the main aspects of PAni based nanomaterials that are part of a sensor device for detecting gases: from different morphological PAni to PAni/metal nanocomposites, other PAni based nanocomposites and PAni-based heterojunctions. The objective of these PAni-based nanomaterials is to obtain better quality parameters (such as sensitivity, selectivity, rapidity, long-term stability, low operation temperature, reproducibility) for gas sensors and to detect more and various gases. It is difficult to compare the performance characteristics of different morphological PAni nanomaterial gas sensors based on different experimental conditions. For PAni nanocomposites, we show the gas sensitivities of these nanocomposites in comparison with pure PAni or another components in the PAni nanocomposites. The PAni nanomaterials with various morphologies, which are prepared by different methods, can exhibit various gas sensitivities to different gases. Functionalization processes are being used to modify the PAni, and the design of heterojuctions composed of PAni and other materials that are applied in preparing gas sensors; both of the two methods are an attempt to increase the selectivity to some target gases, reduce the detection limit, extend the dynamic ranges, or shorten response and recovery times, etc. Other operational parameters such as the cost, reusability and portability of the gas sensors, based on PAni, have seldom been evaluated since the main goals and expectations are the improvement of selectivity and sensitivity. Various mechanisms of gas detection are listed according to the differences between the sensitive materials and the target gas. It may be possible to use PAni-based sensors to efficiently detect various types of gases: alkaline and reduced gases, acidic gases, other inorganic gases, and some organic gases. In the future, it will be possible to minimize the sensitive devices, detect more toxic gases, and optimize the performance parameters.

REFERENCES

[1] Y. Xu, G. Heiland, *J. Transclu. Tech.* 1995, 4, 59.

[2] Y. Xu, G. Heiland, *J. Transclu. Tech.* 1995, 1, 63.

[3] Y. Xu, G. Heiland, *J. Transclu. Tech.* 1995, 2, 56.

[4] Y. Xu, G. Heiland, *J. Transclu. Tech.* 1996, 4, 93.

[5] Y. Xu, G. Heiland, *J. Transclu. Tech.* 1997, 1, 79.

[6] M. Cremer, *Z. Biol.* 1906, 47, 582.

[7] W. H. Brattain, J. Bardeen, *Bell Syst. Tech. J.,* 1953. 32, 1-12.

[8] N. Yamazoe, *Sens. Actuators* B 1991, 5, 7.

[9] W. Gopel, K. D. Schierbaum, *Sens. Actuators* B 1995, 26, 1.

[10] G. Behr, W. Fliegel, *Sens. Actuators* B 1995, 26, 33.

[11] C. H. Kwon, H. K. Hong, D. H. Yun, K. Lee, S. T. Kim, Y. H., Roh, B. H. Lee, *Sens. Actuators B* 1995, 24, 610.

[12] G. Mangamma, V. Jayaraman, T. Gnanasekaran, G. Periaswami, *Sens. Actuators* B 1998, 53, 133.

[13] Y. Yamada, Y. Seno, Y. Masuoka, K. Yamashita, *Sens. Actuators* B 1998, 49 248.

[14] Z. Jiao, S. Wang, L. Bian, J. Liu, *Mater. Res. Bull.* 2000, 35, 741.

[15] P. N. Bartlett, K. Sim, L. Chung, *Sens. Actuators* B 1989, 19, 141.
[16] R. Nohria, R. K. Khillan, Y. Su, R. Dikshit, Y. Lvov, K. Varahramya, *Sens. Actuators* B 2006, 114, 218.
[17] Q. Fang, D. G. Chetwynd, J. A. Covington, C. S. Toh, J. W. Gardner, *Sens. Actuator* B 2002, 84, 66.
[18] D. Xie, Y. Jiang, W. Pan, D. Li, Z. Wu, Y. Li, *Sens. Actuator* B 2002, 81 158.
[19] F. Zee, J. W. Judy, *Sens. Actuator* B 2001, 72, 120.
[20] Y. Cao, P. Smith, A. J. Heeger, *Synth. Met.* 1989, 32, 263.
[21] S. Palaniappan, C. A. Amarnath, *Mater. Chem. Phys.* 2005, 92, 82.
[22] S. Virji, J. Huang, R. B. Kaner, B. H. Weiller, *Nano Lett.* 2004, 4, 491.
[23] J. Huang, S. Virji, B. H. Weiller, R. B. Kaner, *Chem. Eur. J.* 2004, 10, 1314.
[24] K. Domansky, D. L. Baldwin, J. W. Grate, T. B. Hall, J. Li, M. Josowicz, J. Janata, *Anal. Chem.* 1998, 70, 473.
[25] L. L. Miller, J. S. Bankers, A. J. Schmidt, D. C. Boyd, *J. Phys. Org. Chem.* 2003, 13, 808.
[26] V. Svetlicic, A. J. Schmidt, L. L. Miller, *Chem. Mater.* 1998, 10, 3305.
[27] Z. F. Li, F. D. Blum, M. F. Bertino, C. S. Kim, S. K. Pillalamarri, *Sens. Actuator* B 2008, 134, 31.
[28] D. Chowdhury, *J. Phys. Chem.* C 2011, 115, 13554.
[29] S. P. Surwade, S. R. Agnihotra, V. Dua, N. Manohar, S. Jain, S. Ammu, S. K. Manohar, *J. Am. Chem. Soc.* 2009, 131, 12526
[30] M. M. Ayad, G. E. Hefnawey, N. L. Torad, *Journal of Hazardous Materials* 2009, 168, 85.
[31] D. S. Dhawale, D. P. Dubal, V. S. Jamadade, R. R. Salunkhe, S. S. Joshi, C. D. Lokhande, *Sens. Actuator* B 2010, 145, 205.
[32] X. Li, Y. Gao, J. Gong, L. Zhang, L. Qu, *J. Phys. Chem.* C 2009, 113, 69.
[33] J. Gong, Y. Li, Z. Hu, Z. Zhou, Y. Deng, *J. Phys. Chem.* C 2010, 114, 9970.
[34] J. D. Fowler, S. Virji, R. B. Kaner, B. H. Weiller, *J. Phys. Chem.* C 2009, 113, 6444.
[35] H. Liu, J. Kameoka, D. A. Czaplewski, H. G. Craighead, *Nano Lett.* 2004, 4 671.
[36] A. L. Kukla, Yu. M. Shirshov, S. A. Piletsky, *Sens. Actuators B: Chem.* 1996, 37, 135.
[37] N. R. Chiou, A. J. Epstein, *Adv. Mater.* 2005, 17, 1679.
[38] Y. Gao, S. Yao, J. Gong, and L. Y. Qu, *Macromol. Rapid. Commun.* 2007, 28, 286.
[39] X. Zhang, W. J. Goux and S. K. Manohar, *J. Am. Chem. Soc.* 2004, 126, 4502.
[40] H. Xia, D. Cheng, C. Xiao, and H. S. O. Chan, *J. Nanosci. Nanotechnol.* 2006, 6, 3950.
[41] M. K. Park, K. Onishi, J. Locklin, F. Caruso, and R. C. Advincula, *Langmuir* 2003, 19, 8550.
[42] L. H. Meng, Y. Lu, X. Wang, J. Zhang, Y. Duan, C. Li, *Macromolecules* 2007, 40, 2981.
[43] S. P. Armes, M. Aldissi, S. Agnew, S. Gottesfeld, *Langmuir* 1990, 6, 1745.
[44] B. Vincent, J. Waterson, *J. Chem. Soc., Chem. Commun.* 1990, 683.
[45] J. Stejskal, P. Kratochvil, S. P. Armes, S. F. Lascelles, A. Riede, M. Helmstedt, J. Prokes, I. Krivka, *Macromolecules* 1996, 29, 6814.
[46] Z. Wei, M. Wan, *Adv. Mater.* 2002, 14, 1314.
[47] C. R. Martin, *Science* 1994, 266, 1961.
[48] C. R. Martin, *Chem. Mater.* 1996, 8, 1739.
[49] C. G. Wu, T. Bein, *Science* 1994, 264, 1757.

[50] J. Liu, Y. Lin, L. Liang, J. A. Voigt, D. L. Huber, Z. R. Tian, E. Coker, B. Mckenzie, M. J. Mcdermott, *Chem. Eur. J.* 2003, 9, 604.
[51] J. Huang, S. Virji, B. H. Weiller, R. B. Kaner, *J. Am. Chem. Soc.* 2003, 125, 314.
[52] S. Virji, B. H. Weiller, J. Huang, R. Blair, H. Spepherd, T. Faltens, P. C. Haussmann, R. B. Kaner, S. H. Tolbert, *Journal of Chemical Education* 2008, 85, 1102.
[53] J. Chen, J. Yang, X. Yan, Q. Xue, *Synthetic Met.* 2010, 160, 2452.
[54] B. Xue, *Master's thesis of Northeast Normal University*, 2007.
[55] Y. Gao, *Master's thesis of Northeast Normal University*, 2007.
[56] H. Ma, Y. Gao, Y. Li, J. Gong, X. Li, B. Fan, Y. Deng, *J. Phys. Chem.* C 2009, 113, 9047.
[57] H. Ma, Y. Li, F. Cao, Y. Gao, J. Gong, Y. Deng, *Journal of Polymer Science: Part A: Polymer Chemistry,* 2010, 48, 3596.
[58] H. Ma, Y. Li, S. Yang, F. Cao, J. Gong, Y. Deng, *J. Phys. Chem.* C 2010, 114, 9264.
[59] Y. Li, J. Gong, G. He, Y. Deng, *Synthetic Metals* 2011, 161, 56.
[60] Y. Gao, Z. Kang, X. Li, X. Cui, J. Gong, *DOI: 10.1039/c1ce05096f.*
[61] D. S. Sutar, N. Padma, D. K. Aswal, S. K. Deshpande, S. K. Gupta, J. V. Yakhmi, *Sensors and Actuators* B 2007, 128, 286.
[62] K. Crowley, E. O'Malley, A. Morrin, M. R. Smyth, A. J. Killard, *Analyst*, 2008, 133, 391.
[63] C. Liu, K. Hayashi, K. Toko, *Macromolecules* 2011, 44, 2212.
[64] P. P. Sengupta, P. Kar, B. Adhikari, *Thin Solid Films* 2009, 517, 3770.
[65] Y. Gao, X. Li, J. Gong, B. Fan, Z. Su, L. Qu, *J. Phys. Chem.* C 2008, 112, 8215.
[66] Y. Gao, S. Yao, J. Gong, L. Qu, *Macromol. Rapid Commun.* 2007, 28, 286.
[67] L. Y. Yang, W. B. Liau, *Synthetic Metals* 2010, 160, 609.
[68] S. Virji, J. Huang, R. B. Kaner, B. H. Weiller, *Nano Lett.*, 2004, 4, 491.
[69] S. P. Surwade, S. R. Agnihotra, V. Dua, N. Manohar, S. Jain, S. Ammu, S. K. Manohar, *J. Am. Chem. Soc.* 2009, 131, 12528.
[70] G. Li, C. Martinez, S. Semancik, *J. Am. Chem. Soc.* 2005, 127, 4903.
[71] G. Li, M. Josowicz, J. Janata, S. Semancik, *Appl. Phys. Lett.* 2004, 85, 1187.
[72] J. Wang, S. Chan, R. R. Carlson, Y. Luo, G. Ge, R. S. Ries, J. R. Heath, H. R. Tseng, *Nano Lett.*, 2004, 4, 1693.
[73] C. K. Tan, D. J. Blackwood, *Sens. Actuators* B 2000, 71, 184.
[74] A. Choudhury, *Sensors and Actuators* B 2009, 138, 318.
[75] G. M. Neelgund, E. Hrehorova, M. Joyce, V. Bliznyuk, *Polym. Int.* 2008, 57, 1083.
[76] A. Houdayer, R. Schneider, D. Billaud, J. Ghanbaja, J. Lambert, *Synth. Met.* 2005, 151, 165.
[77] J. M. Kinyanjui, N. R. Wijeratne, J. Hanks, D. W. Hatchett, *Electrochim. Acta* 2006, 51, 2825.
[78] J. M. Smith, M. Josowicz, J. Janata, *J. Electrochem. Soc.* 2003, 150, 384.
[79] J. Park, S. Park, A. Koukitu, O. Hatozaki, N. Oyama, *Synth. Met.* 2004, 141, 265.
[80] T. K. Sarma, D. Chowdhury, A. Paul, A. Chattopadhyay, *Chem. Commun.* 2002, 14, 1048.
[81] W. Xue, H. Qiu, K. Fang, J. Li, J. Zhao, M. Li, *Synth. Met.* 2006, 156, 833.
[82] F. Yakuphanoglu, E. Basaran, B. F. Senkal, E. Sezer, *J. Phys. Chem.* B 2006, 110, 16908.
[83] R. Brina, G. E. Collins, P. A. Lee, N. R. Armstrong, *Anal. Chem.* 1990, 62, 2357.

[84] H. Behar-Levy, D. Avnir, *Adv. Funct. Mater.* 2005, 15, 1141.
[85] H. Behar-Levy, G. E. Shter, G. S. Grader, D. Avnir, *Chem. Mater.* 2004, 16, 3197.
[86] I. Yosef, D. Avnir, *Chem. Mater.* 2006, 18, 5890.
[87] Y. Gao, D. Shan, F. Cao, J. Gong, X. Li, H. Ma, Z. Su, L. Qu, *J. Phys. Chem.* C 2009, 113, 15175.
[88] M. D. Shirsat, M. A. Bangar, M. A. Deshusses, N. V. Myung, A. Mulchandani, *Appl. Phys. Lett.* 2009, 94, 083502.
[89] S. Sharma, C. Nirkhe, S. Pethkar, A. A. Athawale, *Sensors and Actuators* B 2002, 85, 131.
[90] Shuang Yao, *Master's thesis of Northeast Normal University*, 2008.
[91] G. Xie, P. Sun, X. Yan, X. Du, Y. Jiang, *Sensors and Actuators* B 2010, 145 373.
[92] J. Zheng, G. Li, X. Ma, Y. Wang, G. Wu, Y. Cheng, *Sensors and Actuators* B 2008, 133, 374.
[93] H. Tai, Y. Jiang, G. Xie, J. Yu, X. Chen, Z. Ying, *Sensors and Actuators* B 2008, 129, 319.
[94] S. Virji, J. D. Fowler, C. O. Baker, J. Huang, R. B. Kaner, B. H. Weiller, *Small* 2005, 1, 624.
[95] S. Virji, R. B. Kaner, B. H. Weiller, *Inorg. Chem.* 2006, 45, 10467.
[96] S. Virji, R. Kojima, J. D. Fowler, R. B. Kaner, B. H. Weiller, *Chem. Mater.* 2009, 21, 3056.
[97] S. Weng, J. Zhou, Z. Lin, *Synthetic Metals* 2010, 160, 1136.
[98] A. Airoudj, D. Debarnot, Bruno Beche, F. Poncin-Epaillard, *Anal. Chem.* 2008, 80, 9188.
[99] L. A. Mashat, K. Shin, K. Kalantar-zadeh, J. D. Plessis, S. H. Han, R. W. Kojima, R. B. Kaner, Dan Li, X. Gou, S. J. Ippolito, Wojtek Wlodarski, *J. Phys. Chem.* C 2010, 114, 16168.
[100] L. Yu, X. Jin, X. Zeng, *Langmuir*, 2008, 24, 11631.
[101] S. Virji, R. Kojima, J. D. Fowler, J. G. Villanueva, R. B. Kaner, B. H. Weiller, *Nano Res.* 2009, 2, 135.
[102] G. A. Sotzing, J. N. Phend, R. H. Grubbs, N. S. Lewis, *Chem. Mater.* 2000, 12, 593.
[103] S. Virji, R. B. Kaner, B. H. Weiller, *Chem. Mater.* 2005, 17, 1256.
[104] N. J. Pinto, R. González, A. T. J. Jr, *Appl. Phys. Lett.* 2006, 89, 033505.
[105] H. Zhang, S. Boussaad, N. Ly, N. J. Tao, *Appl. Phys. Lett.* 2004, 84, 133.
[106] D. S. Dhawale, R. R. Salunkhe, U. M. Patil, K. V. Gurav, A. M. More, C. D. Lokhande, *Sensors and Actuators* B 2008, 134, 988.
[107] S. S. Joshi, T. P. Gujar, V. R. Shinde, C. D. Lokhande, *Sensors and Actuators* B 2008, 132, 349.

In: Trends in Polyaniline Research
Editors: T. Ohsaka, Al. Chowdhury, Md. A. Rahman et al.
ISBN: 978-1-62808-424-5

Chapter 13

ANTICORROSIVE PROPERTIES OF POLYANILINE

Ali Olad*[*] *and Azam Rashidzadeh
Polymer Composite Research Laboratory, Department of Applied Chemistry,
Faculty of Chemistry, University of Tabriz, Tabriz, Iran

ABSTRACT

Corrosion is one of the most serious problems in the industrial world. The most common way to prevent a metal from corrosion is to provide an impervious coating over it. But unfortunately, most of coatings due to the existing pinholes in the coating or by diffusion of oxygen and water through it eventually fail. Therefore finding an ideal material which affords good corrosion protection is highly desirable. Conducting polymers are those desirable materials for corrosion inhibition. Among conducting polymers, polyaniline because of its high electrical conductivity, environmental stability, ease of preparation in large quantities and relatively low cost, has been studied extensively for corrosion protection. It was reported that polyaniline can be used as coating, corrosion inhibitor or anticorrosive additive to improve the resistance of metals against corrosion. However, several strategies have been used to increase the adhesion, mechanical properties, barrier effect of the polyaniline coatings and therefore the efficiency of corrosion protection by the polyaniline coating. The utility of various nanoparticles, such as metal oxides, carbon nanotubes, graphite, natural fibers, nanoclays and zeolites as additives to enhance the mechanical and barrier performance of polyaniline has been reported.

The aim of this chapter is to introduce the anticorrosive property of polyaniline and to focus on the mechanism of corrosion protection of coatings based on polyaniline. Also, this chapter deals with a short review of application of nanomaterials to improve the anticorrosive property of polyaniline coatings reported in published literature in the recent past. Finally some examples of applications of polyaniline as anticorrosive additive to modify the formulation of conventional organic coatings and paints are mentioned.

[*] E-mail: a.olad@yahoo.com; Tel: +98 411 3393164; Fax: +98 411 3340191.

INTRODUCTION

Corrosion is the chemical or electrochemical reaction between a material and its environments which produces a destruction of the material and deterioration of its properties [1].

In other words, extracted metal from its primary ore has a natural tendency to revert to that state under the action of oxygen and water. This action is called corrosion which occurs by an anodic reaction which produces free electrons that pass within the metal to another site on the metal surface (the cathode), where it is consumed by the cathodic reaction. In acid solutions the cathodic reaction is producing hydrogen gas (Figure 1) and in neutral solutions the cathodic reaction involves the consumption of oxygen dissolved in the solution and producing hydroxide ions (Figure 2). Thus the corrosion process occurs at the anode but not at the cathode.

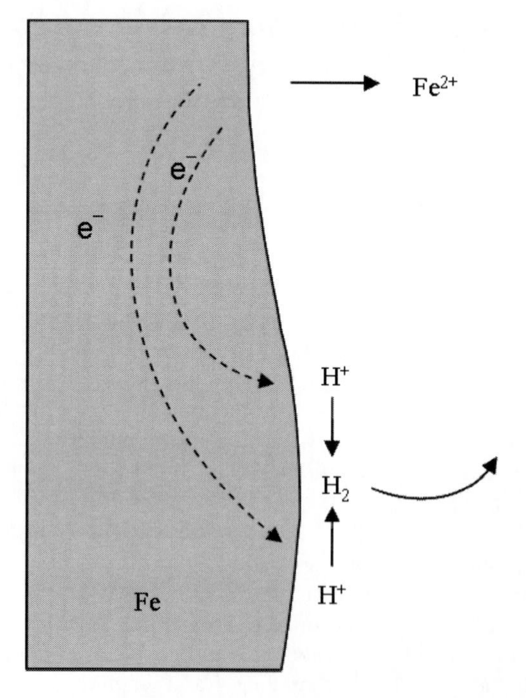

Figure 1. Electrochemical reactions occurring during the corrosion of iron in hydrochloric acid solution.

The anode and cathode in a corrosion process may be close together on the same metal surface as with rusting of steel or on two different metals connected together forming a bimetallic couple.

Economics and safety are two main reasons of corrosion importance. In the economical point of view, corrosion failures of bridges, buildings, aircrafts, automobiles, and gas pipelines as well as and loss of energy is one of the main motivations of scientists to pay high attention to corrosion. Safety is another important factor which is influenced by corrosion of operating equipments.

The most widely used three methods of combating corrosion are protective coatings, cathodic and anodic protection methods.

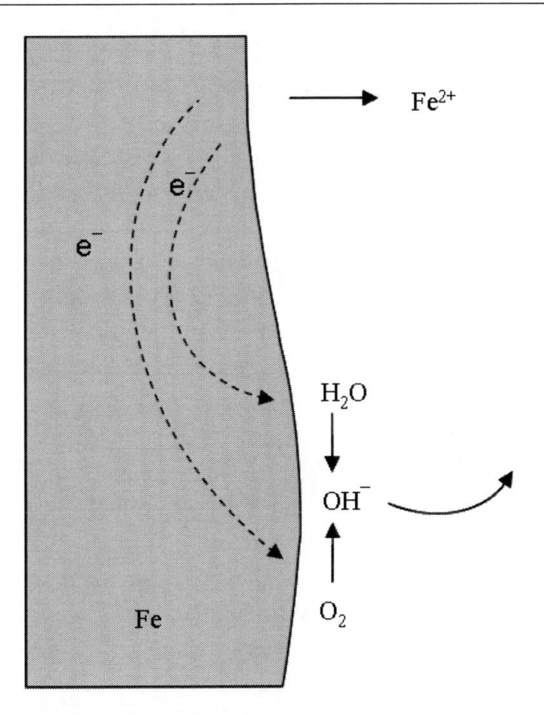

Figure 2. Electrochemical reactions occurring during the corrosion of iron in neutral water.

Protective coatings separate the surfaces of metals that are susceptible to corrosion from the corrosive environment. Thin coatings of metallic, inorganic and organic materials can provide a satisfactory barrier between the metal and its environment. Metalic coatings are applied by electrodeposition, hot dipping, and vapor deposition on the surface of metal. In this method a more noble metal coating is applied on an active metal which provide corrosion resistance of the noble metal. On the other hand applying a more active metal coating takes advantage of the corrosion of the coating preferentially, or sacrificially, to the substrate [2]. Inorganic coatings include oxide, phosphate, chromate, fluoride, or other complex inorganic compounds as well as ceramic coatings, such as carbides and silicides and glass coatings which are applied by oxidizing, phosphatizing, passivation, anodizing and spraying methods [3].

Organic coatings are another class of coatings which apply a relatively thin barrier between substrate material and the environment. These coatings consist of panits, varnishes, lacquers and similar coatings which protect large surfaces of metals than any other method for combating corrosion. Organic coatings provide a barrier that inhibits the penetration of aggressive environmental species. The main goal of these coatings is to prevent the taking place of cathodic reaction ($2H_2O+O_2+4e^- \rightarrow 4OH^-$) beneath the coating. But, over time the barrier coating can fail due to prolonged exposure to the corrosive environment [4]. These organic coatings can develop under coating corrosion which starts from weak spots and then develops into blisters and pitting leading to corrosion. Therefore due to the separation of organic coating at the interface of the substrate, the delamination takes place and the barrier property of the coating fails [5].

Therefore, it is needed to increase the lifetimes of organic coatings through the pretreatment of metal surface. In some cases the use of phosphates, chromates and oxides will extend the lifetimes of barrier coatings through better adhesion of the coatings onto the metal substrate [6]. However because of carcinogenic effect of hexavalent chromium, alternative organic coatings are needed to eliminate chromium from the environment and to protect human health.

Cathodic protection is another electrochemical means of corrosion mitigation. In other words, cathodic protection combats corrosion by use of the same laws which cause the corrosion process. Cathodic protection method reduces the corrosion rate of a metal by reducing its corrosion potential, bringing the metal closer to an immune state. The principle of cathodic protection is using an external anode in connection with the metal to be protected and the passing of an electrical DC current so that all areas of the metal surface become cathodic and therefore do not corrode. There are two basic methods of applying cathodic protection, known as sacrificial (galvanic) anode and impressed current (rectifier) cathodic protection. Effective application of cathodic protection can provide complete protection to any exposed areas for the life of the structure [7]. The combination of an external coating and cathodic protection provides the most economical and effective choice for protection of underground and submerged pipelines. For bare or ineffectively coated existing pipeline systems, cathodic protection often becomes the only practical alternative for corrosion protection. Cathodic protection is commonly applied to a coated structure to provide corrosion control to areas where the coating may be damaged. It may be applied to existing structures to prolong their life.

Anodic protection is based on the formation of a protective film on metals by externally applied anodic currents. The controlled application of anodic currents to the metals causes their passivation and decrease the rate of metal dissolution. Although anodic protection is limited to passive metals and alloys, it can decrease the corrosion rate substantially. The primary advantages of this method are its applicability in extremely corrosive environments and its low current requirements which is preferred to cathodic protection [7].

CONDUCTING POLYMERS

The most common way to prevent a metal from corrosion is to provide an impervious coating over it. If a perfect barrier coating is applied to the surface of metal, then neither oxygen nor water can reach its surface and corrosion will be prevented. But unfortunately, most of coatings are not perfect barrier systems, and all coatings eventually fail due to the existing pinholes in the coating or by diffusion of oxygen and water through it. For this reason, it is highly desirable to find a corrosion protection method involving both chemical and electrochemical techniques, such as chemical inhibitors, cathodic protection, and anodic protection. Conducting polymers capable of providing all three types of corrosion protection are desirable materials for corrosion inhibition.

The use of conducting polymers for corrosion protection was first suggested by MacDiarmid in 1985 [8], and then by Ahmad and MacDiarmid in 1987 [9].Their investigations showed that PANI electrodeposited on passivated steel in a strong acid environment enhanced corrosion protection of the metal. Almost all of the conducting polymers used in corrosion protection fall under the following classes: polyanilines, polyheterocycles and poly (pheny-lene vinylene) s. It has been reported that deposition of conducting polymers onto corrosion-susceptible materials, can inhibit corrosion. This finding opened up the new area in corrosion protection methods in replacing the traditional hexavalent chromium technology.

POLYANILINE

The term polyaniline (PANI) is descriptive of a class of conducting polymers that can be reversibly oxidized and reduced over a wide range of potentials. Unlike other conducting polymers, polyaniline can be doped to a highly conducting form in the acidic media without removing electrons from the polymer backbone which is known as non-oxidative doping. Also, polyaniline can undergo oxidative doping which is a reversible process within the potential range that the polymer is stable. These unique properties enable polyaniline to act as a versatile polymer in corrosion protection studies [10].

Polyaniline and its deravatives have been extensively used to protect metals against corrosion in the form of coating, corrosion inhibitor or anticorrosive additive to improve the resistance of metals against corrosion.

POLYANILINE AS COATING

At least three different methods of using polyaniline as coating have been reported: polyaniline coatings alone; polyaniline coatings as a primer with a conventional polymer topcoat; and polyaniline composite with a conventional polymer coating.

Polyaniline can be synthesized both chemically and electrochemically. Synthesis of polyaniline by electropolymerization method has several advantages in comparison with other coating technologies, such as spin coating. The main one is the ability to form the homogenous and adherent coating polymer at irregular shaped objects [11].

However, it is difficult to electropolymerize aniline monomer on oxidizable or corrosion-susceptible metals, such as iron, mild steels, zinc and aluminium and thereby very poor and non-adherent films with low corrosion-protection properties will form [12].

Polyaniline and polyaniline derivatives are readily grown on active metal electrodes with a suitable choice of electrolyte. Most of the electrochemically polymerization are done in non-aqueous media. The most suitable solvents currently in use include acetonitrile, benzonitrile, and tetrahydrofuran [13-15]. Also, It was reported that it is possible to electropolymerize aniline from an oxalic-acid solution onto iron due to the formation of an iron oxalate complex layer that inhibits dissolution of the iron electrode and facilitates growth of polyaniline over the layer [16], while Bernard et al. [17] have reported the formation of adherent and protective polyaniline layers on iron from a phosphoric acid solution, due to the formation of a stable iron phosphate layer.

DeBerry in 1985 [18] found that the electro-deposited polyaniline on passivated steel in a strong acidic media was enable of providing anodic protection. After that, Wessling proposed that the protection mechanism of steel by PANI is attributed to the formation of a passive layer of metal oxide on steel surface [19]. Investigations with SEM and XPS revealed that the formed oxide layer between the PANI coating and the steel surface was composed of Fe_3O_4/γ-Fe_2O_3 [20, 21]. The doped polyaniline layer deposited over the passive metal oxide film can undergo electron transfer with the metal. This electron transfer may be the responsible of the ability of polyaniline to maintain the passivity of the stainless steel. This protective mechanism was then thought to be anodic protection that maintains the passive film on the metal [22].

Wessling suggested that the formation of the oxide layer on the metal surface might be due to a catalytic redox reaction of PANI [23]. The reaction sequence responsible for the passivation process and the formation of the oxide layer was as follows.

$$2Fe - 4e \rightarrow 2Fe^{2+}$$

$$PANI(ES) + 4H^{+} + 4e \rightarrow PANI(LE)$$

$$1/2O_2 + H_2O + 2e \rightarrow 2OH^{-}$$

$$PANI(LE) \rightarrow PANI(ES) + 4H^{+} + 4e$$

$$O_2 + 2H_2O + 4e \rightarrow 4OH^{-}$$

$$2Fe^{2+} - 2e \rightarrow 2Fe^{3+}$$

$$2Fe^{3+} + 6OH^{-} \rightarrow Fe_2O_3 + 3H_2O$$

Iron is oxidized by the emeraldine salt form of polyaniline (PANI (ES)), which in return is reduced to the leucoemeraldine Salt (PANI (LE)), then Fe^{2+} is further oxidized by oxygen to Fe^{3+}, oxygen also reoxidizes the LE form back to the ES form, which is thereby enabled to continue to passivation, hence is acting as a catalyst. The change in the color of polyaniline after contact with iron in the presence of water from the green to the yellow form due the reduction of polyaniline from the emeraldine salt to the leucoemeraldine salt confirms this mechanism. Finally the formed ferric oxide at the surface of metal acts as a passivating layer. Therefore, the catalytic nature of polyaniline allows one to apply just small amounts of polymer to the surface of the metal and yet produce a substantial metal oxide layer. The ability of polyaniline to be oxidatively regenerated by air seemed to promote corrosion protection even where scratches existed in the protective coating. When the substrate is exposed to corrosive environment as a result of any damage in the coating system, the sacrificial properties of the conducting polymer would inhibit the corrosion process and thereby protect the structure from accelerated deterioration [24].

The emeraldine base form of polyaniline has been used as an anticorrosion undercoat on steel and iron. Emeraldine base polyaniline undercoats were found to offer corrosion protection for both the cold rolled steel and iron samples [25].

In order to reveal a better understanding of the corrosion protection mechanism by polyaniline and confirmation of formation of passive layer on metal surface, further investigations were carried out by Mirmohseni and Olad [10]. They removed polyaniline layer of the samples after treating in corrosive environments NaCl 3.5% Wt. Then they compared anticorrosive behavior of polyaniline removed sample with passivated sample produced at the surface of iron sample by immersing in $KMnO_4$ solution for 2 h. Results showed that the rate of corrosion and corrosion potential for both samples are very similar which supports the idea of formation a passive layer on iron.

Some researchers suggest that formation of passive layer at the metal-coating interface was due to the formation of iron/dopant complex layer. PANI because of its redox capability undergoes a continuous charge transfer reaction at the metal-coating interface in which PANI is reduced from emeraldine salt form (ES) to an emeraldine base (EB). Upon accumulation of excessive corrosive ions, passive layer breakdown takes place [26]. Strength of the passive oxide film is effective on the corrosion protection. Also, the protective behavior of polyaniline depends on the size and charge of the dopant and by increasing the size of the dopant, the strength of the iron/dopant complex film increases, which improves the protective efficiency.

The anticorrosive property of polyaniline/nylon composites coating on steel was also investigated [27]. The obtained results confirmed the passivation of metal mechanism and forming thin, compact, adherent passive layer of Fe_2O_3 at the metal-coating interface. On the other hand, studies on doped polyaniline showed that corrosion is prevented by the generated electric field, which restricts the flow of electrons from the metal to the outside oxidizing species [28, 29].

However, polyaniline coatings suffer from poor barrier, mechanical, brittleness and adhesion properties. Several strategies have been used to increase the effectiveness of polyaniline as anticorrosive coating on metals. It was reported that the mechanical properties of polyaniline coating on metal surface can be increased by formation of polyaniline blend with other insulating polymers or by application of a top coat on the primary pure polyaniline coating [30- 33].

Wessling and Posdorfer [33] used PANI containing primer CORRPASSIV^{TM} sealed with different top coats. Comparisons of the performance of these coatings indicated that the primer/topcoat system shows the most effective corrosion protection property [34].

The type of the dopant anion incorporated into the PANI (emeraldine salt) as well as the top coat type is effective on the corrosion protection property of the coating composed of PANI primer and a barrier topcoat polymer [35]. For example in the case of epoxy, the nature of the amine hardeners used to crosslink the epoxy may convert the ES to the EB which evidenced by color change from green to blue. There was no evidence of such reactions occurring with polyurethane topcoats. Also, the porous nature of the polyurethane topcoat is the reason of the poor corrosion protection provided by the polyurethane systems compared with the epoxy topcoated systems.

Use of large size dopants such as benzosulfonate, salicylate and saccinate, eliminates the penetration of aggressive species and increases the barrier effect of polyaniline coating [36]. Utility of various nanoparticles such as layered silicates like montmorillonite [37], natural clinoptilolite zeolite [38], metal and metal oxide nanoparticles like Zn [39] and ZnO nanoparticles [40], carbon black [41] and carbon nanotubes [42] as additives to enhance the barrier property and anti corrosion performance of polyaniline has been established. Olad and Rashidzadeh [43] had demonstrated the preparation of nanocomposites of polyaniline (PANI) with organophilic montmorillonite (O-MMT) and hydrophilic montmorillonite (Na-MMT) via oxidative in-situ polymerization method and investigated anticorrosion properties of nanocomposite in the form of coating on iron. As shown in Figure 3 and 4, the corrosion current of PANI/MMT nanocomposites coated samples is much lower than for pure polyaniline coated samples. Therefore, it was found that the incorporation of low amount (5 wt %) MMT nanoparticles in polyaniline matrix, promotes the anticorrosive efficiency of PANI/MMT nanocomposites coating on iron samples. The lower corrosion rate of PANI/MMT nanocomposite compared to pure PANI coated samples might resulted from silicate nanolayers of clay dispersed in PANI matrix which increase the tortuosity of diffusion pathway of corrosive agents such as oxygen gas, hydrogen and hydroxide ions (Figure 5).

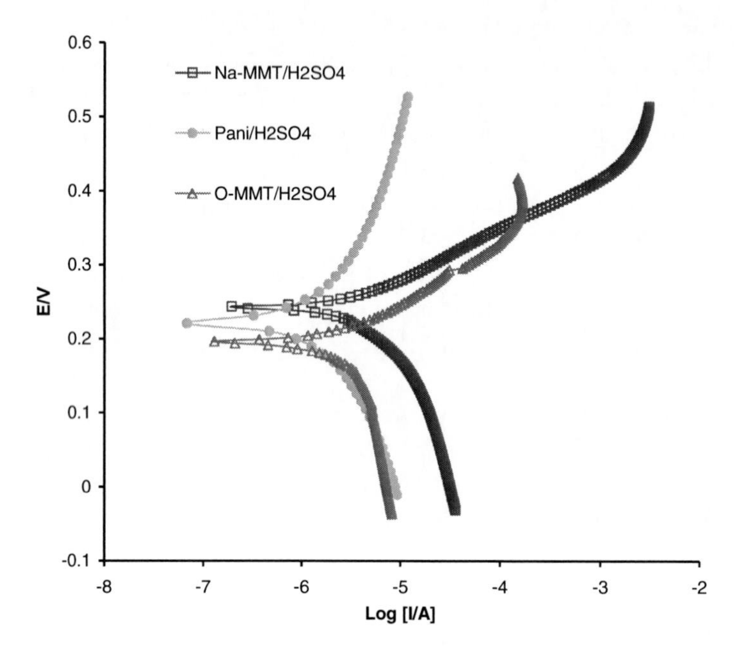

Figure 3. Tafel plots for PANI and PANI/MMT nanocomposite coated iron samples in H_2SO_4 1M.

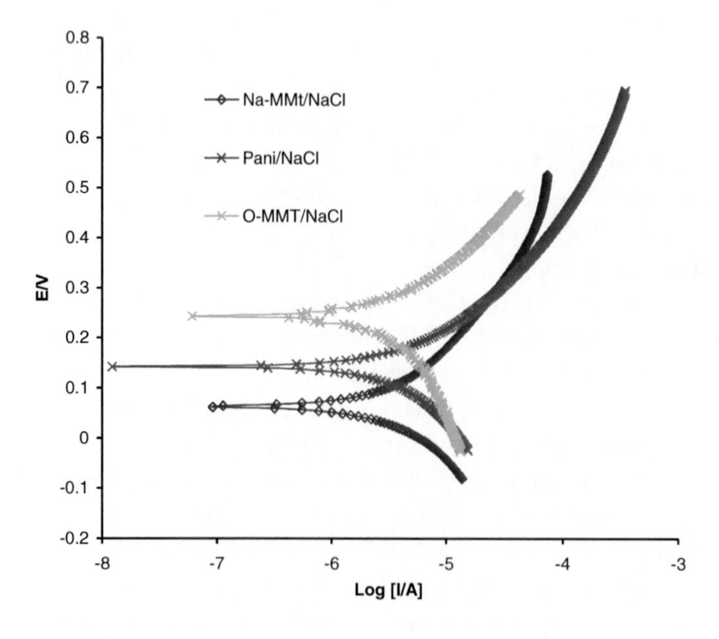

Figure 4. Tafel plots for PANI and PANI/MMT nanocomposite coated iron samples in NaCl 3.5%w/w solution.

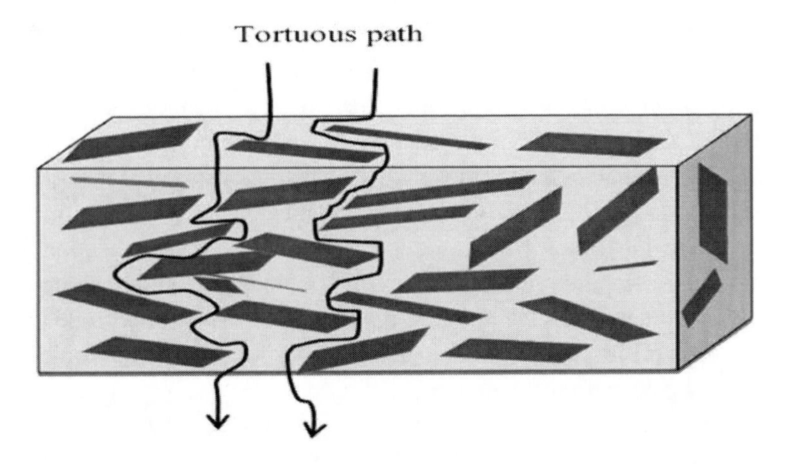

Figure 5. The mechanism of barrier improvement by the addition of clay platelets.

The degree of enhancement in the barrier properties depends on the degree of tortuosity created by clay layers in the diffusion way of molecules trough the polymer film.

On the other hand chemically synthesized polyaniline is difficult to process in common organic solvents which limit its applicability as coating. However these problems may be overcome by synthesis of polyaniline in the presence of aqueous solution of a water soluble polymer like poly(N-vinylpyrrolidone) (PVP) as the steric stabilizers [45]. Poly (N-vinylpyrrolidone) (PVP) acts both as a site for adsorption of oligoaniline initiation centers and as steric stabilizer of the formed colloidal PANI particles. For this purpose, poly (N-vinylpyrrolidone modified polyaniline/MMT nanocomposite was synthesized and applied as anticorrosive coating on iron [46]. The authors concluded that corrosion protection effect of poly (N-vinylpyrrolidone) modified PANI/MMT nanocomposite is effectively higher than that of PANI/MMT nanocomposite. Therefore, introduction of low amount of MMT into the PANI dispersion prepared in poly (N-vinylpyrrolidone) may increase the length of the diffusion pathways for oxygen and water as well as decrease the permeability of the coating, reflecting a significant enhancement in corrosion protection on metallic surface.

Zinc dust and inorganic zinc salts have also, been used as anticorrosive fillers in conducting PANI coating matrices [47] It has been demonstrated by Olad and Rasouli that zinc particles can improve the barrier properties of PANI by the formation of voluminous zinc corrosion products within the pores of PANI coating [48]. Also anticorrosive property of PANI/Zn nanocomposite coating on iron samples is thermodynamically and kinetically much better than pure PANI coating. Therefore, Zn nanoparticles have a good synergetic effect on the corrosion protection behavior of PANI matrix.

In the case of zinc-rich coatings containing PANI, ES form of polyaniline by acting as a catalyst promotes the formation of zinc oxide protective barrier layer [49]. PANI acts as a catalyst, which explains the low concentration of conducting polymer required to protect the metal surface. Moreover, any damage in the coating is catholically protected by corrosion of the zinc pigments, generating the protective zinc oxide layer, which covers the exposed iron, prevent further attack.

However the in situ methods, used for the production of PANI/Zn nanocomposite, are not industrial-friendly methods. For this, in other work, Olad et al. prepared polyaniline/zinc composites and nanocomposites using solution mixing method [50]. According to the results, anticorrosion performances of both PANI/Zn composites and nanocomposites were increased by increasing the zinc loading. Also results showed that the PANI/Zn nanocomposite films and coatings have better corrosion protection effect on iron coupons compared to that of PANI/Zn composite. Zinc has sacrificial oxidation effect in conjunction with iron that makes iron the cathode of a galvanic cell and prevents its corrosion. This behavior of zinc is similar to conducting PANI, and therefore, zinc thermodynamically has a synergetic effect on the corrosion protection behavior of PANI coating. Better corrosion protection of PANI/Zn nanocomposite coating compared to that of PANI/Zn composite coating can be related to the good dispersion of zinc nanoparticles in PANI matrix which may cause to the lower porosity of PANI/Zn nanocomposite coating compared to that of PANI/Zn composite coating containing micro size zinc particles.

In other work, triple hybrid of PANI/epoxy/Zn nanocomposite was prepared using epoxy resin and zinc nanoparticles as additives in the PANI matrix [51]. The addition of these additives was due to improve the mechanical properties of PANI coating and to take advantage of their synergetic effects on the anticorrosion performance of PANI coating.

Several strategies have been used to increase the effectiveness of polyaniline as anticorrosive coating on metals. Encapsulation of polyaniline in the host materials like zeolites is a promising strategy to prepare polymer chains in the nanometer sized of the zeolite channels, to obtain nanocomposite materials with novel special properties. By attention to the unique properties of montmorillonite clay to enhance the anticorrosive property of polyaniline coatings, which has been investigated in our previous work [43], it is conceivable that other layered structures like clinoptilolite can also be tailored to enhance the anticorrosive property of polyaniline coatings due of the ability to promote the barrier property of polyaniline against aggressive species. For getting this purpose, a nanocomposite of polyaniline with natural clinoptilolite (Clino) zeolite was prepared by chemical oxidative polymerization of anilinium cations replaced by protons of acidic clinoptilolite. Then the prepared PANI/Clino nanocomposite was investigated as anticorrosive coating on iron samples and compared with pure polyaniline coating in various corrosive environments [38]. According to the results in corrosive environments PANI/Clino nanocomposite has enhanced corrosion protection effect (lower corrosion current and corrosion rate) in comparison to pure polyaniline coating. They concluded that, clinoptilolite channels would increase the corrosion rate of the iron substrate if the polyaniline were not in the clinoptilolite channels. In fact clinoptilolite channels could act as a pathway for diffusion of corrosive agents. Therefore, it was found that the encapsulation of polyaniline in the clinoptilolite channels and dispersion of clinoptilolite layers in polyaniline matrix, promotes the anticorrosive efficiency of PANI/Clino nanocomposite coating on iron samples. However the improved corrosion protection of PANI/Clino nanocomposite compared to pure polyaniline coated samples might result from dispersion of clinoptilolite nanoparticles in polyaniline matrix which lengthen the *diffusion pathways* for *corrosive species.*

POLYANILINE AS CORROSION INHIBITOR

The use of corrosion inhibitor is one of the most effective methods for protecting metal surfaces against corrosion [52]. Organic compounds having π bonds, heteroatoms (P, S, N, and O), and inorganic compounds, such as chromate, dichromate, nitrite, and so on are the most effective efficient inhibitors [53-57]. However, the use of these chemical compounds due to their negative impacts on the environment has been questioned lately. Therefore, the development of the novel corrosion inhibitors of non-toxic type is more desirable. Inhibitor molecules physically and chemically adsorb on the metal surface and form an adsorption layer that functions as a barrier which protects the metal from the corrosion [58, 59]. During the chemical adsorption, an organic inhibitor by transferring electrons from the organic compounds to the metal forms a coordinate covalent bond and as a result forms a chelate on a metal surface [60].

Organic compounds, containing functional electronegative groups and π-electron in triple or conjugated double bonds, are usually good inhibitors. Heteroatoms, such as sulfur, phosphorus, nitrogen and oxygen having electron pairs, together with aromatic rings in their structure are the major adsorption centers which can be found in heteroaromatic like pyrrole, pyridine, aniline, and their derivatives. [61, 62]. It is generally accepted that the mode of inhibition exhibited by these small molecules when they are dissolved in corrosive environments is predominantly absorption and chelating to the iron atoms exposed on the surface of the metal. This absorption process changes the redox properties of the metal surface as well as protecting it from contact with electrolytes and retarding electron transfer reactions [63]

The availability of π electrons in polyaniline as well as the existence of $-NH_2$ group can enable it to be a good corrosion inhibitor. Hur et al. investigated Corrosion inhibition of stainless steel by polyaniline and halogen substituted polyanilines [64]. Results of their study showed that theses polymers have displayed good protection performance for the corrosion of stainless steel electrode in 0.5M HCl solution by anodic protection and barrier properties. An adsorbed film of inhibitor forms on the metal surface is responsible for the delay of attack of the metal by corrosive environment. In the aniline-halide inhibitor system, aniline primarily inhibits the cathodic reaction, whereas the halide ions seem to strongly inhibit the anodic reaction. The structure-inhibition correlation indicated that Cl substitution on the aniline ring increased inhibition efficiency. It was proposed that π electron density was related to protective ability. Also, N-substituted anilines can act as corrosion inhibitors for aluminum in hydrochloric acid [64]. Alkylaniline oligomers exhibit very good initial inhibition of metal corrosion in aqueous environments [65]. These kinds of amines adsorb on the metal surface through the amino group, and their hydrocarbon chains extend into the aqueous phase to form a protective monolayer film at the metal surface to interfere with either the cathodic or anodic reactions occurring at the adsorption sites.

In the case of polyaniline corrosion inhibition of metals with polyaniline coatings is due to the surface complexation like other inhibitors and formation of passivating metal oxide films that protect the metal surface from further corrosion which undoubtedly plays a more important role in the corrosion protection effect of this polymer.

POLYANILINE AS ADDITIVE

Polyaniline was successfully used as anticorrosive additive to modify the formulation of conventional organic coatings and paints in very low concentration [66]. The anticorrosive property of polymeric blend of camphorsulphonate or phenylphosphonate-doped polyaniline (PANI) and poly-(methyl methacrylate) as a coating on iron was investigated [67]. The results indicated that the protection mechanism of these blends is a two-step mechanism. First, a redox reaction between Fe and PANI takes place leading to PANI reduction and concomitant anion release. Then, iron cations form a passivating complex with the PANI doping anion (camphorsulphonate or phenylphosphonate). This complex by acting as a second physical barrier avoids penetration of aggressive ions. So, it is worth to note that the chemical nature of the dopant is a key factor which determines the protective properties of the coating containing polyaniline. Each time when damage on the polymeric coating is produced, the anions as a consequence of the redox reaction between iron and PANI are released. So, it can be concluded that PANI is a smart paint and anion reservoir, which can release anions in a smart way when damage is produced on the surface of the coating.

Akbarinezhad et al. [48] used PANI as a corrosion inhibitive pigment in an epoxy matrix. Their study revealed that PANI as a pigment in paint showed acceptable protection against the corrosion of carbon steel in 3.5% sodium chloride solution.

Also, polyaniline was found to act as an adhesion promoter. The loss of adherence was significantly higher for the epoxy coating without anticorrosive polyaniline additives which indicated that polyaniline promotes the adherence between the coating and the steel.

Also, the efficacy of the conducting emeraldine salt form of PANI as anticorrosive additive for an epoxy paint based on diglycidyl ether of bisphenol A and polyamide was investigated by Armelin et al. [68]. Their study showed that the addition of PANI-ES improves the resistance of the paint against corrosion with low concentration of polyaniline. The mechanism for corrosion inhibition for the polyaniline is electrochemical in nature which endows the ability to heal the metal surfaces before pitting corrosion due to the reversible redox properties of polyaniline. Laco et al. [69] investigated the anticorrosive property of carbon steel with thermoplastic coatings and alkyd resins containing polyaniline.

Furthermore, the anticorrosive property of an organic coating containing PANI and zinc dust is attributed to anodic corrosion inhibition and barrier effect mechanisms that act together. The protection against corrosion by using epoxy paints modified by the addition of polyaniline and $Zn_3(PO_4)_2$, a conventional corrosion inhibitor, has been investigated and compared [70]. As expected, the incorporation of the $Zn_3(PO_4)_2$ as corrosion inhibitor improves the resistance of the epoxy coating against corrosion. On the other hand, analysis of the panels coated with the epoxy and PANI-EB formulation showed that PANI-EB was providing the highest protection than $Zn_3(PO_4)_2$ and PANI-ES. Thus, the addition of about 0.3% of PANI-EB is very effective for preventing blister formation near the scratch line that exposes the steel surface. This is due to the ability to store charge and electroactivity of polyaniline to be oxidized and reduced in a reversible way.

Moreover, their results indicated that PANI-EB performs better than PANI-ES when they are used as anticorrosive additive of epoxy coatings [25, 71]. This is due to the ability of PANI-EB to store charge and to act as a molecular condenser.

It is evidenced that the undoped form PANI is so or more effective than the doped form of PANI, even although it is generally accepted that the ability to intercept electrons at the metal surface and to transport them typically attributed to CPs is a very effective mechanism to retard corrosion.

The success of PANI-EB as anticorrosive additive is attributed to its intrinsic electroactivity or redox property which can oxidize and reduce in a reversible way. Thus the electrochemical mechanism is combined with a barrier effect of polyaniline against corrosive species. Furthermore, PANI-EB has many having amine groups can form complex with metal ions which may play a role in this process.

Gasparac and Martin [72] reported that the anticorrosion properties of PANI were independent of the doping level, and totally undoped emeraldine base coating was equally capable of passivation of the stainless steel substrate.

In practical applications, PANI is not used alone. The paint consisted from polymeric matrix (a film-forming substance), pigments and other additives required for the creation of a protective film. Kalendova et al. [73] reported that there is a synergic effect between PANI and a corrosion-inhibition pigment and the therefore, anticorrosive property of the paint would be is greater than when using only one type of the inhibition system. The highest anticorrosion efficiency of PANI in the disturbed places on the coating indicates its active effectiveness as anodic inhibitors suppress the anodic reaction of oxidizing Fe to ferrous cations. These inhibitors provide cathodic protection of surface because ferrous cations are prevented from migrating from the lattice. This means that the application of PANI retards the transition of Fe^0 in an elementary lattice to cations Fe^{2+}. The other evidence of the existence of this inhibition protection mechanism is the conductive contact between pigment particles and metal or between the pigment particles alone. This means that the coating must contain a sufficient amount of polyaniline particles because the anodic inhibitors are efficient only if they are concentrated enough and if their concentration is low, corrosion is promoted.

Alam et al. dispersed polyaniline [74] and nano polyaniline/ferrite [75]. In alkyd resin and investigated its corrosion protection. They reported that PANI/alkyd and PANI/ferrite/alkyd coatings act as inhibitors which form a passive protective layer which imparts high resistivity against corrosive ions. Also, the anticorrosive effect of the nanocomposite coatings can be attributed to the presence of ferrite particles. The presence of excessive ferrite particles prevents the reduction of PANI from ES to EB form and maintains PANI in its doped state and prevents metal dissolution. Furthermore, the uniform dispersion of the PANI/ferrite nanocomposite in alkyd coating helps in the formation of a well-adhered, dense, and continuous network-like structure that prevent the penetration of the corrosive ions through to the metal substrate.

Zaarei et al. [76] reported a new epoxy coating by applying emeraldine-base polyaniline in the aminic hardener of these coatings. Their studies showed that presence of low concentration of EB in the hardener resulted in the formulation of an epoxy coating with superior corrosion protection properties. Micro polyaniline and nano polyaniline particles were used as anticorrosion additive in epoxy coating by directly adding to a new water-based hardener (RIPI-W.B.H.) [77]. The results showed the better anticorrosion performance of coating containing nano PANI than the one containing micro PANI. In contrast, Tiitu et al. [78] dissolved emeraldine base (EB) form of PANI in specific aminic hardeners and then added the mixture to the epoxy resin and then investigated the resulted composite as anticorrosive coating.

Ge et al. [79] dispersed polyaniline nanofiber in epoxy resin through mechanical grinding using cone-mill and investigated its anticorrosion property as a coating for Q235 steel. Results showed that the best corrosion protection effect was obtained when the amount of PANI was around 0.5% (wt%). Also, the results of this study showed that different composite coatings of PANI doped by different inorganic acids show different protective abilities in which H_3PO_4-doped PANI showed the best protective effect, followed by H_2SO_4-doped

PANI, and then followed by HNO3-doped PANI, while HCl-doped PANI provided the worst protective effect. These results indicated that morphology and counter-anion are effective on the anticorrosive property of the doped PANI. Thus it can be concluded that polyaniline because of its low cost and easy to synthesize, is undoubtedly highly desirable material for anticorrosive applications in future. Thus, *commercializing polyaniline for* corrosion protection applications needs to be additionally investigated in future research projects.

CONCLUSION

Conducting polymers are always interesting materials due to their continuous usages in many areas. One of the most important applications of conducting polymers is protection of metals against corrosion. Among conducting polymers, PANI and its derivatives are extensively studied as anticorrosive coatings on various metals. On the mechanism of corrosion protection by PANI, it has been well established that PANI has both barrier and electrochemical protection effects. The electrochemical protection is caused by the increase of the corrosion potential and the formation of a protective passive layer on metal surface due to redox catalytic properties of PANI. It has been reported that PANI based coatings can prevent corrosion even in scratched areas where bare metal surface is exposed to the aggressive environment. this is the reason for these coatings being described as smart coatings since PANI can generate on demand a corrosion inhibitor for stopping or slowing the corrosion rate at defects. The anticorrosive property of polyaniline can be improved by addition of various materials like metal oxides, nano clays, zeolites and other additive materials. Also, PANI can be used as additive in organic paints and primer systems to replace chromium containing coatings which have adverse health and environmental concerns. The results of various studies indicated that presence of low concentration of polyaniline in the various conventional organic coatings resulted in the formulation of a coating with superior corrosion protection properties. Thus, polyaniline is undoubtedly a highly *desirable material for anticorrosive applications in future.*

REFERENCES

[1] Yeh, J-M.; Chang, K-C. Polymer/layered silicate nanocomposite anticorrosive coatings. *J. Ind. Eng. Chem.* 2008, *14*, 275–291.

[2] *Handbook of Hot-dip Galvanization;* Maass, P., Peissker, P., Ahner, Eds; WILEY-VCH Verlag Gmbh & Co.: Germany, 2011.

[3] Krishna Reddy, L. *Principles of Engineering Metallurgy*,; New Age International (P) Ltd Publishers: New Delhi, India, 2009.

[4] Grundmeier, G.; Schmidt, W.; Stratmann, M. Corrosion protection by organic coatings: electrochemical mechanism and novel methods of investigation. *Electrochim. Acta* 2000, *45*, 2515–2533.

[5] Frankel, G.S. Pitting corrosion of metals a review of the criticalfactors. *J. Electrochem. Soc.* 1998, *145 (6)*, 2186–2198.

[6] Cape, T.W. Phosphate conversion coatings. In: *ASM Handbook, Corrosion*; Davis, J.R., Ed.; ASM International, Library of Congress, USA, 1987, Vol. 13; p 383.

[7] Fontana, M. G. *Corrosion engineering*, 3rd ed.; McGraw-Hill: Singapore, 1986.

[8] MacDiarmid, *A.G.* Short course on conductive polymers. Suny, New Paltz. New York, 1985, *430*, 243-252.

[9] Ahmad, N.; MacDiarmid, A.G. Inhibition of corrosion of steels using conducting polymers. *Bull. Am. Phys. Soc.* 1987, *32(3)*, 548-555.

[10] Mirmohseni, A.; Oladegaragoze, A. Anti-corrosive properties of polyaniline coating on iron, *Synthetic Met.* 2000, *114*, 105–108.

[11] Pruneanu, S.; Csahok, E.; kertesz, V.; Inzelt, G. Electrochemical quartz crystal microbalance study of the influence of the solution composition on the behavior of poly(aniline) electrodes. *Electrochim. Acta* 1998, *43(16)*, 2305–2323.

[12] Breslin, C.B.; Fenelon, A.M.; Conroy, K.G. Surface engineering: corrosion protection using conducting polymers. Mater. Des. 2005, *26*, 233–237

[13] Petitjean, J.; Aeiyach, S.; Lacroix, J.C.; Lacaze, P.C. Ultra-fast electropolymerization of pyrrole in aqueous media on oxidizable metals in a one-step . *J. Electroanal. Chem.* 1999, *478*, 92-100.

[14] Zaid, B.; Aeiyach, S.; Lacaze, P.C. Electropolymerization of pyrrole in propylene carbonate on zinc electrodes modified by heteropolyanions. *Synthetic Met.* 1994, *65*, 27-34.

[15] Bazzaoui, M.; Bazzaoui, E.A.; Martins, L.; Martins, J.I. Electrochemical synthesis of adherent polypyrrole films on zinc electrodes in acidic and neutral organic media. *Synthetic Met.* 2002, *128*, 103-114.

[16] Lacroix, J. C.; Camalet, J. L.; Aeiyach, S.; Chane-Ching, K.; Petitjean, J.; Chauveau, E.; et al. Aniline electropolymerization on mild steel and zinc in a two-step process. *J. Electroanal. Chem.* 2000, *481*, 76–81.

[17] Bernard, M. C.; Joiret, S.; Hugot-Le, G. A.; Long, P. D. Protection of iron against corrosion using a polyaniline layer – II. Spectroscopic analysis of the layer grown in phosphoric/metanilic solution. *J. Electrochem. Soc.* 2001,*148*, B299–303.

[18] DeBerry, D. Modification of the Electrochemical and Corrosion Behavior of Stainless Steel with an Electroactive Coating. *J. Electrochem. Soc.* 1985, *132*, 1022-1026.

[19] Wessling, B. Passivation of metals by coating with polyaniline: corrosion potential shift and morphological changes. *Adv. Mater.*1994, *6 (3)*, 226–228.

[20] Wessling, B.; Posdorfer, J. Corrosion prevention with an organic metal (polyaniline): corrosion test results. *Electrochim. Acta* 1999, *44*, 2139–2147.

[21] Wessling, B. Corrosion prevention with an organic metal (polyaniline): Surface ennobling, passivation, corrosion test results, *Mater. Corros.* 1996, *47*, 439–445.

[22] Skothenn, T. A.; Elsenbauinei, R. L.; Reynolds, J. R.; Eds.; *Handbook of Conducting Polymers*; Second Ed.; Marcel Dekker, Inc.; New York, 1998.

[23] Wessling, B. In *Handbook of Conducting Polymers*; Skothenn, T. A.; Elsenbauinei, R. L.; Reynolds, J. R.; Eds.; Second Ed.; Marcel Dekker, Inc.: New York, 1998; pp 467-530.

[24] Rout, T.K.; Jha, G.; Singh, A.K.; Bandyopadhyay, N.; Mohanty, O.N. Development of conducting polyaniline coating: a novel approach to superior corrosion resistance. *Surf .Coat. Tech.* 2003, *167*,16–24.

[25] Fahlman, M.; Jasty, S.; Epstein, A.J. Corrosion protection of iron/steel by emeraldine base polyaniline: an X-ray photoelectron spectroscopy study. *Synth. Met.* 1997, *85*, 1323-1326.

[26] Jose, E.; Pereira, S.; Susana, I.; Cordoba, T.; Roberto, M. T. Polyaniline Acrylic Coatings for Corrosion Inhibition: The Role Played by Counter-ions. *Corros. Sci.* 2005, *47 (3)*, 811–822.

[27] Ansari, R.; Alikhani, A. H. Application of polyaniline/nylon composites coating for corrosion protection of steel. *J. Coat. Technol. Res.* 2009, *6 (2)*, 221–227.

[28] Schauer, T.; Joos, A.; Dulog, L.; Eisenbach, C.D. Protection of iron against corrosion with polyaniline primers. *Prog. Org. Coat.* 1998, *33*, 20-27

[29] Sathiyanarayanan, S.; Muthukrishnan, S.; Venkatachari, G.; Trivedi, D.C. Corrosion Protection of Steel by Polyaniline (PANI) Pigmented Paint Coatings. *Prog. Org. Coat.* 2005, *53*, 297-301.

[30] Wei, Y.; Wang, J.; Jia, X.; Yeh, J-M.; Spellane, P. Polyaniline as corrosion protection coatings on cold rolled steel. *Polymer* 1995, *36*, 4535-4537.

[31] Huh, J.H.; Oh, E.J.; Cho, J.H. Investigation of corrosion protection of iron by polyaniline blend coatings. *Synth. Met.* 2003, *137*, 965-966.

[32] Abu, Y.M.; Aoki, K. Corrosion protection by polyaniline-coated latex microspheres. *J. Electrochem. Chem.* 2005, *583*, 133-139.

[33] Alvial, G.; Matencio, T.; Neves, B.R.A.; Silva, G.G. Blends of poly(2,5-dimethoxy aniline) and fluoropolymers as protective coatings. *Electrochim. Acta* 2004, *49*, 3507-3516.

[34] Wessling, B.; Posdorfer, J. Corrosion prevention with an organic metal (polyaniline): corrosion test results. *Electrochim. Acta* 1999, *44*, 2139-2147.

[35] Dominis, A. J.; Spinks, G. M.; Wallace, G. G. Comparison of polyaniline primers prepared with different dopants for corrosion protection of steel. *Prog. Org. Coat.* 2003, *48*, 43–49.

[36] Pud, A.A.; Shapoval, G.S.; Karamchik, P.; Ogurtsov, N.A.; Gromovaya, V.F.; Myronyuk, I.E.; Kontsur, Y.V. Electrochemical behavior of mild steel coated by polyaniline doped with organic sulfonic acids. *Synth. Met.* 1999, *107*, 111-115.

[37] Chang, K-C.; Lai, M-C.; Peng, C-W.; Chen, Y-T.; Yeh, J-M.; Lin, C-L.; Yang, J-C. Comparative studies on the corrosion protection effect of DBSA-doped polyaniline prepared from in situ emulsion polymerization in the presence of hydrophilic Na^+-MMT and organophilic organo-MMT clay platelets. *Electrochim. Acta* 2006, *51*, 5645–5653.

[38] Olad, A.; Naseri, B. Preparation, characterization and anticorrosive properties of a novel polyaniline/clinoptilolite nanocomposite. *Prog. Org. Coat.* 2010, *67*, 233–238.

[39] Patil, R.C.; Radhakrishnan, S. Conducting polymer based hybrid nano-composites for enhanced corrosion protective coatings. *Prog. Org. Coat.* 2006, *57*, 332–336.

[40] Muthirulan, P.; Kannan, N.; Meenakshisundaram, M. In-situ electrochemical fabrication of porous organic-inorganic hybrid nanocomposites on stainless steel for proton exchange membrane fuel cell application, Solid State Physics, Proceedings of the 56th DAE Solid State Physics Symposium 2011. AIP Conference Proceedings, 2012, Vol. 1447, pp 407-408, DOI: 10.1063/1.4710052.

[41] Wu, K.H.; Chang, Y.C.; Yang, C.C.; Gung, Y.J.; Yang, F.C. Synthesis, infrared stealth and corrosion resistance of organically modified silicate–polyaniline/carbon black hybrid coatings. *Eur. Polym. J.* 2009, *45*, 2821–2829.

[42] Martina, V.; De Riccardis, M. F.; Carbone, D.; Rotolo, P.; Bozzini, B.; Mele, C. Electrodeposition of polyaniline–carbon nanotubes composite films and investigation on their role in corrosion protection of austenitic stainless steel by SNIFTIR analysis. *J. Nanopart. Res.* 2011, *13*, 6035-6047.

[43] Olad, A.; Rashidzadeh, A. Preparation and anticorrosive properties of PANI/Na-MMT and PANI/O-MMT nanocomposites. *Prog. Org. Coat.* 2008, *62*, 293–298.

[44] Dispenza, C.; Presti, C.; Belfiore, C.; Spadaro, G.; Piazza, S. Electrically conductive hydrogel composites made of polyaniline nanoparticles and poly(N-vinyl-2-pyrrolidone). *Polymer* 2006, *47*, 961-971.

[45] Olad A.; Rashidzadeh, A. Poly(N-vinylpyrrolidone) Modified Polyaniline/Na^+-Cloisite Nanocomposite: Synthesis and Characterization. *Fiber. Polym.* 2012, *13*, 16-20.

[46] Sathiyanarayanan, S.; Azim, S.S.; Venkatachari, G. Corrosion protection of galvanized iron by polyaniline containing wash primer coating. *Prog. Org. Coat.* 2009, *65*,152–157.

[47] Kumar, S.A.; Meenakshi, K.S.; Sankaranarayanan, T.S.N.; Srikanth, S. Corrosion resistant behavior of PANI–metal bilayer coatings. *Prog. Org. Coat.* 2008, *62*, 285–292.

[48] Olad, A.; Rasouli, H. Enhanced Corrosion Protective Coating Based on Conducting Polyaniline/Zinc Nanocomposite. *J. Appl.Polym. Sci.* 2010, *115*, 2221–2227.

[49] Armelin, E.; Pla, R.; Liesa, F.; Ramis, X.; Iribarren, J.I.; Alemán, C. Corrosion protection with polyaniline and polypyrrole as anticorrosive additives for epoxy paint, *Corros. Sci.* 2008, *50*, 721–728.

[50] Olad, A.; Barati, M.; Shirmohammadi, H. Conductivity and anticorrosion performance of polyaniline/zinc composites Investigation of zinc particle size and distribution effect. *Prog. Org. Coat.* 2011, *72*, 599–604.

[51] Olad, A.; Barati, M.; Behboudi, S. Preparation of PANI/epoxy/Zn nanocomposite using Zn nanoparticles and epoxy resin as additives and investigation of its corrosion protection behavior on iron. *Prog. Org. Coat.* 2012, *74*, 221–227.

[52] Foad El Sherbini, E.E. Effect of some ethoxylated fatty acids on the corrosion behavior of mild steel in sulphuric acid solution. *Mater. Chem. Phys.* 1999, *60*, 286–290.

[53] Leçe, H.D.; Emregül, K.C.; Atakol, O. Difference in the inhibitive effect of some Schiff base compounds containing oxygen, nitrogen and sulfur donors. *Corros. Sci.* 2008, *50*,1460–1468.

[54] Mu, G.; Li, X.; Qu, Q.; Zhou, J. Molybdate and tungstate as corrosion inhibitors for cold rolling steel in hydrochloric acid solution. *Corros. Sci.* 2006, *48*, 445–459.

[55] Samiento-Bustos, E.; González Rodriguez, J.G.; Uruchurtu, J.; Dominguez-Patiño, G.; Salinas-Bravo, V.M. Effect of inorganic inhibitors on the corrosion behavior of 1018 carbon steel in the LiBr + ethylene glycol + H_2O mixture. *Corros. Sci.* 2008, *50*, 2296–2303.

[56] Bastos, A.C.; Ferreira, M.G.; Simões, A.M. Corrosion inhibition by chromate and phosphate extracts for iron substrates studied by EIS and SVET. *Corros. Sci.* 2006, *48*, 1500–1512.

[57] Sahin, M.; Gece, G.; Karcı, F.; Bilgiç, S. Experimental and theoretical study of the effect of some heterocyclic compounds on the corrosion of low carbon steel in 3.5% NaCl medium. J. *Appl. Electrochem.* 2008, *38*, 809–815.

[58] Elayyachy, M.; Hammouti, B.; El Idrissi, A. New telechelic compounds as corrosion inhibitors for steel in 1M HCl. *Appl. Surf. Sci.* 2005, *249*, 176–182.

[59] Bouklah, M.; Hammouti, B.; Lagrenee, M.; Bentiss, F. Thermodynamic properties of 2,5-bis(4-methoxyphenyl)-1,3,4-oxadiazole as a corrosion inhibitor for mild steel in normal sulfuric acid medium. *Corros. Sci.* 2006, *48*, 2831–2842.

[60] Ajmal, M.; Mideen, A.S.; Quaraishi, M.A. 2-Hydrazino-6-methyl-benzothiazole as an effective inhibitor for the corrosion of mild steel in acidic solutions. *Corros. Sci.* 1994, *36*, 79–84.

[61] Fang, J.; Li, J. Quantum chemistry study on the relationship between molecular structure and corrosion inhibition efficiency of amides. *J. Mol. Struct. (Theochem)* 2002, *593*, 179–185.

[62] Quraishi, M.A.; Sharma, H.K. 4-Amino-3-butyl-5-mercapto-1,2,4-triazole: a new corrosion inhibitor for mild steel in sulphuric acid. *Mater. Chem. Phys.* 2002, *78*, 18–21.

[63] Lu, W-K.; Basak, S.; Elsenbaumer, R. L. In *Handbook of Conducting Polymers*; Skothenn, T. A.; Elsenbauinei, R. L.; Reynolds, J. R.; Eds.; Second Ed.; Marcel Dekker, Inc.: New York, 1998; pp 881-921.

[64] Hur, E.; Bereket, G.; Sahin, Y. Corrosion inhibition of stainless steel by polyaniline, poly(2-chloroaniline), and poly(aniline-co-2-chloroaniline) in HCl. *Prog. Org. Coat.* 2006, *57*, 149–158.

[65] Bacskai, R.; Schroeder, A. H.; Young, D. C. Hydrocarbon-soluble alkylaniline/formaldehyde oligomers as corrosion inhibitors. *J. Appl. Polym. Sci.* 1991, *42*, 2435–2441.

[66] Armelin, E.; Oliver, R.; Liesa, F.; Iribarren, J.I.; Estrany, F.; Alemán, C. Marine paint fomulations: Conducting polymers as anticorrosive additives. *Prog. Org. Coat.* 2007, *59*, 46 –52.

[67] Pereira da Silva, J. E.; Cordoba de Torresi, S. I.; Torresi, R. M. Polyaniline acrylic coatings for corrosion inhibition: the role played by counter-ions. *Corros. Sci.* 2005, *47*, 811–822.

[68] Akbarinezhad, E.; Ebrahimi, M.; Faridi, H.R. Corrosion inhibition of steel in sodium chloride solution by undoped polyaniline epoxy blend coating. *Prog. Org. Coat.* 2009, *64*, 361–364.

[69] Iribarren Lacoa, J. I..; Cadena Villotab, F.; Liesa Mestres, F.; Corrosion protection of carbon steel with thermoplastic coatings and alkyd resins containing polyaniline as conductive polymer as conductive polymer. *Prog. Org. Coat.* 2005, *52*, 151–160.

[70] Armelin, E.; Alemán, C.; Iribarren, J. I. Anticorrosion performances of epoxy coatings modified with polyaniline: A comparison between the emeraldine base and salt forms. *Prog. Org. Coat.* 2009, *65*, 88–93.

[71] Lu, W-K.; Elsenbaumer, R.L.; Wessling, B. Corrosion protection of mild steel by coatings containing polyaniline. *Synth. Met.* 1995, *71*, 2163-2166.

[72] Gasparac, R.; Martin, C.R. The Effect of Protic Doping Level on the Anticorrosion Characteristics of Polyaniline in Sulfuric Acid Solutions. *J. Electrochem. Soc.* 2002, *149*, B409-B413.

[73] Kalendova, A.; Vesely, D.; Stejskal, J. Organic coatings containing polyaniline and inorganic pigments as corrosion inhibitors. *Prog. Org. Coat.* 2008, *62*, 105–116.

[74] Alam, J.; Riaz,U.; Ahmad, Sh.; High performance corrosion resistant polyaniline/alkyd ecofriendly coatings. *Curr. Appl. Phys.* 2009, *9*, 80–86.

[75] Alam, J.; Riaz, U.; Ashraf, S. M.; Ahmad, Sh. Corrosion-protective performance of nano polyaniline/ferrite dispersed alkyd coatings. *J. Coat. Technol. Res.* 2008, *5 (1)*, 123–128.

[76] Zaarei, D.; Sarabi, A. A.; Sharif, F.; Moazzami Gudarzi, M.; Kassiriha, S. M. A new approach to using submicron emeraldine-base polyaniline in corrosion-resistant epoxy coatings. *J. Coat. Technol. Res.* 2012, *9 (1),* 47–57.

[77] Bagherzadeh, M.R.; Ghasemi, M.; Mahdavi, F.; Shariatpanahi, H. Investigation on anticorrosion performance of nano and micro polyaniline in new water-based epoxy coating. *Prog. Org. Coat.* 2011,*72*, 348–352.

[78] Tiitu, M.; Talo, A.; Forsen, O.; Ikkala, O. Aminic epoxy resin hardeners as reactive solvents for conjugated polymers: polyaniline base/epoxy composites for anticorrosion coatings. *Polymer* 2005, *46*, 6855–6861.

[79] Ge, C. Y.; Yang, X. G.; Hou, B. R. Synthesis of polyaniline nanofiber and anticorrosion property of polyaniline–epoxy composite coating for Q235 steel. *J. Coat. Technol. Res.* 2012, *9*, 59-69.

[80] Mansfeld, F. The use of electrochemical impedance spectroscopy for the evaluation of the properties of passive films and protective coatings. *ACH-Model Chem.* 1995, *132 (4)*, 619–631.

[81] Zarras, P.; Anderson, N.; Webber, C.; Irvin, D.J.; Irvin, J.A.; Guenthner, A.; Stenger-Smith, J.D. Progress in using conductive polymers as corrosion-inhibiting coatings. *Rad. Phys. Chem.* 2003, *68*, 387–394.

In: Trends in Polyaniline Research
Editors: T. Ohsaka, Al. Chowdhury, Md. A. Rahman et al.
ISBN: 978-1-62808-424-5

Chapter 14

POLYANILINE AND REMOVAL OF POLLUTANTS

Ali Olad*[*] *and Azam Rashidzadeh
Polymer Composite Research Laboratory, Department of Applied Chemistry,
Faculty of Chemistry, University of Tabriz, Tabriz, Iran

ABSTRACT

The increasing amount of various pollutants discarded into the environment has been become a serious problem. In the past decade, many efforts have been done to overcome water pollution problem. The unique chemical, electrochemical and ion exchange properties of polyaniline has made it as one of the affordable materials that exhibit high capability to remove various pollutants like heavy metals and dye molecules from effluents.

This chapter deals with application of polyaniline as a potential material for removal of metal ions and dye molecules from aqueous solutions. Then the review of the works on the capability of polyaniline composites and nanocomposites for removal of metal ions and dyes from wastewater has been discussed. Also, the mechanism of removal of pollutants using polyaniline and its composites and nanocomposites has been elucidated.

INTRODUCTION

Various pollutants affect the quality, health and property of drinking water, rivers, lakes and oceans all over the world. Domestic sewage, industrial wastewater, nuclear waste and oil spill are the main sources of water pollution. Different forms of pollutants like heavy metals, dye molecules, microbial pollutants, radioactive substances and organic materials influence the health in different ways.

The term heavy metal refers to any metallic chemical element that has a relatively high density and is highly toxic or poisonous at low concentration, such as cadmium, chromium, lead and mercury. Heavy metals are natural components of the Earth crust. They cannot be degraded or destroyed. Heavy metals from industrial processes, accumulated in nearby lakes,

[*] E-mail: a.olad@yahoo.com; Tel: +98 411 3393164; Fax: +98 411 3340191.

rivers and soils, can enter in animal and human bodies and their tendency to accumulate in living organisms makes them more dangerous. Long-term exposure to these materials may result in slowly progressing physical, muscular, and neurological degenerative processes. Besides that, continuous releasing of dye stuffs and dye wastewater can cause pollution of lakes and rivers. Also microbial pollutants from sewage results in infectious diseases are a major problem in the developing world. The unregulated release of these pollutants has created environmental pollution as well as medical and aesthetic problems associated with human health and agriculture, thus from a public health standpoint, treatment of contaminated water is of prime importance.

In the recent decades, new and useful ways have been used to solve the environmental problems. Much has been said and written about environmental problems, but in this chapter, we focus on the application of polyaniline (PANI) in water and wastewater treatment. There are many conventional processes for removing heavy metals from wastewater. The ones that have received the most publicity in recent years include chemical precipitation, ion exchange, adsorption, membrane filtration, coagulation, flotation, and electrochemical deposition [1]. Among these processes, PANI removes pollutants mainly via ion exchange, adsorption and membrane filtration methods.

POLYANILINE

Conducting polymers are new class of materials which have been studied extensively during the past three decades. Their unique properties such as electrical conductivity, electrochemical properties and ease of processing made them useful in wide area of applications. Among conducting polymers, polyaniline is a potential material for various commercial applications because of its high electrical conductivity, redox properties, environmental stability, ease of preparation in large quantities and relatively low cost.

As we know, Polyaniline (PANI) can occur in a range of well defined oxidation states. These different states range from the fully reduced leucoemeraldine (LE), emeraldine base (EB) to emeraldine salt (ES) and pernigraniline (PA) (Figure 1). The only conducting form among the four is the green protonated emeraldine base form, which has both the oxidized iminium and reduced amine nitrogens, in equal amounts. The imine sites can be protonated to the bipolaron emeraldine salt form. However, this form undergoes a further rearrangement to form the delocalized polaron lattice, which gives a conductive nature to polymer. Thus, the blue insulating emeraldine form can be transformed into the conducting form by lowering the pH of the media and vice-versa [2].

PANI in the conducting state has a positive charge delocalized over the backbone of the polymer which endows ion-exchange property to PANI. The ion-exchange property of PANI is derived from the positive charge of radical cation form. Ion-exchange property of PANI differs in several ways from those of conventional ion-exchange resins. The reason may be attributed to the charge delocalization. A number of counterions can be incorporated into PANI. Thus PANI can transport ions through continuous charging and discharging, with concomitant diffusion of ions into and from the polymer [3].

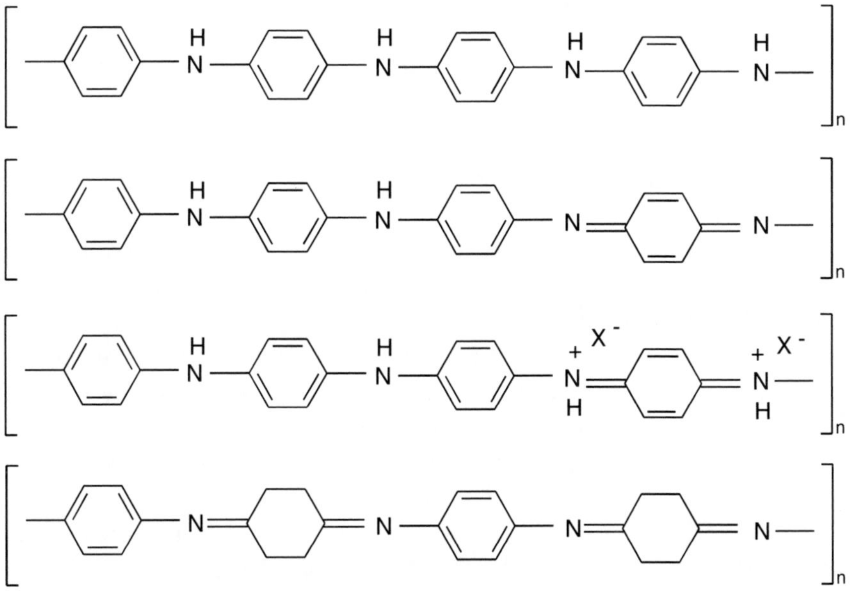

Figure 1. Different oxidation states of polyaniline: leucoemeraldine (LE), emeraldine base (EB), emeraldine salt (ES) and pernigraniline (PA).

Polyaniline via its nitrogen atoms in amine derivatives forms co-ordinate bond with positive charge of heavy metal and dye ions and adsorb them due to the presence of electron in s^2p^3 orbital of nitrogen. The adsorption property of polyaniline is highly dependent to pH of solution. Under low pH in acidic conditions the surface of polyaniline is highly protonated. The protonated form of polyaniline can form electrostatic attraction with solution anions like chromate and dichromate ions.

Furthermore, polyaniline due to its unique electrochemical properties and by undergoing oxidation or reduction can be used to reduce highly toxic materials to their less toxic states. Moreover, it was reported that the conducting polymer obtained from 1,8-diaminonaphthalene may collect some heavy metal ions from the solution [4]. Also, this polymer can collect Ag^+ ions and these ions might be exchanged by Hg^{2+} and other ions such as Cu^{2+}, Mg^{2+}, Cd^{2+} and Pb^{2+}. Polyaniline as well as other conducting polymers can form complexes with heavy metals [5]. Results of these studies lead to the possibility of application of polyaniline in abatement of some toxic materials from the environment [6, 7].

ADSORPTION

Adsorption is one of the cheap and effective technologies used for removal of pollutants from water and wastewater. In adsorption process a substance is transferred from the liquid phase to the surface of a solid adsorbent material [8]. This process is sometimes reversible and adsorbents by suitable desorption process can be regenerated and used again in adsorption cycle.

Activated carbon [9], carbon nanotubes [10], clays [11], agricultural wastes [12], industrial byproducts and wastes [13, 14], natural zeolites [15] and biosorbents [16] are common adsorbents used in waste water treatment.

Polyaniline because of its environmental stability, controllable quality, and good processability have attracted great interest for removal of pollutants from aqueous solution. As mentioned previously, it is believed that PANI can be existed in three major oxidation forms. In acidic media, PANI chains become protonated or oxidized and therefore, the positive charges can be localized over the polymeric backbone. The positive charges can be counterbalanced by anions that are absorbed by the PANI chains. On the other hand, in basic media, the negative charges are produced on polymer backbone which has great affinity for cations. Thus, the deprotonated PANI(EB) adsorbs cationic molecules due to the electrostatic interactions between the cationic groups and the amine and imine nitrogen in PANI(EB) structure [17]. Removal of heavy metal and dye ions from aqueous solutions with PANI is very dependant to the pH value of the solution. In acidic pH values, polyaniline is changed into ES (-N groups are protonated), so the polymer cannot act as a ligand or chelating agent, therefore, the metal uptake process is very poor. In contrast, with increasing the pH of the solution, the polymer is changed into its undoped form (EB), then free amine or imine groups on the polymer backbone will be available for metal chelating, so the sorption of heavy metal ions increases considerably. However at higher pH values (alkaline media), precipitation of heavy metal ions may occur as well.

Beside heavy metal ions, polyaniline was used to remove dye molecules from aqueous solutions. There are various forms of dyes including anionic (direct, acid and reactive) dyes, cationic (basic) dyes and non-ionic (dispersed) dyes. As told before, protonated PANI emeraldine salts (ES) and deprotonated PANI emeraldine base (EB), can be used to adsorp preferentially anionic and cationic dyes from aqueous solution, respectively [18]. Emeraldine base preferentially removes the cationic dyes like methylene blue while the anionic dyes like procion red can predominately be removed by the polyaniline salt [18]. PANI doped with functionalized protonic acids such as p-toluenesulfonic acid and camphorsulfonic acid selectively adsorbed anionic dyes [19]. Different polymerization conditions such as initiator/oxidant amount, dopant acid, polymerization time, and the type of surfactant can affect the removal rate of dyes like direct blue 78 by PANI [20]. In the case of anionic dye removal by PANI, the electrostatic interaction is expected to be the dominant, owing to the existence of –$NH^{+\cdot}$ – centers on the polaron/bipolaron forms of PANI salt and the presence of dye anions generated from the dissociation of the dye molecule in solution. As soon as the dye anions reach to the surface of PANI, they approach to the $NH^{+\cdot}$ centers and interact with them. The suggested mechanism of binding the anionic dye with PANI matrix was shown in Figure 2.

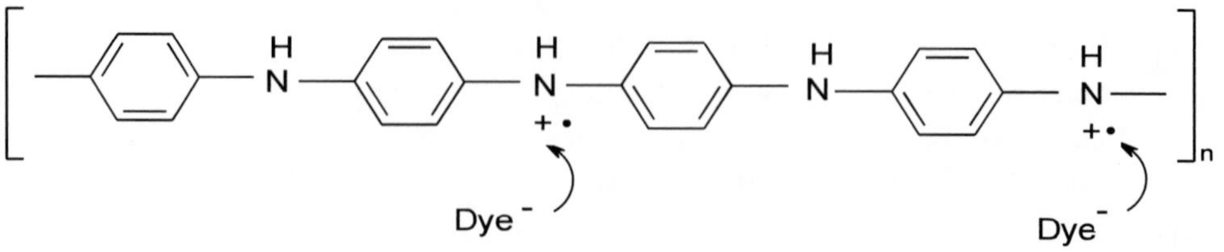

Figure 2. Interaction of anionic dye with polyaniline.

As shown in this figure, each unit structure of PANI adsorbs only two dye molecules. In other words, each $NH^{+\cdot}$ centers on PANI can only interact with one anionic dye molecule. The acidic media is more preferable for adsorption of anionic dyes. Because in acidic pH, PANI exists in the doped state and the positively charged $NH^{+\cdot}$ centers on polymer backbone facilitate the interaction with the dye anions. Whereas, in basic solutions, the ES form of PANI is dedoped into the EB form and therefore the positively charged sites on the ES form are no longer available.

Mechanisms of pollutants removal by adsorption involve several interactions including acid–base interactions, hydrogen bonding, ion exchange, coordination/chelation, complexation, precipitation, physical adsorption, and electrostatic interactions [21]. Mahanta et al. reported that the removal of anionic dyes from aqueous solutions is due to the chemical interaction between dye molecules and polyaniline [22]. They also stated that in acidic conditions, due to high concentration of H^+, the surface of the adsorbent is positively charged so electrostatic attraction between the adsorbent and the anionic dye is enhanced. In alkaline conditions, due to the presence of hydroxyl ions on the surface of adsorbents which compete with the dye for adsorption sites, lower adsorption of dye is occurred. In higher pH conditions, because PANI is in the form of EB, no Cl^- ions or other counter ions existed to be exchanged with anionic dye molecules, then the removal of anionic dyes in neutral or alkaline conditions is lower than acidic solutions [23].

In contrast the cationic dyes can also be adsorbed by basic form of PANI(EB). This process was shown in Figure 3. Basic dyes or cationic dyes molecules upon dissolution release colored dye cations into the solution. The adsorption of these charged cationic dye groups onto the adsorbent surface is primarily influenced by the surface charge on the adsorbent which is in turn influenced by the solution pH [24]. In acidic media the basic dye will become protonated and the positive charge on the adsorbent molecules at low pH, resulted to a lower sorption degree. In contrast in higher pH values, the negative charge density on the surface of the adsorbent increases which favors the sorption of cationic dyes. Therefore, increasing the solution pH, increases the extent of dye removal. At highly acidic conditions, lower adsorption of cationic dye on the adsorbent is probably due to the presence of high concentration of H^+ ions on the surface of adsorbent competing with a cationic dye for adsorption sites on the adsorbent. With increasing the solution pH, the electrostatic repulsion between the positively charged methylene blue and the surface of adsorbent is lowered and consequently removal efficiency is increased.

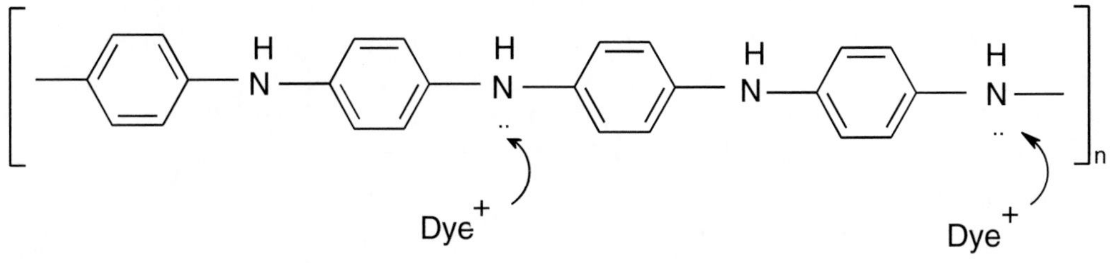

Figure 3. Interaction of a cationic dye with polyaniline.

Several methods have been applied to increase the adsorption capacity of pollutants by polyaniline. Riahi and coworkers investigated the effect of synthesis condition on PANI capability for removal and recovery of chromium from aqueous solution [25]. They synthesized polyaniline in various solvents including water, mixture of solvents containing water, acetonitrile and ethyl acetate and then investigated its effect on removal efficiency. They also concluded that the kind of solvents affects surface morphology of synthesized polyanilines. The variation of polyanilines surfaces affects the capacity of prepared polyaniline for chromium removal due to changes of interface between polyanilines and chromium solutions.

One of the methods to improve the removal efficiency of pollutants through adsorption is synthesis of large-surface-area PANIs. The synthesis of PANI nanocomposite with CNTs is a fascinating way to increase the removal efficiency of both materials through the synergetic effect between materials. CNTs, because of their high adsorption capacity, have been considered as one of the very suitable materials in the removal of organic and inorganic pollutants from aqueous solutions [26]. The mechanisms of adsorption of metal ions onto CNTs are via electrostatic attraction, sorption-precipitation and chemical interaction between the metal ions and the surface functional groups of CNTs. In the case of CNTs, the major mechanism is chemical interaction between the pollutant ions and the surface functional groups of CNTs, by which the protons of carboxylic and phenolic groups in the CNTs is exchanged with the metal ions in aqueous phase. But because of difficulty of separation of CNTs from aqueous solutions by centrifugation and filtration, and their low thermal stability, it is needed to modify them with other materials.

Besides that, polyaniline due to its high stability and the large amount of amine and imine functional groups, which have strong affinity with metal ions, is an attractive polymer to modify CNTs. Nanocomposites composed of PANI and multiwalled carbon nanotubes (MWCNT) and singlewalled carbon nanotubes (SWCNT) can be prepared through in-situ chemical polymerization of aniline in a dispersion of CNT [27] , electrochemical polymerization [28] and plasma induced polymerization technique [29].

Zeng et al. [30] applied PANI/CNT nanocomposite for removal of malachite green dye molecule. They stated that the high adsorption capacity of the CNT/PANI composites is attributed to the large surface area of the composites and the strong π-π^* interactions and improved charge transfer between PANI and CNTs. Du et al. [31] reported that MWCNT/PANI composite film coated platinum wire can be used to remove some phenolic compounds through headspace solid-phase micro extraction method.

Clays are another class of materials used for removal of pollutants from aqueous solutions [32]. Clays are aluminosilicate materials containing exchangeable ions on their surface which play important role in the environment in removal of pollutants either through ion exchange or adsorption or both. These natural materials due to their easy availability, low cost, and good sorption properties could be a good substitute for activated carbon as an adsorbent. The layered structure of clay is considered as host materials for the adsorbents and counter ions. Vermiculite, bentonite, kaolinite, diatomite and montmorillonite are among the clays used for removal of metals and dyes [33-35].

The adsorption efficiency of clays generally results from a net negative on the structure of minerals clays [36]. This negative charge gives clay the capability to adsorb positively charged species. Clays along with polyaniline has been used as adsorbent for removal of metals and dyes from aqueous solutions. These kind of polyaniline/clay nanocomposites have high adsorption capacity compared neat clay. Wang et al. [37] investigated the removal of humic acid (HA) from aqueous solution using polyaniline/attapulgite (ATP–PANI) composite. Humic acid which is formed by biodegradation of dead organic matter in the environment, may cause serious environmental and health problems. It may change the color and taste of water. Also, it may bind heavy metals and results in increase their concentrations in water. So to minimize and remove humic acid from drinking water is very essential. Attapulgite is a kind of crystalline hydrated magnesium aluminum silicate mineral with fibrillar structure, large specific surface area, porous structure and moderate cation exchange. These properties made it an adsorbent for the removal of heavy metal and organic pollutants [38]. However because of low adsorption capacities of attapulgite, modification of attapulgite with small molecules or polymers has been investigated [39]. As told before, PANI because of its large amount of amine and imine nitrogens, has been frequently used in wastewater treatment to eliminate the inorganic and organic pollutants. In low pH, the strong electrostatic forces between protonated adsorbent and disassociated HA molecules lead to the enhanced HA adsorption. But, in high pH media, electrostatic repulsive force increases which may decrease the HA adsorption on ATP–PANI [40].

In another work PANI/palygorskite composite was used for the removal of hazardous heavy metal ions from wastewater [41]. As a crystalline hydrated magnesium silicate, palygorskite because of its fibrous morphology, large specific surface area, and moderate cationic exchange capacity, has been used for adsorption of organic substances and heavy metal ions from solutions [42]. In this work, Kong and coworkers stated that the adsorption mechanism of metal ions removal was through chelation, ionic exchange, and electrostatic attraction. In the case of palygorskite, the removal of the metal ions was assigned to ionic exchange and electrostatic attraction between the metal ions and palygorskite because of a net negative charge of palygorskite. On the other hand, the strong electrostatic attraction between the positive heavy metal ions and the palygorskite particles would lead to adsorption of metal ions [43]. When the PANI/palygorskite composite was used as the adsorbent, the amine and secondary amino groups in PANI chains acted as linkages between the benzene rings and the metal ions which resulted to the third removal mechanism of chelation.

Besides that, natural zeolites due to their large surface area, valuable ion exchange capability and low cost have gained significant interest for adsorption of different species from aqueous solutions. Among the most frequently studied natural zeolites, clinoptilolite because of its availability, high adsorption capacity and selectivity for certain heavy metal ions, has received extensive attention. Zeolites have a net negative structural charge which makes them suitable for ion exchange process and selective for certain cations. This makes zeolites of interest for use in the treatment of nuclear, municipal and industrial wastewaters [44]. The negative charge also causes natural zeolites to have little or no affinity for pollutant anions like chromate, nitrate and anionic dyes. Modification of zeolite surface by certain surfactants like polymers has been emerged as a useful method to make them applicable for adsorption and removal of anions. The modification of these natural minerals with polymers can improve their heavy metals removal capability. By attention to the unique properties of polyaniline and its high potential for reduction of toxic Cr(VI), Olad et al. [45] modified the surface of clinoptilolite with polyaniline and applied the prepared nanocomposite for removal of Cr (VI). The modification of clinoptilolite nanoparticles by polyaniline was carried out by the polymerization of aniline monomer within the clinoptilolite nanoparticles. Results of their studies showed that the polyaniline/clinoptilolite (PANI/Clino) nanocomposite is an efficient material for removal of Cr(VI) from aqueous solutions especially at lower concentrations of Cr(VI). Also the encapsulation of polyaniline in the zeolite channels was lead to its enhanced stability in contact to the higher concentrations of Cr(VI) solutions with respect to pure polyaniline.

To tackle the poor processibility of PANI, Rashidzadeh and Olad [46] synthesized polyaniline/clinoptilolite nanocomposite in the presence of water soluble poly vinyl alcohol. The prepared polyaniline/ poly vinyl alcohol/clinoptilolite (PANI/PVA/Clino) nanocomposite was then applied for the removal of methylene blue dye. The results of their studies showed that the prepared PANI/PVA/Clino nanocomposite had high removal efficiency to remove methylene blue dye from aqueous solutions. They concluded that this behavior may be due to the different components of the adsorbent. As schematically shown in figure 4, PANI via its nitrogen atoms in amine derivatives, forms coordinate bonds with positive charge of methylene blue dye. On the other hand, the hydroxyl groups on PVA can serve to adsorb cationic methylene blue dye molecules. Moreover, methylene blue dye with positive charge adsorb on Clino via electrostatic attraction force.

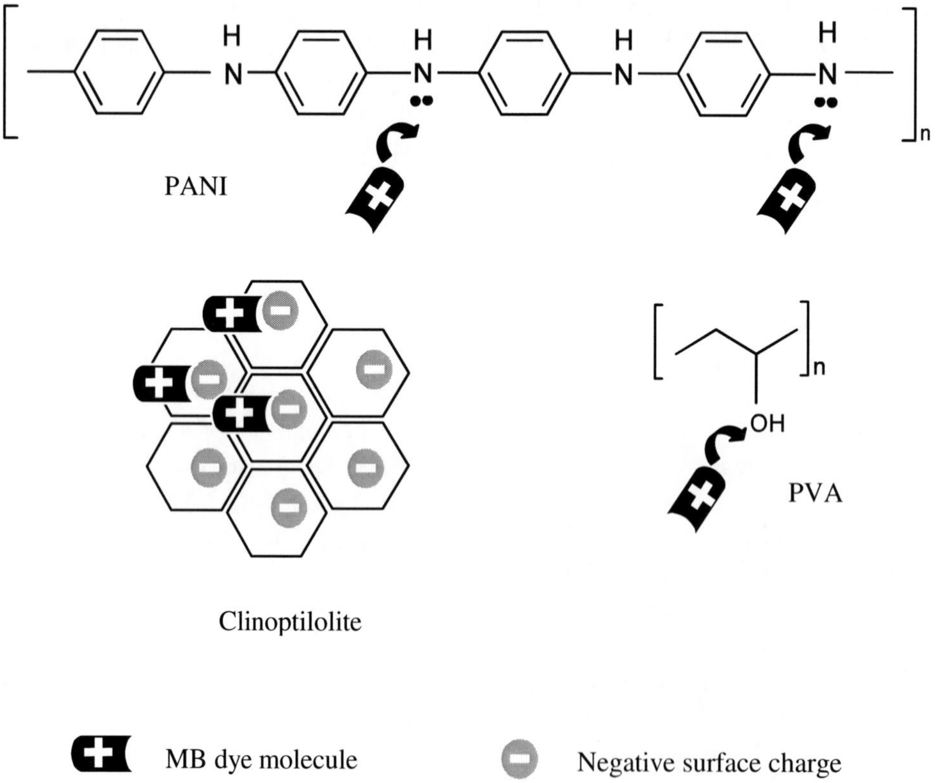

Figure 4. Schematic illustration of adsorption of cationic methylene blue dye on the PANI/PVA/Clino nanocomposite.

Polyaniline coated sawdust was applied as an adsorbent for the removal of acid dye (Acid Violet 49) from aqueous solutions by Baseri et al. [47]. They suggested that the high rate of anionic dye removal is due to the ion exchange mechanism between mobile Cl^- ions and anionic dye molecules. Ansari and coworkers investigated polyaniline coated onto wood sawdust (PANI/SD) for the removal of methyl orange dye [48], acid green 25 [49] and eosin

Y molecules [50] from aqueous solutions. They proposed that the removal of anionic dye in neutral condition might be mostly due to the strong intermolecular interactions (e.g. H-bonding) of anionic dye with the polymer. In acidic conditions because of existence of protonated adsorbent, electrostatic attraction between adsorbent and adsorbate (anionic dye) is enhanced. In contrast in alkaline condition due to the presence of OH^- ions on the surface of adsorbent, which compete with anionic dye, lower adsorption occurred. Polyaniline coated on sawdust was also investigated for removal of lead ion [51], Mercuric Ion [52] and Cd (II) ions [53] from aqueous solutions.

Ahmed and coworkers [54] investigated the removal of anionic sulphonated dye from aqueous solution using nano-polyaniline and baker's yeast. Baker's yeast due to its low price, easily grown ability, readily availability and high yields of biomass, was used as biosorbent for heavy metals [55] and textile dyes [56].

Besides utilizing natural materials composite with polyaniline, a number of wastes or by-products which are generated by many industries have also been used for removal of contaminants from water. Ghorbani et al. [57] applied polyaniline nanocomposite coated on rice husk ash for removal of Hg(II) from aqueous media.

Kumar and coworkers investigated short-chain polyaniline coated on jute fiber for the removal of chromium in batch modes [58]. They stated that in acidic pH, $-NH_2$ sites of PANI-jute was highly protonated and therefore repulsion of Cr(III) ions can occurred. By increasing pH to 6–9, almost 70% removal of Cr(III) was achieved by PANI-jute. Also, the removal of Cr(III) at pH between 7 and 9 was attributed due to precipitation of Cr(III). By further increasing in pH above 9, Cr(III) became soluble again due to its amphotheric nature. However they also studied the removal of hexavalent chromium by PANI-jute in continuous mode [59]. Reduction of Cr(VI) to Cr(III) in acidic pH followed by adsorption of the reduced Cr(III) by forming complex with adsorbent was reported as the main mechanism of Cr(VI) removal. In acidic medium, Cr(VI) was in the form of acid chromate ($HCrO_4^-$). At low pH, amine group (NH_2) of PANI-jute gets protonated and exists as ammonium ($-NH_3^+$) form. So, electrostatic attraction between the protonated ammonium and the negative chromate ions is the mechanism for the removal of Cr(VI) from solution. On the other hand, in strong acidic media and low pH value, Cr(VI) is reduced to Cr(III). Therefore, positively charged Cr(III) were probably repulsed by protonated PANI-jute and remained in solution, but by increasing the pH value, reduction of Cr(VI) to Cr(III) was much less and anionic adsorption of $HCrO_4^-$ ion with PANI-jute was the predominating mechanism.

Zheng et al. reported the preparation of kapok fiber oriented-polyaniline nanofibers and its application for removal of Cr(VI) [60]. Their findings indicated that kapok fiber can guide the growth orientation of PANI which resulted in accelerated adsorption rate of Cr(VI).

To improve the removal efficiency of polyaniline, its composites with other polymers for removal of heavy metals and dyes has also been investigated. For example a composite of polyaniline with nylon and polyurethane was used to remove Cr(VI) from aqueous solutions [61].

To minimize processing costs, the use of low-cost adsorbents such as chitosan as well as polyaniline has been reported. The hydrophilic nature, biocompatibility, biodegradability, non-toxicity, low-cost and fast adsorption kinetics have increased the utilization of chitosan in removal of pollutants from water and wastewater. Also, existence of the amino and hydroxyl groups in the chitosan can serve as coordination and reaction site for the adsorption of pollutants.

Yavuz et al. [62] have used alkyl-substituted polyaniline/chitosan composites as adsorbent materials for the removal of Cr (VI) from aqueous solutions. Also, application of polyaniline/chitosan composites was investigated to remove various dyes of congo red, coomassie brilliant blue, remazol brilliant blue R, and methylene blue dye molecules from aqueous solution [63].

Furthermore, the removal efficiency of Cd(II) [64] and Hg(II) ions [65] from aqueous solution using polyaniline/polystyrene nanocomposite was investigated. Polyaniline/poly ethylene glycol (PANI/PEG) composite was also applied for removal of chromium compounds from solutions [66]. Poly ethylene glycol as a stabilizing agent affects the morphology of prepared composite. The morphology of composite influences the capacity of the composite via the specific surface and adsorption sites on the surface of composites. The removal mechanism of chromium is the combination of surface adsorption and reduction of chromium (VI). The surface adsorption takes place because of the nitrogen atoms in PANI/PEG composite which form coordinate bond with positively charged metals. Moreover, under acidic conditions the protonated nitrogen atoms of PANI/PEG can form electrostatic attraction with solution anions (chromate and dichromate).

Moreover, the removal of chromium (VI) from aqueous solution has been carried out using polypyrrole-polyaniline nanofibers [67]. Incorporation of PANI into the growing polymer chain of polypyrrole provides an increase in surface area and sorption sites that enhances the Cr(VI) adsorption compared to its polypyrrole homopolymer counterpart. Increased removal efficiencies at lower solution pH values is due to the exchange between doped Cl^- ions in the adsorbent and the $HCrO_4^-$ ions in the solution. In contrast, lower removal efficiency at alkaline solutions is mainly due to the competition between CrO_4^{2-} and OH^- ions for the anion exchange sites of the adsorbent. By further increasing the pH value the reduction of Cr(VI) to Cr(III) takes place which causes to increase of removal efficiency.

Chemical and Electrochemical Reduction

For the first time in 1993, the use of conducting polymers for the reduction of very toxic Cr(VI) to less toxic Cr(III) has been reported by Rajeshwar and co-workers [68]. They reported that complete reduction of Cr(VI) by electrosynthesized polypyrrole films can occur. After that in 2004, Breslin reported the use of polyaniline films for the reduction of toxic Cr(VI) species in solution [69]. Ruotolo et al. reported the use of thin polyaniline films deposited on a large surface area substrate reticulated vitreous carbon (RVC) for reduction of Cr(VI) to Cr(III) by electrochemical method [70]. They concluded that the polyaniline-modified electrode is a promising material for use in Cr(VI) reduction because it remained stable after several cycles.

Olad and Nabavi studied the efficiency of Cr(VI) reduction by various forms of polyaniline [71]. Polyaniline powder as emeraldine base (EB) form was used for Cr(VI) reduction. The emeraldine base form of polyaniline can be oxidized to pernigraniline (PA) which is accompanied by reduction of Cr(VI) in solution:

$$3PANI(EB) + 6A^- - 6e^- \rightarrow 3PANI(A)_2(PA) \quad (1)$$

$$Cr_2O7_2^- + 14H^+ + 6e^- \rightarrow 2Cr^{3+} + 7H_2O \quad (2)$$

where PANI represents one tetramer and A^- is the anion present in solution (OH^- in distilled water and HSO_4^- in acidic H_2SO_4 solution). They also investigated the effect of synthesis method, film or powder form, oxidation state, polymer film thickness, initial Cr(VI) concentration and solution pH on the kinetics and performance of Cr(VI) reduction. Their investigations showed that the solution pH value affects the reduction efficiency of Cr (VI) and the reaction kinetics increases by increasing the acidity of Cr(VI) solutions. However the reduction efficiency after ~16min of exposure is almost similar for all three solutions with different pH values. It was found that by increasing the solution acidity, the overoxidation and degradation of polyaniline, occurs in lower concentrations of Cr(VI) solutions. Therefore it can be concluded that the reduction kinetics is higher in acidic solutions.

The authors, to make a comparison between EB powder and EB film in Cr(VI) reduction, used the same weight of films in Cr(VI) reduction. They found that, the reduction efficiency decreases by increasing of the polymer film thicknesses. This is due to the increasing of surface area of the film by decreasing its thicknesses. According to their results, even in very thinner polymer films, the total reduction efficiency of the EB film was lower than the reduction efficiency of EB powder with the same weight.

The electrochemically synthesized polyaniline film has also been used for the reduction of Cr(VI) solutions [71]. The effect of oxidation state of polyaniline on Cr(VI) reduction was investigated by using both the emeraldine base and leuocoemeraldine forms of electrochemically synthesized polyaniline films. LE because of having more amine nitrogens in comparison with EB, is a stronger reducing agent. Thus it has higher reducing capacity than EB form. The advantages of polyaniline for Cr(VI) reduction compared to commonly used reducing agents, is the reversibility of reduction process and possibility of recycling of polymer for further treatments.

In another work, Olad and coworkers applied polyaniline nanofibers for electro catalytic reduction of nitrate ions [72]. Polyaniline nanofibers were synthesized using ethanol as soft template via cyclic voltammetry technique. Their studies showed that conducting polymer-modified electrodes have better performance than classic metal-modified electrodes for nitrate reduction. The better performance is due to the redox and ion-exchange properties of polyaniline and also higher effective surface areas which which display improved mass transport and higher catalytic effect compared to that of electrodes modified with PANI microstructures [73].

Ion Exchange

Ion exchange technique because of its simplicity and low operation cost has been successfully used in the industry for the removal of pollutants from effluents [74]. An ion exchanger is a material capable of exchanging either anions or cations from the surrounding media. [75]. The ion exchange capacities of conducting polyaniline were well understood and it was found to depend on the polymerization conditions, the type and size of the dopants incorporated during the polymerization process as well as the ions present in the electrolyte solution, the polymer thickness and ageing of the polymer [76]. Small dopants such as Cl^-,

ClO_4^-, NO_3^- due to their high mobility in the polymer matrix mainly exhibit anion-exchanging behavior. While large dopants like polyvinylsulfonate and polystyrenesulfonate due to immobility of these ions in the polymer matrix, act as cation exchange materials [77].

Katal and Pahlavanzadeh investigated the removal of Zn(II) ions from aqueous solution by using polyaniline composite. They showed that PANI in the doped state (possessing releasable dopants such as Cl^-), can easily exchange the dopant anions with Zn(II) present in aqueous solutions [78].

Furthermore, PANI as an organic ion-exchanger was incorporated into an inorganic ion exchanger to provide a class of hybrid ion exchangers with a high ion exchange capacity, high thermal stability, good reproducibility and selectivity for heavy metals. Polyaniline Ce(IV) molybdate composite was applied as Cd(II) ion selective cation-exchanger by Alam and co workers [79, 80]. The high uptake of cadmium ion demonstrates that composite has both ion-exchange properties and sorption and ion-selective characteristics.

Polyaniline Sn(IV) arsenosphophate has been found to possess high selectivity for lead [81]. Khan et al. have used polyaniline Sn(IV) tungstoarsenate [82] and polyaniline zirconium titanium phosphate [83] for the removal of cadmium, mercury and lead. Polyaniline titanotungstate was synthesized by incorporation of polyaniline into the inorganic precipitate of titanotungstate which showed high selectivity for cesium and was used for the removal of cesium ions from aqueous solutions [84].

Membrane and Filtration Technology

Membrane filtration technology because of high efficiency, easy operation and space saving, shows great promise for contaminant removal. Polymer enhanced membrane, has increased metal removal efficiency in comparison with using membrane alone. Among these kinds of membranes, amine functionalized polymeric membranes show a promising potential for contaminant removal.

Sankır et. al [85] reported the preparation of novel nanocomposite membranes from poly (acrylonitrile)-co-poly (methylacrylate) copolymer and polyaniline for chromium (VI) removal. They observed that PANI had a great impact on the chromium removal. Higher removal efficiency of chromium was possibly due to the adsorption of mono and divalent chromium on the membrane surface because of doped polyaniline causing better filtration of (VI) and ease of water transport.

PANI was also used in ion-exchange membranes. PANI is polymerized in the membrane matrix by chemical oxidation. Nature of the dopant as well as doping level of the polymer affects the separation properties by PANI. With PANI, separation is possible based on the size of the molecules or ions and their charge. The hydrophobic and hydrophilic nature of ion-exchange membranes can be modified by using PANI [86].

Photocatalysis

Photocatalysis is a rapid and efficient method for destruction of environmental pollutants. Various types of semiconductors like TiO_2, ZnO, CeO_2, CdS, ZnS, etc have been used in this process. Formation of electron–hole pairs in the conduction and the valence band of the semiconductor is responsible of reduction or oxidation of species in solution. Recombination of these electron–hole pairs must be avoided by hydroxyl ions and oxygen species which form radicals [87]. It has been reported that the surface properties of TiO_2 [88], ZnO [89], CeO_2 [90], CdS [91], and ZnS [92] particles and nanoparticles can be improved by modification with conducting polyaniline.

Olad et al. modified the surface of ZnO nanoparticles with polyaniline to form a core-shell nanocomposite. They applied this nanocomposite for degradation of methylene blue dye [93] and ampiciline [94]. Application of conducting PANI to the surface of ZnO nanoparticles causes a red shift in the spectroscopic absorption peak of ZnO nanoparticles toward the visible light region (Figure 5). The authors investigated the effect of light source, photocatalyst loading and photocatalyst type on the degradation efficiency of methylene blue dye. Also the excitation wavelength of PANI coated on the ZnO nanoparticles is in the visible light region (Figure 5).

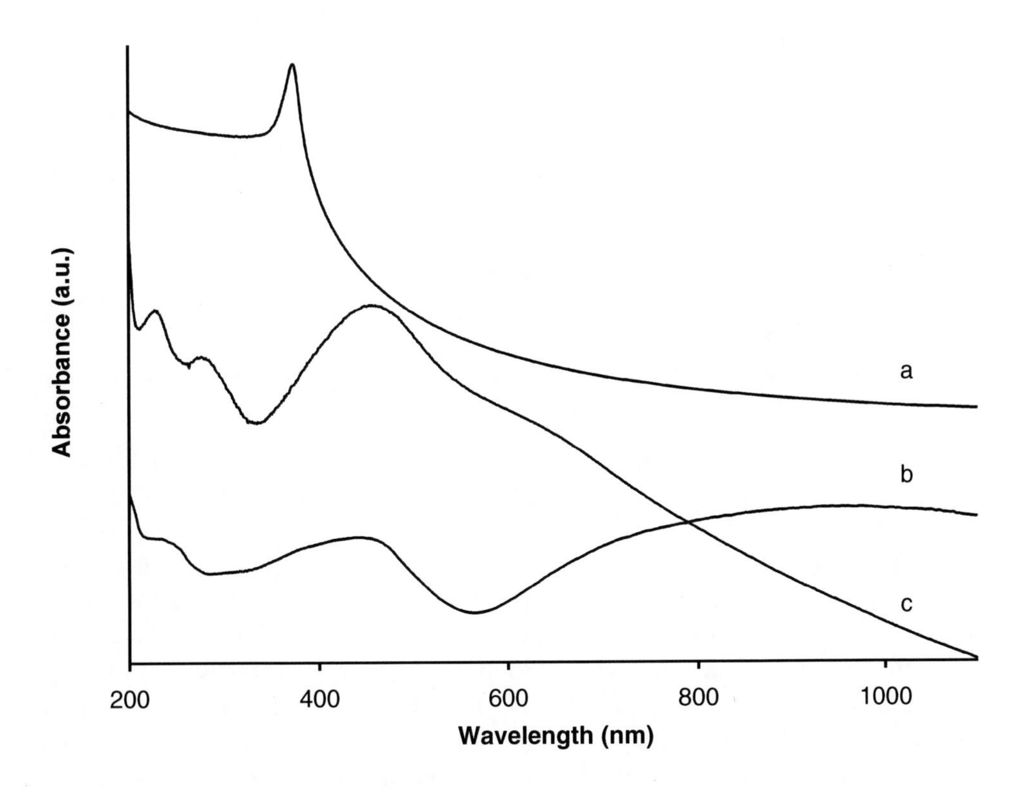

Figure 5. UV-visible spectra of ZnO nanoparticles (a), Pristine PANI (b), and PANI/ZnO nanocomposite (c).

Comparison of pristine ZnO, pure PANI and PANI/ZnO nanocomposite as photocatalyst showed that ZnO nanoparticles have no photocatalytic activity under visible light irradiation. This is because of the large band gap of ZnO, and therfore high energy is required for excitation of electrons from the valence band to the conduction band, which cannot be provided by visible light irradiation. Pure PANI has photocatalytic activity under visible light irradiation in the degradation of MB molecules. However, the efficiency of photocatalytic degradation of MB using PANI/ZnO nanocomposite as photocatalyst under visible light irradiation was much more than that of pristine PANI at the same contact time (figure 6).

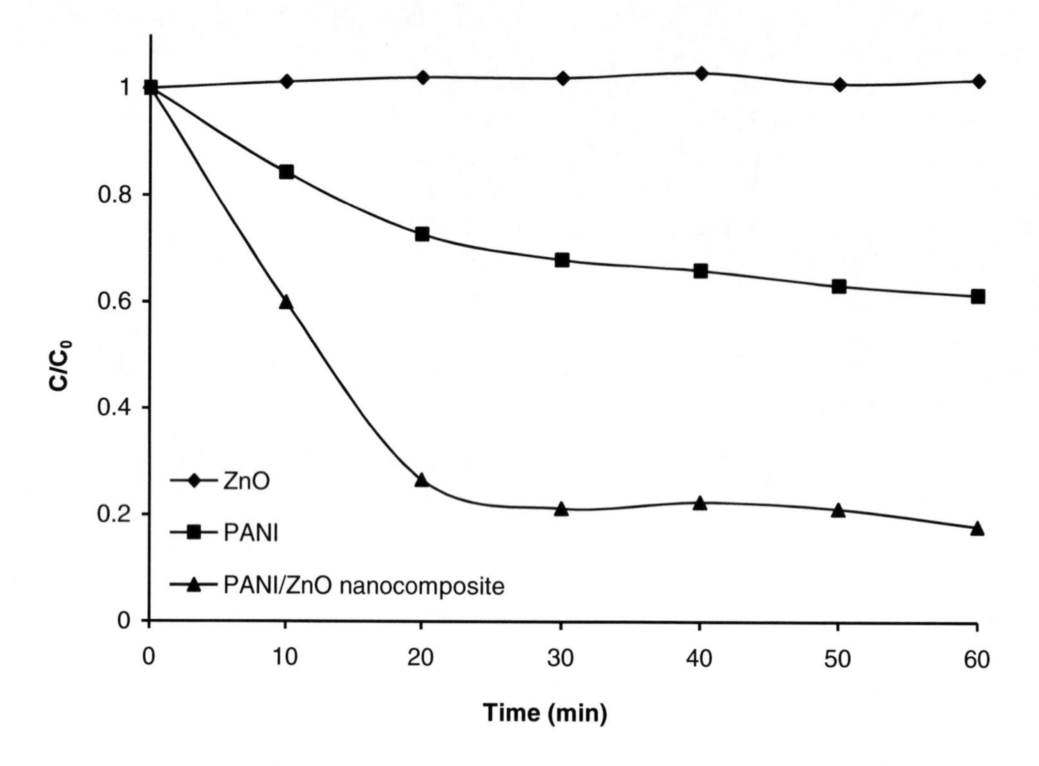

Figure 6. Concentration of MB in solution against contact time during photocatalytic degradation by use of the different photo catalysts ZnO nanoparticles, pristine PANI, and PANI/ZnO nanocomposite under visible light irradiation.

The basic mechanism of photocatalysis over irradiated ZnO was well established [95]. ZnO nanoparticles are irradiated with UV light to generate electron–hole pairs, which can react with water to yield hydroxyl and super-oxide radicals to oxidize and mineralize the organic and inorganic molecules. However, because of large band gap of ZnO, only UV light can excite the ZnO nanoparticles to generate electron–hole pairs. So, it is needed to use a dye with narrow band gap as a sensitizer to increase the photocatalytic efficiency of ZnO. PANI, having a narrower band gap than that of ZnO (2.81 eV), is used as a photo sensitizer to ZnO [96]. Therefore, PANI/ZnO nanocomposite can be excited to produce more electron hole pairs under natural sunlight irradiation which could result in efficient photocatalytic activity. This is because of the synergetic effect between ZnO and PANI on the photocatalytic degradation of dyes in PANI/ZnO nanocomposite. In other word, by irradiation of PANI/ZnO nanocomposite under natural sunlight, both ZnO and PANI absorb the photons at their interface, and then charge separation occurs at the interface. The electrons generated by conducting PANI can be transferred to the conduction band of ZnO, enhancing the charge separation and in turn promoting the photocatalytic ability of the photocatalyst. This mechanism was schematically shown in Figure 7. Also, Eskizeybek and coworkers [97] applied PANI/ZnO nanocomposite for degradation of methylene blue and malachite green dyes under UV and sun light irradiations. They showed that the kinetics of photocatalytic degradation of organic pollutants on the nanocomposite can be described well by the pseudo first-order reaction.

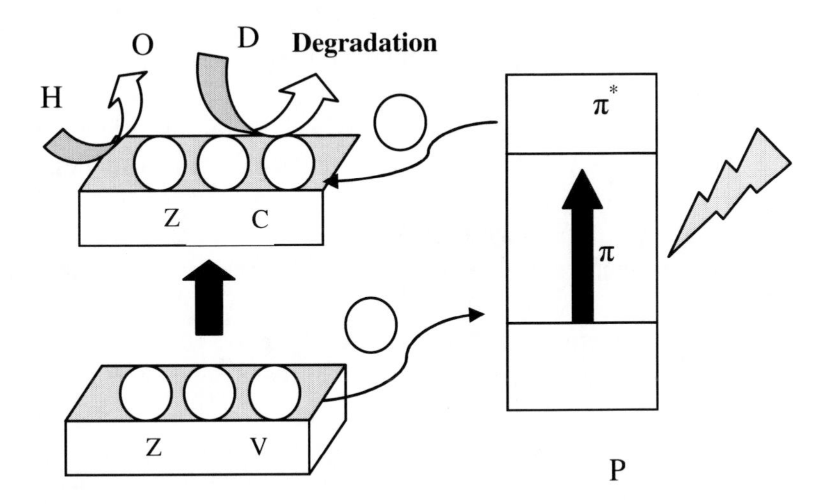

Figure 7. Mechanism of photocatalytic activity of surface modified ZnO nanoparticles with polyaniline for degradation of dye molecule.

CONCLUSION

Polyaniline with the unique properties like its electrochemical properties as well as its ion exchange property and its interactions with pollutant ions is one of the affordable materials to remove various pollutants like heavy metals and dye molecules from effluents. Polyaniline as a reducing agent, in various forms such as powder, free standing film or film on electrodes, can reduce some heavy metals to their less toxic lower oxidation states. Also PANI because of its unique doping properties has been successfully utilized for the fabrication of ion-exchange membranes. Besides that due to the presence of charges on the polymer backbone in different media, polyaniline has been used to adsorb different pollutant molecules. Using other efficient adsorbents like clay, zeolites, metal and metal oxides, development of polyaniline composites and nanocomposites with high removal efficiency has gained great interest.

REFERENCES

[1] Fenglian, F.; Qi, W. Removal of heavy metal ions from wastewaters: A review. *J. Environ. Manage.* 2011, *92*, 407-418.

[2] Wallace, G. G.; Spinks, G. M.; Kane-Maguire, L. A. P. *Conductive Electroactive Polymers: Intelligent Polymer Systems*, 3rd ed.; CRC Press, Taylor & Francis Group, 2009.

[3] Bidan, G.; Ehui, B. One-step electrosynthesis and characterization of poly (aniline)–Nafion and poly (3-methylthiophene)–Nafion composite films. *J. Chem. Soc. Chem. Commun.* 1989, *20,* 1568–1570.

[4] Lee, J. W.; Park, D. S.; Shim, Y. B.; Park, S. M. Electrochemical characterization of poly (1,8-diaminonaphthalene): a functionalized polymer. *J. Electrochem. Soc.* 1992, *139*, 3507-3514.

[5] Pałys, B.J.; Skompska, M.; Jackowska, K. Sensitivity of poly 1,8-diaminonaphthalene to heavy metal ions — electrochemical and vibrational spectra studies. J. *Electroanal. Chem.* 1997, *433*, 41-48.

[6] Wampler, W.A.; Basak, S.; Rajeshvar, K. Composites of polypyrrole and carbon-black 4, use in environmental-pollution abatement of hexavalent chromium. *Carbon*, 1996, *34*,747–757.

[7] Alatorre, M.A.; Gutierrez, S.; Paramo, U.; Ibanez, J.G. Reduction of hexavalent chromium by polypyrrole deposited on different carbon substrates. *J. Appl. Electrochem.* 1998, *28*, 551–557.

[8] Kurniawan, T.A.; Babel, S. *A research study on Cr(VI) removal from contaminated wastewater using low-cost adsorbents and commercial activated carbon.* Proceeding of the Second International Conference on Energy Technology towards a Clean Environment (RCETE), Phuket, Thailand, February, 12-14, 2003.

[9] Huttenloch, P., Roehl, K.E.; Czurda, K. Sorption of nonpolar aromatic contaminants by chlorosilane surface modified natural minerals. *Environ. Sci.Technol.* 2001, *35*, 4260-4264.

[10] Kabbashi, N.A., Atieh, M.A., Al-Mamun, A., Mirghami, M.E.S., Alam, M.D.Z.; Yahya, N. Kinetic adsorption of application of carbon nanotubes for Pb(II) removal from aqueous solution. *J. Environ. Sci.* 2009, *21*, 539-544.

[11] Vengris, T., Binkiene, R.; Sveikauskaite, A. Nickel, copper, and zinc removal from wastewater by a modified clay sorbent. *Appl. Clay. Sci.* 2001, *18*, 183–190.

[12] Sud, D., Mahajan, G.; Kaur, M.P. Agricultural waste material as potential adsorbent for sequestering heavy metal ions from aqueous solutions: a review. *Bioresour. Technol.* 2008, *99*, 6017-6027.

[13] Alinnor, J., 2007. Adsorption of heavy metal ions from aqueous solution by fly ash. Fuel 86, 853–857.

[14] Lee, T.Y., Park, J.W.; Lee, J.H. Waste green sands as a reactive media for the removal of zinc from water. *Chemosphere* 2004, *56*, 571–581.

[15] Kocaoba, S., Orhan, Y.; Akyüz, T. Kinetics and equilibrium studies of heavy metal ions removal by use of natural zeolite. *Desalination* 2007, *214*, 1-10.

[16] Kavamura, V.N.; Esposito, E. Biotechnological strategies applied to the decontamination of soils polluted with heavy metals. *Biotechnol. Adv.* 2010, *28*, 61-69.

[17] Ayad, M. M.; Abu El-Nasr, A. Adsorption of Cationic Dye (Methylene Blue) from Water Using Polyaniline Nanotubes Base. *J. Phys. Chem. C* 2010, *114*, 14377-14383.

[18] Chowdhury, A. N.; Jesmeen S. R.; Hossain, M. M. Removal of dyes from water by conducting polymeric adsorbent, *Polym. Adv. Technol.* 2004, *15*, 633–638.

[19] Mahanta, D.; Madras, G.; Radhakrishnan, S.; Patil, S. Adsorption and desorption kinetics of anionic dyes on doped polyaniline. *J. Phys. Chem. B*, 2009, *113*, 2293–2299.

[20] Salem, M.A. The role of polyaniline salts in the removal of direct blue 78 from aqueous solution: A kinetic study. *React. Funct. Polym.* 2010, *70*, 707–714.

[21] Crini, G. Recent developments in polysaccharide-based materials used as adsorbents in wastewater treatment. *Prog. Polym. Sci.* 2005, *30*, 38–70.

[22] Mahanta, D., Madras, G., Radhakrishnan, S.; Satish, P. Adsorption of sulfonated dyes by polyaniline emeraldine salt and its kinetics. *J. Phys. Chem. B* 2008, *112 (33)*, 10153–10157.

[23] Mahanta, D., Giridhar, M., Radhakrishnan, S.; Satish, P. Adsorption and desorption kinetics of anionic dyes on doped polyaniline. *J. Phys. Chem. B* 2009, *113 (8)*, 2293–2299.

[24] Banimahd Keivani, M.; Zare, K.; Aghaie, H.; Ansari, R. Removal of methylene blue dye by application of polyaniline nano composite from aqueous solutions. *J. Phys. Theor. Chem. IAU Iran*, 2009, *6 (1)*, 50-56.

[25] Riahi Samani, M.; Borghei, S. M.; Olad, A.; Chaichi, M. J. Influence of Polyaniline Synthesis Conditions on its Capability for Removal and Recovery of Chromium from Aqueous Solution. *Iran. J. Chem. Chem. Eng.* 2011, *30(3)*, 97-100.

[26] Rao, G.P., Lu, C.; Su, F. Sorption of divalent metal ions from aqueous solution by carbon nanotubes: a review. *Sep. Purif. Technol.* 2007, *58*, 224-231.

[27] Suckeveriene, R.Y.; Zelikman, E.; Mechrez, G.; Narkis, M. Literature review: conducting carbon nanotube/polyaniline nanocomposites. *Rev. Chem. Eng.* 2011, *27*, 15-21.

[28] Baibarac, M.; Baltog, I.; Godon, C.; Lefrant, S.; Chauvet, O. Covalent functionalization of single-walled carbon nanotubes by aniline electrochemical polymerization. *Carbon* 2004, *42*, 3143-3152.

[29] Nastase, C.; Nastase, F.; Vaseashta, A.; Stamatin, I. Nanocomposites based on functionalized nanotubes in polyaniline matrix by plasma polymerization. *Prog. Solid. State. Chem.* 2006, *34*, 181–189.

[30] Zeng, Y.; Zhao, L.; Wu, W.; Lu, G.; Xu, F.; Tong, Y.; Liu, W.; Du, J. Enhanced Adsorption of Malachite Green onto Carbon Nanotube/Polyaniline Composites. *J. Appl. Polym. Sci.* 2012, DOI: 10.1002/APP.37947.

[31] Wei, D.; Faqiong, Z.; Baizhao, Z. Novel multiwalled carbon nanotubes–polyaniline composite film coated platinum wire for headspace solid-phase microextraction and gas chromatographic determination of phenolic compounds, *J. Chromatogr. A.* 2009, *1216*, 3751–3757.

[32] Jiang, M.Q., Jin, X.Y., Lu, X.Q.; Chen, Z.L. Adsorption of Pb(II), Cd(II), Ni(II) and Cu(II) onto natural kaolinite clay. *Desalination* 2010, *252*, 33-39.

[33] Hong, S.; Wen, C.; He, J.; Gan, F.; Ho, Y.S. Adsorption thermodynamics of methylene blue onto bentonite, *J. Hazard. Mater.* 2009, *167*, 630–633.

[34] Hajjaji, M.; Alami, A.; El Bouadili, A. Removal of methylene blue from aqueous solution by fibrous clay minerals, *J. Hazard. Mater.* 2006, *B135*, 188–192.

[35] Almeida, C.A.P.; Debacher, N.A.; Downs, A.J.; Cottet, L.; Mello, C.A.D. Removal of methylene blue from colored effluents by adsorption on montmorillonite clay. *J. Colloid Interface Sci.* 2009, *332*, 46–53.

[36] McKay, G., Otterburn, M.S.; Aga, J.A. Fuller earth and fired clay as adsorbents for dyestuffs – equilibrium and rate studies. *Water. Air. Soil. Pollut.* 1985, *24*, 307–322.

[37] Wanga, J.; Han, X.; Ma, H.; Ji, Y.; Bi, L. Adsorptive removal of humic acid from aqueous solution on polyaniline/attapulgite composite. *Chem. Eng. J.* 2011, *173*, 171–177.

[38] Zhang, J.; Xie, S.; Ho, Y.S. Removal of fluoride ions from aqueous solution using modified attapulgite as adsorbent, *J. Hazard. Mater.* 2009, *165*, 218–222.
[39] Pang, C.; Liu, Y.; Cao, X.; Hua, R.; Wang, C.; Li, C. Adsorptive removal of uranium from aqueous solution using chitosan-coated attapulgite, *J. Radioanal. Nucl. Chem.* 2010, *286*, 185–193.
[40] Wang, Y.; Zeng, L.; Ren, X.; Song, H.; Wang, A. Removal of Methyl Violet from aqueous solutions using poly (acrylic acid-co-acrylamide)/attapulgite composite. *J. Environ. Sci.* 2010, *22*, 7–14.
[41] Kong, Y.; Wei, J.; Wang, Z.; Sun, T.; Yao, C.; Chen, Z. Heavy Metals Removal from Solution by Polyaniline/Palygorskite Composite. *J. Appl. Polym. Sci.* 2011, *122*, 2054–2059.
[42] Chen, H.; Wang, A. Q. Kinetic and isothermal studies of lead ion adsorption onto palygorskite clay. *J. Colloid. Interface. Sci.* 2007, *307*, 309-316.
[43] Kadirvelu, K.; Thamaraiselvi, K.; Namasivayam, C. Adsorption of nickel(II) from aqueous solution onto activated carbon prepared from coirpith. *Sep. Purif. Technol.* 2001, *24*, 497-505.
[44] Wanga, S.; Peng, Y. Natural zeolites as effective adsorbents in water and wastewater treatment. *Chem. Eng. J.* 2010, *156*, 11–24.
[45] Olad, A.; Khatamian, M.; Naseri, B. Removal of Toxic Hexavalent Chromium by Polyaniline Modified Clinoptilolite Nanoparticles. *Iran. Chem. Soc.* 2011, *8*, S141-S151.
[46] Rashidzadeh, A.; Olad, A. Novel polyaniline/poly (vinyl alcohol)/clinoptilolite nanocomposite: dye removal, kinetic, and isotherm studies. *Desalin. Water Treat.* 2013, DOI: 10.1080/19443994.2013.766904
[47] Raffiea Baseri, J.; Palanisamy, P.N.; Sivakumar, P. Application of Polyaniline Nano Composite for the Adsorption of Acid Dye from Aqueous Solutions, *E-Journal of Chemistry* 2012, *9(3)*, 1266-1275.
[48] Ansari, R.; Mosayebzadeh, Z. Application of polyaniline as an efficient and novel adsorbent for azo dyes removal from textile wastewaters, *Chem. Pap.* 2011, *65 (1)*, 1–8.
[49] Ansari, R.; Alaie S.; Mohammad-Khan, A. Application of polyaniline for removal of acid green 25 from aqueous solutions. *J. Sci. Ind. Res.* 2011, *70*, 804-809.
[50] Ansari, R.; Mosayebzadeh, Z. Removal of Eosin Y, an Anionic Dye, from Aqueous Solutions Using Conducting Electroactive Polymers, *Iran. Polym. J.* 2010, *19 (7)*, 541-551.
[51] Ansari, R.; Raofie, F. Removal of Lead Ion from Aqueous Solutions Using Sawdust Coated by Polyaniline, *E-Journal of Chemistry*, 2006, *3 (1)*, 49-59.
[52] Ansari, R.; Raofie, F. Removal of Mercuric Ion from Aqueous Solutions Using Sawdust Coated by Polyaniline, *E-Journal of Chemistry*, 2006, *3 (1)*, 35-43.
[53] Mansour, M. S.; Ossman, M. E.; Farag, H. A. Removal of Cd (II) ion from waste water by adsorption onto polyaniline coated on sawdust. *Desalination* 2011, *272*, 301–305.
[54] Ahmed, S. M.; El-Dib, F. I.; El-Gendy, N. Sh.; Sayed, W. M.; El-Khodary, M. A kinetic study for the removal of anionic sulphonated dye from aqueous solution using nano-polyaniline and Baker's yeast. Arab. J. Chem. 2012, In Press, Corrected Proof.
[55] Goksungur, Y., Uren, S.; Guvenc, U. Biosorption of cadmium and lead ions by ethanol treated waste baker's yeast biomass. *Bioresour. Technol.* 2005, *96*, 103–109.

[56] Wang, B.; Guo, X. Reuse of waste beer yeast sludge for biosorptive decolorization of reactive blue 49 from aqueous solution. *World J. Microbiol. Biotechnol.* 2011, *27 (6)*, 1297–1302.

[57] Ghorbani, M.; Soleimani Lashkenari, M.; Eisazadeh, H. Application of polyaniline nanocomposite coated on rice husk ash for removal of Hg(II) from aqueous media, *Synthetic. Met.* 2011, *161*, 1430–1433.

[58] Kumar, P.A.; Ray, M.; Chakraborty, S. Removal and recovery of chromium from wastewater using short chain polyaniline synthesized on jute fiber. *Chem. Eng. J.* 2008, *141*, 130–140.

[59] Kumar, P.A.; Chakraborty, S. Fixed-bed column study for hexavalent chromium removal and recovery by short-chain polyaniline synthesized on jute fiber. *J. Hazard. Mater.* 2009, *162*, 1086–1098.

[60] Zheng, Y.; Wang, W.; Huang, D.; Wang, A. Kapok fiber oriented-polyaniline nanofibers for efficient Cr(VI) removal. *Chem. Eng. J.* 2012, *191*, 154– 161.

[61] Ansari, R. Application of Polyaniline and its Composites for Adsorption/Recovery of Chromium (VI) from Aqueous Solutions. *Acta Chim. Slov.* 2006, *53*, 88–94.

[62] Yavuz, A. G.; Dincturk-Atalay, E.; Uygun, A.; Gode, F.; Aslan, E. A comparison study of adsorption of Cr(VI) from aqueous solutions onto alkyl-substituted polyaniline/chitosan composites. *Desalination.* 2011, *279*, 325–331.

[63] Janaki, V.; Oh, B-T.; Shanthi, K.; Leeb, K-J.; Ramasamya, A.K.; Kamala-Kannan, S. Polyaniline/chitosan composite: An eco-friendly polymer for enhanced removal of dyes from aqueous solution. *Synthetic. Met.* 2012, *162*, 974–980.

[64] Taghipour Kolaei, Z.; Tanzifi, M.; Yousefi, A.; Eisazadeh, H. Removal of Cd(II) from aqueous solution by using polyaniline/polystyrene nanocomposite. *J.Vinyl. Addit. Technol.* 2012, *18*, 52-56.

[65] Gupta, R.K.; Singh, R.A.; Dubey, S.S. Removal of mercury ions from aqueous solutions by composite of polyaniline with polystyrene. *Sep. PuriF. Technol.* 2004, *38*, 225–232.

[66] Riahi Samani, M.; Borghei, S. M.; Olad, A.; Chaichi, M. J. Removal of chromium from aqueous solution using polyaniline –Poly ethylene glycol composite. *J. Hazard. Mater.* 2010, *184*, 248–254.

[67] Bhaumik, M.; Maity, A.; Srinivasu, V.V.; Onyango, M. S. Removal of hexavalent chromium from aqueous solution using polypyrrole-polyaniline nanofibers. *Chem. Eng. J.* 2012, *181–182*, 323–333.

[68] Wei, C.; German, S.; Basak, S.; Rajeshwar, K. Reduction of hexavalent chromium in aqueous solutions by polypyrrole. *J. Electrochem. Soc.* 1993, *140*, L60–L62.

[69] Farrell, S.T.; Breslin, C.B. Reduction of Cr(VI) at a polyaniline film: influence of film thickness and oxidation state. *Environ. Sci. Technol.* 2004, *38*, 4671–4676.

[70] Ruotolo, L.A.M.; Gubulin, J.C. Chromium(VI) reduction using conducting polymer films. *React. Funct. Polym.* 2005, *62*, 141–151.

[71] Olad, A.; Nabavi, R. Application of polyaniline for the reduction of toxic Cr(VI) in water. *J. Hazard. Mater.* 2007, *147*, 845–851.

[72] Olad, A.; Farshi, F.; Ettehadi, J. Electrocatalytic reduction of nitrate ions from water using polyaniline nanofibers modified gold electrode. *Water. Environ. Res.* 2012, *84*, 144-149.

[73] Kazimierska, E.; Smyth, M. R.; Killard, A. J. Size-Dependent Electrocatalytic Reduction of Nitrite at Nanostructured Films of Hollow Polyaniline Spheres and Polyaniline–Polystyrene Core–Shells. *Electrochim. Acta.* 2009, *54*, 7260–7267.

[74] Alyüz, B.; Veli, S. Kinetics and equilibrium studies for the removal of nickel and zinc from aqueous solutions by ion exchange resins. *J. Hazard. Mater.* 2009, *167*, 482-488.

[75] Gode, F.; Pehlivan, E. Removal of chromium (III) from aqueous solutions using Lewatit S 100: the effect of pH, time, metal concentration and temperature. *J. Hazard. Mater.* 2006, *136*, 330-337.

[76] Ghorbani, M.; Esfandian, H.; Taghipour, N.; Katal, R. Application of polyaniline and polypyrrole composites for paper mill wastewater treatment. *Desalination* 2010,*263*, 279–284.

[77] Weidlich, C.; Mangold, K.M.; Juttner, K. Conducting polymers as ion-exchangers for water purification. *Electrochim. Acta* 2001, *47*, 741–745.

[78] Katal, R.; Pahlavanzadeh, H. Zn(II) Ion Removal From Aqueous Solution By Using a Polyaniline Composite. *J. Vinyl. Addit. Techn.* 2011, *17*, 138-145.

[79] Inamuddin, Z. A.; Nabi, S. A. Synthesis and characterization of a thermally stable strongly acidic Cd(II) ion selective composite cation-exchanger: Polyaniline Ce(IV) molybdate. *Desalination*. 2010, *250*, 515–522.

[80] AL-Othmana, Z. A.; Inamuddinb; Naushad, M. Forward (M^{2+}–H^+) and reverse (H^+–M^{2+}) ion exchange kinetics of the heavy metals on polyaniline Ce(IV) molybdate: A simple practical approach for the determination of regeneration and separation capability of ion exchanger. *Chem. Eng. J.* 2011, *171*, 456– 463.

[81] Niwas, R.; Khan, A.A.; Varshney, K.G. Synthesis and ion exchange behavior of polyaniline Sn(IV) arsenophosphate: a polymeric inorganic ion exchanger, *Colloid. Surface. A.* 1999, *150*, 7–14.

[82] Khan, A. A.; Alam, M. M. Synthesis, characterization and analytical applications of a new and novel 'organic–inorganic' composite material as a cation exchanger and Cd(II) ion-selective membrane electrode: polyaniline Sn(IV) tungstoarsenate. *React. Funct. Polym.* 2003, *55*, 277–290.

[83] Khan, A. A.; Paquiza, L. Characterization and ion-exchange behavior of thermally stable nano-composite polyaniline zirconium titanium phosphate: Its analytical application in separation of toxic metals. *Desalination* 2011, *265*, 242–254.

[84] El-Naggar, I.M.; Zakaria, E.S.; Ali, I.M.; Khalil, M.; El-Shahat, M.F. Kinetic modeling analysis for the removal of cesium ions from aqueous solutions using polyaniline titanotungstate, *Arab. J. Chem.* 2012, *5*, 109–119.

[85] Sankır, M.; Bozkır, S.; Aran, B. Preparation and performance analysis of novel nanocomposite copolymer membranes for Cr(VI) removal from aqueous solutions. *Desalination.* 2010, *251*, 131–136.

[86] Conklin, J.A.; Su, T.M.; Huang, S.C.; Kaner, R.B. in: Handbook of Conducting Polymers; Akotheim, T.A.; Elsenbaumer, R.L.; Reynolds J.,(Eds.; second ed.; Dekker: New York, 1998.

[87] Zhang, F.S.; Itoh, H. Photocatalytic oxidation and removal of arsenite from water using slag–iron oxide–TiO_2 adsorbent. *Chemosphere* 2006, *65 (1)*, 125–131.

[88] Li, X.; Wang, D.; Cheng, G.; Luo, Q.; An, J.; Wang, Y. Preparation of polyaniline modified TiO_2 nanoparticles and their photocatalytic activity under visible light illumination. *Appl. Catal. B. Environ.* 2008, *81 (3–4)*, 267–273.

[89] Yang, S.; Ishikawa, Y.; Itoh, H.; Feng, Q. Fabrication and characterization of core/shell structured TiO_2/polyaniline nanocomposite. *J. Colloid. Interface. Sci.* 2011, *356*, 734-740.

[90] Chuang, F-Y.; Yang, S-M. Cerium dioxide/polyaniline core–shell nanocomposites. *J. Colloid. Interface. Sci.* 2008, *320 (1)*, 194–201.

[91] He, K.; Li, M.; Guo, L. Preparation and photocatalytic activity of PANI-CdS composites for hydrogen evolution. *Int. J. Hydrogen. Energ.* 2012, *37*, 755-759.

[92] Pant, H. C.; Patra, M. K.; Negi, S. C.; Bhatia, A.; Vadera, S. R.; Kumar, N. Studies on conductivity and dielectric properties of polyaniline–zinc sulphide composites. *Bull. Mater. Sci.* 2006, *29 (4),* 379–384.

[93] Olad, A.; Nosrati, R. Preparation, characterization, and photocatalytic activity of polyaniline/ZnO nanocomposite. *Res. Chem. Intermed.* 2012, *38*, 323-336.

[94] Nosrati, R.; Olad, A.; Maramifar, R. Degradation of ampicillin antibiotic in aqueous solution by ZnO/polyaniline nanocomposite as photocatalyst under sunlight irradiation. Environ. Sci. Pollut. R. 2012, 19 (6), 2291-2299.

[95] Li, Q.; Zhang, C.; Li, J. Photocatalysis and wave-absorbing properties of polyaniline/TiO_2 microbelts composite by in situ polymerization method. *Appl. Surf. Sci.* 2010, *257*, 944–948.

[96] Xiong, S.X.; Wang, Q.; Xia, H.S. Template synthesis of polyaniline/TiO_2 bilayer microtubes. *Synthetic Met.* 2004, *146*, 37–42.

[97] Eskizeybek, V.; Sarı, F.; Gulce, H.; Gulce, A.; Avcı, A. Preparation of the new polyaniline/ZnO nanocomposite and its photocatalytic activity for degradation of methylene blue and malachite green dyes under UV and natural sun lights irradiations. *Appl. Catal. B-Environ.* 2012, *119– 120*, 197–206.

In: Trends in Polyaniline Research
Editors: T. Ohsaka, Al. Chowdhury, Md. A. Rahman et al.
ISBN: 978-1-62808-424-5

Chapter 15

BIOMEDICAL APPLICATIONS OF POLYANILINE

Ali Olad* and Fahimeh Farshi Azhar
Polymer Composite Research Laboratory,
Department of Applied Chemistry,
Faculty of Chemistry, University of Tabriz,
Tabriz, Iran

ABSTRACT

Polyaniline (PANI) is considered as one of the most promising intrinsically conducting polymers (ICPs) because of its unique properties such as high electrical conductivity, environmental stability, easy polymerization and low cost of the monomer. Attractive fields for potential applications of polyaniline are in rechargeable batteries, sensors, switchable membranes, anticorrosive coatings, electronic and more recently in biomedical fields. The fact that several tissues and cells are responsive to electrical fields and stimulus has made PANI attractive for a number of biological and medical applications. This chapter provides information on favorable PANI properties specific to biomedical applications and how PANI has been optimized to be responsible for these properties. The chapter first introduces the advantages of PANI for different applications in biomedical fields. Specific informations are provided by detail on utilizing PANI in tissue engineering, biosensors, and drug delivery devices. Various modification processes have been discussed to improve the usage of PANI in these fields. With attendance to this chapter, evidences can be found on the starting of the use of other ICPs as well as PANI in the biomedical applications.

INTRODUCTION

Intrinsically conducting polymers (ICPs) as novel organic materials have electrical properties similar to metals and inorganic semiconductors. Among the available ICPs, polyaniline (PANI) is found to be the most promising material because of its ease of

* E-mail: a.olad@yahoo.com; Tel: +98 411 3393164; Fax: +98 411 3340191.

synthesis, low cost monomer, and better environmental stability compared to other ICPs. Combination of these unique properties has given PANI a wide range of applications including in rechargeable batteries, electromagnetic interface shielding (EMI), anticorrosive coatings, sensors and actuators, toxic metal recovery, catalysis, fuel cells and more recently in the biomedical fields [1].

Research on ICPs for biomedical applications was started with the discovery in the 1980s which showed that these materials are compatible with many biological molecules such as those used in biosensors. By the mid-1990s, ICPs were also shown that have modulate cellular activities such as cell adhesion and migration, DNA synthesis, and protein secretion due to their electrical stimulation [2-5].

Polyaniline, one of the most promising ICPs, offers several important advantages for biomedical applications over other materials such as metals and semiconductors, including biocompatibility, controllable entrap and release of biological molecules (i.e., reversible doping), charge transfer capability from a biochemical reaction, and ability to change the electrical, chemical, and physical properties in specific applications. These unique characteristics are useful in many biomedical applications, such as drug delivery systems [6], biosensors [7], tissue engineering scaffolds [8], and neural probes [9].

In detail, the advantages of PANI in these applications arise from the fact that PANI can permit control over the level and duration of electrical stimulation for tissue engineering applications. Furthermore, it can be precisely deposited on metal electrodes for biosensors, and can be closely interfaced with biomolecules for a more effective transduction mechanism in drug delivery devices [1]. However, in biomedical fields like as other applications of PANI, various improvement methods have been used to enhance its properties.

1. Polyaniline in Tissue Engineering Applications

Tissue engineering (TE) is a multidisciplinary field of science focused on the development and application of knowledge in chemistry, physics, engineering, life and clinical sciences to solve the critical medical problems such as tissue loss and organ failure [10]. It involves the fundamental understanding of structure-function relationships in normal and pathological tissues and the development of biological substitutes that restore, maintain or improve tissue function [11]. Tissue engineering as a new medical therapy is a rapidly developing area of science, which generally requires the use of scaffolds as three dimensional supports for initial cell attachment, cellular growth, proliferation, and support for new tissue formation [12, 13]. Natural tissues are three-dimensional (3-D) structures composed of cells surrounded by extracellular matrix (ECM). The ECM forms the supporting matrix for residing the cells and the cell–cell and cell–ECM contacts play an important role in maintaining cell differentiation and function. The approach of tissue engineering is utilizing a cell-matrix construct to develop functioning tissues [14].

Extensive studies have been performed to develop materials that can be used as viable scaffolds for tissue engineering. Several parameters should be taken into consideration to choose materials for scaffold fabrication depending on the intended application. Biocompatibility, biodegradability, good mechanical and surface properties, interconnected three-dimensional structures, and feasible fabrication techniques are of essential *requirements* in the design of *scaffolds* [15].

A wide variety of natural and synthetic biomaterials, such as polymers, ceramics, and their combinations have been studied as scaffold materials to fulfill the different needs in tissue engineering applications. The most commonly used synthetic polymers are biodegradable polymers such as poly(glycolic acid) (PGA), poly(lactic acid) (PLA), poly(lactic-co-glycolic acid) (PLGA), poly(ε-caprolactone) (PCL) and poly (hydroxyl butyrate) (PHB).

Naturally derived polymers, such as collagen, gelatin, chitosan, alginate and starch have also been used for scaffold fabrication [16]. Recently developed technologies for preparation of scaffolds need biomaterials that not only can act as physically support for tissue growth, but also they have to be electrically conductive and thus be able to stimulate specific cell functions or activate cell responses. For these reasons, in recent years, using of conducting polymers such as polyaniline (PANI) has become attractive as biocompatible tissue support structures [17].

The use of conducting polymers would allow delivering of local electrical stimulations providing a physical template for cell growth and tissue repair as well as allowing the precise external control over the level and duration of stimulation [18]. These materials show good capacity to support and modulate the growth of various cells, such as nerve and bone cells, and are widely used in biological systems [19]. Hence the importance of conducting polymer composites is based on the hypothesis that such composites can be used as host materials for the growth of the cells, while electrical stimulations can be applied in-situ directly to the cells through the composite [20, 21].

1.1. Modification of Conducting Polymers for Tissue-Engineering Applications

Although conducting polymers offer many advantages over other materials in tissue engineering applications due to their electrical properties, further optimization of these materials is necessary [1]. In the field of biomaterials, the design of bioactive surfaces is of particular interest because that the biological systems interact with biomaterials via the interface.

A variety of surface modification and immobilization methods have been developed to create surfaces having bioactive ligands to interact with biomolecules and cells [22]. Several attempts have been made to combine the conductivity and biocompatibility for conducting polymers, such as doping by biological dopants, incorporation of bioactive molecules for increasing cell adhesion and proliferation, patterning of conductive scaffolds for improvement of their surface topography and surface modification of conducting polymeric materials with biological moieties [22-26]. Figure 1 summarizes the different modification methods that have been carried out for the modification of conducting polymers in tissue engineering applications.

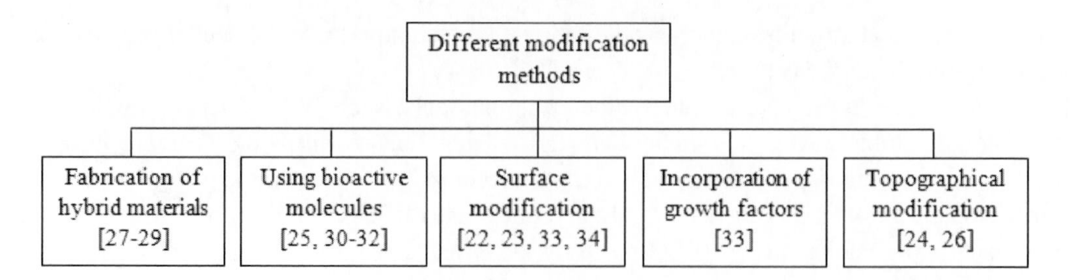

Figure 1. Different modification methods for conducting polymers in tissue engineering applications.

1.1.1. Modification of Conducting Polymers by Bioactive Molecules

Natural tissues are three-dimensional (3-D) structures composed of cells surrounded by extracellular matrix (ECM). In the modification of polymers, usually ECM molecules are chosen as dopants to create polymers with higher binding affinity which causes to the improvement of their adhesion to the cell molecules [35]. Bioactive molecules can be used as dopants to modify the polymers for specific functionality through the incorporation of proteins, peptides or ECM components [35, 36]. Doping of conducting polymers by bioactive molecules such as heparin, dextran sulphate, hyaluronic acid, chitosan, collagen, growth factors, oligodeoxyguanylic acids and ATP are being carried out through their modification [25, 30-32]. However, the doping by biomolecules has some limitations, such as low loading and decrease in conductivity, and in the case of induced release, with polymer reduction the supply becomes limited and the release occurs faster [37]. Gelmi et al. [35] and Gilmore et al. [36] also showed that doping with a bioactive polymer could cause an inverse effect on the physical properties depending on the kind of doping molecules such as surface roughness of the final composite material. To overcome these limitations, surface immobilizations of biomolecules have been investigated. Immobilization of biomolecules into conducting scaffolds can be performed via methods such as adsorption, entrapment and covalent binding [22]. In covalent modification method, a reactive group must be present on the polymer, which is not always the case in many polymers [33].

Addition of bioactive agents to conducting polymers is considered as a major strategy to improve cell–tissue interactions [37]. Electrically conductive and biologically active scaffolds are desirable for enhancing the adhesion, proliferation and differentiation of various cell types. The incorporation of bioactive molecules such as neuroactive molecules into conductive scaffolds has also been applied for modification of conducting scaffolds [33]. However, direct covalent attachment of the factors to the backbone of a conducting polymer usually has an adverse effect on the electrical properties of the polymer [38].

1.1.2. Modification of Conducting Polymers with Biopolymers

Fabrication of hybrid materials has also been investigated as a modification method to enhance the properties of conducting polymers in tissue engineering applications. The biocompatibility of electroactive polymers is commonly improved by their blending with natural polymers [39]. Ghasemi-Mobarakeh et al. [28] carried out the electrospinning of a blend of poly(ε-caprolactone) (PCL), gelatin and PANI to obtain conductive nanofibrous scaffolds for nerve tissue engineering. The scaffold offers the advantages of utilizing both synthetic polymer, PCL, which provides mechanical strength and natural polymer, gelatin, which supplies the cell adhesion and proliferations. Moreover, the incorporation of PANI caused to the making of a nanofibrous scaffold and also it is conducting material suitable for the electrical stimulation required for enhanced nerve regeneration. On the other hand, Li et al. [29] blended PANI and gelatin to fabricate conducting scaffolds and investigate the attachment and proliferation of H9c2 cardiac myoblasts on PANI–gelatin scaffolds. Results showed that the incorporation of PANI into gelatin had the advantages of inducing electrical conductivity to the scaffolds with improved cell attachment and proliferation due to the presence of gelatin.

1.1.3. Modification of Conducting Polymers by Topographical Improving Agents

The topographical characteristics of scaffolds have significant effects in tissue engineering and it has been demonstrated that the surface roughness of scaffolds affects cell behaviors, including cell attachment, proliferation and differentiation. Moreover, the topographical features are described to have a positive effect on axonal orientation, where the physical guidance of axons is a vital parameter in nerve repair. It was reported that hydrophilic and protein rich surfaces are favorable for cell adhesion. To increase hydrophilicity and achieve better cell adhesion, sodium hydroxide and plasma treatments are used. Researchers have also tried to use extracellular matrix proteins, like collagen. In this case, the motivation in cell adhesion is arisen from enriched surface by ECM proteins that mimic the biochemical composition of physiological ECM. Apart from biochemistry, the topology of a biomaterial surface is also important for cell adhesion. In the case of native ECM, the extracellular macromolecules are assembled into an organized fibrillar meshwork. This unique topology has critical implications, namely the maintenance of cellular adhesion and function. Enhancements in protein synthesis and subsequent cell adhesion were also reported. Recently, modern technologies have been developed to emulate the fibrillar topology using an electrospinning process where high voltage electric field is used to spin various biomaterials into nanofibers [39-45].

1.2. Polyaniline in Tissue Engineering Scaffolds

There is a growing interest in the use of PANI as a novel intelligent material in tissue engineering scaffolds because of its unique conductive properties that increases cell attachment, proliferation, migration and differentiation [45]. Past studies have demonstrated that the electrical charge plays an important role in stimulating either the proliferation or differentiation of various cell types, and electrically conducting or electroactive polymers provide interesting surfaces for cell culture. Some properties of the scaffolds such as surfaces charge, wetability, conformation and dimension can be altered reversibly by chemical or electrochemical oxidation or reduction of conducting polymers [46]. Also PANI has ability to locally delivering electrical stimuli at the site of damaged tissue to promote wound healing [47].

It has been suggested that perhaps the compatibility of PANI is specific to some particular cells. But the cell compatibility of different cell lines such as cardiac myoblasts, PC12, bone, skin and nerve cells with PANI has been approved [1]. Studies with PANI have shown that it supports adhesion and proliferation of H9c2 cardiac myoblasts and enhances in vitro neurite extension [29, 48, 49]. PC-12 cells differentiated into neural-like cells upon the electrical stimulation and addition of nerve growth factor on PANI surfaces. For this, the multiblock copolymer PLAAP was designed and synthesized with the condensation polymerization of hydroxyl-capped poly(L-lactide) (PLA) and carboxyl-capped aniline pentamer (AP). The PLAAP copolymer exhibited excellent electroactivity, solubility, and biodegradability. At the same time, as one scaffold material, PLAAP copolymer possesses certain mechanical properties with the tensile strength of 3 MPa, tensile Young's modulus of 32 MPa, and breaking elongation rate of 95%. Also the compatibility of PLAAP copolymer was confirmed in vitro and proved that the electroactive PLAAP copolymer was innocuous, biocompatible, and helpful for the adhesion and proliferation of rat C6 cells. Moreover, the PLAAP copolymer stimulated by electrical signals was demonstrated as accelerating the differentiation of rat neuronal pheochromocytoma PC-12 cells [50]. Li et al. [29] showed that rat cardiac muscle cells are able to attach, migrate, and proliferate on PANI-gelatin electrospun fibers. SEM analysis of these blend fibers containing less than 3% PANI in total weight, revealed uniform fibers with no evidence for phase segregation, as also confirmed by DSC. Data indicated that with increasing the amount of PANI, the average fiber size was reduced and the tensile modulus increased. To test the usefulness of PANI-gelatin blends as a fibrous matrix for supporting cell growth, H9c2 rat cardiac myoblast cells were cultured on fiber-coated glass cover slips. Cell cultures were evaluated in terms of cell proliferation and morphology. Results indicated that all PANI-gelatin blend fibers supported H9c2 cell attachment and proliferation to a similar degree as the control tissue culture-treated plastic (TCP) and smooth glass substrates. Depending on the concentrations of PANI, the cells initially displayed different morphologies on the fibrous substrates, but after 1 week all cultures reached confluence of similar densities and morphologies [29]. Huang et al. [46] combined PANI with poly(D,L-lactide) (PDLA) to make a PLA-PANI-PLA block copolymer that was electroactive and supported C6 glioma cell attachment and proliferation. The in-vitro biodegradation and biocompatibility experiments proved the copolymer is biodegradable and biocompatible. Moreover, this new block copolymer showed good solubility in common organic solvents, leading to excellent processability of the system. Therefore these electroactive biodegradable PAP copolymers possess the properties that are potentially needed for scaffold materials in neuronal or cardiovascular tissue engineering [46]. Later, a PLA-PANI multiblock was synthesized to generate a scaffold with better mechanical properties [50]. Wang et al. [51] investigated the in-vivo tissue responses to PANI and found no distinctive features resulting from tissue incompatibility after PANI implantation. It was observed that inflammation associated with the various forms of PANI was minimal after 50 weeks. Histological examinations of tissues, 24 weeks after the implantation, revealed that the pristine emeraldine form of PANI film was encapsulated by several loosely arranged fibrous tissues. The collagen-immobilized emeraldine film, on the other hand, showed no features resulting from tissue incompatibility near the implant [51]. Kamalesh et al. [52] also studied the biocompatibility of PANI in different states and found that these polymers are sufficiently biocompatible to be used for biomedical applications. The biocompatibility was assessed through subcutaneous implantation of PANI different forms into the dorsal skin of the male Sprague-Dawley rats, for a period ranging from 19 to 90 weeks. Histological examination, interstitial pressure measurement, and X-ray photoelectron spectroscopy (XPS) were employed to determine the biocompatibility of these polymers. The polymers did not provoke inflammatory responses in the subcutaneous tissues over the entire implantation period. Characteristics features associated with tissue-implant incompatibility were not evident near the implantation. Interstitial pressure was measured to evaluate the development of tissue.

Low interstitial pressure readings on the region of implantation confirmed the biocompatibility of these polymer types [51]. Bidez et al. [48] examined the adhesion and proliferation properties of H9c2 cardiac myoblasts on a conducting PANI substrate. Both the non-conductive emeraldine base and the conductive salt forms of PANI were found to be biocompatible and to support cell attachment and proliferation. Scaffaro [53] reported for the first time, the design and development of surface modified ethylene-acrylic acid copolymer (EAA) films with PANI in skin tissue engineering, with the aim of inducing electrical conductivity and potentially enable the electronic control of a range of physical and chemical properties of the surface tin film cells. In nerve tissue engineering, electrospun PLLA/PANI scaffolds were found to possess the nanoscale features of native ECM, had the mechanical and electrical properties suitable for nerve tissue engineering. In-vitro nerve stem cell culture on composite conductive scaffolds demonstrated cell biocompatibility of electrospun conductive PLLA/PANI scaffolds similar to those observed on biodegradable PLLA scaffolds. Moreover the electrical stimulation of nerve stem cells on PLLA/PANI scaffolds showed higher neurite extensions, which facilitate the regeneration of nerve [54].

However like other conducting polymers as mentioned before, utilization of PANI in commercial tissue engineering applications is limited by poor processability and intractability. A number of attempts have been made to improve the processability of PANI, including doping with functionalized protonic acids [55], preparing PANI composites with thermoplastic polymers [56], and synthesizing copolymers of aniline and substituted anilines, e.g. with alkyl (electron-donating) [57], carboxylic acid [58] and sulfonic acid (electron-withdrawing) [59] substituents. Acid functionalized polyaniline copolymers such as poly (aniline-co-o/m-aminobenzoic acid) (P(ANI-co-o/m-ABA)) and sulfonic acid ring-substituted polyaniline (SPAN) have been prepared chemically and electrochemically [60]. These copolymers show quite different properties from those of PANI homopolymer including lower conductivity, but improved solubility in common organic solvents. An interesting property of such copolymers is that the acid group in the polymer backbone can act as a "self dopant" instead of requiring an external dopant as in the case for polyaniline [61]. The advantage of self-doping is that the conductivity and electrochemical behavior of the polymers show less pH dependence, and the polymers retain electroactive even in basic solutions.

To enhance the biocompatibility of polyaniline, it was blended with natural polymers such as collagen [51] and gelatin [29] or covalently grafted with oligopeptides such as Tyr–Ile–Gly–Ser–Arg (YIGSR) [62]. The solubility of polyaniline could be enhanced by covalently grafting side groups or polymers, such as poly(ethylene glycol) [63] and poly(acrylic acid) [64], on the backbone of polyaniline. Despite of all the above-mentioned success, one of the most important issues related to the applications of PANI as tissue engineering scaffolds is the lack of biodegradability, which prevents its in-vivo applications. Keeping polyaniline in the body for a long time may induce chronic inflammation and require surgical removal. Therefore, inducing biodegradability to PANI is a very important and challenging task. In this regard, preparation of a block copolymer having biodegradable polymer segments, covalently bonded to polyaniline, may be useful and could induce biodegradability to the PANI [65].

2. POLYANILINE IN BIOSENSOR APPLICATIONS

Application of biosensors in clinical diagnostics has recently received much attention. This is because of that biosensors are portable, cost-effective, yield quickly specific information about desired analytes, and can be used by semi-skilled operators.

There is a considerable interest towards the use of conducting polymers in biosensors because of their many interesting properties such as biocompatibility, redox properties and the possibility of direct electron transfer between electrode and active sites of biomolecules [66]. The first biosensing device was created by joining an enzyme into an electrode [67], and since that time, much progress has been made in monitoring and diagnosing metabolites (e.g., glucose, hormones, neurotransmitters, antibodies, antigens) for clinical purposes. A biosensor is composed of a sensing element (i.e., biomolecule) and a transducer. The sensing element interacts with the analyte producing a chemical signal that is transferred to the transducer, which ultimately transforms the input into an electrical signal (Figure 2). Conducting polymers are extensively used as transducers that integrate the signals produced by biological sensing elements such as enzymes. Depending on how the chemical signal is sensed and transmitted, biosensors can be divided into several categories: amperometric (measures current), potentiometric (measures potential), conductometric (measures change in conductivity), optical (measures light absorbance or emission), calorimetric (measures change in enthalpy), and piezoelectric (measures mechanical stress) [1].

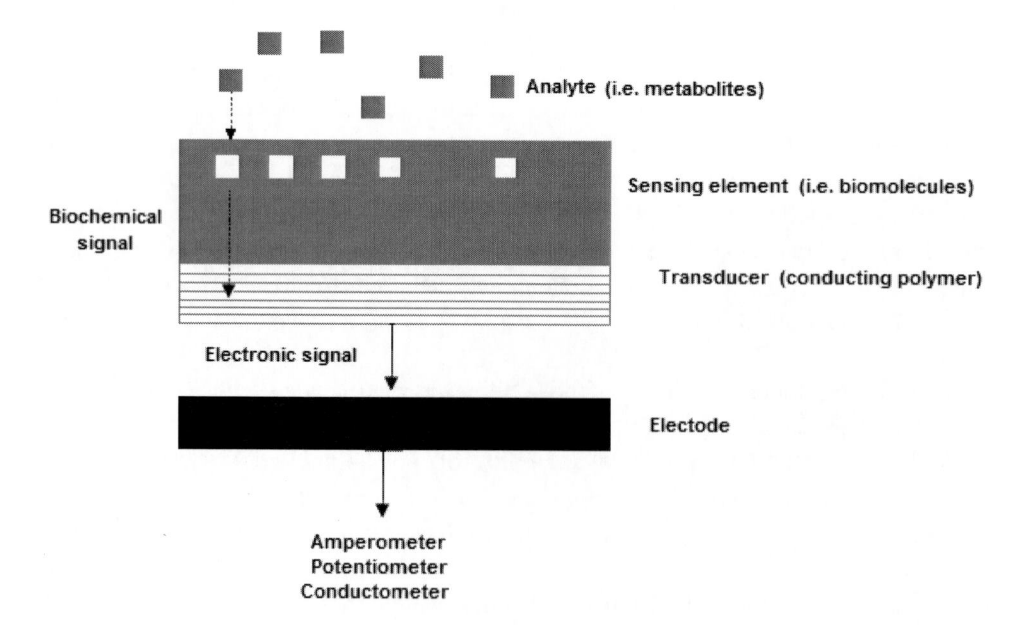

Figure 2. Schematic of a biosensor.

The most common types of transducers are amperometric and potentiometric. An amperometric biosensor measures the current produced when a specific product is oxidized or reduced (e.g., redox reaction of a substrate in an enzyme) at a constant applied potential [68] (Figure 3). The conducting polymer facilitates the electron transfer (e.g., via hydrogen peroxide) between an enzyme, such as an oxidase or dehydrogenase, and the final electrode.

Redox mediators such as ferrocene, viologen, Prussian Blue, or their derivatives are used to improve electron transfer from the biochemical reaction to the conducting polymer and therefore improve sensor sensitivity and selectivity. These redox mediators can be entrapped, incorporated as dopants, or chemically conjugated to the monomer [69]. Potentiometric biosensors use ion-selective electrodes as physical transducers. For example, detection of urea by ureases is performed via the production of NH_3, which interacts with conducting polymer to produce an electrical signal. This signal could be a product of a change in pH and the subsequent ion mobility in the polymer matrix started by an equilibration of the dopants with the free ions in solution [70].

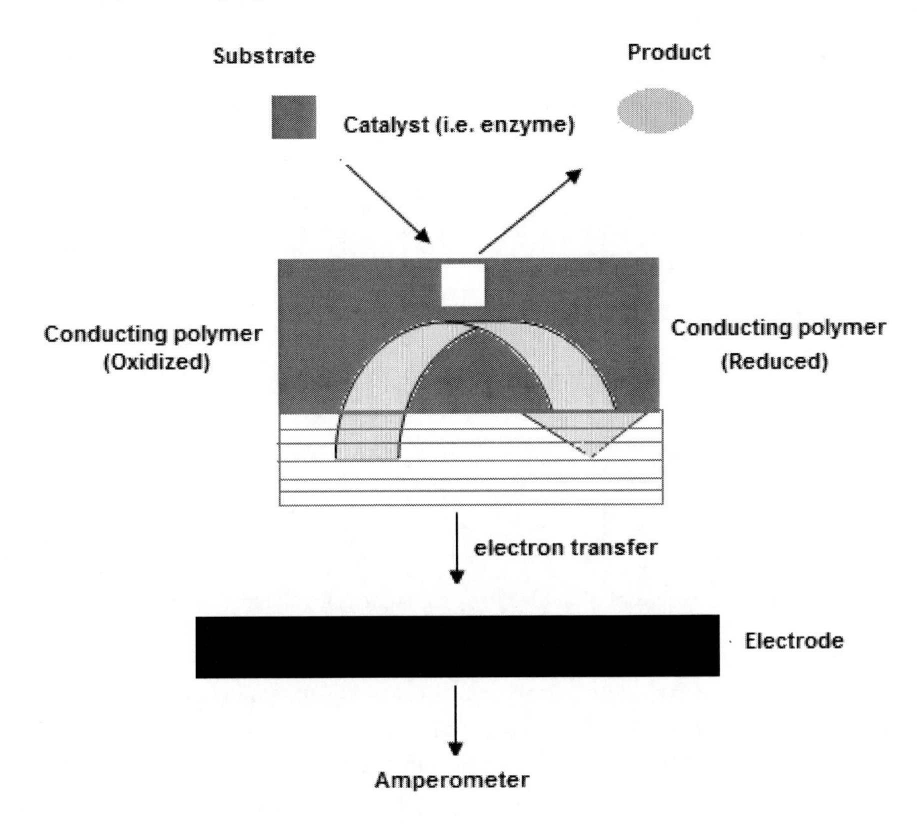

Figure 3. Schematic of electron transfer in an amperometric biosensor.

2.1. Modification of Conducting Polymers for Biosensor Applications

A key aspect in biosensor applications is the integration of the electrical component (i.e., conducting polymer) with the biological recognition components. The immobilization of bioactive macromolecules in or on conducting polymers has been extensively explored in an effort to provide close contact between these two elements [71]. Different available techniques are needed for the immobilization of biologically active molecules on conducting polymers. For this immobilization, it is critical to maintain the activity of the molecules, increase stability, and ensure accessibility of the analyte to perform biological events such as hybridization of complementary oligonucleotides, antigen-antibody binding, or enzyme-catalyzed reactions. Figure 4 summarizes the main categories of immobilization techniques of biological sensing elements on conducting polymers.

Two main classes are distinguished: non-covalent and covalent modification methods. Non-covalent techniques include adsorption, physical entrapment, and affinity binding. Covalent immobilization includes all techniques that create a covalent bond between the conducting substrate and the biomolecule via functional groups. Physical adsorption is the simplest method of immobilization and one of the first approaches used for biosensors. As an example, glucose oxidase has been adsorbed onto conducting polymer for an amperometric sensor and was shown to detect glucose over a wide range of concentrations (2.5–30 mM) using dimethylferrocene as an electron transfer mediator [72].

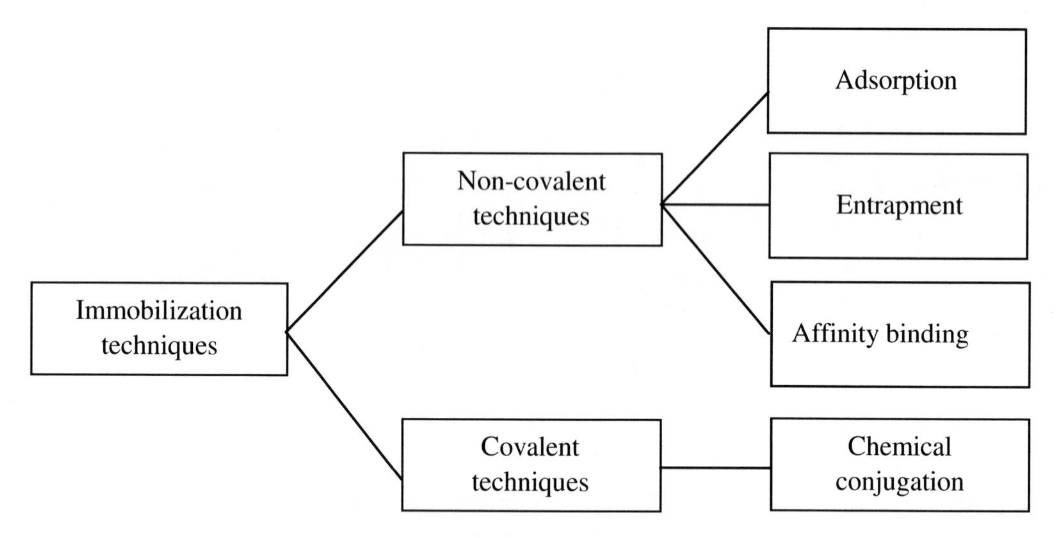

Figure 4. Immobilization techniques of biomolecules on conducting polymers for biosensor devices.

DNA biosensors have also been created using adsorption techniques. DNA has been indirectly adsorbed on the conducting polymer surface via mercapto-oligonucleotide probe immobilization of Au–Ag nanocomposites adsorbed onto conducting polymer [73]. Although adsorption is simple, controlling the concentration of the immobilized compound is difficult and immobilization is not stable because of the weak non-covalent forces involved, which decrease the lifetime of the biosensor [74]. Another drawback is that compound adsorption occurs as a monolayer, which limits the quantity of sensing element.

An alternative to adsorption is physical entrapment of the desired biomolecule during electropolymerization, which is one of the most extensively used techniques. During this process, monomer, dopant, and biomolecules are mixed in a single solution used for electrochemical polymerization. This process is usually performed under mild conditions (i.e., neutral pH, low oxidation potentials) without chemical reactions that could alter the activity of proteins, and only requires a single step for both polymerization and molecule immobilization. For this reason conducting polymer are frequently used to entrap biomolecules [75]. Many early applications successfully entrapped glucose oxidase (GOx) in conducting polymer films [2]. Improvements continue to be made on glucose biosensors, which also serve as models for other biosensors. Wang and coworkers [76] discovered that the sensitivity of biosensors utilizing entrapped enzymes is enhanced with increasing microscopic roughness of the electrode's surface. Other enzymes such as horseradish peroxidase, phosphorylases, polyphenol oxidase, lactate dehydrogenase, and deaminases have also been entrapped in conducting polymer films for biosensors. For example, a PANI-derivative polymer was mixed with polylysine to give it water-insoluble for aqueous-based biosensors and then subsequently modified with horseradish peroxidase (via entrapment) [77]. Entrapment methods have also been used for immobilization of antibodies [78] and DNA [79]. Most entrapment methods involve co-deposition during electrochemical synthesis. An alternative to this method is sol–gel encapsulation, which has been performed using silica and polythiophene (PT) or silica and PANI to entrap glucose oxidase for glucose sensing [80]. Entrapment of living cells within conducting polymer has also been reported to create novel biosensors. Detection of dopamine was achieved by entrapment of cells extracted from banana pulp in conducting polymer films in which the enzyme polyphenol oxidase catalyzes the conversion of dopamine to quinine with a corresponding consumption of oxygen [81]. This study was based on the finding that the browning reaction of the banana tissue was produced by the conversion of dopamine to quinine by polyphenol oxidase [82]. Although entrapment is a popular immobilization technique, it has some important limitations. For example, the hydrophobic nature of the polymer damages the quaternary structure of proteins and decreasing their biological activity. To overcome this limitation, new alternatives have focused on creating more hydrophilic polymers using modified monomers with long hydrophilic chains [83]. Also, entrapment methods require a high concentrations of the biomolecules (~0.2–3.5 mg/mL), which is not always available and increases the cost of the process. Finally, the entrapment procedure diminishes the accessibility of analytes to the sensing element and thus affects on the affinity of complex formation (e.g., antibody-antigen, hybridization of nucleotides). As a result, other immobilization techniques, such as affinity binding and covalent modification, have been developed to overcome these limitations. Affinity binding methods are based on immobilizing molecules on the surface of conducting polymer via strong non-covalent interactions. As a conventional approach in this category, the use of the avidin-biotin complex presents advantages over other immobilization techniques because of the extremely specific and high-affinity interactions between biotin and the glycoprotein avidin ($K_a=10^{15}$ mol^{-1} L) [84]. This technique allows control over orientation of the immobilized molecules by adjusting the location of the binding elements, which increases the activity and accessibility of the biological sensing elements. In contrast with conventional grafting, this approach can also be applied to prepare assemblies containing multilayers of biological molecules.

An attractive alternative to affinity binding for biomolecule attachment involves the introduction of appropriate functional groups into conducting polymer backbones or the surface modification of the polymers, followed by covalent bonding (i.e., grafting) of bioactive macromolecules to the surface. In comparison to adsorption, entrapment, and affinity binding immobilization, this method is typically more strong and stable to external environmental factors, allows high loading, and increases biosensor lifetime; however, it is usually more complex and sometimes requires reaction conditions not suitable for biomolecules. Compared to entrapment methods, surface chemical conjugation increases the accessibility of the analytes and enhances the formation of affinity interactions. Conducting copolymers containing covalently substituted monomers have been fabricated as a means to facilitate the immobilization of biomolecules. One strategy for this type of immobilization is to functionalize conducting polymer monomers prior to polymerization. Another conjugation method is chemical grafting after polymerization of unmodified conducting polymer. One common method for post-conducting polymer modification is the use of glutaraldehyde (often in conjunction with bovine serum albumin). Glutaraldehyde crosslinking techniques have been used to immobilize enzymes, such as glucose oxidase, glutamate oxidase, lactate oxidase, xanthine oxidase, choline oxidase, alcohol oxidase, uricase, trypsin, and acetylcholinesterase on conducting polymers [85]. Cholesterol esterase, cholesterol oxidase, and peroxidase have all been immobilized on PANI using glutaraldehyde crosslinking method [86]. Conducting polymers modified in this fashion retained relatively high conductivity. These surface-modified polymers typically have higher conductivities compared to most bulk-modified polymers. For example, polymers with substituted monomers exhibit a significant decrease in conductivity (3–5 orders of magnitude) [24].

In addition to the methods described above, there are other novel modifications of conducting polymers for sensing applications. Molecular imprinting techniques were used by electropolymerization of conducting polymer in the presence of analyte and removing the template analyte molecules at the end of the polymerization via washing. This allows the formation of unique three-dimensional sites used for the selective recognition of analyte [87]. Another unique biosensor technique involves coating gold electrodes, upon which aptamers have been covalently immobilized, with ferrocene-modified conducting polymer. In this instance, the bioactive molecule (the aptamer) is not immobilized directly to the conducting polymer; however, the conducting polymer surrounds the bioactive molecule allowing the conducting polymer to act as an electrical transducer [88].

Recent advances in carbon nanotubes (CNTs) include the incorporation of CNTs into a number of conducting polymer-based biosensors. For example, preliminary studies have been performed exploring the properties of PANI/CNT devices as pH sensors [89]. Also DNA-doped conducting polymer was used in conjunction with CNTs for label-free detection of DNA. In particular, the unique properties of the conducting polymer-CNTs allowed the detection of hybridization reactions with complementary DNA sequences via a decrease in impedance [90]. CNTs have also been incorporated into biosensors as nanotube arrays onto which enzymes, such as GOx, can be immobilized along with a conducting polymer [91]. In general, the presence of CNTs tends to increase the overall sensitivity and selectivity of biosensors.

For enhancing the sensing performance, the incorporation of metal nanoparticles into PANI composites was developed. For example Guo et al. [92] synthesized Pt nanoparticles on PANI, and applied it for electrochemical detection of hydrogen peroxide. In another study hybrid PANI nanofibers with integrated Pt nanoflowers are used as urea biosensor [93].

Uniform deposition of the preformed polymer on a suitable sensor surface has been another major challenge in biosensor fabrication. Though techniques such as drop casting and spin casting are easy and can be used for producing films from preformed polymers, but deposition of preformed polymers and nanostructures using an electrophoretic deposition (EPD) technique has shown promise for yielding uniform and dense films. EPD is a simple and cost-effective technique, amenable to scaling-up to large dimensions, and applicable to a great variety of materials. In EPD, charged molecules dispersed in a suitable solvent move towards an electrode of opposite polarity, and get deposited as homogeneous and highly dense films via particle coagulation. [94-96]. However, a continuing challenge in the wider use of biosensors has been the sensitivity and stability of the surface bond biosensing molecules. The sensitivity and selectivity is crucially dependent on the matrix used for their immobilization in the sensor. The biosensing molecules need to be integrated with a redox active system in a bio-compatible environment to obtain an optimum output from the electrochemical biosensor. In electrochemical biosensors, conducting polymers and their composites have attracted much interest as suitable materials because of their biomolecule binding ability, enhanced stability, speed, and sensitivity [97-100].

2.2. Polyaniline in Biosensor Devices

PANI has become a most interesting material in biosensor devices. PANI might be greatly applied in many biosensors because of its special properties such as:

(i) stability and ease of synthesis,
(ii) high surface area and chemical specificities,
(iii) redox conductivity and polyelectrolyte characteristics and
(iv) direct and easy deposition on the electrode. Moreover, its electrical properties can be reversibly controlled by both charge-transfer doping and protonation [101,102].

It possesses a large number of amino groups that can be used for easy binding of biosensing molecules.

Like other conducting polymers such as polypyrrole, PANI has found applications as immobilization matrix in the design of conductometric [103], potentiometric [104] and amperometric [105] biosensors. In the designing of amperometric biosensors, PANI has been used as a matrix for covalent enzyme immobilization [106]. Amperometric enzyme biosensors based on PANI nanoparticles have also been reported [107].

Additionally, composite form of PANI with carbon nanotubes (CNTs), gold nanoparticles (AuNPs), and other materials has been developed for improvement of redox stability and electro-activity of PANI based systems [108-113]. The presence of nanomaterials such as AuNP in the composite results in large surface area to volume ratio together with PANI providing the functional groups, thus the sensitivity of nanocomposite is improved strongly. Dhand et al. [109] described the fabrication of a PANI–CNT composite for biosensor applications. They observed that the presence of CNTs in the matrix acts as a dopant for PANI, and stabilizes the system during electrochemical measurements. It was observed that this led to a steady and reversible behavior in the composite. Feng et al. [114] investigated a PANI–AuNP matrix based electrochemical DNA biosensor, and showed that the use of this composite resulted in enhanced immobilization of the DNA probe and detection sensitivity of the target DNA. In another study, Ozdemir et al. [115] described a pyranose oxidase modified AuNP–PANI/AgCl hybrid nanocomposite matrix based glucose biosensor. They stated that the use of a nanocomposite matrix improved the bioactivity and stability of the bound enzyme in operational conditions. Liu et al. [116] demonstrated a highly electroactive polystyrene–PANI–Au composite based glucose biosensor. They showed that the composite exhibited improved electrical conductivity and showed a well-defined redox peak and high catalytic activity for the enzyme bound electrode. Yang et al. [117] described PANI–Au composite formation via oxidative polymerization of aniline, using $HAuCl_4$ as the oxidant and 1-butyl-3-methylimidazolium hexafluorophosphate as the growth media. They observed that an electrode fabricated using a composite possesses high conductivity, a large specific surface area, and excellent electroactivity, and thus can serve as an excellent matrix for enzyme immobilization and electrocatalysis.

3. Polyaniline in Drug Delivery Applications

Drug delivery technology has now appeared as a truly interdisciplinary science with aim of improving human health. The basic goal of a drug delivery system is to release a biologically active molecule at a desired rate for a desired duration, so to maintain the drug level in the body appropriate with the therapeutic amounts.

Based on the nature of the carrier, drug delivery systems can be broadly classified into liposomal, electromechanical, and polymeric delivery systems. But the polymeric drug delivery systems are the most investigated ones. In polymeric drug delivery systems the drugs are incorporated in a polymer matrix. The rate of the release of drugs from such a system depends on a multitude of parameters such as nature of the polymer matrix, matrix geometry, properties of the drug, initial drug loading, and drug–matrix interaction.

Most of the polymers investigated for drug delivery applications are non-biodegradable polymers such as poly(dimethylsiloxane) (PS), polyurethanes (PU), and poly(ethylene-co-vinyl acetate) (EVA) and biodegradable polymers such as poly(glycolic acid) (PGA), poly(lactic acid) (PLA) and poly(lactic-co-glycolic acid) (PLGA) [14]. Also in recent years conducting polymers were investigated for drug-delivery mediators and actuators [1]. This section briefly describes the most representative advances in these areas.

Many of the drug delivery and actuator approaches are based on the reduction of conducting polymers, which activates expulsion of ions (i.e., de-doping) and an associated change in volume [118]. When a conducting polymer like polyaniline is electrochemically oxidized or reduced, simultaneous changes in the polymeric charge and chain conformation result in the modification of the conductivity, color, volume and other polymer parameters. Electromechanical properties are linked to the movement of ions and solvent molecules inside and outside the polymeric matrix and conformational changes along the polymeric chains, driving the opening and closure of the polymeric entanglements [119, 120]. For example polymeric electroactive blends formed by electropolymerized aniline inside a non-conductive polyacrylamide porous matrix were studied as suitable materials for the electrocontrolled release of model compounds like safranin with this mechanism [121].

With respect to drug delivery applications, electrical stimulation of CPs has been used to release a number of therapeutic proteins and drugs including, for example, nerve growth factor (NGF) [122], dexamethasone [123], and heparin [124]. An early investigation demonstrated entrapment and electrically stimulated release of bovine serum albumin and NGF from PPy doped with polyelectrolytes (e.g., dextran sulfate) [122]. The use of polyelectrolytes induced high water content in the CP, allowing an easy release of the entrapped protein. For this release, conducting polymer was reduced with negative potential, producing a rapid expulsion of anions in less than one minute. The retained activity of the released NGF was confirmed with a neurite extension assay with PC12 cells. Although protein entrapment in CPs often affects the protein's folding and activity because of the hydrophobicity of the polymer, the use of polyelectrolytes and high water content might have overcome this limitation. Also NGF was exclusively released when conducting polymer was electrically stimulated. In particular, a 3V pulse for 150 s resulted in a release of 22 ng/cm^2 of NGF from the surface of the polymer. The activity of the released protein was confirmed by observing PC12 cell neurite extension in the presence of the released NGF. The release of heparin from hydrogels immobilized onto conducting polymer films can also be started by electrical stimulation [124]. PVA hydrogels were covalently immobilized onto conducting polymer films via grafting of aldehyde groups to conducting polymer and chemical reaction of these with hydroxyl groups from the hydrogel. An accelerated release of heparin from the hydrogel was reported when conducting polymer was electrically stimulated.

Another research demonstrated the use of conducting polymer nanotubes for drug release, in which conducting polymer was polymerized on top of electrospun PLGA fibers (~100nm diameter) loaded with dexamethasone [125]. PLGA was subsequently removed to produce conducting polymer nanotubes encapsulating small molecules. The conducting polymer nanotubes released the drug in a controlled fashion upon electrical stimulation, probably as a consequence of expansion/reduction of polymer cavities produced by the expulsion of anions. Hua et al. [126] developed a nontoxic drug nanocarrier containing carboxyl groups by mixing magnetic nanoparticles (MNPs) of Fe_3O_4 with the water-soluble polyaniline derivative, poly[aniline-co-sodium N-(1-one-butyric acid) aniline] (SPAnNa), and doping with HCl aqueous solution to form SPAnH/MNPs shell/core. SPAnH/MNPs could be used to effectively immobilize the hydrophobic drug paclitaxel (PTX), thus enhancing the drug's thermal stability and water solubility.

CONCLUSION

Polyaniline (PANI) has usage in a diverse array of applications and more recently in biomedical fields. The fact that several tissues and cells are responsive to electrical fields and stimulus has made PANI attractive in biological and medical applications. In addition to this highly desirable property, the ease of preparation and modification of PANI has made it a popular choice for biomedical applications. PANI offers several important advantages for biomedical applications over other materials such as metals and semiconductors, including biocompatibility, entrap and controllably release of biological molecules (i.e., reversible doping) in drug delivery devices, transfer charge from a biochemical reaction in biosensors, and ability to control over cell proliferation and differentiation in tissue engineering. Despite of the vast amount of research already conducted on PANI for biomedical applications, the field is still growing.

REFERENCES

[1] Guimard, N.K.; Gomez, N.; Schmidt, C.E. Conducting polymers in biomedical engineering. *Progr. Polym. Sci.* 2007, *32*, 876-921.

[2] Foulds, N.C.; Lowe, C.R. Enzyme entrapment in electrically conducting polymers. *J. Chem. Soc. Faraday. Trans.* 1986, *82*, 1259-1264.

[3] Umana, M.; Waller, J. Protein modified electrodes: the glucose/oxidase/polypyrrole system. *Anal. Chem.* 1986, *58*, 2979-2983.

[4] Wong, J.Y.; Langer, R.; Ingber, D.E. Electrically conducting polymers can noninvasively control the shape and growth of mammalian cells. *Proc. Natl. Acad. Sci. USA*. 1994, *91*, 3201-3204.

[5] Shi, G.; Rouabhia, M.; Wang, Z.; Dao, L.H.; Zhang, Z. A novel electrically conductive and biodegradable composite made of polypyrrole nanoparticles and polylactide. *Biomaterials*. 2004, *25*, 2477-2488.

[6] Fan, Q.; Sirkar, K.K.; Michniak, B. Iontophoretic transdermal drug delivery system using a conducting polymeric membrane. *J. Membrane. Sci.* 2008, *321*, 240-249.

[7] Shin, Y.J.; Kameoka, J. Amperometric cholesterol biosensor using layer-by-layer adsorption technique onto electrospun polyaniline nanofibers, *J. Ind. Eng. Chem.* 2012, *18*, 193-197.

[8] Ghasemi-Mobarakeh, L.; Prabhakaran, M.P.; Morshed, M.; Nasr-Esfahani, M.H.; Baharvand, H.; Kiani, S.; Al-Deyab, S.S.; Ramakrishna, S. Application of conductive polymers, scaffolds and electrical stimulation for nerve tissue engineering, *J. Tissue. Eng. Regen. Med.* 2011, *5*, 17-35.

[9] Di, L.; Wang, L.P.; Lu, Y.N.; He, L.; Lin, Z.X.; Wu, K.J.; Ren, Q.S.; Wang, J.Y. Protein adsorption and peroxidation of rat retinas under stimulation of a neural probe coated with polyaniline, *Acta. Biomater.* 2011, *7*, 3738-3745.

[10] Langer, R.; Vacanti, J.P. Tissue engineering. *Science*. 1993, *260*, 920-926.

[11] Shalak, R.; Fox, C.F. Preface. In: Shalak, R., Fox, C.F. *Tissue engineering*. Alan R. Liss: N.Y., 1988, pp. 26-29.

[12] Yang, F.; Murugan, R.; Ramakrishna, S.; Wang, X.; Ma, Y.X.; Wang, S. Fabrication of nano-structured porous PLLA scaffold intended for nerve tissue engineering. *Biomaterials*.2004, *25*, 1891-1900.

[13] Subramanian, A.; Krishnan, U.M.; Sethuraman, S. Development of biomaterial scaffold for nerve tissue engineering: biomaterial mediated neural regeneration. *J. Biomed. Sci.* 2009, *16*, 108-119.

[14] Lakshmi, S.N.; Laurencin, C.T. Polymers as Biomaterials for Tissue Engineering and Controlled Drug Delivery. *Adv. Biochem. Eng. Biotechnol.* 2006, *102*, 47-90.

[15] Huinan, L.; Webster, T.J. Bioinspired Nanocomposites for Orthopedic Applications, 2007, 1-51.

[16] Armentano, I.; Dottori, M.; Fortunati, E.; Mattioli, S.; Kenny, J.M. Biodegradable polymer matrix nanocomposites for tissue engineering: A review. *Polym. Degrad. Stabil.* 2010, *95*, 2126-2146.

[17] Abdul Rahman, N.; Gizdavic-Nikolaidis, M.; Ray, S.; Easteal, A.J.; Travas-Sejdic, J. Functional electrospun nanofibres of poly(lactic acid) blends with polyaniline or poly(aniline-co-benzoic acid, *Synthetic. Met.* 2010, *160*, 2015-2022.

[18] Skotheim, T.A.; Reynolds, J.R. *Handbook of Conducting Polymers,* 3rd ed., CRC Press: Gainesville, 2007.

[19] Zhang, Q.; Yan, Y.; Li, S.; Feng, T. The synthesis and characterization of a novel biodegradable and electroactive polyphosphazene for nerve regeneration. *Mater. Sci. Eng. C.* 2010, *30*, 160-166.

[20] Bettinger, C.J.; Bruggeman, J.P.; Misra, A.; Borenstein, J.T.; Langer, R. Biocompatibility of biodegradable semiconducting melanin films for nerve tissue engineering. *Biomaterials*. 2009, *30*, 3050-3057.

[21] Schmidt, C.E.; Shastri, V.R.; Vacanti, J.P.; Langer, R. Stimulation of neurite outgrowth using an electrically conducting polymer. *Proc. Natl. Acad. Sci. USA.* 1997, *94*, 8948-8953.

[22] Cen, L.; Neoh, K.G.; Kang, E.T. Surface functionalization of electrically conductive polypyrrole film with hyaluronic acid. *Langmuir.* 2002, *18*, 8633-8640.

[23] Lee, J.W.; Serna, F.; Nickels, J.; Schmidt, C.E. Carboxylic acid-functionalized conductive polypyrrole as a bioactive platform for cell adhesion. *Biomacromolecules.* 2006, *7*, 1692-1695.

[24] Song, H.K.; Toste, B.; Ahmann, K.; Hoffman-Kim, D.; Palmore, G.T. Micropatterns of positive guidance cues anchored to polypyrrole dopedwith polyglutamic acid: a new platform for characterzing neurite extension in complex environments. *Biomaterials.* 2006, *27*, 473-484.

[25] Stauffer, W.R.; Cui, X.T. Polypyrrole doped with two peptide sequences from laminin. *Biomaterials.* 2006, *27*, 2405-2413.

[26] Gomez, N.; Lee, J.Y.; Nickels, J.D.; Schmidt, C.E. Micropatterned polypyrrole: a combination of electrical and topographical characteristics for the stimulation of cells. *Adv. Funct. Mater.* 2007, *17*, 1645-1653.

[27] Huang, J.; Hu, X.; Lu, L.; Ye, Z.; Zhang, Q.; Luo, Z. Electrical regulation of Schwann cells using conductive polypyrrole/chitosan polymers. *J. Biomed. Mater. Res. A.* 2010, *93*, 164-174.

[28] Ghasemi-Mobarakeh, L.; Prabhakaran, M.P.; Morshed, M., Nasr-Esfahani, M.H.; Ramakrishna, S. Electrical stimulation of nerve cells using conductive nanofibrous scaffolds for nerve tissue engineering. *Tissue. Eng. A.* 2009, *15*, 3605-3619.

[29] Li, M.; Guo, Y.; Wei, Y.; MacDiarmid, A.G.; Lelkes, P.I. Electrospinning polyaniline-contained gelatin nanofibers for tissue engineering applications. *Biomaterials.* 2006, *27*, 2705-2715.

[30] Ateh, D.D.; Vadgama, P.; Navsaria, H.A. Culture of human keratinocytes on polypyrrole-based conducting polymers. *Tissue. Eng.* 2006, *12*, 645-655.

[31] Meng, S.; Rouabhia, M.; Shi, G.; Zhang, Z. Heparin dopant increases the electrical stability, cell adhesion, and growth of conducting polypyrrole/poly(L,L-lactide) composites. *J. Biomed. Mater. Res. A.* 2008, *87*, 332-344.

[32] Richardson, R.T.; Thompson, B.; Moulton, S.; Newbold, C.; Lum, M.G.; Cameron, A.; Wallace, G.; Kapsa, R.; Clark, G.; O'Leary, S. The effect of polypyrrole with incorporated neurotrophin-3 on the promotion of neurite outgrowth from auditory neurons. *Biomaterials.* 2007, *28,* 513-523.

[33] Lee, J.Y.; Lee, J.W.; Schmidt, C.E. Neuroactive conducting scaffolds: nerve growth factor conjugation on active esterfunctionalized polypyrrole. *J. R. Soc. Interface.* 2009, *6,* 801-810.

[34] Cen, L.; Neoh, K.G.; Li, Y.; Kang, E.T. Assessment of in vitro bioactivity of hyaluronic acid and sulfated hyaluronic acid functionalized electroactive polymer. *Biomacromolecules.* 2004, *5,* 2238-2246.

[35] Gelmi, A.; Higgins, M.J.; Wallace, G.G. Physical surface and electromechanical properties of doped polypyrrole biomaterials. *Biomaterials.* 2010, *31,* 1974-1983.

[36] Gilmore, K.J.; Kita, M.; Han, Y.; Gelmi, A.; Higgins, M.J.; Moulton, S.E.; Clark, G.M.; Kapsa, R.; Wallace, G.G. Skeletal muscle cell proliferation and differentiation on

polypyrrole substrates doped with extracellular matrix components. *Biomaterials.* 2009, *30*, 5292-5304.

[37] Green, R.A.; Lovell, N.H.; Wallace, G.G.; Poole-Warren, L.A. Conducting polymers for neural interfaces: challenges in developing an effective long-term implant. *Biomaterials.* 2008, *29,* 3393-3399.

[38] Thompson, B.C.; Richardson, R.T.; Moulton, S.E.; Evans, A.J.; O'Leary, S.; Clark, G.M.; Wallace, G.G. Conducting polymers, dual neurotrophins and pulsed electrical stimulation– dramatic effects on neurite outgrowth. *J. Control. Release.* 2010, *141,* 161-167.

[39] Chen, F.; Lee, C.N.; Teoh, S.H. Nanofibrous modification on ultra-thin poly (ε-caprolactone) membrane via electrospinning. *Mater. Sci. Eng. C. Biomimet. Supramol. Syst.* 2007, *27,* 325-332.

[40] Naji, A.; Harmand, M.F. Study of the effect of the surface state on the cytocompatibility of a Co–Cr alloy using human osteoblasts and fibroblasts. *J. Biomed. Mater. Res.* 1990, *24,* 861-871.

[41] Meyer, U.; Szulczewski, D.H.; Moller, K.; Heide, H.; Jones, D.B. Attachment kinetics and differentiation of osteoblasts on different biomaterials. *Cell. Mater.* 1993, *3,* 129-140.

[42] Steele, J.G.; Mcfarland, C.; Dalton, B.A.; Johnson, G.; Evans, M.D.M.; Rolfe Howlett, C.; Underwood, P. AnneAttachment of human bone cells to tissue culture polystyrene and to unmodified polystyrene: the effect of surface chemistry upon initial cell attachment. *J. Biomater. Sci. Polym. Ed.* 1993, *5,* 245-257.

[43] Chu, C.F.; Lu, A.; Liszkowski, M.; Sipehia, R. Enhanced growth of animal and human endothelial cells on biodegradable polymers. *Biochim. Biophys. Acta.* 1999, *1472,* 479-485.

[44] Xu, C.; Yang, F.; Wang, S.; Ramakrishna, S. In vitro study of human vascular endothelial cell function on materials with various surface roughness. *J. Biomed. Mater. Res. A.* 2004, *71A,* 154-161.

[45] McKeon, K.D.; Lewis, A.; Freeman, J.W. Electrospun Poly(D,L-lactide) and Polyaniline Scaffold Characterization, *J. Appl. Polym. Sci.* 2010, *115*, 1566-1572.

[46] Huang, L.; Hu, J.; Lang, L.; Wang, X.; Zhang, P.; Jing, X.; Wang, X.; Chen, X.; Lelkes, P.I.; MacDiarmid, A.G.; Wei, Y. Synthesis and characterization of electroactive and biodegradable ABA block copolymer of polylactide and aniline pentamer, *Biomaterials.* 2007, *28*, 1741-1751.

[47] Gizdavic-Nikolaidis, M.; Ray, S.; Bennett, J.R.; Easteal, A.J.; Cooney, R.P. Electrospun Functionalized Polyaniline Copolymer-Based Nanofibers with Potential Application in Tissue Engineering, *Macromol. Biosci.* 2010, *10*, 1424-1431.

[48] Bidez, P.R.; Li, S.; MacDiarmid, A.G.; Venancio, E.C.; Wei, Y.P.; Lelkes, I. Polyaniline, an electroactive polymer, supports adhesion and proliferation of cardiac myoblasts. *J. Biomater. Sci., Polym. Ed.* 2006, *17*, 199-212.

[49] Guo, Y.; Li, M.; Mylonakis, A.; Han, J.; MacDiarmid, A.D.; Chen, X. Electroactive oligoaniline-containing self-assembled monolayers for tissue engineering applications. *Biomacromolecules.* 2007, *8*, 3025-3034.

[50] Huang, L.H.; Zhuang, X.L.; Hu, J.; Lang, L.; Zhang, P.B.; Wang, Y.S.; Chen, X.S.; Wei, Y.; Jing, X.B. Synthesis of Biodegradable and Electroactive Multiblock

Polylactide and Aniline Pentamer Copolymer for Tissue Engineering Applications. *Biomacromolecules*. 2008, *9*, 850-858.

[51] Wang, C.H.; Dong, Y.Q.; Sengothi, K.; Tan, K.L.; Kang, E.T. In vivo tissue response to polyaniline. *Synthetic. Met.* 1999, *102,* 1313-1314.

[52] Kamalesh, S.; Tan, P.; Wang, J.; Lee, T.; Kang, E.T.; Wang, C.H. Biocompatibility of electroactive polymers in tissues. *J. Biomed. Mater. Res.* 2000, *52*, 467-478.

[53] Scaffaro, R.; Re, G.L.; Dispenza, C.; Sabatino, M.A.; Armelao, L. A new route for the preparation of flexible skin–core poly(ethylene-co-acrylic acid)/polyaniline functional hybrids. *React. Func. Polym.* 2011, *71*, 1177-1186.

[54] Prabhakaran, M.P.; Ghasemi-Mobarakeh, L.; Jin, G.; Ramakrishna, S. Electrospun conducting polymer nanofibers and electrical stimulation of nerve stem cells, *J. Biosci. Bioeng*. 2011, *112*, 501-507.

[55] Su, S.J.; Kuramoto, N. Synthesis of processable polyaniline complexed with anionic surfactant and its conducting blends in aqueous and organic system. *Synthetic. Met.* 2000, *108*, 121-126.

[56] Laska, J.; Zak, K.; Pron, A. Conducting blends of polyaniline with conventional polymers. *Synthetic. Met.* 1997, *84*, 117-118.

[57] Leclerc, M.; Guay, J.; Dao, L.H. Synthesis and characterization of poly(alkylanilines) *Macromolecules*. 1989, *22*, 649-653.

[58] Rao, P.S.; Sathyanarayana, D.N. Synthesis of electrically conducting copolymers of aniline with o/m-amino benzoic acid by an inverse emulsion pathway. *Polymer*. 2002, *43*, 5051-5058.

[59] Yue, J.; Epstein, A.J. Synthesis of self-doped conducting polyaniline. *J. Amer. Chem. Soc.* 1990, *112*, 2800-2801.

[60] Rivas, B.L.; Saınchez, C.O. Poly(2-) and (3-aminobenzoic acids) and their copolymers with aniline: Synthesis, characterization, and properties. *J. Appl. Polym. Sci.* 2003, *89*, 2641-2648.

[61] Malinauskas, A. Self-doped polyanilines. *J. Power. Sources*. 2004, *126*, 14-220.

[62] Guterman, E.; Cheng, S.; Palouian, K.; Bidez, P.; Lelkes, P.I.; Wei, Y. Peptide-modified electroactive polymers for tissue engineering applications. *Polymer. Prepr.* 2002, *43*, 766-767.

[63] Wang, P.; Tan, K.L. Synthesis and characterization of poly(ethylene glycol)-grafted polyaniline. *Chem. Mater.* 2001, *13*, 581-587.

[64] Chen, Y.; Kang, E.T.; Neoh, K.G.; Tan, K.L. Chemical modification of polyaniline powders by surface graft copolymerization. *Polymer*. 2000, *41*, 3279-3287.

[65] Chao, D.; Lu, X.; Chen, J.; Zhao, X.; Wang, L.; Zhang, W.; Wei, Y. New method of synthesis of electroactive polyamide with amine-capped aniline pentamer in the main chain. *J. Polym. Sci. Pol. Chem.* 2006, *44*, 477-482.

[66] Khan, R.; Pratima R.S., Kaushik, A.; Singh, S.P.; Ahmad, S.; Malhotra, B.D. Cholesterol biosensor based on electrochemically prepared polyaniline conducting polymer film in presence of a nonionic surfactant, *J. Polym. Res.* 2009, *16*, 363-373.

[67] Clark, L.C.; Lyons, C. Electrode systems for continuous monitoring in cardiovascular surgery. *Ann. NY. Acad. Sci* .1962, *102*, 29-45.

[68] Gerard, M.; Chaubey, A.; Malhotra, B.D. Application of conducting polymers to biosensors. *Biosen. Bioelectron*. 2002, *17*, 345-359.

[69] Li, J.P.; Gu, H.N. A selective cholesterol biosensor based on composite film modified electrode for amperometric detection. *J. Chinese. Chem. Soc. (Taipei, Taiwan).* 2006, *53*, 575-582.

[70] Pandey, P.C.; Mishra, A.P. Conducting polymer-coated enzyme microsensor for urea. *Analyst.* 1988, *113*, 329-331.

[71] Bakker, E.; Telting-Diaz, M. Electrochemical sensors. *Anal. Chem.* 2002, *74*, 2781-2800.

[72] Tamiya, E.; Karube, I.; Hattori, S.; Sizuki, M.; Yokoyama, K. Micro glucose sensors using electron mediators immobilized on a polypyrrole-modified electrode. *Sensor. Actuat.* 1989, *18*, 297-307.

[73] Fu, Y.; Yuan, R.; Chai, Y.; Zhou, L.; Zhang, Y. Coupling of a reagentless electrochemical DNA biosensor with conducting polymer film and nanocomposite as matrices for the detecion of the HIV DNA sequences. *Anal. Lett.* 2006, *39*, 467-482.

[74] Ahuja, T.; Mir, I.A.; Kumar, D.; Rajesh. Biomolecular immobilization on conducting polymers for biosensing applications. *Biomaterials.* 2007, *28*, 791-805.

[75] Cosnier, S. Biosensors based in electropolymerized films: new trends. *Anal. Bioanal. Chem.* 2003, *377*, 507-520.

[76] Wang, J.; Myung, N.V.; Yun, M.; Monbouquette, H.G. Glucose oxidase entrapped in polypyrrole on high-surface area Pt electrodes: a model platform for sensitive electroenzymatic biosensors. *J. Electroanal. Chem.* 2005, *575*, 139-146.

[77] Ngamna, O.; Morrin, A.; Moulton, S.E.; Killard, A.J.; Smyth, M.R.; Wallace, G.G. An HRP based biosensor using sulphonated polyaniline. *Synthetic. Met.* 2005, *153*, 185-188.

[78] Li, C.M.; Sun, C.Q.; Song, S.; Choong, V.E.; Maracas, G.; Zhang, X.J. Impedance labelless detection-based polypyrrole DNA biosensor. *Front. Biosci.* 2005, *10*, 180-186.

[79] Chen, Y.; Elling; Lee Y.l.; Chong, S.C. A fast sensitive and label free electrochemical DNA sensor. *J. Phys. Conf. Ser.* 2006, *34*, 204-209.

[80] Yamagishi, F.G.; Stanford, T.B.; Jr.; van Ast, C.I. Biosensors from conducting polymer transducers and sol–gel encapsulated bioindicator molecules. *Proc. Electrochem. Soc.* 2001- 2001, *18*, 213-223.

[81] Deshpande, M.V.; Hall, E.A. An electrochemically grown polymer as an immobilization matrix for whole cells: applications in an amperometric dopamine sensor. *Biosen. Bioelectron.* 1990, *5*, 431-448.

[82] Sidwell, J.S.; Rechnitz, G.A. Bananatrode, an electrochemical biosensor for dopamine. *Biotechnol. Lett.* 1985, *7*, 419-422.

[83] Mousty, C.; Galland, B. Cosnier S. Electrogeneration of a hydrophilic cross-linked polypyrrole film for enzyme electrode fabrication: application to the Amperometric detection of glucose. *Electroanal.* 2001, *13*, 186-190.

[84] Wilchek, M.; Bayer, E.A. The avidin–biotin complex in bioanalytical applications. *Anal. Biochem.* 1988, *171*, 1-32.

[85] Gade, V.K.; Shirale, D.J.; Gaikwad, P.D.; Savale, P.A.; Kakde, K.P.; Kharat, H.J.; Shirsat, M.D. Immobilization of GOD on electrochemically synthesized Ppy–PVS composite film by cross-linking via glutaraldehyde for determination of glucose. *React. Funct. Polym.* 2006, *66*, 1420-1426.

[86] Singh, S.; Solanki, P.R.; Pandey, M.K.; Malhotra, B.D. Cholesterol biosensor based on cholesterol esterase, cholesterol oxidase and peroxidase immobilized onto conducting polyaniline films. *Sensor. Actuat. B.* 2006, *115*, 534-541.

[87] Ebarvia, B.S.; Cabanilla, S.; Sevilla III, F. Biomimetic properties and surface studies of a piezoelectric caffeine sensor based on electrosynthesized polypyrrole. *Talanta.* 2005, *66*, 145-152.

[88] Le, F.F.; Ho, H.A.; Leclerc, M. Label-free electrochemical detection of protein based on ferrocene-bearing cationic polythiophene and aptamer. *Anal. Chem.* 2006, *78*, 4727-4731.

[89] Ferrer-Anglada, N.; Kaempgen, M.; Roth, S. Transparent and flexible carbon nanotube/polypyrrole and carbon nanotube/ polyaniline pH sensors. *Phys. Stat. Sol. B.* 2006, *243*, 3519-3523.

[90] Cai, H.; Xu, Y.; He, P.G.; Fang, Y.Z. Indicator Free DNA Hybridization detection by impedance measurement based on the DNA-doped conducting polymer film formed on the carbon nanotube modified electrode. *Electroanal.* 2003, *15*, 1864-1870.

[91] Qu, L.; He, P.; Li, L.; Gao, M.; Wallace, G.; Dai, L. Aligned/ micropatterned carbon nanotube arrays: surface functionalization and electrochemical sensing. *Proc. SPIE Int Soc. Opt. Eng.* 2005, *5732*, 84-92.

[92] Guo, S.J.; Dong, S.J.; Wang, E.K. Polyaniline/Pt Hybrid Nanofibers: High-Efficiency Nanoelectrocatalysts for Electrochemical Devices. *Small.* 2009, *5*, 1869-1876.

[93] Jia, W.; Su, L.; Lei, Y. Pt nanoflower/polyaniline composite nanofibers based urea biosensor, *Biosen. Bioelectron.* 2011, *30*, 158-164.

[94] Dhand, C.; Malhotra, B.D. *Organic Electronics in Sensors and Biotechnology*. McGraw Hill, 2009, pp. 361-394.

[95] Dhand, C.; Singh, S.P.; Arya, S.K.; Datta, B.; Malhotra, B.D. Cholesterol Biosensor Based on Electrophoretically Deposited Conducting Polymer Film Derived from Nano-Structured Polyaniline colloidal suspension. *Anal. Chim. Acta.* 2007, *602*, 244-251.

[96] Murphy-Perez, E.; Arya, S.K.; Bhansali, S. Vapor-liquid-solid grown silica nanowire based electrochemical glucose biosensor. *Analyst.* 2011, *136*, 1686-1689.

[97] Wang, J. Electrochemical biosensors: towards point-of-care cancer diagnostics. *Biosens. Bioelectron.* 2006, *21 (10)*, 1887-1892.

[98] Koh, W.C.A.; Son, J.I.; Choe, E.S.; Shim, Y.B. Electrochemical Detection of Peroxynitrite Using a Biosensor Based on a Conducting Polymer–Manganese Ion Complex. *Anal. Chem.* 2010, *82 (24)*, 10075-10082.

[99] Malhotra, B.D.; Chaubey, A.; Singh, S.P. Prospects of conducting polymers in biosensors. *Anal. Chim. Acta.* 2006. *578 (1)*, 59-74.

[100] Wang, G.F.; Huang, H.; Zhang, G.; Zhang, X.J.; Fang, B.; Wang, L. Dual Amplification Strategy for the Fabrication of Highly Sensitive Interleukin-6 Amperometric Immunosensor Based on Poly-Dopamine. *Langmuir.* 2011, *27 (3)*, 1224-1231.

[101] Zhong, H.; Yuan, R.; Chai, Y.; Li, W.; Zhong, X.; Zhang, Yu. In situ chemo-synthesized multi-wall carbon nanotube-conductive polyaniline nanocomposites: Characterization and application for a glucose amperometric biosensor, *Talanta.* 2011, *85*, 104-111.

[102] Dhand, C.; Das, M.; Datta, M.; Malhotra, B.D. Recent advances in polyaniline based biosensors. *Biosens. Bioelectron.* 2011, *26 (6)*, 2811-2821.

[103] Ajay, A.K.; Srivastava, D.N. Microtubular conductometric biosensor for ethanol detection. *Biosens. Bioelectron.* 2007, *23*, 281-284.

[104] Qaisar, A.; Adeloju, S.B. Development of a potentiometric catechol biosensor by entrapment of tyrosinase within polypyrrole film. *Sensor. Actuat. B.* 2009, *140*, 5-11.

[105] Xu, L.; Zhu, Y.; Yang, X.; Li, Ch. Amperometric biosensor based on carbon nanotubes coated with polyaniline/dendrimer-encapsulated Pt nanoparticles for glucose detection. *Mater. Sci. Eng. C.* 2009, *29*, 1306-1310.

[106] Singh, S.; Solanki, P.R.; Pandey, M.K.; Malhotra, B.D. Covalent immobilization of cholesterol esterase and cholesterol oxidase on polyaniline films for application to cholesterol biosensor. *Anal. Chim. Acta.* 2006, *568*, 126-132.

[107] Morrin, A.; Orawan, N.; Killard, A.; Moulton, S.; Smyth, M.; Wallace, G. An Amperometric Enzyme Biosensor Fabricated from Polyaniline Nanoparticles. *Electroanal.* 2005, *17*, 423-430.

[108] Dhand, C.; Arya, S.K.; Datta, M.; Malhotra, B.D. Polyaniline–carbon nanotube composite film for cholesterol biosensor. *Anal. Biochem.* 2008, *383 (2)*, 194-199.

[109] Dhand, C.; Arya, S.K.; Singh, S.P.; Singh, B.P.; Datta, M.; Malhotra, B.D. Preparation of polyaniline/multiwalled carbon nanotube composite by novel electrophoretic route. *Carbon.* 2008, *46 (13)*, 1727-1735.

[110] Feng, X.; Mao, C.; Yang, G.; Hou, W.; Zhu, J.J. Polyaniline/Au Composite Hollow Spheres: Synthesis, Characterization, and Application to the Detection of Dopamine. *Langmuir.* 2006, *22 (9)*, 4384-4389.

[111] Gajendran, P.; Saraswathi, R. Polyaniline-carbon nanotube composites. *Pure. Appl. Chem.* 2008, *80 (11)*, 2377-2395.

[112] Spain, E.; Kojima, R.; Kaner, R.B.; Wallace, G.G.; O'Grady, J.; Lacey, K.; Barry, T.; Keyes, T.E.; Forster, R.J. High Sensitivity DNA Detection Using Gold Nanoparticle Functionalized Polyaniline Nanofibres. *Biosens. Bioelectron.* 2011, *26 (5)*, 2613-2618.

[113] Tian, S.J.; Liu, J.Y.; Zhu, T.; Knoll, W. Polyaniline/Gold Nanoparticle Multilayer Films: Assembly, Properties, and Biological Application. *Chem. Mater.* 2004, *16 (21)*, 4103-4108.

[114] Feng, Y.Y.; Yang, T.; Zhang, W.; Jiang, C.; Jiao, K. Enhanced sensitivity for deoxyribonucleic acid electrochemical impedance sensor: Gold nanoparticle/polyaniline nanotube membranes. *Anal. Chim. Acta.* 2008, *616 (2)*, 144-151.

[115] Ozdemir, C.; Yeni, F.; Odaci, D.; Timur, S. Electrochemical glucose biosensing by pyranose oxidase immobilized in gold nanoparticle-polyaniline/AgCl/gelatin nanocomposite matrix. *Food Chem.* 2010, *119 (1)*, 380-385.

[116] Liu, Y.G.; Feng, X.M.; Shen, J.M.; Zhu, J.J.; Hou, W. Fabrication of a novel glucose biosensor based on a highly electroactive polystyrene/polyaniline/Au nanocomposite. *J. Phys. Chem. B.* 2008, *112 (30)*, 9237-9242.

[117] Yang, W.R.; Liu, J.Q.; Zheng, R.K.; Liu, Z.W.; Dai, Y.; Chen, G.N.; Ringer, S.; Braet, F. Ionic Liquid-assisted Synthesis of Polyaniline/Gold Nanocomposite and Its Biocatalytic Application. *Nanoscale Res. Lett.* 2008, *3 (11)*, 468-472.

[118] Entezami, A.A.; Massoumi, B. Artificial muscles, biosensors and drug delivery systems based on conducting polymers: a review. *Iran. Polym. J.* 2006, *15*, 13-30.

[119] Bay, L.; Jacobsen, T.; Skaarup, S.; West, K. Mechanism of actuation in conducting polymers: osmotic expansion, *J. Phys. Chem. B.* 2001, *105*, 8492-8497.

[120] Low, L.M.; Seetharaman, S.; He, K.Q.; Madou, M.J. Microactuators toward microvalves for responsive controlled drug delivery. *Sensor. Actuat. B.* 2000, *67*, 149-160.

[121] Luiz, M.; Cordoba, L.S.I. de Torresi. Polymeric electro-mechanic devices applied to antibiotic-controlled release. *Sensor. Actuat. B.* 2008, *130*, 638-644.

[122] Hodgson, A.J.; John, M.J.; Campbell, T.; Georgevich, A.; Woodhouse, S.; Aoki, T.; Ogata, N.; Wallace, G.G. Integration of biocomponents with synthetic structures-use of conducting polymer polyelectrolyte composites. *Proc. SPIE Int. Soc. Opt. Eng.* 1996, *2716*, 164-176.

[123] Wadhwa, R.; Lagenaur, C.F.; Cui, X.T. Electrochemically controlled release of dexamethasone from conducting polymer polypyrrole coated electrode. *J. Control. Release*. 2006, *110*, 531-541.

[124] Li, Y.; Neoh, K.G.; Kang, E.T. Controlled release of heparin from polypyrrole–poly(vinyl alcohol) assembly by electrical stimulation. *J. Biomed. Mater. Res. A.* 2005, *73A*, 171-181.

[125] Abidian, M.R.; Kim, D.H.; Martin, D.C. Conducting polymer nanotubes for controlled drug release. *Adv. Mater.* 2006, *18*, 405-409.

[126] Hua, M.Y.; Yang, H.W.; Chuang, C.K.; Tsai, R.Y.; Chen, W.J.; Chuang, K.L.; Chang, Y.H.; Chuang, H.C.; Pang, S.T. Magnetic nanoparticle-modified paclitaxel for targeted therapy for prostate cancer. *Biomaterials*, 2010, *31*, 7355-7363.

INDEX

B

C

D

E

F

G

H

I

J

K

L

M

N

O

P

T

U

V

W

X

Y

Z